Lecture Notes in Computer Science 8787

Commenced Publication in 1973
Founding and Former Series Editors:
Gerhard Goos, Juris Hartmanis, and Jan van Leer

T0183380

Preface

Welcome to the proceedings of the 15th International Conference on Web Information Systems Engineering (WISE 2014), held in Thessaloniki, Greece in October 2014. The series of WISE conferences aims to provide an international forum for researchers, professionals, and industrial practitioners to share their knowledge in the rapidly growing area of Web technologies, methodologies, and applications. The first WISE event took place in Hong Kong, China (2000). Then the trip continued to Kyoto, Japan (2001); Singapore (2002); Rome, Italy (2003); Brisbane, Australia (2004); New York, USA (2005); Wuhan, China (2006); Nancy, France (2007); Auckland, New Zealand (2008); Poznan, Poland (2009); Hong Kong, China (2010); Sydney, Australia (2011); Paphos, Cyprus (2012), and Nanjing, China (2013). This year, for a fifth time, WISE was held in Europe, in Thessaloniki, supported by Aristotle University, the largest University in Greece.

WISE 2014 hosted three well-known keynote and invited speakers: Prof. Krishna P. Gummadi, Head of the Networked Systems Research Group, at Max Planck Institute for Software Systems, Germany who gave a talk on "Understanding Information Exchange in Social Media Systems", Dr Mike Fisher, Chief Researcher in the Research and Innovation Department of British Telekom, UK who gave an industrial focus lecture on "Connected Communities"; and Prof. Santo Fortunato faculty of Complex Systems at the Department of Biomedical Engineering and Computational Science of Aalto University, Finland who gave a talk on "Detecting Communities in Networks". Moreover, four tutorials were presented on the topics : "Blocking Techniques for Web-Scale Entity Resolution", "Community Detection and Evaluation in Social and Information Networks", "Navigating the choices for Similarity Operators", and "Extensions on Map-Reduce".

A total of 196 research papers were submitted to the conference for consideration, and each paper was reviewed by at least two reviewers. Finally, 52 submissions were selected as full papers (with an acceptance rate of 26% approximately), plus 16 as short papers. The program also featured 14 poster papers and 1 WISE challenge summary paper which outlines the WISE challenge succeeded submissions, presented in the WISE challenge separate Workshop. The research papers cover the areas of semantic Web; Web mining, modeling, and classification; Web querying and searching; Web recommendation and personalization; social online networks; Web technologies and frameworks; Software Architectures, techniques, and platforms; and Web innovation and applications.

We wish to take this opportunity to thank the industry program co-chairs, Dr. Shengbo Guo, Dr. Nikos Laoutaris, and Dr. Hamid Motahari; the tutorial and panel co-chairs, Prof. Evimaria Terzi and Prof. Ernestina Menasalvas; the WISE challenge program co-chairs Prof. Grigoris Tsoumakas, Prof. Apostolos

Papadopoulos and Prof. Weining Qian; the Workshop co-Chairs, Prof. Armin Haller and Prof. Barbara Catania; the publication chair Prof. Xue Li; the Local Organizing Committee chairs Prof. Nick Bassiliades and Prof. Eleftherios Angelis; the publicity co-chairs, Prof. Fang Li, the Prof. Roger Whitaker and the Prof. George Pallis; and the WISE society representative, Xiaofang Zhou. The editors and chairs are grateful to the Web site and social media master, Dr. Ioannis Karydis for his continuous active support and commitment.

In addition, special thanks are due to the members of the International Program Committee and the external reviewers for a rigorous and robust reviewing process. We are also grateful to the Department of Informatics of the Aristotle University, and the International WISE Society for supporting this Conference.

We expect that the ideas that have emerged in WISE 2014 will result in the development of further innovations for the benefit of scientific, industrial and societal communities.

October 2014

Azer Bestavros
Boualem Benatallah
Yannis Manolopoulos
Athena Vakali
Yanchun Zhang

Organization

General Chairs

Yannis Manolopoulos   Aristotle University, Greece
Yanchun Zhang   Victoria University, Australia

Program Committee Chairs

Boualem Benatallah   University of New South Wales, Australia
Azer Bestavros   Boston University, USA
Athena Vakali   Aristotle University, Greece

Industry Program Chairs

Shengbo Guo   Samsung Information Systems America, USA
Nikos Laoutaris   Telefonica Research, Spain
Hamid Motahari   IBM, USA

Demo Co-chairs

Srdjan Krco   Ericsson/University of Belgrade, Serbia
Joao Fernandes   Alexandra Institute, Denmark

Tutorial and Panel Co-chairs

Ernestina Menasalvas   Universidad Politecnica de Madrid, Spain
Evimaria Terzi   Boston University, USA

WISE Challenge Program Chairs

Weining Qian   East China Normal University, China
Apostolos Papadopoulos   Aristotle University, Greece
Grigorios Tsoumakas   Aristotle University, Greece

Workshops Co-chairs

Barbara Catania   University of Genoa, Italy
Armin Haller   CSIRO, Australia

Publication Chair

Xue Li   University of Queensland, Australia

Local Organizing Committee Chairs

Nick Bassiliades   Aristotle University, Greece
Eleftherios Angelis   Aristotle University, Greece

Publicity Co-chairs

Fang Li   Shanghai Jiaotong University, China
Roger Whitaker   Cardiff University, UK
George Pallis   University of Cyprus, Cyprus

WISE Society Representative

Xiaofang Zhou   University of Queensland, Australia

Web and Social Media Master

Ioannis Karydis   Ionian University, Greece

Program Committee

Karl Aberer   Ecole Polytechnique Federale de Lausanne,
                                   Switzerland
Divy Agrawal   University of California at Santa Barbara, USA
Marco Aiello   University of Groningen, The Netherlands
Anastasia Ailamaki   Ecole Polytechnique Federale de Lausanne,
                                   Switzerland
Virgilio Almeida   Federal University of Minas Gerais, Brazil
Eleftherios Angelis   Aristotle University of Thessaloniki, Greece
Leonidas Anthopoulos   TEI of Thessaly, Greece
Demetrios Antoniades   Georgia Institute of Technology, USA
Costin Badica   University of Craiova, Romania
Ricardo Baeza-Yates   Yahoo! Labs, Spain
Alistair Barros   Queensland University of Technology, Australia
Ladjel Bellatreche   Laboratoire d'Informatique et d'Automatique
                                   pour les Systemes, France
Boualem Benatallah   University of New South Wales, Australia
Salima Benbernou   University Paris Descartes, France
Christos Berberidis   International Hellenic University, Greece

Arun Iyengar   IBM TJ Watson, USA
Peiquan Jin   University of Science and Technology of China,
                                          China
George Kakarontzas   TEI of Thessaly, Greece
Georgia Kapitsaki   University of Cyprus, Cyprus
Helen Karatza   Aristotle University of Thessaloniki, Greece
Ioannis Katakis   University of Athens, Greece
Dimitrios Katsaros   University of Thessaly, Greece
Vasiliki Kazantzi   TEI of Thessaly, Greece
Anastasios Kementsietsidis   IBM T.J. Watson Research Center, USA
Fotis Kokkoras   TEI of Thessaly, Greece
Yiannis Kompatsiaris   CERTH, ITI, Greece
Ioannis Kompatsiaris   CERTH, ITI, Greece
Efstratios Kontopoulos   Centre of Research & Technology, Greece
Makrina Viola Kosti   Aristotle University of Thessaloniki, Greece
Manolis Koubarakis   University of Athens, Greece
Srdjan Krco   TEI of Thessaly, Greece
Fang Li   Jiao Tong University, China
Xue Li   School of ITEE, Australia
Xue Li   The University of Queensland, Australia
Xuemin Lin   The University of New South Wales, Australia
Georgios Meditskos   Centre of Research & Technology, Greece
Ernestina Menasalvas   Technical University of Madrid, Spain
Natwar Modani   IBM India Research Lab, India
Hamid Motahari   University of New South Wales, Australia
Hamid Motahari   IBM Almaden, USA
Luis Munoz   University of Cantabria, Spain
Miyuki Nakano   University of Tokyo, Japan
Wilfred Ng   The Hong Kong University of Science &
                                          Technology, China
Mara Nikolaidou   Harokopio University, Greece
Kjetil Norvag   Norwegian University of Science and
                                          Technology, Norway
Alexandros Ntoulas   Zynga, USA
George Pallis   University of Cyprus, Cyprus
Apostolos Papadopoulos   Aristotle University of Thessaloniki, Greece
Stelios Paparizos   Microsoft Research, USA
Josiane Xavier Parreira   DERI - National University of Ireland, Ireland
Sophia Petridou   Aristotle University of Thessaloniki, Greece
Evaggelia Pitoura   University of Ioannina, Greece
Dimitris Plexousakis   University of Crete, Greece
Weining Qian   East China Normal University, China
Misha Rabinovich   Case Western Reserve University, USA
Sandra Patricia Rojas Berrio   National University of Colombia, Colombia

Keynote Lectures

Prof. Krishna Gummadi

Krishna Gummadi is a tenured faculy member and head of the Networked Systems research group at the Max Planck Institute for Software Systems (MPI-SWS) in Germany. He received his Ph.D. (2005) and M.S. (2002) degrees in Computer Science and Engineering from the University of Washington. He also holds a B.Tech (2000) degree in Computer Science and Engineering from the Indian Institute of Technology, Madras. Krishna's research interests are in the measurement, analysis, design, and evaluation of complex Internet-scale systems. His current projects focus on understanding and building social Web systems. Specifically, they tackle the challenges associated with protecting the privacy of users sharing personal data, understanding and leveraging word-of-mouth exchanges to spread information virally, and finding relevant and trustworthy sources of information in crowds. Krishna's work on online social networks, Internet access networks, and peer-to-peer systems has led to a number of widely cited papers and award (best) papers at ACM/Usenix's SOUPS, AAAI's ICWSM, Usenix's OSDI, ACM's SIGCOMM IMW, and SPIE's MMCN conferences.

"Understanding Information Exchange in Social Media Systems", Monday Oct. 13th, 2014

The functioning of our modern knowledge-based societies depends crucially on how individuals, organizations, and governments exchange information. Today, much of this information exchange is happening over the Internet. Recently, social media systems like Twitter and Facebook have become tremendously popular, bringing with them profound changes in the way information is being exchanged online. In this talk, focus is on understanding the processes by which social media users generate, disseminate, and consume information. Specifically, the trade-offs between relying on the information generated by (i.e., wisdom of) crowds versus experts and the effects of information overload on how users consume and disseminate information are investigated. Limitations of our current understanding is highlighted and arguing that an improved understanding of information exchange processes is the necessary first step towards designing better information retrieval (search or recommender) systems for social media.

 **Dr. Mike Fisher**

Mike Fisher is a Chief Researcher in the Research and Innovation Department of British Telekom, UK. Following a PhD in Physics from University of Surrey, he joined BT and worked on "blue sky" research projects investigating semiconductor optical materials and devices. He later moved into distributed systems where his research interests have included policy-based management, active networks, Grid computing, Cloud computing and most recently the Internet of Things. Mike has had a strong involvement in collaborative projects on these topics at national and European level. He was involved in the establishment of the NESSI European Technology Platform and was the Chairman of the ETSI Technical Committee responsible for Grid and Cloud. His current focus is on information-centric network services that can enable improved exchange of information, and the value that these can deliver.

"Connected Communities", Monday Oct. 13th, 2014

Any process or activity can be improved by timely access to better information. The long-held vision of a connected world is now becoming a reality as technological advances make it increasingly cost-effective to publish, find and use a huge variety of data. New ways of managing information offer the potential for transformational change, with the network as the natural point of integration. In this talk some of BT recent work exploring technologies to promote sharing in communities unified by an interest in similar information is highlighted. This includes experiences in a number of sectors including transport, supply chain and future cities.

 **Prof. Santo Fortunato**

Santo Fortunato is Professor of Complex Systems at the Department of Biomedical Engineering and Computational Science of Aalto University, Finland. Previously he was director of the Sociophysics Laboratory at the Institute for Scientific Interchange in Turin, Italy. Prof. Fortunato got his PhD in Theoretical Particle Physics at the University of Bielefeld In Germany. He then moved to the field of complex systems. His current focus areas are network science, especially community detection in graphs, computational social science and science of science. His research has been published in leading journals, including Nature, PNAS, Physical Review Letters, Reviews of Modern Physics, Physics Reports and has collected over 10,000 citations (Google Scholar). His review article Community detection in graphs (Physics Reports 486, 75-174, 2010) is the most cited paper on networks of the last years. He is the recipient of the Young Scientist Award in Socio- And Econophysics 2011 from the German Physical Society.

"Detecting Communities in Networks", Tuesday Oct. 14th, 2014

Finding communities in networks is crucial to understand their structure and function, as well as to identify the role of the nodes and uncover hidden relationships between nodes. In this talk I will briefly introduce the problem and then focus on algorithms based on optimization. I will discuss the limits of global optimization approaches and the potential advantages of local techniques. Finally I will assess the delicate issue of testing the performance of methods.

Tutorials (Abstracts)

Blocking Techniques
for Web-Scale Entity Resolution

George Papadakis and Themis Palpanas

Institute for the Management of Information Systems
- Athena Research Center, Greece
gpapadis@imis.athena-innovation.gr
Paris Descartes University, France
themis@mi.parisdescartes.fr

Abstract. Entity Resolution constitutes one of the cornerstone tasks for the integration of overlapping information sources. Due to its quadratic complexity, a bulk of research has focused on improving its efficiency so that it can be applied to Web Data collections, which are inherently voluminous and highly heterogeneous. The most common approach for this purpose is blocking, which clusters similar entities into blocks so that the pair-wise comparisons are restricted to the entities contained within each block.

In this tutorial, we elaborate on blocking techniques, starting from the early, schema-based ones that were crafted for database integration. We highlight the challenges posed by today's heterogeneous, noisy, voluminous Web Data and explain why they render inapplicable the early blocking methods. We continue with the presentation of the latest blocking methods that are crafted for Web-scale data. We also explain how their efficiency can be improved by meta-blocking and parallelization techniques.

We conclude with a hands-on session that demonstrates the relative performance of several, state-of-the-art techniques, and enables the participants of the tutorial to put in practice all the topics discussed in the theory.

Community Detection and Evaluation in Social and Information Networks

Christos Giatsidis, Fragkiskos D. Malliaros, and Michalis Vazirgiannis

Ecole Polytechnique, France
http://www.lix.polytechnique.fr/~giatsidis
Ecole Polytechnique, France
http://www.lix.polytechnique.fr/~fmalliaros
Ecole Polytechnique, France
http://www.lix.polytechnique.fr/~mvazirg
Athens University of Economics and Business, Greece

Abstract. Graphs (or networks) constitute a dominant data structure and appear essentially in all forms of information (e.g., social and information networks, technological networks and networks from the areas of biology and neuroscience). A cornerstone issue in the analysis of such graphs is the detection and evaluation of communities (or clusters) - bearing multiple and diverse semantics. Typically, the communities correspond to groups of nodes that tend to be highly similar sharing common features, while nodes of different communities show low similarity. Detecting and evaluating the community structure of real-world graphs constitutes an essential task in several areas, with many important applications. For example, communities in a social network (e.g., Facebook, Twitter) correspond to individuals with increased social ties (e.g., friendship relationships, common interests). The goal of this tutorial is to present community detection and evaluation techniques as mining tools for real graphs. We present a thorough review of graph clustering and community detection methods, demonstrating their basic methodological principles. Special mention is made to the degeneracy (k-cores and extensions) approach for community evaluation, presenting also several case studies on real-world networks.

Extensions on Map-Reduce

Himanshu Gupta, L. Venkata Subramaniam,
and Sriram Raghavan

IBM Research, India
http://researcher.watson.ibm.com/researcher/view.php?person=in-higupta8
http://researcher.watson.ibm.com/researcher/view.php?person=in-lvsubram
http://researcher.watson.ibm.com/researcher/view.php?person=
in-sriramraghavan

Abstract. This tutorial will present an overview of various systems and algorithms which have extended the map-reduce framework to improve its performance. The tutorial will consist of four parts. The tutorial will start with an introduction of the map-reduce framework along-with its strengths and limitations. The goal of this tutorial is to explain how research has attempted to overcome these limitations.

The first part will look at systems which focus on (1) processing relational data on map-reduce and (2) on providing indexing support on map-reduce. The second part will then move on to systems which provide support for various classes of queries like join-processing, incremental computation, iterative and recurring queries etc. The third part will present an overview of systems which improve the performance of map-reduce framework in a variety of ways like skew-management, data-placement, reusing the results of a computation etc. The fourth part will finally look at various initiatives in this space currently underway within IBM across all global labs.

Similarity Search: Navigating the Choices for Similarity Operators

Deepak S. Padmanabhan and Prasad M. Deshpande

http://researcher.watson.ibm.com/researcher/view.php?person=
in-deepak.s.p
http://researcher.watson.ibm.com/researcher/view.php?person=
in-prasdesh

Abstract. With the growing variety of entities that have their presence on the web, increasingly sophisticated data representation and indexing mechanisms to retrieve relevant entities to a query are being devised. Though relatively less discussed, another dimension in retrieval that has recorded tremendous progress over the years has been the development of mechanisms to enhance expressivity in specifying information needs; this has been affected by the advancements in research on similarity operators. In this tutorial, we focus on the vocabulary of similarity operators that has grown from just a set of two operators, top-k and skyline search, as it stood in the early 2000s. Today, there are efficient algorithms to process complicated needs such as finding the top-k customers for a product wherein the customers are to be sorted based on the rank of the chosen product in their preference list. Arguably due to the complexity in the specification of new operators such as the above, uptake of such similarity operators has been low even though emergence of complex entities such as social media profiles warrant significant expansion in query expressivity. In this tutorial, we systematically survey the set of similarity operators and mechanisms to process them effectively. We believe that the importance of similarity search operators is immense in an era of when the web is populated with increasingly complex objects spanning the entire spectrum, though mostly pronounced in the social and e-commerce web.

Table of Contents – Part II

Social Online Networks

Predicting Elections from Social Networks Based on Sub-event
Detection and Sentiment Analysis   1
  Sayan Unankard, Xue Li, Mohamed Sharaf, Jiang Zhong,
  and Xueming Li

Sonora: A Prescriptive Model for Message Authoring on Twitter   17
  Pablo N. Mendes, Daniel Gruhl, Clemens Drews, Chris Kau,
  Neal Lewis, Meena Nagarajan, Alfredo Alba, and Steve Welch

A Fuzzy Model for Selecting Social Web Services   32
  Hamdi Yahyaoui, Mohammed Almulla, and Zakaria Maamar

Insights into Entity Name Evolution on Wikipedia   47
  Helge Holzmann and Thomas Risse

Assessing the Credibility of Nodes on Multiple-Relational Social
Networks   62
  Weishu Hu and Zhiguo Gong

Result Diversification for Tweet Search   78
  Makbule Gulcin Ozsoy, Kezban Dilek Onal,
  and Ismail Sengor Altingovde

WikipEvent: Leveraging Wikipedia Edit History for Event Detection . . .   90
  Tuan Tran, Andrea Ceroni, Mihai Georgescu,
  Kaweh Djafari Naini, and Marco Fisichella

Feature Based Sentiment Analysis of Tweets in Multiple Languages 109
  Maike Erdmann, Kazushi Ikeda, Hiromi Ishizaki, Gen Hattori,
  and Yasuhiro Takishima

Incorporating the Position of Sharing Action in Predicting Popular
Videos in Online Social Networks 125
  Yi Long, Victor O.K. Li, and Guolin Niu

An Evolution-Based Robust Social Influence Evaluation Method in
Online Social Networks 141
  Feng Zhu, Guanfeng Liu, An Liu, Lei Zhao, and Xiaofang Zhou

A Framework for Linking Educational Medical Objects:
Connecting Web2.0 and Traditional Education 158
  Reem Qadan Al Fayez and Mike Joy

An Ensemble Model for Cross-Domain Polarity Classification
on Twitter   168
    Adam Tsakalidis, Symeon Papadopoulos, and Ioannis Kompatsiaris

A Faceted Crawler for the Twitter Service   178
    George Valkanas, Antonia Saravanou, and Dimitrios Gunopulos

Diversifying Microblog Posts   189
    *Marios Koniaris, Giorgos Giannopoulos, Timos Sellis,
    and Yiannis Vassiliou*

Software Architectures and Platforms

MultiMasher: Providing Architectural Support and Visual Tools
for Multi-device Mashups   199
    *Maria Husmann, Michael Nebeling, Stefano Pongelli,
    and Moira C. Norrie*

MindXpres: An Extensible Content-Driven Cross-Media Presentation
Platform   215
    Reinout Roels and Beat Signer

Open Cross-Document Linking and Browsing Based on a Visual Plug-in
Architecture   231
    Ahmed A.O. Tayeh and Beat Signer

Cost-Based Join Algorithm Selection in Hadoop   246
    *Jun Gu, Shu Peng, X. Sean Wang, Weixiong Rao, Min Yang,
    and Yu Cao*

Consistent Freshness-Aware Caching for Multi-Object Requests   262
    Meena Rajani, Uwe Röhm, and Akon Dey

ε-*Controlled-Replicate*: An Improved *Controlled-Replicate* Algorithm
for Multi-way Spatial Join Processing On Map-Reduce   278
    Himanshu Gupta and Bhupesh Chawda

REST as an Alternative to WSRF: A Comparison Based
on the WS-Agreement Standard   294
    Florian Feigenbutz, Alexander Stanik, and Andreas Kliem

Web Technologies and Frameworks

Enabling Cross-Platform Mobile Application Development:
A Context-Aware Middleware   304
    Achilleas P. Achilleos and Georgia M. Kapitsaki

GEAP: A Generic Approach to Predicting Workload Bursts
for Web Hosted Events 319
  Matthew Sladescu, Alan Fekete, Kevin Lee, and Anna Liu

High-Payload Image-Hiding Scheme Based on Best-Block Matching
and Multi-layered Syndrome-Trellis Codes.......................... 336
  Tao Han, Jinlong Fei, Shengli Liu, Xi Chen, and Zhu Yuefei

Educational Forums at a Glance: Topic Extraction and Selection 351
  *Bernardo Pereira Nunes, Ricardo Kawase, Besnik Fetahu,
  Marco Antonio Casanova,
  and Gilda Helena Bernardino B. de Campos*

PDist-RIA Crawler: A Peer-to-Peer Distributed Crawler for Rich
Internet Applications 365
  *Seyed M. Mirtaheri, Gregor V. Bochmann,
  Guy-Vincent Jourdan, and Iosif-Viorel Onut*

Understand the City Better: Multimodal Aspect-Opinion
Summarization for Travel 381
  Ting Wang and Changqing Bai

Event Processing over a Distributed JSON Store: Design and
Performance.. 395
  Miki Enoki, Jérôme Siméon, Hiroshi Horii, and Martin Hirzel

Cleaning Environmental Sensing Data Streams Based on Individual
Sensor Reliability 405
  Yihong Zhang, Claudia Szabo, and Quan Z. Sheng

Managing Incentives in Social Computing Systems with PRINGL 415
  Ognjen Scekic, Hong-Linh Truong, and Schahram Dustdar

Consumer Monitoring of Infrastructure Performance in a Public
Cloud 425
  *Rabia Chaudry, Adnene Guabtni, Alan Fekete, Len Bass,
  and Anna Liu*

Business Export Orientation Detection through Web Content
Analysis 435
  *Desamparados Blazquez, Josep Domenech, Jose A. Gil,
  and Ana Pont*

Web Innovation and Applications

Towards Real Time Contextual Advertising 445
  Abhimanyu Panwar, Iosif-Viorel Onut, and James Miller

On String Prioritization in Web-Based User Interface Localization 460
   Luis A. Leiva and Vicent Alabau

Affective, Linguistic and Topic Patterns in Online Autism
Communities 474
   Thin Nguyen, Thi Duong, Dinh Phung, and Svetha Venkatesh

A Product-Customer Matching Framework for Web 2.0 Applications ... 489
   Qiangqiang Kang, Zhao Zhang, Cheqing Jin, and Aoying Zhou

Rapid Development of Interactive Applications Based on Online Social
Networks 505
   Ángel Mora Segura, Juan de Lara, and Jesús Sánchez Cuadrado

Introducing the Public Transport Domain to the Web of Data 521
   Christine Keller, Sören Brunk, and Thomas Schlegel

Measuring and Mitigating Product Data Inaccuracy in Online
Retailing.. 531
   Runhua Xu and Alexander Ilic

Challenge

WISE 2014 Challenge: Multi-label Classification of Print Media Articles
to Topics 541
   Grigorios Tsoumakas, Apostolos Papadopoulos, Weining Qian,
   Stavros Vologiannidis, Alexander D'yakonov, Antti Puurula,
   Jesse Read, Jan Švec, and Stanislav Semenov

Author Index 549

Table of Contents – Part I

Web Mining, Modeling and Classification

Coupled Item-Based Matrix Factorization 1
Fangfang Li, Guandong Xu, and Longbing Cao

A Lot of Slots – Outliers Confinement in Review-Based System 15
Roberto Di Pietro, Marinella Petrocchi, and Angelo Spognardi

A Unified Model for Community Detection of Multiplex Networks 31
Guangyao Zhu and Kan Li

Mining Domain-Specific Dictionaries of Opinion Words 47
*Pantelis Agathangelou, Ioannis Katakis, Fotios Kokkoras,
and Konstantinos Ntonas*

A Community Detection Algorithm Based on the Similarity
Sequence.. 63
Hongwei Lu, Qian Zhao, and Zaobin Gan

A Self-learning Clustering Algorithm Based on Clustering Coefficient ... 79
MingJie Zhong, ZhiJun Ding, HaiChun Sun, and PengWei Wang

Detecting Hierarchical Structure of Community Members by Link
Pattern Expansion Method 95
Fengjiao Chen and Kan Li

An Effective TF/IDF-Based Text-to-Text Semantic Similarity Measure
for Text Classification 105
Shereen Albitar, Sébastien Fournier, and Bernard Espinasse

Automatically Annotating Structured Web Data Using a SVM-Based
Multiclass Classifier 115
Daiyue Weng, Jun Hong, and David A. Bell

Mining Discriminative Itemsets in Data Streams 125
Majid Seyfi, Shlomo Geva, and Richi Nayak

Modelling Visit Similarity Using Click-Stream Data:
A Supervised Approach.. 135
*Deepak Pai, Abhijit Sharang, Meghanath Macha,
and Shradha Agrawal*

BOSTER: An Efficient Algorithm for Mining Frequent Unordered
Induced Subtrees 146
Israt J. Chowdhury and Richi Nayak

Web Querying and Searching

Phrase Queries with Inverted + Direct Indexes 156
  Kiril Panev and Klaus Berberich

Ranking Based Activity Trajectory Search 170
  Wei Chen, Lei Zhao, Xu Jiajie, Kai Zheng, and Xiaofang Zhou

Topical Pattern Based Document Modelling and Relevance Ranking 186
  Yang Gao, Yue Xu, and Yuefeng Li

A Decremental Search Approach for Large Scale Dynamic
Ridesharing 202
  Ali Shemshadi, Quan Z. Sheng, and Wei Emma Zhang

Model-Based Search and Ranking of Web APIs across Multiple
Repositories 218
  Devis Bianchini, Valeria De Antonellis, and Michele Melchiori

Common Neighbor Query-Friendly Triangulation-Based Large-Scale
Graph Compression 234
  Liang Zhang, Chen Xu, Weining Qian, and Aoying Zhou

Continuous Monitoring of Top-k Dominating Queries over Uncertain
Data Streams.. 244
  Guohui Li, Changyin Luo, and Jianjun Li

Keyword Search over Web Documents Based on Earth Mover's
Distance 256
  *Jiangang Ma, Quan Z. Sheng, Lina Yao, Yong Xu,
  and Ali Shemshadi*

iPoll: Automatic Polling Using Online Search 266
  *Thin Nguyen, Dinh Phung, Wei Luo, Truyen Tran,
  and Svetha Venkatesh*

Web Recommendation and Personalization

Comparing the Predictive Capability of Social and Interest Affinity
for Recommendations... 276
  Alexandra Olteanu, Anne-Marie Kermarrec, and Karl Aberer

End-User Browser-Side Modification of Web Pages 293
  *Oscar Díaz, Cristóbal Arellano, Iñigo Aldalur, Haritz Medina,
  and Sergio Firmenich*

Mobile Phone Recommendation Based on Phone Interest 308
  Bozhi Yuan, Bin Xu, Tonglee Chung, Kaiyan Shuai, and Yongbin Liu

Two Approaches to the Dataset Interlinking Recommendation
Problem   324
    Giseli Rabello Lopes, Luiz André P. Paes Leme,
    Bernardo Pereira Nunes, Marco Antonio Casanova,
    and Stefan Dietze

Exploiting Perceptual Similarity: Privacy-Preserving Cooperative
Query Personalization   340
    Christoph Lofi and Christian Nieke

Identifying Explicit Features for Sentiment Analysis in Consumer
Reviews   357
    Nienke de Boer, Marijtje van Leeuwen, Ruud van Luijk,
    Kim Schouten, Flavius Frasincar, and Damir Vandic

Facet Tree for Personalized Web Documents Organization   372
    Róbert Móro, Mária Bieliková, and Roman Burger

Mobile Web User Behavior Modeling   388
    Bozhi Yuan, Bin Xu, Chao Wu, and Yuanchao Ma

Effect of Mood, Social Connectivity and Age in Online Depression
Community via Topic and Linguistic Analysis   398
    Bo Dao, Thin Nguyen, Dinh Phung, and Svetha Venkatesh

A Review Selection Method Using Product Feature Taxonomy   408
    Nan Tian, Yue Xu, and Yuefeng Li

Semantic Web

A Genetic Programming Approach for Learning Semantic Information
Extraction Rules from News   418
    Wouter IJntema, Frederik Hogenboom, Flavius Frasincar,
    and Damir Vandic

Ontology-Based Management of Conflicting Products in Pixel
Advertising   433
    Ferry Boon, Sabri Bouzidi, Raymond Vermaas, Damir Vandic,
    and Flavius Frasincar

Exploiting Semantic Result Clustering to Support Keyword Search
on Linked Data   448
    Ananya Dass, Cem Aksoy, Aggeliki Dimitriou, and
    Dimitri Theodoratos

Discovering Semantic Mobility Pattern from Check-in Data   464
    Ji Yuan, Xudong Liu, Richong Zhang, Hailong Sun,
    Xiaohui Guo, and Yanghao Wang

An Offline Optimal SPARQL Query Planning Approach to Evaluate
Online Heuristic Planners 480
    Achille Fokoue, Mihaela Bornea, Julian Dolby,
    Anastasios Kementsietsidis, and Kavitha Srinivas

Agents, Models and Semantic Integration in Support of Personal
eHealth Knowledge Spaces 496
    Haridimos Kondylakis, Dimitris Plexousakis, Vedran Hrgovcic,
    Robert Woitsch, Marc Premm, and Michael Schuele

Probabilistic Associations as a Proxy for Semantic Relatedness 512
    Shahida Jabeen, Xiaoying Gao, and Peter Andreae

A Hybrid Model for Learning Semantic Relatedness Using
Wikipedia-Based Features 523
    Shahida Jabeen, Xiaoying Gao, and Peter Andreae

An Ontology-Based Approach for Product Entity Resolution
on the Web 534
    Raymond Vermaas, Damir Vandic, and Flavius Frasincar

Author Index 545

Predicting Elections from Social Networks Based on Sub-event Detection and Sentiment Analysis

Sayan Unankard[1], Xue Li[1], Mohamed Sharaf[1], Jiang Zhong[2], and Xueming Li[2]

[1] School of Information Technology and Electrical Engineering,
The University of Queensland, Brisbane QLD 4072, Australia
[2] Key Laboratory of Dependable Service Computing in Cyber Physical Society
Ministry of Education, Chongqing 400044, China
{uqsunank,m.sharaf}@uq.edu.au, xueli@itee.uq.edu.au,
{zhongjiang,lixuemin}@cqu.edu.cn

Abstract. Social networks are widely used by all kinds of people to express their opinions. Predicting election outcomes is now becoming a compelling research issue. People express themselves spontaneously with respect to the social events in their social networks. Real time prediction on ongoing election events can provide feedback and trend analysis for politicians and news analysts to make informed decisions. This paper proposes an approach to predicting election results by incorporating sub-event detection and sentiment analysis in social networks to analyse as well as visualise political preferences revealed by those social network users. Extensive experiments are conducted to evaluate the performance of our approach based on a real-world *Twitter* dataset. Our experiments show that the proposed approach can effectively predict the election results over the given baselines.

Keywords: election prediction, event detection, sentiment analysis, micro-blogs.

1 Introduction

Micro-blog services such as *Twitter* generate a large amount of messages carrying event information and users' opinions over a wide range of topics. The events discussed on social networks can be associated with topics, locations, and time periods. The events can be in a variety, such as celebrities or political affairs, local social events, accidents, protests, or natural disasters. Messages are posted by users after they have experienced or witnessed the events happening in the real world and they want to share their experiences immediately. For a long-running event like a nation-wide election which usually has fixed start and end times, users may want to monitor sub-events (i.e., hierarchically nested events that break down an event into more refined parts) such as the debate or campaign-launch speech. Alternatively, policy-makers may want to know the feeling of users during the course of an election. The new research in computer science, sociology and political science shows that data extracted from social media platforms yield

B. Benatallah et al. (Eds.): WISE 2014, Part II, LNCS 8787, pp. 1–16, 2014.

accurate measurements of public opinion. It turns out that what people say on *Twitter* is a very good indicator of how they would vote in an election [1,2,3,4].

Existing studies have focused on counting of preferences or sentiment analysis on a party or candidate. They neglect the fact that the voters' attitudes and opinions of people may be different depending on specific political topics and in different geographic areas. Moreover, the same voters participating in different discussions may have different political preferences. In this paper, we are interested in predicting the result of elections from micro-blog data by incorporating sub-event detection and sentiment analysis to detect their political preferences and predict the election results at a state as well as a national level.

The main contributions of this paper are as follows. (1) We present an approach to forecast the vote of a sample user based on the analysis of his/her micro-blog messages and count the votes of users to predict the election results. (2) Sub-event detection and sentiment analysis are incorporated to predict the vote of users as different level of sub-events user engaged in the discussions will affect the prediction results. We evaluate our proposed approach with a real-world *Twitter* data posted by Australia-based users during the 2013 Australian federal election.

The rest of the paper is organized as follows. First, we describe the related work in Section 2. Second, the proposed approach is presented in Section 3. Third, we present the experimental setup and results in Section 4. Finally, the conclusions are given in Section 5.

2   Related Work

2.1   Election Prediction on Social Networks

Twitter is a micro-blog service that has been attracting growing attention from researchers in Data Mining and Information Retrieval. Recently, extensive research has been done on social networks in election prediction [1,2,3,4].

O'Connor et al. in [1] presented the feasibility of using *Twitter* data as a substitute and supplement for traditional polls. Subjectivity lexicon is used to determine opinion scores (i.e., positive and negative scores) for each message in the dataset. Then, the authors computed a sentiment score. Consumer confidence and political opinion are analysed and found to be correlated with sentiment word frequencies in *Twitter* data. However, they do not describe any prediction method. Tumasjan et al. in [2] examined whether *Twitter* can be seen as a valid real-time indicator of political sentiment. The authors also found that the mere number of messages reflects the election result and comes close to traditional election polls. Sang et al. in [3] analysed *Twitter* data regarding the 2011 Dutch Senate elections. The authors presented that improving the quality of the document collection and performing sentiment analysis can improve performance of the prediction. However, the authors need to manually annotate political messages to compute sentiment weight and only the first message of every user is taken into account. In addition, the method relies on polling data to correct for demographic bias. Makazhanov et al. in [4] proposed political preference prediction models based on a variety of contextual and behavioural features. The authors extract all interactions of the candidates, group them on a per-party basis, and build a feature vector for each group. Both a decision tree-based J48 and Logistic regression classifiers are utilized for each party. However, this method needs labelling of training examples for each user. The labelling of training set based on a set of users whose political preferences are known based on the explicit statements (e.g., *"I voted XXX today!"*) made on the Election Day or soon after. Moreover, it does not predict the election outcomes.

However, there are several works presented the problems on election prediction using *Twitter* data. Jungherr et al. in [5] presented that a lack of well-grounded rules for data collection and the choice of parties and the correct period in particular can cause the problems. Metaxas et al. in [6] concluded that *Twitter* data is only slightly better than chance when predicting elections. However, the authors described three necessary standards for predicting elections using *Twitter* data: (1) it should be a clearly defined algorithm, (2) it should take into account the demographic differences between *Twitter* and the actual population, and (3) black-box methods should be avoided. Gayo-Avello has criticized several flaws in [7]. For example, there is not a commonly accepted way of counting votes in *Twitter*. Sentiment analysis is applied as a black-box and demographics are neglected. Nevertheless, the author has outlined some of the research lines for future works in this topic. For example, researches need to clearly define which are a vote and the ground truth; sentiment analysis is a core task and researches should acknowledge demographic bias.

2.2   Sub-Event Detection from Social Networks

There are a few research works on search and retrieval of relevant information from social networks [8,9]. Abel et al. in [8] introduced *Twitcident*, a framework for filtering, searching and analysing information about real-world incidents or crises. Given an incident, the system automatically collects and filters relevant information from *Twitter*. However, this work focuses on how to enrich the semantics of *Twitter* messages to improve the incident profiling and filtering rather than detecting sub-events. A research which is similar to our work is presented by Marcus et al. in [9]. A system for visualizing and summarizing events on *Twitter* in real-time, namely *TwitInfo*, is proposed. The system detects sub-events and provides an aggregate view of user sentiment. Sub-events are extracted by identifying temporal peaks in message frequency and by using weighted moving average and variance to detect an outlier as a sub-event. The *Naïve Bayes* classifier is used to analyse the sentiment of messages into positive and negative via *unigram* features. Training datasets are generated for the positive and negative classes using messages with happy and sad *emoticons*. *Emoticon* is a representation of a facial expression such as a smile or frown, formed by various combinations of keyboard characters and used in electronic communications to convey the writer's feelings or intended tone.

2.3   Sentiment Analysis on Social Networks

There are several research papers discussing sentiment analysis via lexicon-based approaches [10,11]. Meng et al. in [10] presented an entity-centric topic-based opinion summarization framework in *Twitter*. Topic is detected from *hashtags* – human annotated tags for providing additional context and metadata to messages. Target-dependent sentiment classification is used to identify the sentiment orientation of a message. Recent researches in the field of political sentiment analysis are presented by Wang et al. in [11] and Ringsquandl et al. in [12]. A similar work to our approach is introduced in [12]. This work studies the application of the Pointwise Mutual Information measure to extract relevant topics from *Twitter* messages. Unsupervised sentiment classification is proposed; the semantic orientation of word is the most probable class (positive, negative, neutral) of each opinion word according to *synsets* (i.e., synonym) in *WordNet*[1]. The final aspect-level sentiment is determined by a simple aggregation function which sums the semantic orientation of all words in the message that mentions the specific aspect.

3   Proposed Approach

In order to understand whether the activity on *Twitter* can serve as a predictor of the election results, we propose an approach to incorporate sub-event detection and sentiment analysis for each sub-event for predicting user's political preference. The proposed approach consists of three main components: sub-event detection, sentiment analysis and the prediction model. We collected the *Twitter* messages related to the 2013 Australian federal election event to demonstrate our approach. The following information provides details of each component.

3.1   Sub-Event Detection for a Particular Event

The notion of event detection was proposed in our recent work [13] for location-based hotspot emerging events. However, the problem that we address in this paper is how to group a set of micro-blog messages into a cluster (or sub-event) for a particular longer-running event (i.e., an election). The user defines an event by specifying a keyword query. For example, search keywords such as *"election"*, *"Kevin Rudd"*, *"Tony Abbott"*, *"#ausvote" and "#auspol"* are used to collect the data of the 2013 Australian federal election. In the following, we brief the techniques for sub-event detection.

It has three steps as we are not consider the emergence of event. Firstly, the pre-processing was designed to ignore common words that carry less important meaning than keywords and to remove irrelevant data e.g., *re-tweet* keyword, web address and message-mentioned username. Slang word and extensions like "booooored" are replaced by proper English words. The stop words are removed

---

[1] http://wordnet.princeton.edu

and all words are stemmed by using *Lucene 3.1.0 Java API*[2]. Message location identification is conducted in order to understand users' opinions in particular areas. We firstly extract message location from the *geo-tagged* (latitude/longitude) information. If *geo-tagged* information is not available we extract user location in the user profile to query the Australia *Gazetteer* database for acquiring the location's address. Then, if neither of them is available we set user location equal to "Australia".

Secondly, for clustering step, we consider a set of messages where each message is associated with a sub-event. With the number of sub-events being unknown in advance, we applied event detection using hierarchical clustering from our previous work [13] with some modifications. We use a sliding window to divide the messages. The size of the sliding window is defined in time intervals (i.e., one day for our experiment). According to our experiment, the clustering method performs well when using the augmented normalized term frequency and cosine similarity function. The cosine similarity function is used to calculate the similarity between the existing cluster and the new message. Every message is compared with all previous cluster's centroids. The algorithm creates a new cluster for the message if there is no cluster whose similarity to the message is greater than the threshold (α). In order to find the most suitable value for the threshold, we conducted the clustering experiments with different threshold values. Our tests show that when $\alpha = 0.30$ it renders the best performance. The mean is used to represent the centroid of the cluster, which trades memory use for speed of clustering.

Finally, after the clustering is performed, all clusters cannot be assigned as event clusters because they can be private conversations, advertisements or others. A cluster can be considered as sub-event if there is strong correlation between the event location (i.e., location mentioned in the messages) and the user location. For event location identification, we find all terms or phrases which reference geographic location (e.g., country, state and city) from message contents. We simply extract the message-mentioned locations via *Named Entity Recognition (NER)*. We use the *Stanford Named Entity Recognizer* [14] to identify locations within the messages. We also use the *Part-of-Speech Tagging* for *Twitter* which is introduced in [15] to extract proper nouns. We use an extracted terms query into the *Gazetteer* database to obtain candidate locations of the event. We find the most probable location of the event using the frequency of each location in the cluster. The location which has the highest frequency is assigned as the event location. In order to understand what the sub-event cluster is about, we find the set of keywords to represent the sub-event topic. To extract the set of co-occurring keywords, firstly we create a directed, edge-weighted graph. We adopt the smoothed correlation weight function, to calculate the semantic correlation weight between terms. We identify the sub-event topic by extracting the *Strongly Connected Components (SCCs)* from the graph. The details of our algorithm are presented in [13].

---

[2] http://lucene.apache.org

3.2 Political Sentiment Analysis

In general, opinions can be expressed about anything, such as a product, service, person, topic or event and by any person or organization. Entity is used to denote the opinion target. For example, the targets/entities of messages likes *"As much as you dislike XXX please Australia...Hate YYY more! I beg you"* are *"XXX"* and *"YYY"*. Sentiment analysis can be a supervised approach or an unsupervised approach or a combination of the two. In the supervised approach, the process of labelling training datasets requires considerable time and effort. Collecting training datasets for all application domains is very time consuming and difficult. In this paper, we focus on a lexicon-based approach to perform sentiment classification. However, spotting the target/entity in a microblog message is not the focus of this paper. Our method has two steps. First, an opinion lexicon is constructed and then, the opinion is classified, based on a statistical calculation.

For sentiment analysis, the pre-processing is conducted. We performed the part of speech (*POS*) processing from the original messages. We use *Twitter NLP and Part-of-Speech Tagging* proposed by Gimpel et al. in [15] for tagging the messages. Moreover, the *emoticons* are extracted from the messages. Finally, all messages after being tagged are stored in the database.

1) Opinion Lexicon: We used the lexicon dictionary which was introduced in [16]. It consists of 4,783 negative and 2,006 positive, distinct words. However, micro-blog messages are informally written and often contain slang words and abbreviations. The traditional lexicon dictionary does not cover opinion words in micro-blogs. In order to expand the lexicon dictionary, we manually annotated the Internet slang dictionary, downloaded from http://www.noslang.com, into 262 positive and 903 negative slang words. *Emoticons*[3] are also grouped into happy and unhappy facial expressions.

2) Lexicon-based Algorithm: Our algorithm assigns the messages into positive, negative and neutral classes. Given a message, the tasks are divided into three steps: word-level sentiment, aspect-level sentiment and sarcasm identification.

Word-level sentiment: This step aims to mark all opinion words or phrases in the message. Each positive word is assigned an opinion score of +1 while each negative word is assigned the score of −1. We extracted adjectives, adverbs, verbs, nouns, interjections and *hashtags* to assign the opinion score. Also, the happy emoticon is assigned the opinion score of +1 and vice versa. In order to detect a phrase, we applied natural language rules which are shown in Table 1.

In this step, it is important to deal with complex linguistic constructions, such as negation, intensification, diminishes and modality because of their effect on the emotional meaning of the text. Negation and modality are computed in the same way. We defined the rules for negation and intensification as follows. For negation (e.g., "no", "not" and "never"), there are three cases to compute an opinion score (*OS*) of a given phrase.

---

[3] http://en.wikipedia.org/wiki/List_of_emoticons

Table 1. Natural language rules for phrase detection

Rule	Example
Adverb + Adjective	not good, very sad
Comparative Adverb + Adjective	more offensive, more sincere
Adverb + Verb	not vote, never truth
Intensifier/Diminishes + Adverb	really good, slightly nervous
Modals Verb + Verb	can't promise, can't believe

(1) Negation + Neg. e.g., "not bad"; $OS = +1$
(2) Negation + Pos. e.g., "not good"; $OS = -1$
(3) Negation + Neu. e.g., "not work"; $OS = -1$
Intensifiers (e.g., "very", "really" and "extremely") increase the semantic intensity of a neighbouring lexical item, whereas diminishes (e.g., "quite", "less", "slightly") decrease it. The opinion score of a phrase is computed as follows.
(1) Intensifier + Neg. e.g., "very bad"; $OS = -1.5$
(2) Intensifier + Pos. e.g., "very good"; $OS = 1.5$
(3) Diminishes + Neg. e.g., "slightly mad"; $OS = -0.5$
(4) Diminishes + Pos. e.g., "quite good"; $OS = 0.5$

Aspect-Level Sentiment: In this step we aim to compute the opinion orientation for each aspect/target. For the message likes *"As much as you dislike XXX please Australia...Hate YYY more! I beg you"*, we want to extract a pair of opinion word and the aspect such as { *"dislike"* and *"XXX"*} and { *"hate"* and *"YYY"*} then we can calculate the aspect-level score. We applied an opinion aggregation function to assign the final opinion orientation for each aspect in the message. Each aspect has many names that refer to it, even within the same message and clearly, across messages. For example, { *"Tony Abbott"*, *"Abbott"* and *"TonyAbbottMHR"*} refer to the same person who is one of the candidates of the 2013 Australian federal election. As extracting the aspect/target in microblog messages is not the focus of this paper, we simply set the aspects of our experiments to two sets of keywords as follows:
A_1 = { *"Tony Abbott"*, *"Abbott"*, *"TonyAbbottMHR"*},
A_2 = { *"Kevin Rudd"*, *"Rudd"*, *"KRudd"*, *"KRuddPM"*}
Every word opinion score is computed related to its distance to the aspect. The number of words between the current word and the aspect (i.e., the matched keywords in the aspect keyword set) is assigned as the distance of the current word to the aspect. The aspect-level score is computed as:

$$asp\_score(m, A) = \sum_{w_i \in m} \frac{opinion\_score_{w_i}}{min(distance(w_i, a)), a \in A} \qquad (1)$$

where m is the message, A is the set of aspect keywords, w_i is the word in the messages m and a is the aspect keyword in A. The aspect sentiment is positive

Table 2. The statistical information of sarcasm messages

List	Kevin Rudd	Tony Abbott
No. of messages	1,481	3,254
No. of users	959	1,737
No. of users who posted sarcastic messages	48	114
% of users who posted negative sarcasm	100.00%	100.00%
% of users who have the same opinions in every message for a given topic/event	89.58%	92.98%
% of users who have both positive and negative messages for a given topic/event	10.42%	7.02%

if $asp\_score(m, A) > 0$, and is negative when $asp\_score(m, A) < 0$. Otherwise, the aspect sentiment is neutral.

Sarcasm Identification: In addition, micro-blog messages also contain extensive use of irony and sarcasm, which are particularly difficult for a machine to detect [17]. Sarcasm transforms the polarity of the message into its opposite. Negative sarcasm is a message that sounds positive but is intended to convey a negative attitude. Positive sarcasm is a message that sounds negative but is apparently intended to be understood as positive. Watching people's faces while they talk is a good way to pick up on sarcasm. However, it is very difficult to detect sarcasm in writing due to lack of intonation and facial expressions.

In order to understand the sarcastic messages in micro-blogs, we conducted statistical studies. We manually labelled 5,735 messages sent by users around Australia related to one sub-event (i.e., the first debate of the 2013 Australian federal election between *Kevin Rudd* and *Tony Abbott* on 11 August 2013 from 6pm to 9pm). There are 1,481 and 3,254 messages which discussed *Kevin Rudd* and *Tony Abbott*, respectively. The messages are annotated with the polarity being positive, negative or neutral and are also marked as sarcastic messages where applicable. The statistical information for sarcasm is shown in Table 2.

As we can see from Table 2, most users hold negative views on sarcastic messages. Our interest in this task is to mark off whether a message is intended to be sarcastic and assign the polarity of the message. Considering a single message, it is very difficult to classify sarcasm, even for humans. In general, a message like *"XXX: Road is the future of transport! Brilliant."* will be considered as a positive opinion; however, some people in developed countries might think this is a sarcastic message as they have too many roads now. Therefore, the message itself cannot be effective to predict sarcastic message. The previously messaged opinions of the author may help to classify whether the current message tends to be sarcastic or not. However, some people may have different opinions on different topics/sub-events. Based on our observation on sarcasm in micro-blogs we found that most of the micro-blog users have only one opinion on a specific topic or event (89.58% and 92.98% of messages related to *Kevin Rudd* and *Tony Abbott* respectively).

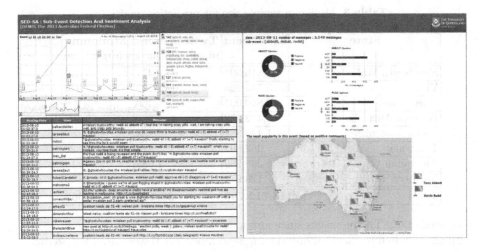

Fig. 1. A dashboard to display sub-event and sentiment of two specific candidates

Therefore, a reasonable ways to classify sarcastic messages are to consider a specific facial expression (i.e., *emoticon* expression) and to compare them with the author's previous messages in the same topic or sub-event. To address this issue, the *emoticon* expression will be compared with the message polarity. All messages accompanied with an *emoticon* are computed as follows.

(1) Pos. message + Neg. *emoticon*; *polarity* = −1

(2) Neg. message + Pos. *emoticon*; *polarity* = +1

If there are no *emoticons* in the message, we compare the aspect opinion score of the current message with the previous messages of the author in the same sub-event and within the same interval of time (i.e., the size of the sliding window). If the current message opinion differed from the overall opinions of previous messages in the same sub-event (i.e., greater than 90%), we change the aspect opinion score of the current message to the same as that for the previous message's opinion. However, if the opinions of the previous messages are divergent, the current message opinion is not changed because it is surmised that this author tends to have different opinions on the same sub-event.

For usability and understanding issues of visualizing the model, we designed a dashboard to display sub-event and sentiment of two specific candidates. Sub-events are presented via *Annotated Time Line Chart* as show in Figure 1 (left) for each day (represented by letters A to Z). The sub-event name is represented by a keywords list described in 3.1. Figure 1 (right) displays how people feel about specific opinion targets for a given sub-event.

3.3   Election Prediction Model

In order to predict the election results, we learn from the professional pollsters. Our prediction model can be divided into two parts; sampling process and user's

Table 3. Minimum sample size for prediction model

State	Enrolment	Twitter users in our dataset	Minimum sample size
New South Wales	4,816,991	13,471	349
Victoria	3,715,925	12,233	270
Queensland	2,840,091	5,360	206
Western Australia	1,452,272	2,630	105
South Australia	1,130,388	2,234	82
Tasmania	362,892	314	26
Australian Capital Territory	265,269	1,683	19
Northern Territory	128,971	268	10
Total	**14,712,799**	**38,193**	**1,067**

vote prediction. The messages since announce Election day (i.e., 4 August 2013) until the day before Election day (i.e., 6 September 2013) were used for predicting the results. Also, we decided to predict the two-party-preferred vote as in Australian politics the candidates will be from the two major parties.

Sampling Process: Since no one can be sure that who will actually vote, the prediction can be approximated by sampling those who will likely to vote. The most important aspect of correct prediction is the selection of a representative. We need to decide who is a particular sample of our prediction and how many people we need to predict. Almost all surveys rely on sampling. This paper analyses a sample of *Twitter* users in Australia. A user account which has *username* contains the words *"news"* and *"TV"* is removed (e.g., "abcnews", "abctv" and etc.) as it is news media account. We compute our sample size by using Cochran's sample size formula [18]. We want to estimate sample size (ss) with 95% confidence and the margin of error no larger than 3%. The formulas used in our sample size calculator are shown as follows:

$$n = \frac{Z^2 p(1-p)}{e^2}, \quad ss = \frac{n}{1 + (n-1)/P} \tag{2}$$

where Z is $Z-score$ corresponds to confidence level ($Z = 1.96$ for confidence level 95%), p is the maximum possible proportion (50% is the most conservative assumption), e is the acceptable margin of error (i.e., the amount of error that you can tolerate) and P is the population size. The minimum sample size (ss) for our experiments is 1,067 people. We randomly select the sample users according to the numbers of enrolment by State[4] as shown in Table 3. We only determine the locations of users because *Twitter* users are not required to specify the age and gender in their profile.

User's Vote Prediction: According to the voters' attitudes and opinion may be different depending on the specific political topic and the voters participating

---

[4] http://results.aec.gov.au/17496/Website/
GeneralEnrolmentByState-17496.htm

in different discussion events may have different political preference, our predicting model were computed based on the significance of sub-event topics and sentiment scores. The sub-event score is calculated to evaluate the significance of each sub-event topic. The sub-event topic will have a high score if there is a lot of a message of them and many users discussing about it. In this work, sub-event score (SE_e) for a given event topic (e) is defined as:

$$SE_e = \frac{NoOfMessages_e}{NoOfTotalMessages} \times \frac{NoOfUsers_e}{NoOfTotalUsers} \tag{3}$$

All sub-events are ranked based on sub-event scores. In order to determine the voter preference among the candidates, for a given user we compute sentiment score for each candidate (i.e., "Abbott" and "Rudd"). For a given user, Aspect Sentiment (AS) scores are defined as Eq. 4 and 5 for "Tony Abbott" and "Kevin Rudd" respectively.

$$AS_{Abbott} = \frac{\sum_{m=1}^{pos}(asp\_score(m, Abbott) \times SE_m)}{\sum_{m=1}^{neg}(|asp\_score(m, Abbott)| \times SE_m)} \times \frac{C_{Abbott}}{C_{Abbott} + C_{Rudd}} \tag{4}$$

$$AS_{Rudd} = \frac{\sum_{m=1}^{pos}(asp\_score(m, Rudd) \times SE_m)}{\sum_{m=1}^{neg}(|asp\_score(m, Rudd)| \times SE_m)} \times \frac{C_{Rudd}}{C_{Abbott} + C_{Rudd}} \tag{5}$$

where $asp\_score(m, A)$ is the aspect-level score of message m, pos is the number of positive messages, neg is the number of negative messages, C_x is the number of both positive and negative messages of aspect x. If a given user posts only positive messages, we assign the summation of negative messages equal to 1. On the other hand, we assign the summation of positive messages equals to 1 when a user posts only negative messages. The voter preference is defined as the highest score out of the two candidates. If the scores are equal, we randomly selected the user vote. In addition, there is another possibility that people has negative sentiment while he still favour to the candidate however it is very difficult to identify.

$$UserVote_u = \begin{cases} \text{``Abbott''} & \text{if } AS_{Abbott} > AS_{Rudd} \\ \text{``Rudd''} & \text{if } AS_{Abbott} < AS_{Rudd} \\ Random(\text{``Abbott''}, \text{``Rudd''}) & \text{otherwise} \end{cases} \tag{6}$$

4   Experiments and Evaluation

In this section, we firstly assess sub-event detection and sentiment analysis methods because both components may affect the final prediction results of our approach. Next, we evaluate our prediction results by computing the Mean Absolute Error (MAE) between the actual and predicted outcomes.

4.1   Dataset and Experimental Setting

A collection of messages posted by Australia-based users (given latitude, longitude and radius) via the *Twitter Search API* service from 4 August 2013 to 8

September 2013 with 808,661 messages with the user's initial event query is used for our experiments. We define an event by specifying the keyword query (i.e., "#ausvotes13", "#election2013", "#AusVotes", "#auspol", "Kevin Rudd" and "Tony Abbott"). We decided to choose this period because the election date is announced on 4 August 2013 and people started to discuss about this event. Also, we decided to choose the keywords related to the two candidates because as in Australian politics the candidates will be from the two major parties. Therefore, in this work we will predict the two-party-preferred vote.

For sub-event detection evaluation, we download the ground truth from *The Sydney Morning Herald* website in *Federal Politics* section[5]. It contains 115 real-world events during 4 August 2013 to 8 September 2013.

For sentiment analysis evaluation, we manually labelled 5,735 messages sent by users in Australia related to the first debate event of the 2013 Australian federal election, between *Kevin Rudd* and *Tony Abbott* on 11 August 2013 from 6pm to 9pm. There are 1,481 messages related to *Kevin Rudd* and 3,254 messages referring to *Tony Abbott*. The messages are annotated with a polarity score (positive, negative or neutral) and sarcasm by three local persons who have political knowledge. We assigned the message polarity score which was determined by the majority view of the three annotators.

For prediction evaluation, the messages since announce election day (i.e., 4 August 2013) until the day before election day (i.e., 6 September 2013) were used for predicting the results. We download the election results from *Australian Electoral Commission* website[6]. The two-party-preferred results for all states and territories as a national summary are compared. The four different national opinion polls are also compared with our results.

4.2 Baseline Approaches

In order to evaluate our approach for detecting sub-events in a collection of *tweets*, we compare our approach performance with temporal peaks detection approach in [9]. The authors bin the messages into a histogram by time (i.e., one hour in this paper). Then, the authors calculate a historically weighted running average of message rate and identify rates that are significantly higher than the mean message rate. A window surrounding the local maximum is identified. Finally, top five frequent terms are presented as event name of each peak.

To evaluate our sentiment analysis method, we compare the performance of our method with aspect-based opinion summarization on *Twitter* data in the domain of politics introduced by Ringsquandl et al. in [12] which is the most similar work to ours. Researchers used the opinion lexicon which is presented in [19]. Semantic orientation of a word is the most probable class (positive, negative, neutral) of each opinion word according to *synsets* in *WordNet*. The final aspect-level sentiment is determined by a simple aggregation function which sums the semantic orientation of all words in the message that mentions the specific aspect.

---

[5] http://www.smh.com.au/federal-politics/the-pulse-live
[6] http://www.aec.gov.au/Elections/Federal_Elections/2013/

Table 4. The performance of sub-event detection

Method	# of detected events	# of real-life events	# of distinct real-life events	Precision (%)	Recall (%)	F1-Score (%)
Peak detection	19	14	14	73.68	12.17	20.89
Our approach	542	229	79	42.25	68.70	**52.32**

Table 5. The performance of sentiment analysis

Aspect	Polarity	No. of Messages	Baseline (%)			Our approach (%)		
			Prec.	Rec.	F1	Prec.	Rec.	F1
Kevin Rudd (ALP)	Positive	327	32.72	37.15	34.80	70.34	47.92	**57.00**
	Negative	726	18.87	70.98	29.82	54.41	83.51	**65.89**
	Neutral	428	79.44	34.00	47.62	67.76	54.92	**60.67**
Tony Abbott (LNP/Coalition)	Positive	334	38.92	22.03	28.14	62.28	31.09	**41.48**
	Negative	1,624	22.84	72.89	34.79	59.05	74.75	**65.98**
	Neutral	1,296	76.47	45.99	57.43	62.89	62.60	**62.74**

Finally, we evaluate our prediction by comparing the performance of our approach with counting-based approaches [2] for our first baseline. For a second baseline, we adopt the idea from [3] by counting the number of tweets one week before the election day and using only the first message of each user for the prediction. However, we do not incorporate polls data in the second baseline. The third baseline is based on sentiment analysis only. We use the same size of our sample and the same algorithm of our sentiment analysis. We use the sum of sentiment scores for each aspect to predict the user votes. The third baseline is compared in order to see how well the combination between sub-event detection and sentiment analysis improve our results.

4.3 Evaluation

In this section, we evaluate the performance of our sub-event detection, sentiment analysis and the prediction approaches. For sub-event detection, we compare the precision, recall and F1-score against the peak detection baseline.

$$Precision_{event} = \frac{\#detect\_realworld\_events}{\#total\_detect\_events}, \tag{7}$$

$$Recall_{event} = \frac{\#distinct\_detect\_realworld\_events}{\#total\_realworld\_events} \tag{8}$$

There is more than one detected event can relate to the same real-world event, then they are considered correct in terms of precision but only one event is considered in counting recall. In order to evaluate the performance of our sentiment analysis method, we compare the the *Precision*, *Recall* and *F1-Score* of each polarity category against the aspect-based baseline.

$$Precision_{opinion} = \frac{T}{C}, \quad Recall_{opinion} = \frac{T}{L} \tag{9}$$

Table 6. MAE for comparing election results with three baselines (%)

State	Election result		Baseline1		Baseline2		Baseline3		Our method	
	ALP	LNP	ALP	MAE	ALP	MAE	ALP	MAE	ALP	MAE
NSW	45.65	54.35	37.94	7.71	44.81	0.84	42.60	3.05	43.11	2.54
VIC	50.20	49.80	37.11	13.09	40.41	9.79	39.00	11.20	38.48	11.72
QLD	43.02	56.98	42.44	0.58	51.35	8.33	45.05	2.03	45.56	2.54
WA	41.72	58.28	37.11	4.61	44.80	3.08	41.38	0.34	41.02	0.70
SA	47.64	52.36	33.21	14.43	46.88	0.76	40.94	6.70	42.38	5.26
TAS	51.23	48.77	26.35	24.88	35.00	16.23	35.11	16.12	38.40	12.83
ACT	59.91	40.09	38.23	21.68	46.58	13.33	42.61	17.30	45.54	14.37
NT	49.65	50.35	35.11	14.54	58.06	8.41	38.08	11.57	42.74	6.91
Average				12.69		7.60		8.54		**7.11**
National	46.51	53.49	37.23	9.28	55.64	9.13	41.69	4.82	42.08	**4.43**

Table 7. MAE for comparing election results (National) with opinion polls (%)

Firm	Date	ALP	LNP	MAE	Remark
Morgan (multi) [20]	4-6 Sep 2013	46.50	53.50	1.01	
ReachTEL [21]	5 Sep 2013	47.00	53.00	0.49	
Newspoll [22]	3-5 Sep 2013	46.00	54.00	0.51	excludes Northern Territory
Essential [23]	1-4 Sep 2013	48.00	52.00	1.49	
Our approach		42.08	57.92	4.43	

where T is the number of correct classified messages in one opinion category, C is the number of messages classified in one opinion category and L represents the number of the true labelled messages in one opinion category. Finally, we evaluate our prediction results by computing the Mean Absolute Error (MAE) between the actual and predicted outcomes.

Table 4 shows the *Precision, Recall* and *F1-Score* of the sub-event detection of our approach against the peak detection baseline. In Table 4, we can observe that our approach can effectively detect real-world events which is significantly larger than the baseline. The baseline can detect smaller number of events because it considers only the temporal peaks in *tweet* frequency. Some events might not be frequently posted on social networks. On the other hand, our approach detects many duplicated events such as the first debate event. There are many different topics discussed during the debate which can cause many clusters when we perform the clustering process. However, our approach outperforms the baseline method by 31.43%. Table 5 represents the performance of the sentiment analysis of our approach against the baseline. It can be seen that our approach can effectively classify the micro-blog messages with a *F1-Score* which is significantly higher than the baseline in the same domain of politics.

Table 6 illustrates the performance of our prediction method against the three baselines. It can be seen that by incorporating sub-event detection and sentiment analysis can effectively improve the prediction accuracy in both state and national levels. In addition, it can correctly predict five out of eight states and territories with smallest error and only 4.43% error for national level. Table 7 presents the performance of our approach against the four different national opinion polls. It can observe that our method comes close to traditional polls with the same trend.

In our study, the incorporating between sub-event detection and sentiment analysis achieved better prediction results than the three baselines. It might suggest that the discussions of sub-event topics that user had engaged in influenced their voting. Also, it can be seen that *Twitter* is able to reflect underlying trend in a political campaign. Even if people who use social media are not completely representative of the public, the amount of attention paid to an issue is an indicator of what is happening in society. Our approach allows researchers to surface user opinions of the social sphere at different time points to determine a view of sentiment for a given event. Also, it turns out that what people say on *Twitter* is a very good indicator of how they will vote.

5   Conclusions

In this paper, we studied a problem of predicting elections based on publicly available data on social networks, like *Twitter*. An effective method of predicting election results is proposed. An approach to detecting sub-events and performing sentiment analysis over micro-blogs in order to predict user preferences is also presented. Extensive experiments are conducted to have evaluated the performance of our approach on a real-world *Twitter* dataset. The proposed approach is effective in predicting election results against the given baselines and comes close to the results of traditional polls. In future work, we will further consider the sarcasm identification and analysis. More studies on the credibility will be conducted in order to remove disinformation and spamming.

Acknowledgement. This paper is partially supported by the Australian Research Council Discovery Project ARC DP140100104.

References

1. O'Connor, B., Balasubramanyan, R., Routledge, B.R., Smith, N.A.: From tweets to polls: Linking text sentiment to public opinion time series. In: ICWSM, pp. 122–129 (2010)
2. Tumasjan, A., Sprenger, T.O., Sandner, P.G., Welpe, I.M.: Predicting elections with twitter: What 140 characters reveal about political sentiment. In: ICWSM, pp. 178–185 (2010)
3. Sang, E.T.K., Bos, J.: Predicting the 2011 dutch senate election results with twitter. In: EACL Workshop on Semantic Analysis in Social Media, pp. 53–60 (2012)
4. Makazhanov, A., Rafiei, D.: Predicting political preference of twitter users. In: ASONAM, pp. 298–305 (2013)

5. Jungherr, A., Jurgens, P., Schoen, H.: Why the pirate party won the german election of 2009 or the trouble with predictions: A response to tumasjan, a., sprenger, t. o., sander, p. g., & welpe, i. m. "predicting elections with twitter: What 140 characters reveal about political sentiment". Soc. Sci. Comput. Rev. 30(2), 229–234 (2012)
6. Metaxas, P.T., Mustafaraj, E., Gayo-Avello, D.: How (not) to predict elections. In: SocialCom/PASSAT, pp. 165–171 (2011)
7. Gayo-Avello, D.: I wanted to predict elections with twitter and all i got was this lousy paper - a balanced survey on election prediction using twitter data. CoRR abs/1204.6441, 1–13 (2012)
8. Abel, F., Hauff, C., Houben, G.J., Stronkman, R., Tao, K.: Semantics + filtering + search = twitcident. exploring information in social web streams. In: HT, pp. 285–294 (2012)
9. Marcus, A., Bernstein, M.S., Badar, O., Karger, D.R., Madden, S., Miller, R.C.: Twitinfo: aggregating and visualizing microblogs for event exploration. In: CHI, pp. 227–236 (2011)
10. Meng, X., Wei, F., Liu, X., Zhou, M., Li, S., Wang, H.: Entity-centric topic-oriented opinion summarization in twitter. In: KDD, pp. 379–387 (2012)
11. Wang, H., Can, D., Kazemzadeh, A., Bar, F., Narayanan, S.: A system for real-time twitter sentiment analysis of 2012 u.s. presidential election cycle. In: ACL (System Demonstrations), pp. 115–120 (2012)
12. Ringsquandl, M., Petkovic, D.: Analyzing political sentiment on twitter. In: AAAI Spring Symposium: Analyzing Microtext, pp. 40–47 (2013)
13. Unankard, S., Li, X., Sharaf, M.A.: Emerging event detection in social networks with location sensitivity. World Wide Web, 1–25 (2014)
14. Jurafsky, D., Martin, J.H.: Speech and Language Processing: An Introduction to Natural Language Processing, Computational Linguistics, and Speech Recognition. Prentice Hall (2000)
15. Gimpel, K., Schneider, N., O'Connor, B., Das, D., Mills, D., Eisenstein, J., Heilman, M., Yogatama, D., Flanigan, J., Smith, N.A.: Part-of-speech tagging for twitter: Annotation, features, and experiments. In: ACL, pp. 42–47 (2011)
16. Hu, M., Liu, B.: Mining and summarizing customer reviews. In: KDD, pp. 168–177 (2004)
17. González-Ibáñez, R., Muresan, S., Wacholder, N.: Identifying sarcasm in twitter: A closer look. In: ACL, pp. 581–586 (2011)
18. Cochran, W.G.: Sampling techniques. Wiley, New York (1977)
19. Wilson, T., Wiebe, J., Hoffmann, P.: Recognizing contextual polarity in phrase-level sentiment analysis. In: HLT/EMNLP, pp. 347–354 (2005)
20. RoyMorgan: Two party preferred voting intention (%).,
    http://www.roymorgan.com/morganpoll/federal-voting/
    2pp-voting-intention-recent-2013-2016 (accessed: July 7, 2014)
21. ReachTel: Two party preferred result based on (2010), election distribution,
    https://www.reachtel.com.au/blog/7-news-national-poll-5september13
    (accessed: July 7, 2014)
22. NewsPoll: Two party preferred,
    http://polling.newspoll.com.au.tmp.anchor.net.au/image_uploads/130922
    (accessed: July 7, 2014)
23. Essential: Two party preferred, federal politics – voting intention,
    http://essentialvision.com.au/documents/essential_report_130905.pdf
    (accessed: July 7, 2014)

Sonora: A Prescriptive Model
for Message Authoring on Twitter

Pablo N. Mendes, Daniel Gruhl, Clemens Drews, Chris Kau, Neal Lewis,
Meena Nagarajan, Alfredo Alba, and Steve Welch

IBM Research, USA

Abstract. Within social networks, certain messages propagate with more ease
or attract more attention than others. This effect can be a consequence of several
factors, such as topic of the message, number of followers, real-time relevance,
person who is sending the message etc. Only one of these factors is within a user's
reach at authoring time: how to phrase the message. In this paper we examine how
word choice contributes to the propagation of a message.

We present a prescriptive model that analyzes words based on their historic
performance in retweets in order to propose enhancements in future tweet performance. Our model calculates a novel score (SONORA SCORE) that is built on
three aspects of diffusion - volume, prevalence and sustain.

We show that SONORA SCORE has powerful predictive ability, and that it complements social and tweet-level features to achieve an F1 score of 0.82 in retweet
prediction. Moreover, it has the ability to prescribe changes to the tweet wording
such that when the SONORA SCORE for a tweet is higher, it is twice as likely to
have more retweets.

Lastly, we show how our prescriptive model can be used to assist users in
content creation for optimized success on social media. Because the model works
at the word level, it lends itself extremely well to the creation of user interfaces
which help authors incrementally – word by word – refine their message until
its potential is maximized and it is ready for publication. We present an easy to
use iOS application that illustrates the potential of incremental refinement using
SONORA SCORE coupled with the familiarity of a traditional spell checker.

Introduction

It is estimated that 72% of online adults use social media sites [11]. This percentage
is even higher within the subgroup of young adults. Perhaps more surprisingly, the
presence of senior citizens has roughly tripled in recent years. As the usage of social
networking websites become routine for adults of all ages, these platforms represent
an ever increasing opportunity for content sharing for virtually any content-producing
professional or institution.

Authoring popular content for social media is challenging, especially considering the
many variables that contribute to the "uptick" of a message [8,19,21]. Nagarajan et al.
[17] show that presence or absence of attribution can largely dictate how retweetability
is observed in a diffusion network. Hansen et al. [6] show that negative sentiment enhances virality in the news segment, but not in the non-news segment. Of these many

B. Benatallah et al. (Eds.): WISE 2014, Part II, LNCS 8787, pp. 17–31, 2014.

attributes that enhance a message's tendency to propagate, word choice is the only one controllable at the time of writing.

Given parlance variation among different demographics and communities, it is intuitive that word choice impacts an audience's reception of content. This variation is quite pronounced in the "New Media" age [12]. For instance, the word choice of middle age professionals discussing their product goals most certainly differs from teenage students discussing their music interests. This highlights a fundamental reason why word choice is important: by speaking the wrong vernacular one can not only distort the core of a message but also its reach.

In this paper we primarily concentrate upon the impact of word or phrase selection on the propagation of a message throughout its intended audience. To this end, we have developed a measure we call SONORA SCORE to prescribe word changes for an uptick in retweetability. SONORA SCORE estimates how well the language used in a tweet has performed, based on the observation of past data. The intuition behind SONORA SCORE is that certain words may 'resonate' better within a community. Based on this sound-related metaphor, we introduce measures of how 'loud' a word sounds, how prevalent it sounds within a time period, and for how long it sounds.

Authors, if well instructed, can modify their word choices to increase their retweetability. Although it would be beneficial to have a 'spell checker'-equivalent system to help improve a message for highest impact, no such solution has yet been disclosed in the literature. SONORA SCORE can be used in realtime to assist authors by prescribing word changes. In this paper, we highlight the impact of SONORA SCORE on 'words' in a tweet. However the proposed SONORA SCORE is easily extended to n-grams of any length and to other kinds of data such as email, blogs, news articles etc. A mobile application prototype using the SONORA SCORE is shown in Figure 1. The figure illustrates the process of scoring words and helping the user identify those words that have potential for improvement. If the user wishes to change a word the prototype proposes alternative words which our model sees as more effective.

In the remainder of this paper, we start by discussing related research in this field, and providing a high level look at how SONORA SCORE works. Our datasets and experiments are described next, exploring the predictive features of retweetability and the prescriptive function of SONORA SCORE. We also include an application section to illustrate the usage of our score in practice. Lastly, we present our conclusions and propose future work.

Related Work

Most of the analysis of message success in the Twitter-sphere has focused on asking questions of "global" features that might help us understand this phenomenon [21] – do tweets with URLs tend to get shared more? Do past retweets of an author boost future retweetability? What role does the topic of the tweet play? etc. Our focus is somewhat different. The system presented in this work aims to identify **wording-based features** that can be used to prescribe more successful features at message authoring time.

Recently, several studies have used measures of retweets, replies, and likes on Twitter as proxies to measure message virality. Suh et al. [21] found that the age of a user's account on a medium, number of friends they had, and the presence of URLs and hashtags

Fig. 1. Screenshot of prescriptive system for iterative content refinement. A potential tweet is entered, and the system dynamically computes the SONORA SCORE of each word. Words below a threshold are colored to suggest that better options are available. Selecting a word reveals alternatives with their respective SONORA SCORE. The user can choose from those suggestions or enter an alternative.

in the content had strong relationships with retweetability. This finding is supported by Petrović and Osborne [19], who concluded that the number of followers and the number of times a user was added to 'lists', along with content features were strong predictors of message retweetability. Yang et al. [23] found that the author's rate of being mentioned by other people predicts how fast a tweet spreads; including links in tweets often generates more number of tweets; and greater number of posts and mentions for a user are better predictors of longer diffusion hops.

Bigonha et al. [3] found that the readability quality of a message was useful (along with network features) in identifying the most influential members of a network. This stream of work has also been applied outside Twitter. Utilizing linguistic categories, [5] show that the presence of certain stylistic features (such as using assertive words and fewer tentative words) and better readability of abstracts correlated positively with the viral capability of a scientific article. Virality was defined as number of article downloads, citations, and bookmarks received by an article. These studies essentially support the motivation behind our work – content features have undeniable effects on message diffusion and author perception.

Our work is motivated in part by psycholinguistic analysis and readability tests that explore the effect of word and language choice [2], [4]. It addresses limitations of the above conclusions in their lack of prescribing what an author can do to write a more effective message. While we acknowledge that the multitude of features on a medium are central to understanding message "uptake", the goal of our work in contrast to that of our predecessors is not to achieve a global optimization of the problem (how to achieve the most retweets for any tweet) but rather a local optimization problem (what word choices will increase the likelihood of tweet retweetability).

Lakkaraju, McAuley and Leskovec [14] study resubmissions of images on the social network Reddit, and find that wording features can be predictive of whether a title will be a successful resubmission. It is unclear from their evaluation if their language model retains its success without features resubmission-specific features – e.g. words that have been previously successful with a given image. A contemporary paper to ours by Tan, Lee and Pang [22] (their results were not published when we conducted our research) has also confirmed that words have strong predictive ability. They report, for example, that words such as 'rt', 'retweet', 'win', and 'official' are strong predictors of retweet-ability. However, their discussion concentrates on globally successful words – with the danger of including words that are prevalent in spam messages (e.g. 'win'). They do not discuss prescribing word substitutions, which is the main interest in our work.

Approach

The algorithm consists of two disjoint steps: firstly selecting a set of historical tweets and extracting a few pieces of metadata. Secondly, combining metadata into a single score. In pseudo-code, the first part of the algorithm works as follows: (1) Select the word you wish to evaluate. Say, "popcorn".[1] (2) From the corpus of tweets defining your target audience, select all tweets that mention the word "popcorn" (note: while we are looking at a single token here, we support n-grams such as "buttered popcorn") (3) Examine for each of these tweets the following (a) What is its root tweet[2]; (b) How often was that root tweet retweeted; (c) How long did the retweeting go on for; (d) When did the root tweet occur. Computing the SONORA SCORE can be defined as an operation on this set of posts and metadata. In the remainder of this section we will formally define these calculations.

Definitions

While it is possible to focus on different outcomes the most generic form of SONORA SCORE is calculated as a combination of three sub scores that can be understood in an analogy to sounds – i.e. given a word, how well will it sound in the social network – in terms of: VOLUME, PREVALENCE, SUSTAIN.

Posts. Let D be a subset of posts of size $n = |D|$, selected via the mechanism above from a universe (e.g. tweets) as a representative sample of a community c. Hereafter, we will use $D(w)$ as the shorthand notation for the subset of all posts in D that contain the word w.

Retweets. Let $RT(t)$ be the observed *retweetability* of a tweet t where $RT(t)$ is the number of times a tweet has been forwarded by a user through the retweet[3] function on

---

[1] While we use the example of a word here, any attribute of a tweet can define a set. e.g., author, class of influencers, time of tweet, geographic location of tweet, etc.

[2] We use 'root tweet' to refer to the tweet that was the initial source of all retweets.

[3] https://support.twitter.com/articles/77606-faqs-about-retweets-rt

Twitter™. Let NZRT (non-zero retweets) be the set of retweeted posts, and ZRT (zero retweets) be the set of non-retweeted posts. Similarly, NZRT(w) and ZRT(w) are the corresponding subsets of posts containing the word w.

Word Volume. VOLUME of a word w captures the intuition of how 'loud' w 'resonates' in the subset of posts. It is represented by the sum of the retweet counts of all posts in $D(w)$ that have a non-zero retweet count: $V(w) = \sum_{\forall t_i \in \text{NZRT}(w)} RT(t_i)$.

Word Amplitude. AMPLITUDE of a word w is a variant of Volume that models the difference in volume for a word w in a subset of posts that were retweeted versus a subset of posts that were not retweeted: $A(w) = \phi(|\text{NZRT}(w)|) - \phi(|\text{ZRT}(w)|)|$. Here we use $\phi(x \in X) = x - mean(X)$ as a mean-centering transformation function to center the counts around 0.

Word Prevalence. PREVALENCE captures the notion that some words are more common than others over a timespan of interest. It is computed based on the number of elements in $D(w)$ that occur for each day in the timespan $\{\tau-, \tau+\}$ of interest. $P(w) = \sum_{d \in \{\tau-, \tau+\}} DC(w, d)$, and the daily count $DC(w, d)$ is the cardinality of the set of posts t_i in day d that contain the word w: $DC(w, d) = |\{\forall t_i \in D(w) | date(t_i) \in d\}|$.

Word Sustain. SUSTAIN of a word captures the notion of how long a word 'resonates' in a subset. Let us define the r_s score to be the sum of the number of hours from first to last tweet of all t_i where this number is non zero: $S(w) = \sum_{\{\forall t_i \in D(w) | \text{HR}(t_i) > 0\}} \text{HR}(t_i)$. The number of hours (HR) is calculated from the first time a tweet id appeared in the dataset, until the last time it appeared.

Tweet Scores from Word-Level Features. So far, we have defined the word-level scores $V(w)$, $P(w)$ and $S(w)$ as functions over words, i.e. functions of the type $g : w \to \mathbb{R}$. It is also possible to define a corresponding score $V_f(t)$ as an aggregation of word scores $V(w_i)$ for a tweet $t = \{w_1, ..., w_n\}$ according to an aggregation function $f : \{x_1, ..., x_n\} \in \mathbb{R} \to \mathbb{R}$. Corresponding definitions of aggregation functions apply to $P_f(t)$ and $S_f(t)$. Possible choices for f are mean, min, max and stdev.

We can now define $Sonora(t, c)$ for a tweet t and community c as an estimate of how well the tweet's words $t = \{w_1, ..., w_n\}$ resonate with the community c. The SONORA SCORE of a tweet is computed through an aggregation of the $V_f(t)$, $P(t)$ and $S(t)$ scores based on its words: $Sonora(t) = \alpha \cdot V_f(t) + \beta \cdot P_f(t) + \gamma \cdot S_f(t)$. Here α, β, γ are mixture weights to control the influence of each component on the final SONORA SCORE, according to the desired outcome ($\alpha + \beta + \gamma = 1$).

Evaluation

For experiments in this work, we collected roughly 225M tweets from an unfiltered 1% feed from Twitter over a three month period, with a few days missing due to network connectivity challenges. To test the predictive nature of SONORA SCORE and other features, the the sample was split in training/testing sets. We chose an arbitrary point τ in time (Apr 01 06:59:59), and divided the data into two sets: $D\tau+$ and $D\tau-$ with all tweets before and after τ, respectively. A total of 3.8% of the tweets in $D\tau+$ has either been tweeted or retweeted in $D\tau-$. The remaining 96.2% are unseen tweets in $D\tau-$.

In order to control for the effect that spam may have on our analysis, we have a collected a set of words that commonly appear in spam messages [1]. We then removed from our dataset every tweet that contained a word from this list. After the spam removal, from an evaluation on 300 tweets, we observed that none of the highly retweeted messages ($\geq$1000) were spam, while 3.85% of the mid-retweet and 3.23% of the low-retweet messages were considered to be spam ($\pm$ 5.66, confidence level 95%). The ranges for defining low-, mid- and high-retweets used were $[0, 10)$, $[10, 1000)$, $[1000, +\infty)$. We consider this an acceptable level of noise for the experiments in this paper. In future work we plan to evaluate the SONORA SCORE's robustness to noise.

We have sampled $100, 000$ tweets from $D\tau+$ that were written in English. We chose English as initial focus, as it is a common language between the authors, but the methods we present are not limited to any particular language. We used *langid.py* [15], an open source automated language detection tool with reported accuracy of 94% on Twitter text [15]. We performed an informal evaluation in our dataset with a random 100 tweets and verified that 87% were correctly detected as English. Tokenization was performed with a state-of-the-art tweet tokenizer from [20] that is aware of Twitter entities such as users, URLs and hashtags. Therefore, barring tokenization errors, occurrences of users, hashtags and URLs can also be treated as "words" for the features described in our approach section.

Features

Our experiments evaluate three categories of features: social features, tweet-level features and word-level features. The **social** features include characteristics related to the Twitter user network and are intended to help model a tweet's prior probability of getting retweeted based only on who is tweeting it to whom. Here, we use the number of followers and number of friends of a user[4]. The **tweet-level** features are intended to describe the tweet's *a priori* likelihood of getting retweeted without looking at the individual words they include or the message they convey. In this work we include the number of hashtags, number of URLs, number of user mentions and the number of stopwords present in a tweet. The **word-level** features, which are the novel contributions of this work and are formally defined in our approach section, focus on the mentions of words in tweets within a period or community of interest. The volume captures how frequently and to what extent tweets containing a word have been retweeted, the prevalence seeks to capture how steadily the word has appeared, and sustain captures for how long a "discussion" continues in which the word appears. For a comprehensive list of features considered by prior art and their predictive abilities please refer to the related work section.

Note that social and tweet-level features are observations, while word-level features are estimates. For instance, the number of hashtags is counted directly from each tweet being evaluated, and so is the number of followers. Meanwhile, the word-level features compute scores estimated from past tweets in $D\tau-$, i.e. not in the $D\tau+$ set from which the testing examples were sampled.

---

[4] Twitter API's 'Get Friend' request returns a collection of users the specified user is following. The 'Get Followers' request returns a collection of users following the specified user.

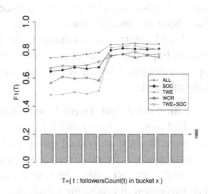

(a) The majority of users have less then 500 followers. Histogram plot clipped at 5000 for presentation purposes.

(b) Prediction F1 performance (y) varies with the number of followers (x). Lines represent different feature sets. Bars represent the sizes of each bin

Fig. 2. F1 improvement obtained from including SONORA SCORE features is higher for buckets with lower number of followers

Prediction Evaluation

In order to validate the effectiveness of the SONORA SCORE, we look to prior work that treats retweetability as an information retrieval problem – i.e. retrieve, from a collection of tweets, those that were retweeted. First, our datasets were preprocessed, assigning a label of 1 to the tweets where $RT(t) > 1$, and 0 otherwise. Under this experimental setting, different approaches can be compared by their ability to correctly retrieve as many of the retweeted posts as possible (high recall) while making as few classification mistakes as possible (high precision).

We selected 80% of the data for training, and performed testing with the remaining 50,000 tweets. Social features and tweet features were directly extracted from each tweet being tested. For the word features, the SONORA SCORES were computed from word occurrences in $D\tau-$, in order to avoid any bias. We experimented with three different popular models for predicting usefulness of features. In all cases, the models have been tested on the same sample from $D\tau+$.

First, we were interested in learning optimal weights to linearly combine our features into one SONORA SCORE that optimizes the likelihood a message will get retweeted. For that purpose, we have trained several Generalized Linear Models (GLM) [18] using subsets of our features. An advantage of GLMs is that they are transparent and friendly for user-interaction – the weights can be displayed as knobs or sliders on an interface, giving the users the freedom to disagree with our model's suggestions.

To explore non-linear combinations of our features, we also trained Conditional Inference Tree (CIT) models [9]. CITs learn relationships between features and the retweetability labels by recursively performing binary partitions in the feature space until no significant association between features and the labels can be stated.

Finally, we also included in our experiments Random Forests (RF) models [7], as they have been repeatedly shown in literature to perform well in a multitude of classification tasks. RFs learn a large number of decision trees that can be used as an ensemble to collectively predict the label at classification time. One disadvantage of RFs is that they operate as a black box, making it harder to take user input into consideration when tweaking the model.

Findings for Social and Tweet-Level Features. Table 1 shows the performance of each method (rows) in terms of F1 for each of the feature sets (columns). The social features have the best individual group performance, reflecting the intuition that the more popular you are, the more retweets you tend to receive. Out of the scores we tested, FOLLOWERS COUNT, USER MENTIONS, NUMBER OF HASHTAGS and NUMBER OF URLs were the most successful in predicting which tweets were retweeted.

Table 1. F1 results on our 50K sample from $D\tau+$using social features (SOC), tweet-level features (TWE), both social and tweet-level features (SOC+TWE), our word-level features (WOR) and a combination of all features (ALL)

	SOC	TWE	SOC+TWE	WOR	ALL
CIT	0.71	0.70	0.73	0.59	0.74
RF	0.76	0.70	0.75	0.70	**0.82**
GLM	0.65	0.69	0.66	0.56	0.68

It is a fact that the prior probability of a message getting retweeted at all, independently of what one is writing about, depends on how many people will see a particular tweet. A person cannot retweet a post they have not seen. The number of people that see a tweet, in turn, depends on a number of factors. Intuitively, the more followers a user has, the higher the likelihood that their tweets will be seen. One the other hand, hashtags are topic markers that are commonly used for searching, therefore could serve as a way to send the message beyond the stream of followers. Similarly, tweet analysis software will notify users when they are mentioned, which is another way to reach out to users that are not followers. Other features that may impact message visibility include time of the day, trendiness of topic, and other variables that are out of the scope of this work.

We argue that the high performance of the social and tweet-level features in our results confirms the intuition that targeting a larger audience increases the chances that someone will retweet a message.

Findings for Word-Level Features. Note that both social and tweet-level features were extracted from the actual tweet being evaluated, while the word-level features were estimated from data from the previous month i.e. $D\tau-$. We find it remarkable that they generalize well over time and offer powerful predictive ability. The best word-level model results (WOR) – i.e. based on RFs – are only .06 F1 points away from the best social (SOC) and tweet-level models (TWE). Moreover, by aggregating our word-level features with social and tweet-level (ALL), we are able to obtain an increase in .06 F1 points over the best social model (SOC), .12 over the best tweet-level model (TWE) and .07 over the combination of social and tweet-level features (SOC+TWE). This is encouraging evidence that word-level features can support the detection of aspects of a tweet's message that lead to better or worse retweetability.

This is encouraging evidence that word-level features can support us detecting aspects of a tweet's message that leads to better or worse retweetability. This characteristic also sets us apart from previous work, as it does not suffice to focus on identifying "good" features, but also identifying "bad" features that when changed might provide better "uptake".

Prescription Evaluation

A prescriptive setting is substantially different from the tweet classification/retrieval setting discussed in previous work and evaluated in our previous section. When guiding users in formulating tweets to garner higher "uptake", our choice of features is limited to what can be changed at tweet authoring time.

For instance, in this setting the social features are fixed. If users want to achieve higher "uptake", they cannot easily enhance their social features instantaneously. Similarly, the tweet-level features that performed well in the predictive setting are rather vague in a prescriptive setting. Although we show that tweets with hashtags often get retweeted more, it is unclear from the tweet-level features which hashtags should be used. On the other hand, word-level features have the potential to help us to prescribe words (or hashtags) that have shown good historical performance in terms of a particular aspect that the user may want to explore.

In summary, we isolate and evaluate the performance of word-level features. We go about it in two ways. First, simulate a prescriptive setting by investigating the performance of each approach when the social features are fixed and tweet-level features are known. Second, we evaluate the likelihood that a suggestion given based on the SONORA SCORE will yield a higher retweet rate.

The effect of fixed social features. The histogram for the number of followers in Figure 2a shows that a large fraction of users in our sample have between 0-500 followers. Examining a particularly dense area (100-200 followers) shows (see Table 2) the F1 performance of their tweets which contain user mentions. This simulates a particular prescriptive setting, where a given user may be required to mention someone, while having at that point in time between 100-200 users. It can be noted that social features loose predictive power in this setting, as their variability decreases. The word-level features, however, not only retain their predictive power but are also prescriptive, allowing us to point out which words in a tweet have a low score.

While Table 2 shows a fixed number of followers and user mentions, Figure 2b shows the performance of each approach across groups of users with distinct follower counts. We start by selecting subsets of data with the same or similar number of followers. For that purpose, we sort the tweets based on their number of followers, sweep the space of FOLLOWERS COUNT, and partition the data into 10 bins. In Figure 2b, each bar on the horizontal axis represents one bin, each containing 2000 tweets. The small vertical axis on the right-hand size of the figure displays the scale of bin sizes.

Note, for users with a large number of followers, it is possible to predict retweetability very well based on social features alone. However, for tweets that were sent to

Table 2. F1 results on a subset with fixed settings: containing user mentions and with number of followers between 100 and 200. Columns refer to social features (SOC), tweet-level features (TWE), word-level features (WOR) and a combination of all features (ALL).

	SOC	TWE	WOR	ALL
CIT	0.63	0.53	0.63	0.45
RF	0.71	0.45	**0.79**	**0.83**
GLM	0.53	0.45	0.56	0.56

less than 800 followers, closer inspection of content is warranted. In those cases, adding word-level and tweet-level features to the mix significantly increases performance.

A similar effect to the followers count is seen with tweet-level features, which underperforms in tweets with lower number of followers and contributes more for popular users. In all cases, adding word-level features has helped increase F1 for all buckets for a given fixed social setting.

Success probability. Finally, we test how often a prescription from our system yields enhancements on a tweet's chance to be retweeted based on a Monte Carlo-style evaluation. We randomly draw (1M times, with replacement) two distinct tweets A and B from the set, and check: if $Sonora(A) > Sonora(B)$ is it also the case that $RT(A) > RT(B)$? In essence, this tests the assertion "If the system tells you one tweet is better than the other, what are the odds it is correct".

Table 3 can be interpreted as follows: Independent of the number of followers the best individual word-level predictor of retweet seems to be the Volume (prior effectiveness of those words). If combined with hashtags (as has been discussed prior) achieves $\approx .65$. This value can be thought of as "if the SONORA SCORE tells you that tweet A has a better phrasing than tweet B, it will be right twice as often as it is wrong.

$$0.574 \cdot A_{min} + 1.51 \cdot A_{max} - .00003 \cdot V_{min} + .0002 \cdot V_{mean}$$
$$-0.0002 \cdot V_{max} + 0.0002 \cdot V_{sum} - 0.000005 \cdot P_{sum}$$
$$-0.002 \cdot S_{mean} + .0003 \cdot S_{max} - 0.00008 \cdot S_{sum} +$$
$$1.35 \cdot \text{nURLs} - 0.637 \cdot \text{nUserMentions} + 1.29$$

Fig. 3. Formula of the best performing GLM (weights truncated for presentation)

We have searched through the space of feature combinations for those that could be used in a prescriptive setting with highest probability of success. Table 3 reports the results. The best performing model relied on A_{min}, A_{max}, V_{min}, V_{mean}, V_{max}, V_{sum}, P_{sum}, S_{mean}, S_{max}, S_{sum} as defined in our approach section, as well as nURLs, and nUserMentions. Figure 3 shows the formula for the best performing GLM, which obtained .69. These numbers are skewed by the "social" features not in this prescriptive model – with them included, the predictor goes to $\approx .8$, or odds are that it will be right four times as often as it is wrong.

In short, there are many factors that impact a tweets "success", but not unlike a spelling or grammar checker, our prescriptive model gives a fair indication of what needs a second look.

Table 3. Relative frequency of a SONORA SCORE recommendation yielding an RT improvement

Model	Feature Sets	P(success)
CIT	Amplitude, Volume, Prevalence, Sustain, nURLs, nUserMentions	72.12%
GLM	Amplitude, Volume, Prevalence, Sustain, nURLs, nUserMentions	69.04%
CIT	Amplitude, Volume, Prevalance and Sustain	68.71%
GLM	Amplitude, Volume, Prevalence and Sustain	66.67%
CIT	Volume, nHashtags	65.44%

Prescribing word substitutes for an audience. In order to prescribe message enhancements at authoring time, our system needs to find word substitutes that convey an equivalent message and that have better performance within an audience. In this work we have used WordNet [16] synonyms as our source of equivalent words. Other lexical databases, thesauri, or techniques for automatic extraction of related words [13] could also be used. To estimate better performance within an audience, we use the aforementioned word-level features.

Table 4. Ranked Wordnet synonyms of 'great' for two datasets ($\rho = 0.74$). Omitting the top 3 suggestions (great, greatest and greater) as they did not vary between groups.

interest	synonym of 'great'
poetry	..., cracking, bully, **neat**, smashing, keen, **swell**, **nifty**, groovy, dandy, bang-up
science	..., **neat**, keen, smashing, **nifty**, cracking, bully, groovy, dandy, bang-up, **swell**

In this setting, two questions come to mind. First, do different words really resonate differently with different audiences? Second, do audiences vary only in their interests for different topics, or do they also differ in their preference for more topic-independent language constructs such as adjectives?

Given a word w_i, we look at how well each of the synonyms of w_i have performed for a target audience. For example, consider two audiences: one interested in poetry and one interested in science. If both groups seem to prefer the same synonyms, then there is little a system can do to prescribe word substitutions to enhance a message's score using this method. However, if there is a difference in preference for different synonyms between the groups, then there is a real opportunity to prescribe message changes to target different audiences.

We collected the set of all synonyms in WordNet for all words mentioned in our dataset from March and April, keeping the synsets with size greater than two. We also selected two focused tweet datasets, filtering for all users mentioning 'poetry' or 'science' in their user profiles. Then, for each set, we ranked the synonyms according to the number of times they appeared in a retweeted message. We then contrasted the rankings with one another. As an example, Table 4 shows synonyms of 'great', and the rank for each of its synonyms in the 'poetry' and 'science' sets. Although the most common forms 'great', 'greater' and 'greatest' seem to be the most common within both groups, there is significant difference in the usage of 'nifty', 'neat' and 'swell'.

We used Spearman's rank correlation coefficient (ρ) to quantify the similarity in ranking between two synonym sets. A $\rho = 1$ indicates that the two groups have identical preference within that synonym set, while a $\rho = -1$ indicates opposite preference. The rankings shown in Table 4 have $\rho = 0.74$.

In order to control for variation in language that may occur in a dataset by chance, we analyzed the distribution of ρ scores for our focused datasets in contrast with a base dataset of retweets in March and in April. Table 5 shows the average ρ values for our focused (poetry-science) and base datasets (mar-apr).

Table 5. Rank correlation ρ between two random samples (mar-apr) is higher than between two samples focused on different audiences (poetry-science)

data sets / ρ	adjectives	nouns	verbs
poetry-science	0.79	0.73	0.73
mar-apr	0.88	0.79	0.76

For all three classes of words, the rank correlation in our base dataset is higher than in our focused data sets. Therefore, different synonyms get retweeted at different rates for the poetry and science audiences(more so than two random samples). This highlighs the importance of word choice for message authoring, and shows that the differences in language go beyond interest in different topics.

Applications

The SONORA SCORE application prototype has been developed to help social media communication professionals message to varied audiences. However, for the sake of the argument, let us pick an intuitive example. Consider two authors: June, a teenage girl, and Augustus, a middle aged man. Both are trying to write a tweet thanking a friend for introducing them to people in a social situation. We select these two social groups as there is a strong difference in common diction between them; a difference that most people are at least somewhat familiar with.

Audience Specific Sonora Score. First, let's consider the two tweets being sent out. June is going to tweet. We highlighted (e.g. [green]) those non-stop words that particularly resonate (i.e., have a high SONORA SCORE) with her target audience: *"give a [shoutout] to my [awesome] BFF for bringing me to her classmate's [kickin] hip-hop party."* Contrast the language with Augustus' tweet: *"[thanks] to my social guru for helping me shamelessly [network] at the [xfactor] party - mucho appreciated."* Both of these authors clearly reflect (or understand) their target audiences and are writing appropriately for their peer groups. Several of the words they are using have very good SONORA SCORE and none of them are particularly poor choices. But consider what would happen if June had sent a tweet phrased like Augustus' to her peer group. We highlighted (e.g. *red*) the words that may be particularly poor choices (i.e., have a low SONORA SCORE): *"thanks to my *social guru* for helping me *shamelessly* network at the *xfactor* party - *mucho appreciated*."* It can be noticed that many of the "good" words for the older target audience are just average with this one, and a number of the words drop to *red* warnings as likely to not "resonate" well for June. The story is not all that much better if Augustus were to use language from the "younger" diction in his tweet: *"give a *shoutout* to my *awesome BFF* for bringing me to her *classmate's kickin hip-hop* party."* Again we see that this language would likely not resonate very well with Augustus' peer group. This ability to analyze tweets for a particular audience provides the opportunity to do "synthetic market studies" of every tweet looking at a specific target audience. This can be a very useful tool for those who are directing their communications to, at times, strongly differing audiences.

Mobile App Prototype. Implemented as a simple Twitter posting client (see Figure 1), the iOS application prototype allows a user to type a proposed tweet (given a target audience), and get a real-time estimate of how well it will resonate. The design of the user interface was guided by 3 main themes established by the iOS Human Interface Guidelines [10] (i) **deference:** the user interface helps the user to understand and interact with the content, but never competes with it; (ii) **clarity:** text is legible at every size, icons are precise and lucid, adornments are subtle and appropriate, and a sharpened focus on functionality motivates the design; (iii) **depth:** visual layers and realistic motion impart vitality and heighten users' delight and understanding. More specifically we wanted the user interface to: (a) provide immediate feedback on the effectiveness of a tweet while composing it (b) display the SONORA SCORE for each word and the complete tweet; (c) provide detailed SONORA SCORES for individual words; (d) present the user with a list of synonyms with higher SONORA SCORES to replace a word with a low score; (e) follow established interaction patterns of the platform.

To achieve the first objective the user interface makes use of color to show the SONORA SCORE for each word as well as for the overall tweet to provide the user with an immediate understanding of the effectiveness of each word in the tweet. Additionally the overall SONORA SCORE for the tweet is shown in numerical representation, as is the number of characters remaining from the 140 character limit for a tweet. Tapping a word twice highlights the word and provides the detailed SONORA SCORE for the selected word. If synonyms are available they are presented in a context menu layered above or beyond the selected word. Tapping one of the synonyms replaces the selected word with the selected synonym. We chose to use a context menu due to it being a familiar interaction pattern for text manipulation on the iOS platform. The screenshot in Figure 1 demonstrates message composition using the above outlined techniques. In this case while the application notes "amazing" resonates slightly better, June feels the nearly as good "kickin" is a better word choice and goes with it. Again, SONORA SCORE may best be thought of as a "grammar checker" that is often right, but still needs a human to make the decisions on best phrasing.

Conclusion

Effective messaging is an integral part of building and engaging relevant communities. By using poorly chosen words and tone, one can distort a message rendering it ineffective or worse, alienating communities which may have a negative impact on brand image. A well "sounding" message not only gets shared more, it also increases the chances that the user will be cited in the future and be considered influential. In this work we presented SONORA SCORE, a new feature based on word distribution that is aimed at helping authors construct better messages on Twitter. We evaluated the score and investigated the role word choice plays in retweetability. SONORA SCORE models three aspects of retweetability - volume, prevalence and sustain - that allow us to predict message popularity and prescribe word choice for message optimization. While past work has focused on identifying features that make for a popular tweet, none of them are prescriptive in nature, i.e. offer suggestions on what to change for better "uptake".

We found that SONORA SCORE serves as a good predictor of retweetability, and complements known predictors such as social and tweet-level features. For users with fewer followers our score's predictive ability is even higher; suggesting that if you cannot change their popularity or their topic, word choice plays a very important role. Based on a Monte Carlo-style evaluation, we found that when SONORA SCORE gives a higher score to a tweet, it is twice as likely to also have higher retweet. Although the experiments presented here focus on tweets, the SONORA SCORE prescriptive model applies to practically all types of messages and networks – emails, blogs, news articles etc., since the basic building block for the SONORA SCORE is at the 'word' level.

This paper has focused on the retweet rate as a measure of success. In future work we plan to study also the length of the retweet cascade, among other aspects of successful tweets.

Acknowledgements. We thank the anonymous reviewers for their careful reading and their many insightful comments and suggestions.

References

1. Alexe, B., Hernandez, M.A., Hildrum, K.W., Krishnamurthy, R., Koutrika, G., Nagarajan, M., Roitman, H., Shmueli-Scheuer, M., Stanoi, I.R., Venkatramani, C., Wagle, R.: Surfacing time-critical insights from social media. In: Proceedings of the 2012 ACM SIGMOD International Conference on Management of Data, SIGMOD 2012, pp. 657–660. ACM, New York (2012)
2. Areni, C.S.: The effects of structural and grammatical variables on persuasion: An elaboration likelihood model perspective. Psychology & Marketing 20(4), 349–375 (2003)
3. Bigonha, C., Cardoso, T.N., Moro, M.M., Gonçalves, M.A., Almeida, V.A.: Sentiment-based influence detection on twitter. Journal of the Brazilian Computer Society 18(3), 169–183 (2012)
4. Bormuth, J.R.: Readability: A new approach. Reading Research Quarterly, 79–132 (1966)
5. Guerini, M., Pepe, A., Lepri, B.: Do linguistic style and readability of scientific abstracts affect their virality? In: ICWSM (2012)

6. Hansen, L.K., Arvidsson, A., Nielsen, F.A., Colleoni, E., Etter, M.: Good friends, bad news-affect and virality in twitter. In: Park, J.J., Yang, L.T., Lee, C. (eds.) FutureTech 2011, Part II. CCIS, vol. 185, pp. 34–43. Springer, Heidelberg (2011)

7. Ho, T.K.: Random decision forests. In: Proceedings of the 3rd International Conference on Document Analysis and Recognition, August 14-16, pp. 278–282. IAPR (1995)

8. Hong, L., Dan, O., Davison, B.D.: Predicting popular messages in twitter. In: Proceedings of the 20th International Conference Companion on World Wide Web, pp. 57–58. ACM (2011)

9. Hothorn, T., Hornik, K., Zeileis, A.: Unbiased recursive partitioning: A conditional inference framework. Journal of Computational and Graphical Statistics 15(3) (2006)

10. Apple Inc. ios human interface guidelines (2013) (Online: accessed October 3, 2013)

11. Smith, A., Brenner, J.: 72% of Online Adults are Social Networking Site Users (2013)

12. Ke, J., Gong, T., Wang, W.S.: Language change and social networks. Communications in Computational Physics 3(4), 935–949 (2008)

13. Kim, J.-K., de Marneffe, M.-C.: Deriving adjectival scales from continuous space word representations. In: EMNLP, pp. 1625–1630. ACL (2013)

14. Lakkaraju, H., McAuley, J.J., Leskovec, J.: What's in a name? understanding the interplay between titles, content, and communities in social media. In: International Conference on Weblogs and Social Media (2013)

15. Lui, M., Baldwin, T.: langid.py: An off-the-shelf language identification tool. In: ACL (System Demonstrations), pp. 25–30. The Association for Computer Linguistics (2012)

16. Miller, G.A.: Wordnet: A lexical database for english. Commun. ACM 38(11), 39–41 (1995)

17. Nagarajan, M., Purohit, H., Sheth, A.P.: A qualitative examination of topical tweet and retweet practices. In: ICWSM (2010)

18. Nelder, J.A., Wedderburn, R.W.M.: Generalized linear models. Journal of the Royal Statistical Society. Series A (General) 135(3), 370–384 (1972)

19. Petrovic, S., Osborne, M., Lavrenko, V.: Rt to win! predicting message propagation in twitter. In: ICWSM (2011)

20. Ritter, A., Clark, S., Mausam, Etzioni, O.: Named entity recognition in tweets: An experimental study. In: EMNLP (2011)

21. Suh, B., Hong, L., Pirolli, P., Chi, E.H.: Want to be retweeted? large scale analytics on factors impacting retweet in twitter network. In: 2010 IEEE Second International Conference on Social Computing (SocialCom), pp. 177–184. IEEE (2010)

22. Tan, C., Lee, L., Pang, B.: The effect of wording on message propagation: Topic- and author-controlled natural experiments on twitter. In: Proceedings of ACL (2014)

23. Yang, J., Counts, S.: Predicting the speed, scale, and range of information diffusion in twitter. In: ICWSM, vol. 10, pp. 355–358 (2010)

A Fuzzy Model for Selecting Social Web Services

Hamdi Yahyaoui[1], Mohammed Almulla[1], and Zakaria Maamar[2]

[1] Computer Science Department, Kuwait University,
P.O. Box 5969, Safat 13060, State of Kuwait
[2] College of Information Technology, Zayed University,
P.O. Box 19282, Dubai, U.A.E
{hamdi,almulla}@ku.edu.kw, zakaria.maamar@zu.ac.ae

Abstract. This paper discusses a fuzzy model to select social component Web services for developing composite Web services. The social aspect stems from the qualities that Web services exhibit at run-time such as selfishness and trustworthiness. The fuzzy model considers these qualities during the selection, which allows users to express their needs and requirements with respect to these qualities. The ranking in this fuzzy model is based on a hybrid weighting technique that mixes Web services' computing and social behaviors. The simulation results show the appropriateness of fuzzy logic for social Web services selection as well as better performance over entropy-based ranking techniques.

Keywords: Fuzzy Logic, Selection, Social Quality, Web Service.

1 Introduction

A rapid survey of Web services topic returns an impressive number of initiatives that examine them from various dimensions including semantic description, security, and composition [4,11,14,16]. This paper is concerned with the selection dimension. Generally speaking, selection arises when competing options are made available for users and, only, one option can be taken at a time. Although Web services are known to be heterogeneous (developed by independent bodies), there is a consensus that their respective functionalities are sufficiently well-defined and homogeneous enough to allow for market competition to happen [2]. Selection also is dependent on matching successfully users' non-functional requirements (*aka* preferences) to Web services' non-functional properties (*aka* QoS). An agreed-upon list of such requirements/properties (e.g., response time) already exists [12], which "eases" the matching. Tremendous efforts are put into this matching through multiple initiatives that consider the *ambiguity* of users' non-functional requirements [1,3,19,20]. Ambiguity is exhibited with terms like *affordable* cost and *rapid* performance. However, terms that stem out of people's social life like *good* friends and *selfish* attitudes are overlooked. From an IT perspective, ambiguity is addressed using *fuzzy logic* that quantifies users' preferences using specific reference scales [23]. In a recent work [21], we assigned successfully *social* qualities like selfishness and fairness to Web services depending on how they interact with peers, sometimes competing peers. In this paper,

B. Benatallah et al. (Eds.): WISE 2014, Part II, LNCS 8787, pp. 32–46, 2014.

we apply fuzzy logic to social Web services (more details on how social Web services differ from regular Web services are available in [6,7,8]) selection, i.e., how do social qualities affect this selection? In addition to expressing their preferences on Web services' *computing behaviors* (i.e., QoS), users now have the opportunity of fine tuning these preferences with respect to these Web services' *social behaviors*. A user avoids *selfish* Web services, i.e., those that are not willing to help out other peers. The user insists on *predictable* Web services, i.e., those that will not suspend or drop her requests because of other appealing requests. Also, the user emphasizes on *trusted* Web services, i.e., those that will not leak her private details.

This paper is at the crossroad of two disciplines: social computing and service-oriented computing. We report in [9] on their blend's great potential. On one hand, social computing builds applications upon the principles of collective action, content sharing, and information dissemination at large. On the other hand, service-oriented computing builds applications upon the principles of service offer and demand, loose coupling, and cross-organization flow. The blend of both disciplines results into *social Web services* that "know" with whom they worked in the past and with whom they would like to work in the future [9]. This paper discusses the design and development of a fuzzy model to select social Web services in compositions. Section 2 presents social Web services. Section 3 reviews briefly the literature. Section 4 discusses a fuzzy model for the selection of social Web services. Experiments are reported in Section 5. Finally, conclusions are drawn in Section 6.

2   Social Web Services

This section is an overview of the basis upon which our social Web services operate namely community, social network, and social quality. More details are given in [21].

• *Web Services communities.* We gather Web services with similar functionalities (e.g., `weatherForecast` and `weatherPrediction`) into communities [10]. Some benefits out of this gathering include: (*i*) communities constitute pockets of expertise so the search for required Web services is restricted to specific communities; (*ii*) communities ease Web services substitution in the case of failure since potential substitutes are already known; and (*iii*) communities' structures can be of different types such as master-slave, alliance, and peer-to-peer.

• *Building social networks of Web services.* Social networks are either intra-community or inter-community. We discuss the former in terms of possible relationships between Web services and means of building networks with nodes and edges corresponding to these Web services and relationships, respectively. Zooming closely into a community, we identify three relationships: (*i*) supervision between the master Web service and the slave Web services; (*ii*) competition between the slave Web services; they all offer the same functionality and only one slave is selected to satisfy a user's request; and (*iii*) substitution between the slave Web services; they all offer the same functionality and can replace each other. Each relationship is mapped onto a specific type of social network of Web services so that elements like supervisors, supervisees, competitors, and substitutes are defined. A supervision social network has two types of nodes to represent the master and slave Web services and one type of edge to connect the nodes.

To evaluate the weight of a supervision edge (Equation 1), which we refer to as Supervision Level ($SupL$) between the master Web service (mws) of a community and a certain slave Web service (sws_i) in this community, we use the Functionality Similarity Level (FSL) between the functionalities of these two Web services, the Trust Level (TL) that shows how confident the master Web service is in the capacity of the slave Web service to satisfy a user's request, and the Responsiveness Level (RL) that shows the acceptance rate of a slave Web service to the demands of the master Web service to satisfy users' requests.

$$SupL_{mws,sws_i} = FSL_{mws,sws_i} \times TL_{mws,sws_i} \times RL_{mws,sws_i} \qquad (1)$$

where: FSL_{mws,sws_i} corresponds to the similarity level between the functionality of the mws and the functionality of the sws_i; $TL_{mws,sws_i} = \frac{SP_{sws_i}}{TP_{sws_i}}$, with TP_{sws_i} as Total number of Participations of sws_i in compositions and SP_{sws_i} as total number of Successful Participations of sws_i in these compositions; and $RL_{mws,sws_i} = \frac{AP_{sws_i}}{TP_{sws_i}}$, with AP_{sws_i} as total number of Accepted Participations of sws_i in these compositions.

As per Equation 1 the weights of edges change regularly as compositions get executed over time. The more the supervision level is close to one, the closest a slave Web service is to a master Web service, which makes this slave a good candidate for satisfying a user's request. By reassessing the trust and responsiveness levels regularly since a slave Web service either accepts or rejects to participate in compositions, it is possible to either promote (getting closer to the master Web service) or demote (getting farther from the master Web service) a slave Web service.

- *Working out the social qualities of Web services.* The social qualities we associate with Web services include *centrality, selfishness, unpredictability,* and *trustworthiness.* We restrict the discussion to selfishness in an intra-community context.

A slave Web service behaves in a selfish way if it does not show a positive attitude towards first, its direct master Web service and second, peers located in its community or other communities. Contrarily, the direct master Web service has high expectations towards this slave Web service and the peers show a positive attitude towards this slave Web service. By high expectations and positive attitude we mean (i) accepting to participate in compositions in response to the invitations of the direct master Web service (using the supervision social network) or other master Web services acting on behalf of their respective slave Web services (using the recommendation social network) or (ii) accepting to substitute others (using the substitution social network).

To establish the selfishness of a slave Web service, we introduced *reward* and *penalty* factors.

- Reward (r_{intra}): a master Web service rewards its slave Web services when they accept its invitation of either taking part in under-development compositions ($r_{intra}(comp)$) or acting as substitutes ($r_{intra}(sub)$) in under-execution compositions. The confirmation of rewards is subject to composition successful completion. Because of the importance of guaranteeing composition execution continuity despite the slave Web services failure, the reward to act as a successful substitute is higher than the reward to act as a successful component ($r_{intra}(sub) > r_{intra}(comp)$). Reward means expense for a master Web service and income for a slave Web service, as per Equations 2 and 3, respectively (n is the number of slave Web services in a community).
- Penalty (p_{intra}): a master Web service penalizes its slave Web services when they either do not take part in compositions at run-time despite their initial acceptance ($p_{intra}(comp)$) or decline continuously to act as substitutes ($p_{intra}(sub)$). Continuous declines can be motivated by the "fear" of not meeting some non-functional properties or inappropriateness for replacing a failing peer as per the competition social network ($ComL$ close to 0). Penalties for composition and substitution are treated equally ($p_{intra}(sub) = p_{intra}(comp)$). Penalty means income for a master Web service and expense for a slave Web service, as per Equations 4 and 5, respectively.

$$
\begin{aligned}
expense_{mws} =&(r_{intra}(comp) \times \sum_{i=1}^{n} SP_{sws_i}) \\
&+ (r_{intra}(sub) \times \sum_{i=1}^{n}\sum_{j=1}^{n}(SR_{sws_i,sws_j}))
\end{aligned}
\tag{2}
$$

$$
\begin{aligned}
income_{sws_i} =&(r_{intra}(comp) \times SP_{sws_i}) \\
&+ (r_{intra}(sub) \times SR_{sws_i,sws_j})
\end{aligned}
\tag{3}
$$

$$
\begin{aligned}
income_{mws} =&(p_{intra}(comp) \times \sum_{i=1}^{n}(TP_{sws_i} - SP_{sws_i})) \\
&+ (p_{intra}(sub) \times \sum_{i=1}^{n}\sum_{j=1}^{n}(TR_{sws_i,sws_j} - SR_{sws_i,sws_j}))
\end{aligned}
\tag{4}
$$

$$
\begin{aligned}
expense_{sws_i} =&(p_{intra}(comp) \times (TP_{sws_i} - SP_{sws_i})) \\
&+ (p_{intra}(sub) \times \sum_{j=1}^{n}(TR_{sws_i,sws_j} - SR_{sws_i,sws_j}))
\end{aligned}
\tag{5}
$$

Definition 1. *Selfishness in intra-community - A community is the host of selfish slave Web slaves when $income_{mws} > \sum_{i=1}^{n} income_{sws_i}$, i.e., the master Web service's income is more than the total income of all its slave Web services. The selfishness level of a slave Web service sws_i is calculated using $\frac{income_{sws_{Ai}}}{expense_{sws_{Ai}}}$ compared to Ts_{intra} that is selfishness threshold set by the community.*

3   Existing Fuzzy Models to Select Web Services

In [19], Wang et al. devise a fuzzy model where q experts use crisp or fuzzy values to rate Web services according to scalability, capability, performance, reliability, and availability criteria. These rates feed a performance matrix, which is used to derive weights for each criterion using the objective weighting Entropy technique. Finally, a score is given to each Web service based on the obtained ratings and computed weights. The scores rank the candidate Web services, which permits to select the best ones. In [20], Xiong and Fan model QoS-based Web services selection as a Fuzzy Multiple Criteria Decision Making (FMCDM) problem. The criteria used in the selection are cost, latency, availability, capacity, robustness, and security. The weights of these criteria, called synthetic weights, are computed with an hybrid technique that combines the objective Entropy and the subjective expert weights. These weights are used together with a judgement matrix to rank Web services. In [3], Chen et al. model Web services selection as a FMCDM problem like in [20], but use triangular fuzzy numbers to represent vague evaluations (based on linguistic terms such as "high", "medium", and "low") of QoS criteria. The objective of Chen et al.'s work is to determine which Web service quality attributes have the highest priority for customers who are invited to evaluate Web services performance. The overall service quality is determined by combining several subjective customers' perception towards each of the evaluated criteria. In [17], Tseng and Wu propose a fuzzy expert system to monitor the performance of Web services. The system detects and resolves potential problems that might affect Web services performance in order to improve their stability and reliability. The proposed expert system is based on a fuzzy table containing a pair of values: rate for each attribute indicating a fuzzy value given by an expert and degree of certainty for giving this rate. The fuzzy values are computed based on a trapezoidal membership function. Then, each column of the fuzzy table is translated into a fuzzy rule, which assesses the overall quality of a Web service. A decision model under fuzzy information is proposed in [18] by Wang. It leverages Intuitionistic Fuzzy Sets (IFS) for Web services selection. The ranking algorithm in this model solves conflicts between expert rankings. The subjective weights provided by experts are used to compute some scores from which a final score is derived. Afterwards, this score is used for ranking Web services. In [13], Peng et al. assess service quality based on first, runtime quality attributes such as price and second, non-runtime quality attributes such as availability, mean response-time, trustworthiness, and performance. A series of partial-ordering models has been developed to rank Web services according to their qualities.

The aforementioned list of works assesses the performance of individual Web services without taking into consideration the quality of interactions that take place over time with users and/or peers. These interactions can reveal different behaviors like honesty and maliciousness. Our main contribution in this paper is the elaboration of a fuzzy model for the selection of Web services based on social qualities namely selfishness, unpredictability, trustworthiness, and centrality, which is to the best of our knowledge not handled in other works. Social qualities are derived from the social networks established in our previous work [21]. The model relies on a weighting technique that considers the distribution of the data rather than its quality in deriving the weight of each social quality. A ranking technique for Web services based on these social qualities is also provided in this paper.

4   A Fuzzy Model to Select Social Web Services

In this section, we first assess Web services' social qualities. Then we transform the values obtained from this assessment into fuzzy values so that we derive the weights of the social qualities based on a new weighting technique. Finally, we compute the crisp values of these qualities so that we derive the utility values that allow ranking social Web services.

4.1   Fuzzification

Fuzzification maps each social quality value onto a fuzzy value based on three triangular membership functions: μ_L, μ_M, and μ_H. They assess low, medium, and high membership values of a social quality, respectively.

$$\mu_L(x) = \begin{cases} 0 & \text{if } x < a \\ \dfrac{x-a}{b-a} & \text{if } a \leqslant x \leqslant b \\ \dfrac{c-x}{c-b} & \text{if } b < x \leqslant c \\ 0 & \text{if } x > c \end{cases} \qquad (6)$$

$$\mu_M(x) = \begin{cases} 0 & \text{if } x < b \\ \dfrac{x-b}{c-b} & \text{if } b \leqslant x \leqslant c \\ \dfrac{d-x}{d-c} & \text{if } c < x \leqslant d \\ 0 & \text{if } x > d \end{cases} \qquad (7)$$

$$\mu_H(x) = \begin{cases} 0 & \text{if } x < c \\ \dfrac{x-c}{d-c} & \text{if } c \leqslant x \leqslant d \\ \dfrac{e-x}{e-d} & \text{if } d < x \leqslant e \\ 0 & \text{if } x > e \end{cases} \qquad (8)$$

It is worth mentioning that constants a, b, c, d, and e are provided by an expert depending on the data. After defining the membership functions, we proceed with the fuzzification of the social quality values in a specific community using Equations 6, 7, and 8 as follow. A community contains a set of (slave) Web services ws_i sharing the same functionality. Each ws_i has four values corresponding to selfishness, unpredictability, trustworthiness, and centrality social qualities. Each value represents an evaluation of a social quality Q_j. The association of Web services with social qualities corresponds to a matrix CV where each row represents a Web service and each column represents a social quality:

$$CV = \begin{bmatrix} CV_{11} & CV_{12} & \dots & CV_{1m} \\ CV_{21} & CV_{22} & \dots & CV_{2m} \\ & & \dots & \\ CV_{n1} & CV_{n2} & \dots & CV_{nm} \end{bmatrix}$$

The fuzzification phase converts each crisp value CV_{ij} into a fuzzy value FV_{ij} $= (L_{ij}, M_{ij}, H_{ij})$ using the aforementioned fuzzy membership functions, i.e.,

$$L_{ij} = \mu_L(CV_{ij}), \; M_{ij} = \mu_M(CV_{ij}), \; H_{ij} = \mu_H(CV_{ij}) \tag{9}$$

As a result, the output of the fuzzification phase is the following matrix:

$$FV = \begin{bmatrix} FV_{11} & FV_{12} & \dots & FV_{1m} \\ FV_{21} & FV_{22} & \dots & FV_{2m} \\ & \dots & & \dots \\ FV_{n1} & FV_{n2} & \dots & FV_{nm} \end{bmatrix}$$

To rank social Web services, we need to compute the weight of each attribute as per the next section.

4.2 Social Quality Weight Computation Phase

Entropy is a well-known objective technique for weight evaluation [15]. Based on a matrix of n candidates and m criteria $(a_{ij})_{n \times m}$, the entropy for each criterion and its relative weight is computed as follows.

Definition 2. *Entropy - The entropy of the jth attribute is defined as $E_j = -e \times \sum_{i=1}^{n} f_{ij} \ln f_{ij}$, where $e = \frac{1}{\ln n}$ and $f_{ij} = \frac{a_{ij}}{\sum_{i=1}^{n} a_{ij}}$.*

Definition 3. *Entropy Weight - The balanced entropy weight of the jth attribute is defined as $EW_j = \frac{1-E_j}{m-H}$, where $H = \sum_{j=1}^{m} E_j$.*

When the evaluated candidates have the same value of a specific attribute, the entropy reaches the maximum value of 1. Therefore, the entropy weight for this attribute is 0, which means that this attribute does not provide any useful information during decision making. Our *objective* weighting technique, which we call Distribution based Ranking (DR), focusses on data distribution rather than quality. In fact, when the data of a social quality has a high standard deviation, this means that the data are spread out over a large range of values and so the quality is a good candidate to discriminate between social Web services. Henceforth, its weight should be high and *vice-versa*.

Based on this fact, we define a weight of a social quality Q_i as its standard δ_i divided by the summation of all the standard deviation values, i.e.,

$$DW_i = \frac{\delta_i}{\sum_{i=1}^{m} \delta_i} \tag{10}$$

It is well-known that objective weighting techniques are very dependent on the data and sometimes do not reflect the real importance of a quality element. Therefore, we resort to the elaboration of a subjective weighting, which takes into account user preferences. Both the subjective and the objective techniques will be used together to derive more reasonable quality weights that permit ranking same Web services.

4.3   Capturing User Requirements

User requirements are generally non-quantifiable and inaccurate. These requirements can be captured based on linguistic terms. We define the association of linguistic terms with fuzzy numbers for all the social qualities in Table 1. These numbers should be provided by experts where low and high values are close to zero and one, respectively.

Table 1. Mapping linguistic terms onto fuzzy numbers for trustworthiness

Linguistic Term	Fuzzy Number
Very Low (VL)	(1,0,0)
Low (L)	(0.25,0.5,0)
Medium (M)	(0,0.5,0.25)
High (H)	(0,0.75,1)
Very High (VH)	(0,0,1)

The user can specify his request based on these linguistic terms, which reflect the relative importance of each social quality to the user. For instance, a user's preferences can be high for trustworthiness, medium for selfishness, and very high for predictability. Each of the fuzzy numbers associated with the user's request linguistic terms are defuzzified to derive one value that represents the subjective weight of each attribute. This defuzzification is explained below.

4.4   Defuzzification Phase

We use the signed distance technique [22] to defuzzy values of the matrix FV. The signed distance of a triangular fuzzy number (L_{ij}, M_{ij}, H_{ij}) is defined by Equation 11:

$$U_{ij} = \frac{1}{4}(L_{ij} + 2 \times M_{ij} + H_{ij}) \tag{11}$$

At the end of the defuzzification process, the obtained matrix is:

$$U = \begin{bmatrix} U_{11} & U_{12} & \dots & U_{1m} \\ U_{21} & U_{22} & \dots & U_{2m} \\ \dots & & \dots & \\ U_{n1} & U_{n2} & \dots & U_{nm} \end{bmatrix}$$

4.5   Ranking Social Web Services

Each of the fuzzy numbers (L_i, M_i, H_i) corresponding to the linguistic terms provided by the user for a social quality O_i is defuzzified into a quantifiable *subjective* weight SW_i using Equation 11. This constitutes the *subjective* part of our weighting technique. Based on the computed objective and subjective weights, we derive a hybrid weight W_i for each social quality O_i, where λ is a parameter in the interval [0,1], which can be fined tuned (Equation 12).

$$W_i = \lambda \times OW_i + (1 - \lambda)SW_i \qquad (12)$$

To rank social Web services, we compute a social score of a Web service based on the defuzzified value and hybrid weight of each social attribute. A weighted average social utility function u is used to compute the score of a Web service ws_j as follows:

$$u_{ws_j} = \frac{\sum_{i=1}^{m} U_{ij} * W_i}{\sum_{i=1}^{m} W_i} \qquad (13)$$

A higher score indicates a better sociable behavior for a Web service. The intuition behind the ranking is that best Web services that are not selfish, predictable, trustworthy, and with good centrality, have higher scores and so it is recommended for the user, who is looking for best candidates that will be engaged for instance in services composition, to select these Web services.

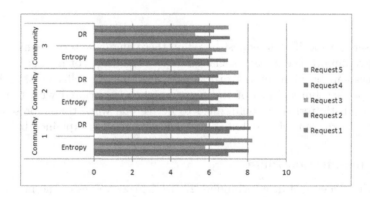

Fig. 1. DR-ranking *versus* Entropy-based ranking for top-20 for $\lambda = 0.25$

5   Experiments

Our dataset contains three communities of Web services and different social networks as per [21]. Each community contains 40 Web services. Four intra-community social qualities are considered namely centrality, unpredictability, selfishness and trustworthiness. The values of these social qualities are computed as explained in Section 2. Our ranking technique is implemented and tested over these communities. It is compared to an entropy-based ranking technique using the Discounted Cumulative Gain (DCG) metrics [5]. The DCG value for top-k ranked Web services is calculated by:

$$DCG_k = \sum_{i=1}^{k} \frac{2^{u_i} - 1}{log(1 + p_i)} \tag{14}$$

Where u_i is the ith Web service's utility value computed using Equation 13 and p_i is the rank of that Web service. The gain is accumulated starting at the top of the ranking and discounted at lower ranks. A large DCG_k value means high utilities of the top-k returned Web services. We created five requests in which the user preferences are different. Table 2 shows these possible user preferences over the social qualities.

For each of the five requests $R_{i,i=1...5}$, the user specifies the importance of each social quality based on linguistic terms (VL,L,M,H,VH). For instance, for R_1, the user considers the importance of the centrality quality attribute as very low, the unpredictability as low, the selfishness as medium, and the trustworthiness as high. We use the different requests in the experiment in order to show the impact of user preferences on the ranking.

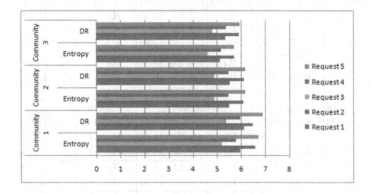

Fig. 2. DR-ranking versus Entropy-based ranking for top-20 for $\lambda = 0.50$

For each of the five requests, a ranking of top-20 is performed on the three communities. The results provided in Table 3 and Figures 1, 2 and 3 show that DCG scores for the DR-ranking are better than those obtained from an Entropy-based ranking for top 20.

For different top-k values, the DR-ranking also provides better results than those of Entropy-based ranking. Table 4 and Figure 4 depict the DCG scores for request R_5 and $\lambda = 0.75$.

From Table 4, it is clear that the DCG grows as the k does. This is obvious since the DCG is a cumulative gain. The gain can reach an increase up to 40%

Table 2. Five user requests

Req.	Centra.	Unpredic.	Selfish.	Trustwor.
R_1	VL	L	M	H
R_2	VL	L	M	VH
R_3	VL	L	L	H
R_4	VL	L	L	VH
R_5	VL	M	M	H

Table 3. DCG scores

$\lambda = 0.25$								
Community$_1$			Community$_2$			Community$_3$		
Request	Entropy	DR	Request	Entropy	DR	Request	Entropy	DR
R_1	6.97	7.05	R_1	6.41	6.43	R_1	5.96	6.05
R_2	8.04	8.13	R_2	7.49	7.5	R_2	6.96	7.06
R_3	5.77	5.86	R_3	5.45	5.45	R_3	5.14	5.23
R_4	6.76	6.86	R_4	6.46	6.46	R_4	6.12	6.22
R_5	8.21	8.30	R_5	7.49	7.49	R_5	6.86	6.98
$\lambda = 0.50$								
Community$_1$			Community$_2$			Community$_3$		
Request	Entropy	DR	Request	Entropy	DR	Request	Entropy	DR
R_1	5.96	6.11	R_1	5.50	5.53	R_1	5.12	5.33
R_2	6.58	6.47	R_2	6.10	6.12	R_2	5.71	5.91
R_3	5.20	5.36	R_3	4.87	4.87	R_3	4.60	4.79
R_4	5.79	5.96	R_4	5.46	5.46	R_4	5.17	5.36
R_5	6.72	6.88	R_5	6.16	6.16	R_5	5.7	5.93
$\lambda = 0.75$								
Community$_1$			Community$_2$			Community$_3$		
Request	Entropy	DR	Request	Entropy	DR	Request	Entropy	DR
R_1	5.01	5.21	R_1	4.62	4.65	R_1	4.33	4.67
R_2	5.24	5.45	R_2	4.85	4.88	R_2	4.55	4.89
R_3	4.64	4.85	R_3	4.32	4.33	R_3	7.29	7.75
R_4	4.87	5.09	R_4	4.55	4.57	R_4	7.49	7.96
R_5	5.38	5.59	R_5	4.97	4.98	R_5	4.63	4.98

Table 4. Top-k results for R_5 with $\lambda = 0.75$

	Community$_1$		Community$_2$		Community$_3$	
Top-k	Entropy	DR	Entropy	DR	Entropy	DR
Top-5	2.63	2.70	2.64	2.63	2.14	2.29
Top-10	3.81	3.93	3.63	3.62	3.21	3.42
Top-15	4.67	4.84	4.36	4.36	3.99	4.29
Top-20	5.38	5.59	4.97	4.98	4.63	4.98

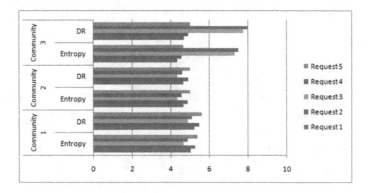

Fig. 3. DR-ranking versus Entropy-based ranking for top-20 for $\lambda = 0.75$

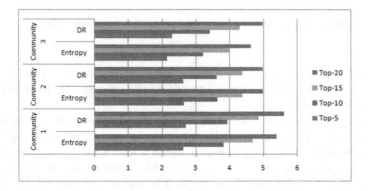

Fig. 4. Top-k DCG scores for request R_5 and $\lambda = 0.75$

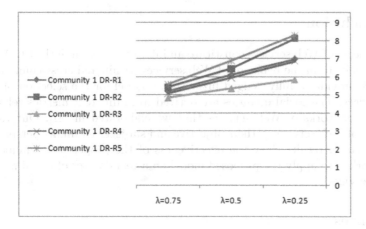

Fig. 5. DR-ranking for different values of λ for Community 1

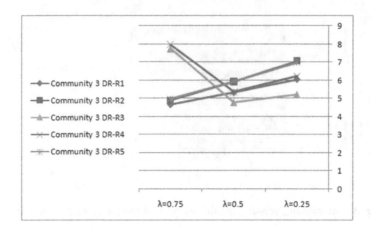

Fig. 6. DR-ranking for different values of λ for Community 3

in some cases. Since the quality weights are derived partially from the data (objective part), the DCG increase depends on the data and so it varies from a community to another.

Fine tuning the value of λ does not provide any conclusion about the superiority of the objective technique with respect to the subjective one or *vice-versa*. For some requests (R_1 and R_2 in Figure 5), as λ increases, the DCG scores decrease, which proves that the subjective part has more impact in this case. For other requests (R_3 and R_4 in Figure 6), as λ increases, no conclusion about the trend for DCG scores can be drawn.

6   Conclusion

Social aspects of Web services constitute an interesting research thrust. A social aspect is built from the behavior of Web services within and cross-communities. Our work provides a fully fledged fuzzy social model for the selection of social Web services. The social qualities are derived from different metrics related to the social networks of Web services that we elaborated in [21]. Our ranking technique of Web services gathers objective and subjective ranking perspectives. The simulation results show the effectiveness of the proposed technique. Our future work is to apply the proposed model on a real dataset of social networks of Web services.

References

1. Amdouni, S., Barhamgi, M., Benslimane, D., Hadjali, A., Benouaret, K., Faiz, R.: Answering Fuzzy Preference Queries over Data Web Services. In: Brambilla, M., Tokuda, T., Tolksdorf, R. (eds.) ICWE 2012. LNCS, vol. 7387, pp. 456–460. Springer, Heidelberg (2012)

2. Bui, T., Gacher, A.: Web Services for Negotiation and Bargaining in Electronic Markets: Design Requirements and Implementation Framework. In: Proceedings of the 38th Hawaii International Conference on System Sciences (HICSS 2005), Big Island, Hawaii, USA (2005)

3. Chen, M.F., Tzeng, G.H., Ding, C.G.: Fuzzy MCDM Approach to Select Service Provider. In: 12th IEEE International Conference on Fuzzy Systems, pp. 572–577. IEEE Computer Society (2003)

4. Chen, W., Paik, I., Hung, P.C.K.: Constructing a Global Social Service Network for Better Quality of Web Service Discovery. IEEE Transactions on Services Computing 99 (2013)

5. Jarvelin, K., Kekalainen, J.: Cumulated Gain-based Evaluation of IR techniques. ACM Transactions on Information System 20, 422–446 (2002)

6. Maamar, Z., Badr, Y., Faci, N., Sheng, Q.Z.: Realizing a social ecosystem of web services. In: Bouguettaya, A., Sheng, Q.Z., Daniel, F. (eds.) Handbook on Web Services - Advanced Web Services Part, Springer (2014)

7. Maamar, Z., Faci, F., Boukadi, K., Sheng, Q.Z., Yao, L.: Commitments to Regulate Social Web Services Operation. IEEE Transactions on Services Computing (forthcoming, 2014)

8. Maamar, Z., Faci, N., Sheng, Q.Z., Yao, L.: Towards a User-Centric Social Approach to Web Services Composition, Execution, and Monitoring. In: Wang, X.S., Cruz, I., Delis, A., Huang, G. (eds.) WISE 2012. LNCS, vol. 7651, pp. 72–86. Springer, Heidelberg (2012)

9. Maamar, Z., Hacid, H., Huhns, M.N.: Why Web Services Need Social Networks. IEEE Internet Computing 15(2) (March/April 2011)

10. Maamar, Z., Subramanian, S., Thiran, P., Benslimane, D., Bentahar, J.: An Approach to Engineer Communities of Web Services - Concepts, Architecture, Operation, and Deployment. International Journal of E-Business Research 5(4) (2009)

11. Medjahed, B., Bouguettaya, A., Elmagarmid, A.K.: Composing Web services on the Semantic Web. The VLDB Journal 12(4) (2003)

12. Menascé, D.A.: QoS Issues in Web Services. IEEE Internet Computing 6(6) (November/December 2002)

13. Peng, D., Chen, Q., Huo, H.: Fuzzy Partial Ordering Approach for QoS based Selection of Web Services. Journal of Software 5, 405–412 (2010)

14. Price, S., Flach, P.A., Spiegler, S., Bailey, C., Rogers, N.: SubSift Web Services and Workflows for Profiling and Comparing Scientists and their Published Works. Future Generation Computer Systems 29(2) (2013)

15. Shanonm, C.: The Mathematical Theory of Communication. Bell System Technical Journal 27, 379–423 (1948)

16. Sheng, Q.Z., Qiaob, X., Vasilakosc, A.V., Szaboa, C., Bournea, S., Xud, X.: Web Services Composition: A Decade's Overview. In: Information Sciences (to appear, 2014)

17. Tseng, J., Wu, C.: An Expert System Approach to Improving Stability and Reliability of Web Service. Expert Systems with Applications 33, 379–388 (2007)

18. Wang, P.: QoS-aware Web Services Selection with Intuitionistic Fuzzy Set under Consumer's Vague Perception. Expert Systems with Applications 36, 4460–4466 (2009)

19. Wang, P., Chao, K., Lo, C., Huang, C., Li, Y.: A Fuzzy Model for Selection of QoS-Aware Web Services. In: Proceedings of the 3rd International Conference on E-Business Engineering (ICEBE 2006), Shanghai, China (2006)

20. Xiong, P., Fan, Y.: QoS-Aware Web Service Selection by a Synthetic Weight. In: Proceedings of the 4th International Conference on Fuzzy Systems and Knowledge Discovery (FSKD 2007), Hainan, China, pp. 632–637. IEEE Computer Society (2007)
21. Yahyaoui, H., Maamar, Z., Lim, E., Thiran, P.: Towards a Community-based, Social Network-driven Framework for Web Services Management. Future Generation Computer Systems 29(6) (2013)
22. Yao, J.S., Wu, K.: Ranking fuzzy numbers based on decomposition principle and signed distance. Fuzzy Sets and Systems 116, 275–288 (2000)
23. Zadeh, L.: A Computational Approach to Fuzzy Quantifiers in Natural Languages. Computer Mathematics with Applications 9 (1983)

Insights into Entity Name Evolution
on Wikipedia*

Helge Holzmann and Thomas Risse

L3S Research Center,
Appelstr. 9a, 30167 Hanover, Germany
{holzmann,risse}@L3S.de
http://www.L3S.de

Abstract. Working with Web archives raises a number of issues caused by their temporal characteristics. Depending on the age of the content, additional knowledge might be needed to find and understand older texts. Especially facts about entities are subject to change. Most severe in terms of information retrieval are name changes. In order to find entities that have changed their name over time, search engines need to be aware of this evolution. We tackle this problem by analyzing Wikipedia in terms of entity evolutions mentioned in articles regardless the structural elements. We gathered statistics and automatically extracted minimum excerpts covering name changes by incorporating lists dedicated to that subject. In future work, these excerpts are going to be used to discover patterns and detect changes in other sources. In this work we investigate whether or not Wikipedia is a suitable source for extracting the required knowledge.

Keywords: Named Entity Evolution, Wikipedia, Semantics.

1 Introduction

Archiving is important for preserving knowledge. With the Web becoming a core part of the daily life, also the archiving of Web content is gaining more interest. However, while the current Web has become easily accessible through modern search engines, accessing Web archives in a similar way raises new challenges. Historical and evolution knowledge is essential to understand archived texts that were created a longer time ago. This is particularly important for finding entities that have evolved over time, such as the US president Barack Obama, the state of Czech Republic, the Finnish phone company Nokia, the city of Saint Petersburg or even the Internet. These entities have not always been as we know them today. There are several characteristics that have changed: their jobs, their roles, their locations, their governments, their relations and even their names. For instance, did you know an early version of the Internet was called Arpanet? Maybe you know, but are future generation going to know this? And how can they find

---

* This work is partly funded by the European Research Council under ALEXANDRIA (ERC 339233).

B. Benatallah et al. (Eds.): WISE 2014, Part II, LNCS 8787, pp. 47–61, 2014.

information about the early Web without knowing the right terms? A name is the most essential parameter for information retrieval regarding an entity. Therefore, we need to know what the entity used to be named in former days. Or even more convenient, search engines should directly take it into account.

Similar to search engines, many data centric applications rely on knowledge from external sources. Semantic Web knowledge bases are built upon ontologies, which are linked among each other. These *Linked Data Sources* can be conceived as the mind of the modern Web. However, other than human minds, it represents just the current state, because none of the popular knowledge bases, like DBPedia[1], Freebase[2], WordNet[3] or Yago [4], provide explicit evolution information. Even though some of them include alternative names of entities, they do not provide further details, e.g. whether or not a name is still valid and when a name was introduced.

Although much of this data has been extracted from Wikipedia and Wikipedia provides evolution information to a certain extent, it is missing in these knowledge bases. The reasons are obvious: Wikipedia provides current facts of an entity in semi-structured info boxes, which are simple to parse and therefore often used for gathering data. Evolution data on the other hand is mostly hidden in the texts.

In this paper we will show how to collect this information by incorporating lists that are available on Wikipedia and provide name evolution information in a semi-structured form (s. Section 3). We wanted to understand whether or not Wikipedia can be used as a resource for former names of entities. Eventually, we will show that it provides a good starting point for discovering evolutions. In future work, these findings can be used for pattern extraction. The patterns in turn can help to discover more names, even in different sources. Using the data from Wikipedia, we might be able to do this in future work in a completely automatic way. By analyzing the structures of Wikipedia, we obtained some interesting insights on how name changes are handled (s. Section 2). The main contributions of this work are:

- We present detailed statistics on entity name changes in Wikipedia's content beyond structural meta data (s. Section 4.2).
- We show that a large majority of more than two-thirds of the name changes mentioned in Wikipedia articles occur in excerpts in a distance of less than three sentences (s. Section 4.3).
- We detect and extract those excerpts without manual effort by incorporating dedicated lists of name changes on Wikipedia (s. Section 3.2).

2   Handling of Name Changes on Wikipedia

Wikipedia does not provide a functionality to preserve historic data explicitly. There is no feature to describe the evolution of an entity in a structured way. The only historic information that is made available is the revision history. However, this information does not describe the evolution of an entity, it rather shows the development of the corresponding article. These two things may correspond but they do not have to. There are a number of reasons for a change of a Wikipedia page, even though the actual name of the corresponding article did not change. One of the reasons is that an article did not follow Wikipedia's naming conventions in the first place and had to be corrected. By exploiting actual revised versions of an article rather than its history, Kanhabua and Nørvåg [5] achieved promising results in discovering former names from anchor texts. However, in terms of the time of evolution, the only information available is the revision date of the article, as stated above.

According to Wikipedia's guidelines, the title of an article should correspond to the current name of the described entity: *"if the subject of an article changes its name, it is reasonable to consider the usage since the change"*[1]. If the official name of an entity changes, the article's name should be changed as well and a redirect from the former name should be created[2]. Hence, the only trace of the evolution is the redirection from the former name. However, important details like the information that this has been done as consequence of an official name change are missing.

Another guideline of Wikipedia advices to mention previous and historic names of an entity in the article[3]: *"Disputed, previous or historic official names should [... be ...] similarly treated in the article introduction in most cases. [...] The alternative name should be mentioned early (normally in the first sentence) in an appropriate section of the article."*. Accordingly, the text is likely to contain name evolution information in an unstructured form, which is hard to extract by automatic methods. An approach to tackle this would be recognizing patterns like *[former name] ... has changed to [new name], because of ... in [year]*. We will investigate if name evolutions are described in such short excerpts of limited length, which are dedicated to the changes, rather than spread through the article.

To explore these excerpts that describe name evolutions we utilized list pages on Wikipedia. These are specialized pages dedicated to list a collection of common data, such as entity name changes, in a semi-structured way. An example of such a list page is the *List of city name changes*[4]. Name changes on this list are presented as arrow-separated names, e.g. *Edo → Tokyo (1868)*. Unfortunately not all lists on Wikipedia have the same format. While the format of this list is easy to parse, others, such as the *List of renamed products*[5], consist of prose describing the changes, like: *"Dime Bar, a confectionery product from Kraft Foods was rebranded Daim bar in the United Kingdom in September 2005 [...]."*

---

[1] http://en.wikipedia.org/wiki/Wikipedia:Naming_conventions#Use_commonly_recognizable_names

[2] http://en.wikipedia.org/wiki/Wikipedia:Move

[3] http://en.wikipedia.org/wiki/Wikipedia:Official_names#Where_there_is_an_official_name_that_is_not_the_article_title

[4] http://en.wikipedia.org/wiki/List_of_city_name_changes

[5] http://en.wikipedia.org/wiki/List_of_renamed_products

As the lists are not generated automatically, also the presentation of the items within a single list can vary. For instance, while some contain a year, other do not, but include additional information instead. Another drawback of the lists is that they are not bound to the data of the corresponding articles. Therefore, a name change that is covered by such a list is not necessarily mentioned in the article, neither the other way around. By analyzing the coverage of name changes in Wikipedia articles more deeply and gathering statistics by focusing on the content rather than Wikipedia's structure, we wanted to answer the question: *Is Wikipedia a suitable seed for automatically discovering name evolutions in texts?*

3   Analysis Method

In the following we define our research questions and describe the procedure we followed during gathering the statistics.

3.1   Research Questions

In our study we want to investigate to what extent Wikipedia provides information about name evolutions and the way these are presented. The goal is to reinforce our hypothesis, that name evolutions are described by short dedicated excerpts (compare Section 2). Based on this hypothesis, we derived more specific research question, which will be answered in following sections.

We started our analysis with the question: *How are name evolutions handled and mentioned on Wikipedia?* This question has already been partially answered in Section 2, based on the structure and guidelines of Wikipedia. As described, name evolutions are not available in a fully structured form, but rather mentioned in prose within the articles of the corresponding entities. Additionally, some name changes are available in a semi-structured manner on specialized list pages, which serve as a starting point for our research. Ultimately, our goal is to use the data from Wikipedia to build up a knowledge base dedicated to name evolution. In order to investigate if Wikipedia and the available lists of name changes can serve a seed for extracting suitable patterns, we require statistical data. Most important is the question: *How many complete name changes, consisting of preceding name, succeeding name and change date, are mentioned in these articles?*.

With the curiosity provoked by these statistics we extracted minimum parts from the articles that include the names of a change together with the date. In future work, we want to use these texts to extract patterns for discovering name evolutions in other Wikipedia articles or texts from other sources. This raised the question: *Do pieces of texts of limited length exist that are dedicated to a name change?*. The availability of a limited length is required for the extraction task. Otherwise, a pattern-based approach would not be applicable and this task would become more complex.

```
# computing min_distance of name change "P -> S (D)"
sentences  = [pseudo: Array with extracted sentences]
preceding  = [pseudo: Array with indices of sentences containing P]. sort
succeeding = [pseudo: Array with indices of sentences containing S]. sort
date       = [preudo: Array with indices of sentences containing D]. sort
components = [preceding, succeeding, date]
min_distance, min_from, min_to = nil, nil, nil
# until no index is available for one component
# (components.first is on that hass been previously shifted)
until components.first.empty?
  components.sort! # sort P, S, D by their next index
  from     = components.first. shift # remove smallest
  to       = components.last. first
  distance = to − from
  # save excerpts bounds for new min_distance
  if min_distance.nil? || distance < min_distance
    min_distance, min_from, min_to = distance, from, to
  end
end
# min excerpt spanning from sentence min_from to min_to
exists = !min_distance.nil?
excerpt = sentences[min_from..min_to].join('_') if exists
```

Listing 1.1. Algorithm in Ruby for computing the sentence distance of the minimum excerpt covering a name change

The final question we want to answer is: *How many sentences do excerpts span that mention a complete name change?* We will analyze percentages of different lengths in order find a reasonable number of sentences for the extraction in other articles later on. This is going to help us validating the initial hypothesis.

3.2 Procedure

The analysis was conducted in two steps. First, we created the dataset as described in Section 4.1. Afterwards, the actual analysis was perfomed by collecting and analyzing the statistical information.

Gathering Data. The lists of name changes from Wikipedia constitute our set of name evolutions. Each item in these lists represents one entity and has the format *Former Name → ... → Preceding Name → Succeeding Name → ... → Current Name.* Every pair of names delimited by an arrow is considered as a name change. Some of the names are followed by parenthesis containing additional data. Mostly, these are the dates of the change. However, there are cases where the brackets contain alternative names, spelling variations (e.g., the name in another language) or additional prose, like "changed", which is redundant in this context and means noise for our purpose. As we are only interested in the dates as well as name variations, we used the brackets containing numbers to extract dates and from the remaining we collected those starting with a capital letter as aliases.

While parsing the lists, we downloaded the corresponding Wikipedia article for each name of an entity by following the links on the names. In case no links were available, we fetched the article with the name or one of the aliases, whatever could be resolved first. For entities with names referring to different articles, we downloaded all of them. If the article redirected to another article on Wikipedia, the redirection was followed. During extraction all HTML tags were removed and we only kept the main article text.

Analysis. In the second step, we gathered statistics from the extracted data. The results are explained in the next section. Since one of the objectives of this analysis is to identify the number of sentences spanned by excerpts that cover a name evolution (s. Section 3.1), we split the articles into sentences. This was done by using the sentence split feature of Stanford's CoreNLP suite [6].

Entity names were identified within sentences by performing a case insensitive string comparison. We required either a match of the name itself or one of its aliases, as extracted from the lists, to consider the name mentioned in a sentence. Due to the different formats that dates can be expressed in, we only matched the year of a date in a sentence.

Finally to analyze the sentence distances of all three components of a name change the algorithm shown in Listing 1.1 has been used. It computes the minimum distance between the first sentence and the last sentence of an excerpt that includes a complete name change within a text. As an example consider the following text about Swindon, a town in South West England, taken from the Wikipedia article of Swindon[6].

*"On 1 April **1997** it was made administratively independent of Wiltshire County Council, with its council becoming a new unitary authority. It adopted the name **Swindon** on 24 April **1997**. The former **Thamesdown** name and logo are still used by the main local bus company of **Swindon**, called **Thamesdown** Transport Limited."*

It describes the name change from Swindon's former name Thamesdown to its current name in 1997, which is mentioned on the Wikipedia page of geographical renaming[7] as *Thamesdown → Swindon (1997)*. The excerpt spans three sentences and includes all three components of the name change twice: Swindon is contained in the sentences with index 1 and 2, Thamesdown is contained twice in sentence 2 and the change date (1997) is mentioned in the sentences 0 and 1. Therefore, one possible sentence distance would be 2, spanning the entire excerpt. However, without the first sentence still all components are included with a shorter distance of 1 (from sentence 1 to sentence 2). This is the minimum sentence distance of this change. If all three components are included in one sentence, the distance would be 0.

---

[6] http://en.wikipedia.org/wiki/Swindon
[7] http://en.wikipedia.org/wiki/Geographical_renaming

4   Analysis Results

In this section we present the results of our analysis, based on the research questions from Section 3.1. After describing the used Wikipedia dataset, we present statistical results and observations we made during the analysis and final discussions.

4.1   Wikipedia Dataset

The data we used for our analysis was collected from the English Wikipedia on February 13, 2014. As starting point we used list pages dealing with name changes. Due to the different formats of these lists we focused on lists of the form that each entity is represented as a single bullet and the name changes are depicted by arrows ($\rightarrow$) (s. Section 2). These lists are easy to parse and therefore, provide a reliable foundation in terms of parsing errors. We found 19 of those lists on Wikipedia. Nine of the lists were identified as fully redundant. After filtering redundant items, we ended up with 10 lists: *Geographical renaming*, *List of city name changes*, *List of administrative division name changes* as well as lists dedicated to certain countries.

A downside of the format constraint is that we only found lists of geographic entities. This will be ignored for the moment and discussed later in Section 5.1 where we manually parsed a list of name changes for different kinds of entities and performed the same analysis again. This allows us to argue general applicability of the results to a certain extend.

The parsed lists contain 1,926 distinct entities with 2,852 name changes. For the found names, we fetched a total of 2,782 articles for 1,898 entities. The larger number of articles compared to entities is a result of 766 entities with names that could be resolved to different articles. For 28 entities we were not able to resolve any name to an article.

4.2   Statistics about Name Changes on Wikipedia List Pages

Table 1 shows a detailed listing of the gathered statistics. The entity with most names in the analysis is *Plovdiv*, the second-largest city of Bulgaria, with 11 changes:

Kendros (Kendrisos/Kendrisia) $\rightarrow$ *Odryssa* $\rightarrow$ *Eumolpia* $\rightarrow$ *Philipopolis* $\rightarrow$ *Trimontium* $\rightarrow$ *Ulpia* $\rightarrow$ *Flavia* $\rightarrow$ *Julia* $\rightarrow$ *Paldin/Ploudin* $\rightarrow$ *Poulpoudeva* $\rightarrow$ *Filibe* $\rightarrow$ *Plovdiv*

The average number of changes per entity is 1.48. Each of the entities that have changed their name has 2.39 different names in average.

Unlike *Plovdiv*, on 708 (36.8%) of the entities we extracted at least one name change was annotated with a date. Overall, 933 of the total 2,852 are annotated with dates (32.7%). Only these were subject of further analysis as the ultimate goal is to identify excerpts that describe a full name change, which we consider consisting of preceding name, succeeding name and change date (s. Section 3.2).

Table 1. Statistics on name evolutions mentioned on Wikipedia. (percentages are in relation to the 100% above)

Subject	Count	Percentage	
Entities	1,926	**100%**	
- annotated with change dates	708	36.8%	
- resolvable to articles	1,898	98.5%	**100%**
- most current name resolvable	1,829	95.0%	96.4%
- linked on a list	1,786	92.7%	94.1%
- with multiple articles	766	39.8%	40.4%
- annotated with change dates	696	36.1%	36.7%
Name changes	2,852	**100%**	
- of entities with articles	2,810	98.5%	
- annotated with dates	933	32.7%	
- of entities with articles, annotated with dates	918	32.2%	**100%**
- mentioned in an article	572	20.1%	62.3%
- mentioned in the most current name's article	551	19.3%	60.0%
Extracted excerpts	572	**100%**	
- sentence distance less than 10	488	85.3%	**100%**
- sentence distance less than 3	389	68.0%	79.7%
- sentence distance 2	45	7.9%	9.2%
- sentence distance 1	118	20.6%	24.2%
- sentence distance 0	226	39.5%	46.3%

An additional requirement in order to extract excerpts is that the entities need to have a corresponding article, which potentially describes the evolution. Out of the 1,926 entities, we found 1,786 entities being linked to an article on one of the lists under consideration. Additionally, we were able to find articles corresponding to names or aliases of 112 entities. In total, 1,898 of the entities could be resolved to Wikipedia articles (98.5%). These compromise 2,810 name changes, which is also 98.5% of the initial 2,852. For 1,829 entities the most current name could be resolved to an article (96.4%). In addition, 114 have a previous name that could be resolved to another article. One reason for this is former names that have a dedicated article, such as *New Amsterdam*[8], the former name of New York City.

The intersection between the entities annotated with change dates and the ones with articles is 696, which is 36.1% of all entities we started with. In terms of name changes, the intersection of name changes with dates and those belonging to entities with articles is 918, which is 32.2% of the initial name changes. These constitute the subject of our research regarding excerpts describing evolutions.

4.3 Statistics about Evolution Mentions in Wikipedia Articles

Proceeding with 1,898 entities with articles, we analyzed 2,782 fetched articles. For entities that were resolved to multiple articles, all articles have been taken

---

[8] http://en.wikipedia.org/wiki/New_Amsterdam

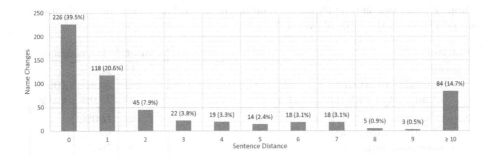

Fig. 1. Distribution of minimum sentence distances

into account in this analysis. To consider a name change being reported in an article, we checked for all three components of a change to be mentioned (i.e., preceding and succeeding name as well as the change date). This holds for 572 out of the 918 name changes with a date available from the entities with articles (62.3%). By taking only articles into account that were resolved for the most current name, 551 (60.0%) complete changes are mentioned.

As described in Section 3.1, we narrowed our objective of the analysis down to the core question: *How many sentences do excerpts span that mention a complete name change?* In order to answer this, we measured the sentence distances of the three components for each of the 572 name changes as described in Section 3.2. Overall, the average sentence distance of the extracted excerpts was 19.9. However, this is caused by a very few, very high distances and is not representative, as indicated by the median of 1. In fact, 488 out of the total 572 excerpts, which is 85.3%, have a distance of less than 10. This shows that for a majority of name changes all three components of a name change occur close together in a text, which positively answers the question if excerpts of limited length cover full name changes (compare Section 3.1). Most likely, those excerpts are dedicated to describe the name evolution of the corresponding entity. As the distribution of sentence distances in Figure 1 shows, a significant majority of 389 changes, around 80% (79.7%) of excerpts with 10 sentences or less, even have a shorter distance of less than three. For our research in terms of extracting evolution patterns, shorter excerpts are more interesting, because they are more likely to constitute excerpts dedicated to describe an evolution. Passages consisting of more than 10 sentences are likely to only cover two independent mentions of preceding and succeeding names of a change. Therefore, in the following we will concentrate on the shortest excerpts with a distance less than three, which constitute the largest amount of all findings.

5   Generalization and Discussions

The found results are supposed to serve as a seed to extract knowledge about name changes of any kind of entities. At this point however, the analyzed lists

Table 2. Statistics on product name evolutions mentioned on Wikipedia. (percentages are in relation to the 100% above)

Subject	Count	Percentage	
Entities	48	**100%**	
- resolvable to articles	45	93.8%	**100%**
- annotated with change dates	36	75.0%	80.0%
Name changes	63	**100%**	
- of entities with articles	59	93.7%	
- annotated with dates	48	76.2%	
- of entities with articles, annotated with dates	45	71.4%	**100%**
- mentioned in an article	36	57.1%	80.0%
Extracted excerpts	36	**100%**	
- sentence distance < 10	33	91.7%	**100%**
- sentence distance less than 3	22	61.1%	66.7%
- sentence distance 2	2	5.5%	6.1%
- sentence distance 1	6	16.7%	18.2%
- sentence distance 0	14	38.9%	42.4%

only consisted of geographic entities. Reliably parsing name changes from different domains on Wikipedia was not possible. In contrast to the semi-structured lists of geographic entities, they are only described in an unstructured way. Thus, we can only carefully make the assumption that our observations hold for entities of other kinds as well. To verify this, we manually parsed a list of renamed products and companies and performed the same analysis in order to generalize the results. In this section we compare and discuss our findings and outline how the results can be used to extend the training set towards general entities.

5.1   Towards Generalization

In addition to the automatically parsed lists of different geographic entities, we manually extracted 47 entities from the list of renamed products[9], which also contains company names. For these we performed the same analysis as described in Section 3. The details of the dataset and statistical results are summarized in Table 2. As before, we only proceeded with entities that were resolvable to Wikipedia articles and annotated with change dates on the list. 36 entities meet these conditions, which is 75%. These consist of 45 (71.4%) name changes out of the total 63 that are listed on the parsed collection. As for geographic entities, we searched for the succeeding and preceding name as well as the date of the changes in the corresponding articles. We discovered all three components for 36 changes. This corresponds to 80%, which is even more than the 62.3% we found for the geographic entities. This result supports our hypothesis that a large amount of name changes are mentioned in the Wikipedia article of the corresponding entity.

---

[9] http://en.wikipedia.org/wiki/List_of_renamed_products

After extracting the minimum excerpts that cover the found changes we computed the sentence distance of each change as described before. We wanted to know if small excerpts of the articles are dedicated to changes, which we consider true if all three components of the majority of changes are mentioned within the shortest distances. On geographic names, 85.3% spanned 10 sentences of less. On products and companies this even holds for 91.7%. Also comparable to our analysis of geographic entities is the fact that the majority of more than two-thirds of these changes are covered by excerpts with three sentences or less. While this includes 79.7% changes on geographic names, we found 66.7% on products and companies.

5.2   Discussion

The analysis was driven by the question about Wikipedia's suitability as a resource for extracting name evolution knowledge. The idea was to extract excerpts from Wikipedia articles that describe name evolutions. This knowledge can be used to build up an entity evolution knowledge base to support other application as well as serve as a ground truth for further research in this field.

Our hypothesis was, that name evolutions are described in short excerpts with a limited length. Based on the statistics, this can be affirmed. Almost 70% of 918 name changes that were available with corresponding articles and dates are mentioned in the Wikipedia articles of their entities. Out of these, 85.3% were found within excerpts of ten sentences or less. Two-thirds even had a distance less than three sentences. Regarding the shape of the excerpts, a closer look at some excerpts (s. Listing 1.2) revealed that most of them contain certain words, such as "became", "rename", "change". The extraction of particular patterns in order to train classifiers for the purpose of identifying evolutions automatically remains for future work.

The generalization of the results leads to the conclusion that a knowledge base that uses Wikipedia as a source for name change information would cover at least 41.7% of all name changes by using only excerpts with a sentence distance of less than three. This is the product of the number of entities with articles (98.5%), the completely mentioned name changes (62.3% of the changes of entities with articles and dates annotated) and the number of excerpts with distance less than three (68%). Certainly, it is not sufficient to use Wikipedia as the only source, but the extracted excerpts constitutes a solid foundation for discovering evolution knowledge on other sources. These could include historical texts, newswire articles and, especially for upcoming changes, social networks and blogs.

In future work we are going to involve more kinds of entities. It is planned to use the geographic name changes as a seed to train classifiers which detect excerpts that follow the same patterns. By setting highest priority on accuracy, we can neglect the fact that our seed only consists of geographic entities. It will definitely lower the recall in the beginning, but as the analysis of products and companies shows, name changes on other kinds are included in Wikipedia in a similar manner. Therefore, we are confident of finding new ones, which will increases our set of name changes for which we can extract new excerpts

---

— *Nyasaland to Malawi in 1964 (current name: Malawi)*
The Federation was dissolved in 1963 and in 1964, Nyasaland gained full
independence and was renamed Malawi.

— *General Emilio Aguinaldo to Bailen in 2012 (current name: Bailen)*
The Sangguniang Panglalawigan (Provincial Board) has unanimously approved
Committee Report 118–2012 renaming General Emilio Aguinaldo, a municipality
in the 7th District of the province, to its original, "Bailen" during the 95th
Regular Session.

— *Western Samoa to Samoa in 1997 (current name: Samoa)*
In July 1997 the government amended the constitution to change the country's
name from Western Samoa to Samoa.

— *Badajoz to San Agustin in 1957 (current name: San Agustin)*
On 20 June 1957, by virtue of Republic Act No. 1660, the town's name of Badajoz
was changed to San Agustin.

— *Bombay to Mumbai in 1996 (current name: Mumbai)*
In 1960, following the Samyukta Maharashtra movement, a new state of
Maharashtra was created with Bombay as the capital. The city was renamed
Mumbai in 1996.

Listing 1.2. Excerpts covering name changes with sentence distance less than 3

afterwards. Under the assumption that these changes are mentioned in different
ways on Wikipedia and other sources, this approach can incrementally extend
the training set.

6   Related Work

Most related to the long-term aim of this work, a knowledge base dedicated to
entity evolution, is YAGO2 [7]. It is an endeavor to extend the original YAGO
knowledge base with temporal as well as spatial information. Most relevant to
us is the temporal data, which YAGO2 incorporates to enhance entities as well
as facts. In contrast to our aim, they do not gather this data by extracting
new knowledge. Instead, they use temporal information which has already been
extracted for YAGO and connect it to the corresponding entity or fact. For
instance, date of birth and date of death are considered as a person's time of
existence. Therefore, dates of name changes are still missing as they are not
present in YAGO either.

   In terms of the process and analysis results presented in this paper, the related
work can be divided into two areas: related research on Wikipedia as well as on
entity evolution.

Research on Wikipedia

A prominent research topic in the context of Wikipedia is prediction of quality flaws. It denotes the task of automatically detecting flaws according to Wikipedia's guidelines, something not to neglect when working with Wikipedia. Anderka et al. [8] have done an impressive work in this field and give a nice overview of the first challenge dedicated to this topic [9]. Another related topic is the research on Wikipedia's revision history and talk pages. This could also serve as an additional resource for name evolutions in the future. Ferschke et al. [10] are working on automatically annotating discussions on talk pages and eventually link these to the corresponding content on Wikipedia articles. Additionally, they provide a toolkit for accessing Wikipedia's history [11].

Besides research on Wikipedia's infrastructure, many analyses on Wikipedia data have been done. Recently Goldfarb et al. [12] analyzed the temporal dimension of links on Wikipedia, i.e., the time distance a link bridges when connecting artists from different eras. However, to the best of our knowledge, no analysis has tackled the question on named entity evolutions in Wikipedia articles before.

Research on Entity Evolution

Most of the prominent research in the field of entity evolution focuses on query translation. Berberich et al. [13] proposed a query reformulation technique to translate terms used in a query into terms used in older texts by connecting terms through their co-occurrence context today and in the past. Kaluarachchi et al. [14] proposed another approach for computing temporally and semantically related terms using machine learning techniques on verbs shared among them. Similarly to our approach, Kanhabua and Nørvåg [5] incorporate Wikipedia for detecting former names of entities. However, they exploit the history of Wikipedia and consider anchor texts at different times pointing to the same entity as time-based synonyms. This is a reasonable approach, however, as we showed, Wikipedia is not complete in terms of name evolutions. As anchor texts can occur in arbitrary contexts, they are not suitable for pattern discovery as we proposed and thus, cannot be used to extend the knowledge by incorporating other sources, either. Not focused on name changes, but the evolution of an entity's context, presented Mazeika et al. [15] a tool for visually analyzing entities by means of timelines that show co-occurring entities. Tahmasebi et al. [16, 17] proposed an unsupervised method for named entity evolution recognition in a high quality newspaper (i.e., New York Times). Similarly to the excerpts that we extracted from Wikipedia, they also aim for texts describing a name change. Instead of taking rules or patterns into account, they consider all co-occurring entities in change periods as potential succeeding names and filter them afterwards using different techniques. Based on that method the search engine *fokas*[18] has been developed to demonstrate how awareness of name evolutions can support information retrieval. We adapted the method to work on Web data, especially on blogs [19]. However, big drawbacks of this approach are that change periods need to be known or detected upfront and the filters do not incorporate characteristics of the texts indicating a name change.

7   Conclusions and Future Work

In our study we investigated how name changes are mentioned in Wikipedia articles regardless of structural elements and found that a large majority is covered by short text passages. Using lists of name changes, we were able to automatically extract the corresponding excerpts from articles. Although the name evolutions mentioned in Wikipedia articles by far cannot be called complete, they provide a respectable basis for discovering more entity evolutions. In future work, we are going to use the excerpts that we found on Wikipedia for discovering patterns and training classifiers to find similar excerpts on other sources as well as unconsidered articles on Wikipedia. The first step on this will be a more detailed analysis of the extracted excerpts, followed by engineering appropriate features. As soon as we are able to identify the components of a name change in the excerpts, we can also incorporate other language versions of Wikipedia. In the long run, we are going to build a knowledge base dedicated to entity evolutions. Such a knowledge base can serves as a source for application that rely on evolution knowledge, like information retrieval systems, especially on Web archives. Furthermore, it constitutes a ground truth for future research in the field of entity evolution, like novel algorithms for detecting entity evolutions on Web content streams.

References

[1] Lehmann, J., Isele, R., Jakob, M., Jentzsch, A., Kontokostas, D., Mendes, P.N., Hellmann, S., Morsey, M., van Kleef, P., Auer, S., Bizer, C.: DBpedia - a large-scale, multilingual knowledge base extracted from wikipedia. Semantic Web Journal (2014)

[2] Bollacker, K.D., Evans, C., Paritosh, P., Sturge, T., Taylor, J.: Freebase: a collaboratively created graph database for structuring human knowledge. In: SIGMOD Conference, pp. 1247–1250 (2008)

[3] Miller, G.A.: Wordnet: A lexical database for english. Commun. ACM, 39–41 (1995)

[4] Suchanek, F.M., Kasneci, G., Weikum, G.: Yago: A Core of Semantic Knowledge. In: 16th International World Wide Web Conference (WWW 2007). ACM Press, New York (2007)

[5] Kanhabua, N., Nørvåg, K.: Exploiting time-based synonyms in searching document archives. In: Proceedings of the 10th Annual Joint Conference on Digital Libraries, JCDL 2010, pp. 79–88. ACM, New York (2010)

[6] The Stanford Natural Language Processing Group. Stanford corenlp - a suite of core nlp tools (2010), http://nlp.stanford.edu/software/corenlp.shtml (accessed February 3, 2014)

[7] Hoffart, J., Suchanek, F.M., Berberich, K., Weikum, G.: Yago2: A spatially and temporally enhanced knowledge base from wikipedia. Artificial Intelligence 194, 28–61 (2013)

[8] Anderka, M., Stein, B., Lipka, N.: Predicting quality flaws in user-generated content: the case of wikipedia. In: SIGIR, pp. 981–990 (2012)

[9] Anderka, M., Stein, B.: Overview of the 1th international competition on quality flaw prediction in wikipedia. In: CLEF (Online Working Notes/Labs/Workshop) (2012)

[10] Ferschke, O., Gurevych, I., Chebotar, Y.: Behind the article: Recognizing dialog acts in wikipedia talk pages. In: Proceedings of the 13th Conference of the European Chapter of the Association for Computational Linguistics (2012)

[11] Ferschke, O., Zesch, T., Gurevych, I.: Wikipedia revision toolkit: Efficiently accessing wikipedia's edit history. In: ACL (System Demonstrations), pp. 97–102 (2011)

[12] Goldfarb, D., Arends, M., Froschauer, J., Merkl, W.: Art history on wikipedia, a macroscopic observation. In: Proceedings of the ACM WebSci 2012, pp. 163–168. ACM (2012)

[13] Berberich, K., Bedathur, S.J., Sozio, M., Weikum, G.: Bridging the terminology gap in web archive search. In: WebDB (2009)

[14] Kaluarachchi, A.C., Varde, A.S., Bedathur, S.J., Weikum, G., Peng, J., Feldman, A.: Incorporating terminology evolution for query translation in text retrieval with association rules. In: CIKM, pp. 1789–1792. ACM (2010)

[15] Mazeika, A., Tylenda, T., Weikum, G.: Entity timelines: Visual analytics and named entity evolution. In: Proc. of the 20th ACM Int. Conference on Information and Knowledge Management, CIKM 2011, pp. 2585–2588. ACM, New York (2011)

[16] Tahmasebi, N., Gossen, G., Kanhabua, N., Holzmann, H., Risse, T.: Neer: An unsupervised method for named entity evolution recognition. In: Proceedings of the 24th International Conference on Computational Linguistics (Coling 2012), Mumbai, India (2012)

[17] Tahmasebi, N.: Models and Algorithms for Automatic Detection of Language Evolution. Towards Finding and Interpreting of Content in Long-Term Archives. PhD thesis, Leibniz Universität Hannover (2013)

[18] Holzmann, H., Gossen, G., Tahmasebi, N.: fokas: Formerly known as - a search engine incorporating named entity evolution. In: Proceedings of the 24th International Conference on Computational Linguistics: Demonstration Papers (Coling 2012), Mumbai, India (2012)

[19] Holzmann, H., Tahmasebi, N., Risse, T.: Blogneer: Applying named entity evolution recognition on the blogosphere. In: Proc. of the 3rd Int. Workshop on Semantic Digital Archives (SDA 2013), in Conjunction with TPDL 2013, Valetta, Malta (September 2013)

Assessing the Credibility of Nodes
on Multiple-Relational Social Networks

Weishu Hu and Zhiguo Gong

Department of Computer and Information Science, University of Macau,
Av. Padre Tomás Pereira Taipa, Macau
{yb07405,fstzgg}@umac.mo
http://www.fst.umac.mo/cis

Abstract. With the development of the Internet, social network is
changing people's daily lives. In many social networks, the relationships
between nodes can be measured. It is an important application to predict
trust link, find the most reliable node and rank nodes. In order to implement those applications, it is crucial to assess the credibility of a node.
The credibility of a node is denoted as the expected value, which can be
evaluated by similarities between the node and its neighbors. That means
the credibility of a node is high while its behaviors are reasonable. When
multiple-relational networks are becoming prevalent, we observe that it
is possible to apply more relations to improve the performance of assessing the credibility of nodes. We found that trust values among one type
of nodes and similarity scores among different types of nodes reinforce
each other towards better and more meaningful results. In this paper,
we introduce a framework that computes the credibility of nodes on a
multiple-relational network. The experiment result on real data shows
that our framework is effective.

Keywords: Trust-based network, credibility, similarity, social network.

1 Introduction

People are used to sharing all kinds of content such as message, images, songs,
video, opinions and blogs on different social networks e.g. Facebook, Twitter
and YouTube. In these social networks, the reputation of a publisher plays an
important role; otherwise, a user may receive some disgusting content such as
virus and Trojan horse. Thus, the trust of a node granted by other users is a
vital property of it. For example, there are explicit opinions on other users as
trust/distrust on Slashdot and Epinions networks.

Unfortunately, online content is not always trustable. And it is no way to
ensure the validity of the information on the Internet. Even worse, different
users usually provide conflicting opinions, as following two examples.

Example 1 (Battery of IPhone). Suppose a user plans to buy an IPhone and
reads the product review from Epionions.com. Among the top 20 opinions, he or
she will find the following comments: three users say "Battery life suffers under

B. Benatallah et al. (Eds.): WISE 2014, Part II, LNCS 8787, pp. 62–77, 2014.

heavy use", four users say "Pretty long battery life", one says "The battery life on the phone drains really easily", and another one says "Increased battery life". Which suggestion should the user adopt?

Example 2 (Definition of Spam). We want to know what is spam? We notice that various definitions from different websites, so we show some of them in Table 1. From the integrity of expression, we found that typepad.com provides the most precise information. In comparison, the information from ask.com is incomplete, and that from Wikipedia is incorrect.

Table 1. Conflicting information about Spam

Web Site	Definition of Spam
Wikipedia	Spam is a canned precooked meat product made by the Hormel Foods Corporation, first introduced in 1937.
about.com	Spam is the practice of purposely deceiving a search engine into returning a result that is unrelated to a users query, or that is ranked artificially high in the result set.
ask.com	Spamming is a fairly easy task which involves a mass sending of a message, for any number of purposes.
webopedia.com	Spam is electronic junk mail or junk newsgroup postings.
typepad.com	Spam is commercial, unsolicited, unanticipated, irrelevant messaging, sent in bulk.

The Credibility Problem of the Internet has been acquainted by current network users. Princeton Survey Research [1] made a survey on the credibility of websites. The conclusion shows no less than 54% online users' trust news sites in most of the time, comparing to only 26% for sales websites and is barely 12% for blogs.

According to authority (or popularity) based on hyperlinks, there are many researches on ranking web pages. The most famous techniques are HITS [2], and PageRank [3] applied in Google.com. These two studies provide high scores to nodes having better connectivity. But unfortunately, authority does not lead to credibility of information. High ranked websites are usually the most popular ones. However, popularity does not equal to credibility. In trust-based networks, a highly nasty node may also have good connectivity but have a low credibility. It means that the credibility of one node depends on the opinions of other nodes, and also depends on how the node makes a fair evaluation about other nodes. In fact, a node with higher credibility should be as trustworthy as another similar node.

Belief propagation (sum-product message passing) is a message passing algorithm, which is used for performing inference on graphical models including Bayesian networks and Markov random fields. It computes the marginal distribution for each unobserved node based on other observed nodes. In addition, beliefs are the estimated marginal probabilities. Belief propagation is mainly applied in information theory and artificial intelligence, which is demonstrated empirical success in numerous applications such as free energy approximation, low-density parity-check codes, turbo codes and satisfaction [4]. Belief propagation is not suitable for creditability problem, because of a lack of observed nodes in advance.

The trust-based network is very different from common network, and it is actually a directed graph. Trust-based network is a special social network having explicit links to express one node trusts/distrusts other nodes. The nodes are individual users, with the relationship "User X trusts User Y" resulting in an edge directed from User X's node to User Y's. Everything happens in some reasons, there is no absolutely independent behavior without any cause. We believe that one user trusts others for some sake, and we can observe that they have similar opinions, common friends or like the same items on these social networks. In a network such as Facebook and Twitter, an explicit link implies that two nodes are close for their frequent communication. However, in a trust-based network, two nodes may be closely connected but the link may show unreasonable. More importantly, a reasonable trust link in trust-based network is the two connected nodes have similar opinions or behaviors (make friends with the same users, focus on the same items). If two users are similar in terms of their opinions or other behaviors, then their trust links are more reasonable. In the other way, users have similar opinions or behaviors, and they don't have to trust each other. For instance, user A, B, C and D have trust relations as Fig. 1. It is intuitive that the trust link from A to B is more reasonable than that from C to D. Because D has no similar behaviors as C (C has two inlinks while D has nothing), the trust link from C to D is not reasonable. A and B have similar behaviors (they are trusting and trusted by the same user C).

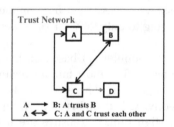

Fig. 1. An Example of Trust-Based Network

Trust is a measure of confidence that an entity or entities will behave in an expected manner [5]. Trust has emerged as a major impediment to the success of electronic markets and communities where interaction with the strangers is the norm [6]. Recently, a novel algorithm known as TrustRank is proposed for combatting web spam. However, the weakness of TrustRank is sensitive to the seed set, which could not be completely involved the different topics on the Internet. Moreover, TrustRank prefers to larger communities with prejudice for a given seed set.

In a rating system such as IMDB, Slashdot and Epinions, users can rate objects. A movie is an object in IMDB, and a product could be an object in Epinions and Slashdot. In other situations, each user can comment on other users and get feedbacks from them. The credibility of a user is depended on the feedbacks from other users, which can be considered as the inlinks of a user received. The attitude of a user about others is based on his comments (outlinks). In an isolate consideration, credibility of a user depends on the quality of inlinks and fairness of a user who has made an opinion towards him, no matter the quantity. In other words, the fairness depends on the opinion he gives in the form of outlinks. If a user is unfair, then his opinion should weigh less. Then, the credibility of another user mainly relies on the links from truthful users. A user that only gives groundless opinions irrespective of the similarities between other users and him is highly injustice. Similarly, a user receiving groundless inlinks from highly unfair users has a lower credibility than a user receiving inlinks from reliable users. In some extreme cases, a user may receive all high credibility but still express his opinion on another user that differs from all other users' in the network.

When the multiple-relational network is coming up all over the world, we observe that it is not only requisite but also advantageous to combine the trust and similarity analysis into one framework because we can apply information from one side to improve the other side. In our case, due to the multiple-relation nature of the network, when computing the similarity of one type of nodes, we should put the trust of the other type of nodes into the formulation. This leads to an asymmetric similarity formulation. This approach has never been studied before. As we analyzed in the above subsection, introducing similarity to trust analysis in a multiple-relational network could be beneficial to many key aspects on the trust analysis. Furthermore, similarity measure gets more customized information from trust analysis side, which should potentially be beneficial too. Our experiment results confirm this mutual beneficial relation.

Similarity and Trust computation can benefit each other, which motivates us to study how to effectively combine them together in one framework. Our technique adopts reinforcement scheme on the top of multiple-relational network decomposing. To be more detail, we firstly define a special bi-typed multiple-relational network as Trust Similarity (TS) network; then we decompose its different types of nodes into two homogeneous networks based on their relations, and we do the analysis of social trust spreading on the Trust network and expectation of similarity measures on the Similarity network. However, the two networks are not totally separated. There is a latent tunnel connecting them for the sake of delivering information back and forth to improve the performance trust and similarity analysis.

In Fig. 2, Alice and Bob trust each other, while Bob trusts Tom and Alice trusts Tom. Alice rates two products (IPad, Bag) and comments on one of Tom's reviews; Bob rates two products (Telephone, IPad) and comments on one of Alice's reviews; and Tom rates three products (Telephone, IPad, and Bag). There are three relations (Trust, Rating, and Comment) in Epinions Network. A reasonable trust between users should base on their similar behaviors. For example,

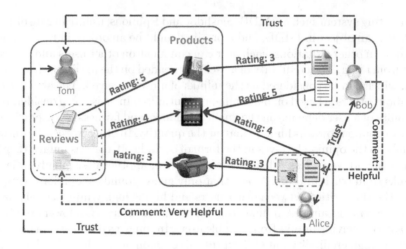

Fig. 2. An Illustration of Multiple-Relational Network of Epinions

they are trusted by the same users, give similar rating to the same products, or make similar comment on the same review. In this example, Alice trusts Tom is more reasonable than Bob trusts Tom, because Alice and Tom rate the same products with the same scores, and Alice makes a comment (very helpful) on Tom's rating, while Bob's rating is very different from Tom's.

We summarize our major contributions as follows.

- We present a model that computes the credibility of nodes in a multiple-relational network. The basic idea is the opinions of trustworthy nodes weigh more. The trustworthiness of a node is computed by the credibility of its neighbors. We observe that the credibility of a node depends on its behaviors, and a reasonable behavior is based on the similarity.
- We propose a new viewpoint of treating trust and similarity computation together. To our best knowledge, our work is leading to explicitly explore how to make use of both two techniques together to analyze a multiple-relational network in a more comprehensive way.
- We study the mutual improvements of trust analysis and similarity computation on each other. An iterative algorithm with optimization on decomposed multiple-relational networks is suitable for this reinforcement relationship.
- We demonstrate its effectiveness through real world social network analysis. Our method outperforms state of the art in trust analysis and similarity computation when they are performed separately.

The paper is organized as follows. We present the related works in Section 2. Section 3 describes the algorithm. Section 4 presents the experimental results before Section 5 concludes.

2   Related Works

Graph theoretic methods for ranking nodes in a network have gained popularity since the application of HITS [2] and PageRank [3] algorithms. As a result, a number of other methods have also been proposed. Most of these methods are usually a variant of eigenvector centrality measure [7]. The algorithm Eigen-Trust [8, 9] removes negative entries by not considering negative ratings. Ranking has been done on trust-based network as well while considering negative links, e.g. PageTrust [10]. The authors of [11] propose an algorithm to compute trust/distrust between two objects. There have been some studies on the social aspects of trust-based networks. One important example is the balance theory [12] that considers relationships of type "enemy of an enemy" as a friend. Another popular theory is status theory [13], where a positive link denotes higher status. These theories have been well evaluated in [14, 15]. [16, 17] compute the bias and prestige of nodes in simple networks where the edge weight denotes the trust score. These methods emphasize on single relation and neglect comprehensive utilization of multiple relations.

For anomaly detection and classification in numerous settings such as calling-card fraud, accounting fraud and cyber-security, guilt-by-association methods [18] derive stronger signal from weak ones. The authors focus on comparing and assessing three very effective algorithms: Random Walk with Restarts, Semi-Supervised Learning and Belief Propagation (BP). Their main contributions are two aspects: firstly, they theoretically prove that the three algorithms are effective in a similar matrix inversion problem; secondly, in practice, they propose a fast convergence algorithm called FaBP, which runs twice as efficient with equal or higher accuracy than BP. They show the advantage in synthetic and real datasets (YahooWeb). Guilt-by-association methods do not apply to Credibility Problem. The premise of guilt-by-association is like attracts like, but unreliable users try their best to get close to reliable ones in trust network, which leads to unreliable users may obtain high credibility. Besides, most of guilt-by-association methods needs supervised or semi-supervised.

TrustRank introduced in [19] is used for web spam, which is also related to Credibility Problem. Web spam is cheating behavior that finds ways to acquire top ranking by using loopholes of search engine ranking algorithms. The basic idea of TrustRank is that good websites seldom point to spam websites and people believe in these good websites. This confidence can be spread through the link topology on the network. Hence, a set of websites with high trustworthiness are picked to make up the seed set and each of them is initialized a non-zero trust value, while all the other websites on the network are assigned an initial value 0. Then a biased PageRank algorithm is used to propagate the initial trust values to their outgoing websites. When convergence, good websites will obtain higher trust values, and spam websites are tend to obtain lower trust values. The results show that TrustRank improves upon PageRank by maintaining good sites in top buckets, while most spam websites are moved to lower buckets. In an enhanced TrustRank [20], the authors propose Topical TrustRank, which applies topical information to partition the seed set and compute trust scores for each different topic respectively. The combination of these trust scores for a website or page is used to decide its ranking. Their experimental results on two large datasets indicate that Topical TrustRank makes a better performance than TrustRank in degrading spam websites or pages. Compared to TrustRank, Topical TrustRank can decrease spam from the top ranked websites by as much as 43.1%. Both TrustRank and Topic TrustRank need seed set selection, and topic TrustRank also needs finding different topics. It is very difficult to gain all the necessary information from a very large network manually in advance, and the size of seed set is also hard to determine. So these methods are not suitable in a large trust network.

Truth discovery is another research involving Credibility Problem. Xiaoxin Yin et al. [21] propose a Truthfinder algorithm to find true facts with conflicting information from different information providers on the network. This approach is applied on certain domain such as book authors and Movie run time. Truthfinder is a fact based search engine, which ranks websites by computing trustworthiness score of each website using the confidence of facts provided by websites. It utilizes the relationships between websites and their information to find the website with accurate information which is ranked at the top. It discovers trustworthy websites better than popular search engines. A new algorithm called Probability of Correctness of Facts(PCF)-Engine [22] is proposed to find the accuracy of the facts provided by the web pages. It uses the Probability based similarity function (SIM), which performs the string matching between the true facts and the facts of web pages to find their probability of correctness. The existing semantic search engines may give the relevant result to the user query, but may not be completely accurate. Their algorithm compute trustworthiness of websites to rank the web pages. Simulation results show that their approach is efficient compared with existing Voting and Truthfinder [21] algorithms. However, these algorithms require to pre-compute the implicit facts in the knowledge base, which is difficult to achieve in large trust networks.

What we present in this paper starts from a unique observation that combining social trust and similarity analysis could benefit each other through information exchange. Credibility Problem is an important research branch on social networks. The task of trust problem is to choose most reliable users in a certain social network [7, 23]. Similarity analysis on social network is usually based on nodes common neighbors or link properties, e.g., [13]. Unlike most of them, we apply more information from social trust side to improve the similarity measure, which is not considered in previous similarity research, such as Reinforced Similarity Integration in Image-Rich Information Networks [24]. And Simrank [25–27] is a famous similarity algorithm in structure network, which is improved in this work.

We have also noticed that there is another work studied a total different relation of social influence and similarity together [12, 28]. They studied how peoples influence and their similarities affect each other. In another word, they consider the same type of nodes similarity and influence in a simple network.

We consider trust of one type of nodes, with the information of similarity of another type of nodes, and vice versa, in a multiple-relational network.

3 Methodology

In this section, we first define the term "credibility" precisely before describing and analyzing an algorithm to compute it. Given a multiple-relational network G_M (e.g., Epinions network as Fig. 2), we present how to convert G_M (only consider trust and rating relations) into two simple graphs $G_T = (V_T, E_T, w_T)$ and $G_S = (V_S, E_S, w_S)$. V and E can be constructed by exploiting one relationship from G_M. For instance, we could construct the vertices and the edges based on the users and their trustiness relationships in Epinions network, respectively. As we discussed in Section 1, the edge weight w can be assigned by analyzing the degree of relation in G_M and the detail is discussed as follows.

3.1 Problem Definition

Formally, let $G = \{V, E, w\}$ be a simple graph, where an edge $e_{ij} \in E$ (directed from node i to node j) has weight $w_{ij} \in [0, 1]$. We say that node i gives the trust-score of w_{ij} to node j.

Let $d^o(u_i)$ denotes the set of all outgoing links from node u_i and likewise, $d^i(u_i)$ denotes the set of all incoming links to node u_i. Credibility of a node is directly proportional to the confidence of all the behaviors provided by it and the implication on it. We firstly introduce one important definition in this paper, the confidence of behavior.

- Confidence of behavior: the confidence of a behavior b (denoted by c(b)) is the probability of b being reliable, according to the best of our knowledge.

Different behaviors about the same object may be conflicting. For example, one user claims that a product is "perfect" whereas another claims that it is "terrible". However, sometimes behaviors may be supportive to each other although they are slightly different. For example, one user claims the product to be "acceptable" and another one claims "not bad" or one user says that a certain mobile phone is 4 inches screen, and another one says 10 cm. If one of such behaviors is reliable, the other is also likely to be reliable. We find that the confidence of behavior is decided by the similarity, so we only consider the similarity to replace the confidence of behavior in the following sections.

Finally, we measure credibility of a node:

- Credibility: This reflects the expected value of each inlink from its neighbor nodes based on their similarities (confidence of their behaviors).

This definition means the credibility of one node depends on its neighbors' credibilities and their similarities. In the next section, we show all steps to calculate the Credibility.

3.2 Trust Similarity

Trust Similarity (TS) network is a special type of heterogeneous network with edge weights of different practical meanings for different edge types. We have noticed that it is generic enough to get important relations for different types of nodes and explore the hidden reinforcement between trust and similarity. Similarity of two nodes in a trust-based network is defined as the summation of similarities between their neighbors. A strong assumption here is that nodes are trusting and trusted by nodes those are similar to them. TS network is a directed multiple-relational network G_M (V, E, w) of two different types of nodes, two types of edges with associated edge features. For ease of presentation, let V_T be the set of nodes we want to study trust on and V_S is the set of the type of nodes for similarity research, where $V = V_T \bigcup V_S$. There are two types of edges E_{TT}, E_{TS} connecting different types of nodes, and $E = E_{TT} \bigcup E_{TS}$. w is a weight vector associated with different types of edges. $w = w_T \bigcup w_S$. $w_T = \{w_T | \forall e_T \in E_{TT}\}$ is a vector of variables, each one of which describes the trust scores between two nodes of an edge e_T. Similarly, $w_S = \{w_S | \forall e_S \in E_{TS}\}$ is another vector of variables for similarity scores on the other type of nodes.

As seen in the above discussion, Fig. 3 and Fig. 4 illustrate the decoupled result of Fig. 2. In the following subsections, we will model the information passing in details in the coordination with the Trust Similarity reinforcement. The similarity of two nodes is the propensity to do the behaviors on other objects. Thus, the propensity or similarity of two nodes can be measured by the difference between the ratings that a node provides to another node. Multiple-relational Network G_M can be converted into two simple G_T and G_S. It is obvious that a higher score similarity of (u_i, u_j) $(Sim^{(\phi+1)}(u_i, u_j))$ implies that the trust relationship of u_i and u_j is more reliable. The similarity of two user nodes u_i, u_j is determined by the similarity in G_T and G_S, using Reinforcement Learning, given by Eq. 1.

$$Sim^{(\phi+1)}(u_i, u_j) = \alpha Sim_T^{(\phi)}(u_i, u_j) + (1 - \alpha) Sim_S^{(\phi)}(u_i, u_j) \tag{1}$$

Where α is a parameter, which is used to weigh the importance of two similarities.

In Trust network like Fig. 3, a user's similarity depends on its neighbors [27], user tends to trust similar users like him, in other words, user's trust similarity $(Sim_T^{(\phi+1)}(u_i, u_j))$ is determined by his inlinks and outlinks, which can be calculated by Eq. 2.

$$
\begin{aligned}
Sim_T^{(\phi+1)}(u_i, u_j) = {} & \frac{\beta}{|d^i(u_i)||d^i(u_j)|} \sum_{u_p \in d^i(u_i)} \sum_{u_q \in d^i(u_j)} Sim_S^{(\phi)}(u_p, u_q) \\
& + \frac{1 - \beta}{|d^o(u_i)||d^o(u_j)|} \sum_{u_p \in d^o(u_i)} \sum_{u_q \in d^o(u_j)} Sim_S^{(\phi)}(u_p, u_q)
\end{aligned}
\tag{2}
$$

Here, β is a parameter, which determining the weight of inlinks and outlinks in G_T.

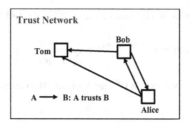

Fig. 3. An Example of Trust Network

In Epinions network, the similarity of two users can be viewed as their ratings on different products. For the sake of calculation, we simply transform the rating graph G_S into a bipartite graph $G_b = (V_1, V_2, E_b)$, where V_1 and V_2 represent two different types of objects. As shown in Fig. 4, V_1 represents a set of users (u_x) and V_2 represents a set of products (o_y) in Epinions. Given such a bipartite graph, we can compute the similarity score for each pair of objects of the same type using Simrank++ [26], which is based on the underlying idea that two objects of one type are similar if they are related to similar objects of the second type. Formally, the similarity score in Simrank++ is computed by the following equations.

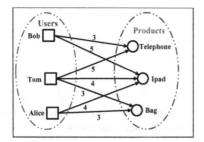

Fig. 4. An Example of Users-Products Bipartite Graph

$$Sim_S^{(\phi+1)}(u_i, u_j) = evidence(u_i, u_j) \cdot \Theta_1 \cdot$$
$$\sum_{(u_i, o_i) \in E_b} \sum_{(u_j, o_j) \in E_b} \mathbf{W}(u_i, o_i) \mathbf{W}(u_j, o_j) Sim_S^{(\phi)}(o_i, o_j) \qquad (3)$$

$$Sim_S^{(\phi+1)}(o_i, o_j) = evidence(o_i, o_j) \cdot \Theta_2 \cdot$$
$$\sum_{(o_i, u_i) \in E_b} \sum_{(o_j, u_j) \in E_b} \mathbf{W}(o_i, u_i) \mathbf{W}(o_j, u_j) Sim^{(\phi)}(u_i, u_j) \qquad (4)$$

Where Θ_1 and Θ_2 are constant. $evidence(x, y)$ and $\mathbf{W}(x, y)$ are defined as follows.

$$evidence(x, y) = \sum_{i=1}^{|N(x) \cap N(y)|} \frac{1}{2^i} \qquad (5)$$

Where $|N(x) \cap N(y)|$ denotes the common neighbors between x and y.

$$\mathbf{W}(x,y) = e^{-variance(y)} \cdot \frac{w(x,y)}{\sum_{(x,z) \in E_b} w(x,z)} \tag{6}$$

Where $variance(y)$ is the variance of the weight of edges that are connected to the node y.

3.3   Credibility

The credibility value of a node represents the true trust of a node. We can use credibility to define true trust. Credibility is the expected value of an incoming link from a reliable node. The credibility value depends on the quality of the inlinks, and not only the quantity: credibility of a node with one high quality inlink is equivalent to a node with many high quality inlinks. This definition differs from the usual random-walk based methods where the number of inlinks matter. For each inlink, we remove the effect of unreliability from the weight and then we compute the mean of all inlinks. The credibility of a node i is given by Eq. 7.

$$Credibility(u_i) = \sum_{u_p \in d^i(u_i)} \frac{Sim(u_p, u_i)}{\sum_{u_q \in d^o(u_p)} Sim(u_p, u_q)} w_{u_p u_i} \tag{7}$$

The credibility value lies in the range $[0, 1]$.

4   Experiment

In this section, we conduct different kinds of experiments to evaluate and analyze our algorithms on the real-world social network.

4.1   Datasets

In our experiment, we apply two real datasets (Epinions and DBLP). The Epinions data are crawled from Epinions.com until April 2010. The three relations in Epinions are, 1) users can trust other users; 2) users can post a review with product rating (from star 1 to star 5) about one product belonging to a certain category; 3) people can vote (very helpful, helpful, somewhat helpful, not helpful and off topic) for someone's review. In this evaluation, we only consider the trust relation and rating relation between users in Epinions. The multiple-relational network can be shown as Fig. 2. And the DBLP data are available from the Citation Network Dataset (http://arnetminer.org/citation). There are two relations in DBLP, 1) coauthor relation; 2) citation relation. The multiple-relational network can be described as Fig. 5. Table 2 shows the various classes of statistics about the two datasets.

   In a small graph, we can expect a high credibility value of a node, while the connections are much reasonable. For example, consider a pair of nodes with just

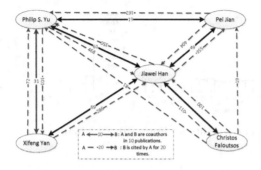

Fig. 5. An Illustration of Multiple-Relational Network of DBLP

Table 2. Detail of Dataset

Dataset	Epinions		DBLP	
	Trust	Rating	Coauthor	Citation
Nodes	98027	273437	595561	116667
Edges	612452	1076051	1311712	500000
Avg. of Degree	6.2478	3.9353	2.2025	4.2857

one directed edge. It is easy to see whether the reliability of one undefined node is reasonable from the other nodes. Similarly, the credibility of one undefined node will be the summation of each edge weight from the other nodes. While random-walk based techniques will give low scores to nodes in such components due to their low connectivity, in our model, they may get high credibility value based on the same connectivity. However, in general, credibility values do not make much sense if the graph is very small.

4.2 Distribution of Credibility

The first set of experiments measure the distribution of credibility values of the nodes. Fig. 6 shows the chart of the credibility value for both the datasets. In both datasets, count of nodes with credibility as 0.6 is very high.

However, the distribution of credibility is smoother due to the removal of the effect of unreliability, especially for Epinions. For DBLP, the distribution is not so smooth because of the presence of too many disconnected components in small size. In such small sized graphs, as previously discussed, if the degree is 1 or 2, the credibility values become close to that as well.

4.3 Comparison of Credibility with Ranking

The next set of experiments compare the ranking of nodes using the credibility values against that produced by the popular ranking algorithms such as PageRank [3] and HITS [2].

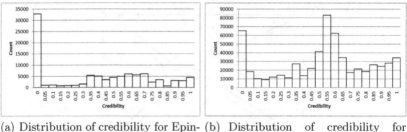

(a) Distribution of credibility for Epinions

(b) Distribution of credibility for DBLP

Fig. 6. Distribution of credibility for two Datasets

Fig. 7 shows the comparison. Note that we have scaled the PageRank and HITS score by multiplying with 1000. One common trend we observe is that nodes with less credibility have low HITS and PageRank score, and those with high credibility have high score. This shows that the ranking determined by the credibility values conform to the perception that more popular nodes have more credibility.

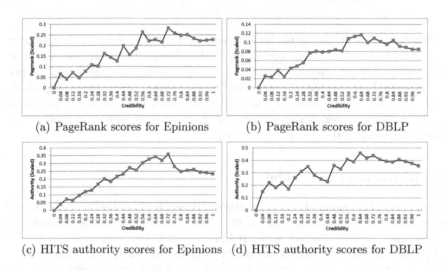

(a) PageRank scores for Epinions

(b) PageRank scores for DBLP

(c) HITS authority scores for Epinions

(d) HITS authority scores for DBLP

Fig. 7. Comparison of credibility with ranking

However, towards the end (when credibility is also almost equal to 1), there is a drop in the scores. This is partly due to our model. Even if a node has few connections but has a high quality inlink, it will attract high reputation. The same is not true for the other two algorithms. Moreover, the two datasets have a large number of strongly connected components and most of them are very small in size. The nodes in these small components have high credibility, but due to their small sizes, they have low scores.

4.4   Connection to Balance Theory

In this evaluation, our method conforms to balance theory. The balance theory includes (i) "a friend's friend is a friend", (ii) "an enemy's friend is an enemy", (iii) "a friend's enemy is an enemy" and (iv) "an enemy's enemy is a friend". More information of balance theory can be found in [12–15].

Assume S is the set of all occurrences of the pattern $i \to j \to k$ where the direct link $i \to k$ exists. We compute the error using the following equation as [16]:

$$\delta = \frac{1}{4|S|} \sum_{\{i,j,k\} \in S} (w_{ij} w_{jk} - w_{ik})^2 \tag{8}$$

Here, the number 4 in the denominator is a normalizing constant.

In this assessment, we evaluate our model with the balance theory. Initially, we compute the conformity of the graph using Eq. 8 on the original network. Secondly, we compute the error removing the bias from each edge as [16]. Finally, we remove unreliability from each edge using Eq. 7, and re-compute the error.

Table 3. Error of conformity with balance theory

Relationship	Epinions			DBLP		
	δ_o	δ_b	δ_u	δ_o	δ_b	δ_u
friend-friend-friend	0.03	0.03	0.03	0.04	0.05	0.04
friend-enemy-enemy	0.55	0.43	0.37	0.61	0.49	0.42
enemy-friend-enemy	0.38	0.34	0.29	0.33	0.28	0.25
enemy-enemy-friend	0.62	0.41	0.27	0.48	0.35	0.22

Table 3 shows the errors of two datasets. Here, δ_o shows the error of the original graph, δ_b shows the error after removing bias [16] and δ_u shows the error after removing unreliability. Our method almost improves the result except in the case of "a friend's friend is a friend" where the error is a small value. Thus, we can conclude that considering the credibility of a node and utilizing it benefits the conformity of the balance theory in the network.

4.5   Case Study

In the final experiment, we want to compare the scores computed by three different algorithms including PageRank (PR), HITS and Credibility (Cred.) for top-10 reliable users from Epinions and DBLP. We select Top-10 reviewers for the most popular authors overall provided by Epinions, and we select Top 10 researchers with highest H-Index in DBLP. In Table 4, we can find that our method produces a better result than others. The credibility scores conform to the manual ranking. PageRank and HITS scores in Table 4 is multiplied by 1000.

Table 4. Top-10 Users' Different Scores in Epinions & DBLP

Epinions				DBLP			
Name	PR	$HITS$	$Cred.$	Name	PR	$HITS$	$Cred.$
jo.com	0.2798	0.3553	0.7431	Herbert A. Simon	0.1183	0.4525	0.6714
dkozin	0.2251	0.3147	0.7251	Anil K. Jain	0.0852	0.4042	0.6542
mkaresh	0.2342	0.3341	0.7209	Scott Shenker	0.0913	0.4228	0.6173
Freak369	0.2739	0.2324	0.7123	Terrence Sejnowski	0.0964	0.3805	0.6065
Bryan_Carey	0.1951	0.3027	0.7037	Hector Garcia-Molina	0.1002	0.4149	0.6042
three_ster	0.2031	0.2232	0.6725	Takeo Kanade	0.0722	0.3623	0.5937
shoplmart	0.2594	0.1851	0.6328	Jiawei Han	0.0773	0.3774	0.5861
dlstewart	0.2161	0.2331	0.6074	Tomaso Poggio	0.0817	0.3811	0.5431
Howard_Creech	0.1973	0.1927	0.5596	Philip S. Yu	0.0753	0.3626	0.5135
ChrisJarmick	0.1912	0.2198	0.5284	David Haussler	0.0761	0.3341	0.4912

5 Conclusion

We observe the benefits of modeling trust and similarity together for ubiquitous
multiple-relational network. We design a method to model and demonstrate the advantages for both sides using a large scale real world data. We believe that analysis
on multiple-relational network has a bright future because social trust and similarity studies are two building blocks for many research interests, such as ranking,
clustering, classification and recommendation. For many applications involving
trust-based networks, it is crucial to assess the credibility of a node. In this paper, we have proposed an algorithm to compute the credibility of nodes in networks
where the edge weight denotes the trust score, using the similarity of nodes in Reinforcement Learning. The experiment result shows that our algorithm significantly
improves the performance than others'. Our model conforms well to other graph
ranking algorithms and social theories such as the balance theory. However, our
algorithm may be misguided by malicious nodes and plotting groups. In the future work, we would like to deal with these problems and other malicious attacks.
Moreover, we will explore a distributed application of our approach.

Acknowledgement. The work was supported in part by Fund of Science and
Technology Development of Macau Government under FDCT/106/2012/A3 and
by University Macau Research Committee under MYRG188-FST11-GZG.

References

1. International, P.S.R.A., WebWatch, C.R.: Leap of Faith: Using the Internet Despite the Dangers: Results of a National Survey for Consumer Reports WebWatch.
   Consumer Reports WebWatch (October 2005)
2. Kleinberg, J.M.: Authoritative sources in a hyperlinked environment. J.
   ACM 46(5), 604–632 (1999)
3. Page, L., Brin, S., Motwani, R., Winograd, T.: The pagerank citation ranking:
   Bringing order to the web. Technical report, Stanford InfoLab (November 1999)
4. Braunstein, A., Mézard, M., Zecchina, R.: Survey propagation: an algorithm for
   satisfiability. CoRR cs.CC/0212002 (2002)

5. Sherchan, W., Nepal, S., Paris, C.: A survey of trust in social networks. ACM Comput. Surv. 45(4), 47 (2013)
6. Orman, L.V.: Bayesian inference in trust networks. ACM Trans. Manage. Inf. Syst. 4(2), 7:1–7:21 (2013)
7. Bonacich, P.: Factoring and weighting approaches to status scores and clique identification. The Journal of Mathematical Sociology 2(1), 113–120 (1972)
8. Kamvar, S.D., Schlosser, M.T., Garcia-Molina, H.: The eigentrust algorithm for reputation management in p2p networks. In: WWW, pp. 640–651 (2003)
9. Richardson, M., Agrawal, R., Domingos, P.: Trust management for the semantic web. In: Fensel, D., Sycara, K., Mylopoulos, J. (eds.) ISWC 2003. LNCS, vol. 2870, pp. 351–368. Springer, Heidelberg (2003)
10. de Kerchove, C., Dooren, P.V.: The pagetrust algorithm: How to rank web pages when negative links are allowed? In: SDM, pp. 346–352 (2008)
11. Guha, R.V., Kumar, R., Raghavan, P., Tomkins, A.: Propagation of trust and distrust. In: WWW, pp. 403–412 (2004)
12. Cartwright, D., Harary, F.: Structural balance: A generalization of heider's theory. Psychological Review 63(5), 277–293 (1956)
13. Heider, F.: Attitudes and cognitive organization. J. Psychology 21, 107–112 (1946)
14. Leskovec, J., Huttenlocher, D.P., Kleinberg, J.M.: Signed networks in social media. In: CHI, pp. 1361–1370 (2010)
15. Leskovec, J., Huttenlocher, D.P., Kleinberg, J.M.: Predicting positive and negative links in online social networks. In: WWW, pp. 641–650 (2010)
16. Mishra, A., Bhattacharya, A.: Finding the bias and prestige of nodes in networks based on trust scores. In: WWW, pp. 567–576 (2011)
17. Li, R.H., Yu, J.X., Huang, X., Cheng, H.: A framework of algorithms: Computing the bias and prestige of nodes in trust networks. CoRR abs/1207.5661 (2012)
18. Koutra, D., Ke, T.-Y., Kang, U., Chau, D.H(P.), Pao, H.-K.K., Faloutsos, C.: Unifying guilt-by-association approaches: Theorems and fast algorithms. In: Gunopulos, D., Hofmann, T., Malerba, D., Vazirgiannis, M. (eds.) ECML PKDD 2011, Part II. LNCS, vol. 6912, pp. 245–260. Springer, Heidelberg (2011)
19. Gyöngyi, Z., Garcia-Molina, H., Pedersen, J.O.: Combating web spam with trustrank. In: VLDB, pp. 576–587 (2004)
20. Wu, B., Goel, V., Davison, B.D.: Topical trustrank: using topicality to combat web spam. In: WWW, pp. 63–72 (2006)
21. Yin, X., Han, J., Yu, P.S.: Truth discovery with multiple conflicting information providers on the web. IEEE Trans. Knowl. Data Eng. 20(6), 796–808 (2008)
22. Srikantaiah, K.C., Srikanth, P.L., Tejaswi, V., Shaila, K., Venugopal, K.R., Patnaik, L.M.: Ranking search engine result pages based on trustworthiness of websites. CoRR abs/1209.5244 (2012)
23. Golub, G.H., Van Loan, C.F.: Matrix computations, 3rd edn. Johns Hopkins University Press, Baltimore (1996)
24. Jin, X., Luo, J., Yu, J., Wang, G., Joshi, D., Han, J.: Reinforced similarity integration in image-rich information networks. IEEE Trans. Knowl. Data Eng. 25(2), 448–460 (2013)
25. Zhao, P., Han, J., Sun, Y.: P-rank: a comprehensive structural similarity measure over information networks. In: CIKM, pp. 553–562 (2009)
26. Antonellis, I., Garcia-Molina, H., Chang, C.C.: Simrank++: query rewriting through link analysis of the click graph. PVLDB 1(1), 408–421 (2008)
27. Jeh, G., Widom, J.: Simrank: a measure of structural-context similarity. In: KDD, pp. 538–543 (2002)
28. Wang, G., Hu, Q., Yu, P.S.: Influence and similarity on heterogeneous networks. In: CIKM, pp. 1462–1466 (2012)

Result Diversification for Tweet Search

Makbule Gulcin Ozsoy, Kezban Dilek Onal, and Ismail Sengor Altingovde

Middle East Technical University, Ankara, Turkey
{makbule.ozsoy,dilek,altingovde}@ceng.metu.edu.tr

Abstract. Being one of the most popular microblogging platforms, Twitter handles more than two billion queries per day. Given the users' desire for fresh and novel content but their reluctance to submit long and descriptive queries, there is an inevitable need for generating diversified search results to cover different aspects of a query topic. In this paper, we address diversification of results in tweet search by adopting several methods from the text summarization and web search domains. We provide an exhaustive evaluation of all the methods using a standard dataset specifically tailored for this purpose. Our findings reveal that implicit diversification methods are more promising in the current setup, whereas explicit methods need to be augmented with a better representation of query sub-topics.

Keywords: Microblogging, Tweet search, novelty, diversity.

1 Introduction

Microblogging sites have recently become world-wide popular platforms for sharing and following emerging news and trending events, as well as expressing personal opinions and feelings on a wide range of topics. In Twitter, a prominent example of such platforms, hundreds of millions of users post 500 million tweets per day. In addition to reading tweets in their own feed, Twitter users often conduct search on the posted content. As of 2014, the number of queries submitted to Twitter per day is reported to be more than two billion [2].

The nature of search in Twitter is different from that of the typical Web search in several ways. Twitter users are more interested in searching other people (especially celebrities) or trending events (usually expressed via hashtags) and more likely to repeat the same query to monitor the changes in the content in time [23]. Given the users' desire on the timely and novel content, earlier works essentially focus on filtering near-duplicates in search results, which might be abundant due to retweeting or posting of the same/similar content by several users in the same time period. A complementary issue, which is mostly overlooked in the literature, is the diversification of the search results, i.e., covering different aspects (sub-topics) of the query topic in the top-ranked results.

In typical web search, diversification of results is usually needed due to the ambiguous or vague specification of queries. In case of Twitter, queries are even shorter (1.64 words on the average [23]), which indicates a similar need for diversification. Furthermore, due to the bursty nature of the microblogging, some

B. Benatallah et al. (Eds.): WISE 2014, Part II, LNCS 8787, pp. 78–89, 2014.

content regarding a particular query aspect can be quickly buried deep in the tweet stream, if the user is not fast enough to see it. For instance, assume a query "WISE2014", with possible different aspects that may discuss the technical coverage of the conference (accepted papers, etc.), logistics (visa issues, travel arrangements, hotels, etc.) and social events during the conference. At the time a user submits this query, it might be possible that the other users are essentially posting about, say, the logistics issues, which are not interest of the searcher if she does not plan to attend. In this case, if she has no patience or time to scroll down tens of tweets, she would miss the earlier messages abut the accepted papers that she is really interested in. Note that, diversification of search results does not only help the end-users to quickly grasp different dimensions of the topic in question, but it may also improve applications that submit queries to the Twitter API and retrieve a few top-ranked results for further processing.

In this paper, we address the result diversification problem for tweet search. To this end, we adopt various methods from the literature that are introduced for text summarization and web search result diversification. In particular, we consider LexRank [8] and Biased LexRank [14] as representative methods from the field of text summarization, and several implicit and explicit diversification methods, namely, MMR [3], Max-Sum [10], Max-Min [10] and xQuAD [19] from web search domain. For comparison, we also include the Sy method proposed in the context of near-duplicate elimination in tweet streams [21,22]. While investigating the performance of these approaches, we analyze the impact of various features in computing the query-tweet relevance and tweet-tweet similarity scores. We evaluate the performance using a benchmark collection, the Tweets2013 corpus [22], recently released for this purpose. To the best of our knowledge, apart from the Sy method [22], none of these methods were employed in the context of tweet search diversification and evaluated in a framework involving a specifically tailored dataset and diversity-aware evaluation metrics.

Our findings reveal that, in contrast to the case of web search, implicit diversification methods perform better than the explicit ones for diversifying tweet search results. This contradictory result is due to the fact that most of the additional terms describing the query aspects do not appear in tweets, which are very short pieces of text. Among the implicit methods, Sy and Max-Sum are found to be the top-performers for different evaluation metrics.

The rest of the paper is organized as follows. In the next section, we present the features utilized for computing the relevance and similarity scores to be used in tweet ranking. In Section 3, we discuss how various diversification methods are adopted for ranking results in tweet search. We present our experimental setup and results in Section 4. In Section 5, we briefly review the related work in the areas of search result diversification and tweet ranking. Finally, we conclude and point to future research directions in Section 6.

2   Features for Tweet Ranking

For our purposes in this paper, we first need to specify how we compute the relevance and similarity scores for the query-tweet and tweet-tweet pairs, respectively. In what follows, we briefly discuss the features and functions employed in our tweet ranking framework.

Similarity function. While computing the similarity of a pair of tweets, we use three types of features, namely, the pure textual content (i.e., terms), hashtags, and tweet time, as follows.

- *Content features.* While extracting the textual content, we remove the media links, urls, mention tags and hashtags in the tweets and reduce the tweet content only to a set of terms. Then, we stem the terms using JWNL (Java WordNet Library). The similarity score between two sets of terms is computed using various well-known functions from the literature. In particular, we employ the traditional Jaccard, Cosine and BM25 functions while computing the similarity of tweets. Whenever needed, IDF values are obtained over the initial retrieval results for a given query.

  Following the practice in the earlier works that employ simpler overlap-based functions in case of tweet ranking (e.g., [21]), we also define a ratio-based similarity function. This function computes the percentage of common terms in tweets t_i and t_j as shown in Eq. 1. In our preliminary analysis, we observed that Twitter users tend to use hashtags also as words to construct a full sentence as in the following example: "Ya Libnan: #Lebanon #credit #rating suffering because of #Syria #crisis"[1]. So, in the experiments, we also employ a variant of this function (denoted as Ratio-H in Section 4) that considers each hashtag as a typical term by stripping of the # sign.

$$S_W(t_i, t_j) = \frac{|Terms(t_i) \cap Terms(t_j)|}{|Terms(t_i)|} \tag{1}$$

- *Hashtag features.* For a given pair of tweets, we compute the Jaccard similarity of the set of hashtags, which is denoted as S_H.
- *Time feature.* Time similarity score between two tweets is based on the difference between their normalized timestamps (using Min-Max Normalization), and computed by Eq. 2.

$$S_T(t_i, t_j) = 1 - |t_{norm}(t_i) - t_{norm}(t_j)| \tag{2}$$

The overall tweet-tweet similarity is computed as a linear weighted function of the content similarity (S_W), hashtag similarity (S_H) and time similarity (S_T) scores, as shown in Eq. 3. In this equation, α_i represents the weight for the similarity score for each feature, where $\sum_i \alpha_i = 1$.

$$sim(t_i, t_j) = \alpha_1 S_W(t_i, t_j) + \alpha_2 S_H(t_i, t_j) + \alpha_3 S_T(t_i, t_j) \tag{3}$$

Relevance function. We compute the relevance of a query to a given tweet, i.e., $rel(q, t)$, using only term features, as hashtags and time features are not available for the queries in our dataset. In our evaluations, we consider all four functions employed for the similarity computation case, namely, Jaccard, Cosine, BM25 and Ratio functions, while computing the relevance scores.

---

[1] The sentence is a tweet collected for topic 7, namely "syria civil war".

3   Tweet Ranking Using Diversification Methods

Let's assume a query q that retrieves a ranked list of tweets C (where $|C| = N$) over a collection of tweets. Our goal in this work is to obtain a top-k ranking S (where $k < N$) that maximizes both the relevance and diversity among all possible size-k rankings S' of C.

To achieve our goal, we employ six different diversification methods that are, to the best of our knowledge, not applied in tweet ranking framework before. In particular, we adopt LexRank [8] and Biased LexRank [14] from text summarization domain, and MMR [3], Max-Sum [10], Max-Min [10], and xQuAD [19] methods from web search domain. All methods except xQuAD are implicit methods, as they only rely on the initial retrieval results for obtaining the top-k diversified ranking. In contrast, xQuAD is a representative of the explicit diversification paradigm, which leverages apriori information regarding the aspects (sub-topics) of the queries during the ranking. Finally, we also include the Sy method proposed in [21,22]. As far as we know, this is the only method directly applied for tweet search diversification in the literature. In what follows, we briefly summarize each of these methods.

Tweet Ranking using LexRank. LexRank [8] is a graph-based multi-document summarization approach that constructs a graph of the input sentences and ranks the sentences performing random walks. The score of the sentences, namely the score vector p, is computed by Eq. 4. In this equation, A is a square matrix with all elements being set to to $1/M$, where M is the number of sentences. As usual in a random walk, A matrix represents the probability of jumping to a random node in the graph. The pairwise similarities of the sentences are captured in the matrix B. When this algorithm converges, the sentences with the highest scores are selected to construct the summary.

$$p = [\lambda A + (1 - \lambda)B]^T p \qquad (4)$$

In this work, instead of sentences, we use tweets that are in the initial retrieval set C for a given query q, and apply the same algorithm to select top-k tweets into S. Note that, some earlier works (e.g.,[20]) also employ LexRank for summarizing tweet streams; however, they do not provide an evaluation based on diversity-aware IR metrics as we do in this paper.

Tweet Ranking using Biased LexRank. Biased LexRank [14] is an extension of LexRank method, which additionally takes into account the relevance of the documents (in our case, tweets) to a given query. The computation is performed using the same formula shown in Eq. 4. However, in this case, A represents the query-tweet relevance matrix. As in the LR method, top-k tweets with the highest stationary probabilities after the convergence are selected into S.

Tweet Ranking using MMR. MMR [3] is a well-known greedy method to combine query relevance and information novelty. In MMR, the set S is initialized with the tweet that has the highest relevance to the query. In each iteration, MMR reduces the relevance score of a candidate tweet by its maximum similarity to already selected tweets in S, and then selects the next tweet based on these discounted scores, i.e., the tweet that maximizes Eq. 5.

$$f_{MMR}(t_i) = \lambda \, rel(t_i, q) - (1 - \lambda) \max_{t_j \in S} sim(t_i, t_j) \tag{5}$$

Note that, in MMR, we involve a trade-off parameter λ to balance the relevance and diversity in the final result set.

Tweet Ranking using Max-Sum and Max-Min. We adopted two of the diversification methods proposed in [10], namely Max-Sum and Max-Min approaches that are based on the solutions for the facility dispersion problem in operations search. In the former method, the objective function aims to maximize the sum of the relevance and diversity (i.e., dissimilarity) in the final result set. This is achieved by a greedy approximation algorithm that selects a pair of tweets that maximizes Eq. 6 in each iteration.

$$f_{MaxSum}(t_i, t_j) = (1 - \lambda)(rel(t_i) + rel(t_j)) + 2\lambda(1 - sim(t_i, t_j)) \tag{6}$$

In the Max-Min method, the objective is maximizing the minimum relevance and dissimilarity of the result set. In this case, the greedy algorithm initially selects the pair of tweets that maximizes Eq. 7. Then, in each iteration, it selects the tweet that maximizes Eq. 8. As in the case of MMR, these methods employ a parameter λ for setting the trade-off between the relevance and diversity.

$$f_{MaxMin}(t_i, t_j) = 1/2(rel(t_i) + rel(t_j)) + \lambda(1 - sim(t_i, t_j)) \tag{7}$$

$$f_{MaxMin}(t_i) = \min_{t_j \in S} f_{MaxMin}(t_i, t_j) \tag{8}$$

Tweet Ranking using xQuAD. xQuAD [19] is an explicit diversification method based on the assumption that aspects of a query can be known apriori. The method aims to maximize the coverage of tweets related to the different aspects of the query and to minimize the redundancy with respect to these aspects. The greedy algorithm selects the tweet that maximizes Eq. 9 in each iteration. In this formula, $P(t_i|q)$ denotes the relevance of t_i to query q, $P(q_i|q)$ denotes the likelihood of the aspect q_i for the query q, $P(t_i|q_i)$ denotes the relevance of t_i to the query aspect q_i, and finally the product term represents the probability of q_i not being satisfied by the tweets that are already selected into S.

$$f_{xQuAD}(t_i) = (1 - \lambda)P(t_i|q) + \lambda \sum_{q_i} \left[P(q_i|q)P(t_i|q_i) \prod_{t_j \in S} (1 - P(t_j|q_i)) \right] \tag{9}$$

Tweet Ranking using Sy In a recent study [21], Tao et al. present a framework for detecting duplicate and near-duplicate tweets and define a large set of features to be employed in this context. Furthermore, they define a simple yet effective diversification method, so-called Sy, leveraging these features. Their method scans the list of initial retrieval results, C, in a top-down fashion. For each tweet t_i, all the succeeding (lower-ranked) tweets that are near-duplicates of t_i (i.e., with a similarity score greater than a pre-defined threshold) are removed.

4   Experiments

Dataset. For our evaluations, we use the Tweets2013 corpus [22] that is specifically built for tweet search result diversification problem. The dataset includes tweets collected between February 1, 2013 and March 31, 2013. There are forty seven query topics and each topic has, on the average, 9 sub-topics.

The owners of the Tweets2013 corpus only share the tweet identification numbers, as the Twitter API licence does not allow users to share the content of the tweets. Using the provided IDs and Twitter API, we attempted to obtain top-100 and top-500 tweets for each topic. Since some of these tweets were erased or their sharing status were changed, we ended up with 81 tweets per topic for top-100 set and 403 tweets per topic for top-500 set, respectively, on the average.

In top-100 tweet collection, we observed that 80% of the tweets are not assigned to any sub-topics. In particular, there are four query topics (with ids 5, 9, 22 and 28) for which the resulting tweets are not related to any of their sub-topics. Besides, there are six more topics (with ids 7, 8, 14, 43, 46 and 47) that retrieve at most 2 relevant tweets among their top-100 results. We removed all of these topics from our query set to avoid misleading or meaningless results.

Similarly, in our top-500 tweet collection, 91% of the tweets are not assigned to any sub-topics. In this case, there are eleven topics (with ids 3, 5, 7, 9, 14, 28, 37, 43, 45, 46 and 47) for which less than 3% of the tweets in top-500 are relevant, and one additonal topic (with id 22) having no relevant tweets at all. These topics are again removed from our query set in the experiments that employ top-500 collection.

Evaluation Metrics. We evaluate diversification methods using the `ndeval` software[2] employed in TREC Diversity Tasks. We report results using three popular metrics, namely, α-nDCG [6], Precision-IA [1], and Subtopic-Recall [27] at the cut-off values of 10 and 20, as typical in the literature.

Results. We present the evaluation results for the methods adopted in this paper, namely LexRank (LR), Biased LexRank (BLR), MMR, Max-Sum, Max-Min and xQuAD, as well as the Sy method that is previously utilized for tweet diversification. We also report the performance for the initial retrieval results (i.e., without any diversification) obtained by a system employing the query-likelihood (QL) retrieval model. Note that, these initial retrieval results were provided in the Tweets2013 corpus, however we re-compute their effectiveness scores based on only those tweets that were still accessible using Twitter API, for the sake of fair comparison. Therefore, the effectiveness of the baseline QL run slightly differs from what is reported in [22].

---

[2] http://trec.nist.gov/data/web10.html

Table 1. Effectiveness of diversification methods for $N = 100$ (We denote content, hashtag and time features that are used in the similarity functions with C, H and T, respectively. Ratio-H function computes the ratio-based similarity using both terms and hashtags).

Method	Rel.	Sim.	Sim. Features	α-nDCG @10	@20	Prec-IA @10	@20	ST-Recall @10	@20
QL	-	-	-	0.303	0.346	0.065	0.058	0.357	0.505
SY	QL	Sy	Syntactic [22]	0.339	0.378	0.080	0.068	0.401	0.529
SY	QL	Jaccard	C,H,T	**0.348**	**0.383**	**0.083**	**0.069**	**0.419**	0.542
LR	BM25	BM25	C, H, T	0.301	0.342	0.066	0.059	0.361	0.486
BLR	BM25	BM25	C, H, T	0.316	0.344	0.067	0.055	0.382	0.473
MMR	Ratio-H	Ratio-H	C	0.341	0.374	0.066	0.056	0.417	0.539
MaxSum	Ratio-H	Cosine	C, H	0.325	0.374	0.064	0.060	0.397	**0.561**
MaxMin	Ratio	Cosine	C	0.322	0.365	0.060	0.057	0.380	0.527
xQuAD	Jaccard	Jaccard	C	0.235	0.263	0.050	0.041	0.302	0.419

In our evaluations, we employed the diversification methods to compute the final top-k ranking S out of N initial retrieval results, where k is 30 and N is from $\{100, 500\}$. In Tables 1 and 2, we report the best-results (based on the α-nDCG@20 scores) for each method when N is 100 and 500, respectively. We also present the functions and features that are used for computing the relevance and similarity scores in each case[3]. For the Sy method, in addition to using the features described in Section 2, we also experimented with its best performing setup reported in a previous study, i.e., employing the syntactic feature set with the associated feature weights for computing the tweet-tweet similarity [22]. For xQuAD, following the practice in [19], we use the official query sub-topics provided in the dataset to represent an ideal scenario.

Table 1 reveals that Max-Sum and Sy are the best diversification strategies for different evaluation metrics when N is set to 100. In particular, Sy (with our features) outperforms all other methods in terms of the Prec-IA metric, whereas Max-Sum achieves the highest score for the ST-Recall@20. Note that, MMR also outperforms the Sy version that incorporates the syntactic features in [22] in terms of the ST-Recall. MMR and both versions of Sy are the best performers for α-nDCG metric and yield comparable results to each other. We also observe that BLR, the query-aware version LR, is slightly better than the original algorithm.

A surprising result that is drawn form Table 1 is that implicit diversification methods outperform xQuAD, an explicit diversification strategy, by a wide margin. This is contradictory to the findings in the case of web search result diversification, where explicit methods are usually the top-performers. For further insight on this finding, we analyzed the occurrence frequency of sub-topic terms in our tweet collection. It turns out that most of the sub-topic terms do

---

[3] We only report the features employed in the similarity functions, as all the relevance functions use just the content (terms) feature.

Table 2. Effectiveness of diversification methods for $N = 500$ (We denote content, hashtag and time features that are used in the similarity functions with C, H and T, respectively. Ratio-H function computes the ratio-based similarity using both terms and hashtags.)

Method	Rel.	Sim.	Sim. Features	α-nDCG @10	@20	Prec-IA @10	@20	ST-Recall @10	@20
QL	-	-	-	0.303	0.346	0.065	0.058	0.357	0.505
SY	QL	Sy	Syntactic [22]	0.339	0.378	0.081	**0.069**	0.402	0.529
SY	QL	Jaccard	C,H,T	**0.348**	**0.382**	**0.082**	0.068	**0.419**	**0.542**
LR	BM25	BM25	C, H, T	0.302	0.341	0.066	0.059	0.361	0.480
BLR	BM25	BM25	C, H, T	0.301	0.340	0.066	0.058	0.362	0.482
MMR	Ratio-H	Ratio-H	C	0.207	0.264	0.043	0.047	0.296	0.467
MaxSum	Ratio-H	Cosine	C, H	0.223	0.287	0.049	0.053	0.311	0.483
MaxMin	Ratio	Cosine	C	0.175	0.238	0.036	0.042	0.270	0.459
xQuAD	Jaccard	Jaccard	C	0.113	0.140	0.0202	0.020	0.142	0.233

Table 3. Effectiveness of diversification methods with the syntactical features in [21,22] and for $N = 100$

Method	α-nDCG @10	@20	Prec-IA @10	@20	ST-Recall @10	@20
LR-Syntactic	0.191	0.243	0.035	0.039	0.271	0.438
BLR-Syntactic	0.201	0.256	0.038	0.042	0.278	0.447
MMR-Syntactic	0.147	0.203	0.033	0.038	0.263	0.407
MaxSum-Syntactic	0.118	0.145	0.022	0.020	0.175	0.254
MaxMin-Syntactic	0.097	0.128	0.019	0.020	0.156	0.272
xQuAD-Syntactic	0.207	0.268	0.046	0.048	0.289	0.469

not appear in the top-100 tweets retrieved for the corresponding queries. More specifically, we find that while tweets lack only 30% of the query terms on the average, they lack 85% of the terms appearing in the sub-topics. We believe that this is due to the way sub-topics are formulated in the Tweets2013 corpus. While defining the sub-topics, human judges seem to use more general expressions that are unlikely to overlap with the terms in the actual tweets (e.g., see the example in [22] for the sub-topics of "Hillary Clinton" query). This implies that there is room for improving the performance of explicit diversification methods, by using external sources such as an ontology or query reformulations from a query log (as in [19]) for a better representation of the sub-topics.

Table 2 reveals that Sy with our similarity features is the best diversification strategy for different evaluation metrics when N is set to 500. LR and BLR scores on top-500 are similar to those using top-100 results. However, a significant decrease in the performance of the algorithms MMR, MaxSum, MaxMin and xQuAD is observed when N is increased to 500. The latter methods seem to trade-off relevance against diversity when the initial tweet set size is increased.

As a final experiment, we incorporate the syntactical features used for the Sy method into all other diversification methods while computing the similarity scores for the tweet pairs. These syntactical features include Levenshtein Distance between tweet contents, overlap in terms, overlap in hashtags, overlap in URLs, overlap in extended URLs and length difference (please refer to [21,22] for details). In this case, query-tweet relevance scores are computed based on the content (term) overlap. In this experiment, we use the best-performing feature weights obtained via logistic regression in [21]. Our results in Table 3 reveal that these features are less useful for the diversification methods we consider in this paper; and usually degrade their performance (cf. Table 1). Note that, we only report the results for $N = 100$ for the sake of brevity.

5   Related Work

5.1   Ranking Tweets

There exists a considerable number of studies which focus on tweet ranking. Relevance to the search query is the major ranking criteria in most of the work on tweet ranking for Twitter search. Jabeur et al. [12] model the relevance of a tweet to a query by a Bayesian network that integrates a variety of features, namely micro-blogger's influence on the query topic, time and content features of tweets. In [29], the authors train machine learning models for ranking tweets against a query. Another study [13] reports that taking the hyper-links in tweet content into account improves the relevance of the retrieved results.

To the best of our knowledge, there are only two earlier studies, namely [18] and [22], that consider novelty as an additional criteria to relevance for ranking tweets. In the first study [18], an approach based on MMR [3] and clustering of tweets is proposed. However, they do not evaluate the proposed approach using diversity-aware evaluation metrics, as we do here. In the second study, Tao et al. [22] introduce Tweets2013 corpus, a data set designed for evaluating result diversification approaches for Twitter search. They also report the diversification performance of their duplicate detection framework introduced in [21] on this corpus. Note that, in this paper, we compare several other approaches to their method, Sy, in a framework that again employs Tweets2013 corpus.

Personalized tweet ranking aims to rank tweets according to the likelihood of being liked by a user. Feng et al. propose a model for personalization of Twitter stream based on the observation that a user is more interested in a tweet if she is likely to retweet it [9]. Therefore, tweets are ranked with respect to the likelihood of being re-tweeted by a user. Retweet likelihood is modeled with a graph that incorporates information from different sources such as the user's profile and interaction history.

Another recent study for personalized tweet ranking is [25]. Vosecky et al. propose a model for delivering personalized and diverse content in response to a search query. In particular, they explicitly represent both a user's and her friends' interests using topic models, and re-rank the search results based on these models. Note that, their evaluation is again based on traditional IR metrics, i.e., they do not explicitly evaluate whether the search results cover different aspects of a given query.

5.2   Diversifying Web Search Results

As web queries are inherently ambiguous and/or underspecified, diversifying the search results to cover the most probable aspects of a query among the top-ranked results (usually, top-10 or -20) arise as a popular research topic. In the *implicit diversification methods*, such query aspects are discovered from the initial retrieval results in various ways that usually involve constructing clusters [11] or topic models [5]. Since finding the optimal diversification is shown to be NP-hard [4], greedy approximation heuristics need to be employed. In this sense, a large number of implicit methods employ greedy best-first search strategy. The representative methods in this category include MMR [3], risk minimization framework proposed by Zhai et al. [28], Greedy Marginal Contribution [24] method that extends the traditional MMR, and Modern Portfolio Theory (MPT) that takes into account the variance of the relevance of the query results over different query aspects [26,17]. Gollapudi and Sharma model the result diversification problem as a bi-criteria optimization problem and then cast it to the well-known obnoxious facility dispersion problem in Operations Research [10]. In this framework, depending on the objective function, it is possible to adopt the greedy heuristics such as the Max-Sum and Max-Min approaches. In contrast, Zuccon et al. cast the diversification problem to the desirable facility dispersion problem and apply greedy local search strategy to find an approximate solution [30]. Vieira et al. also consider a semi-greedy strategy based on local search to obtain diversified query results [24].

In the *explicit diversification methods*, we assume that query aspects are known apriori, i.e., discovered from a taxonomy or query log. To this end, in one of the earliest studies, Radlinski and Dumais utilize query re-formulations [16]. In contrast, IA-Select strategy assumes the existence of a taxonomy that can be used to assign category labels to queries and retrieved results, and exploit these labels for diversification [1]. The xQuAD strategy again makes use of the query reformulations to discover the aspects and proposes a probabilistic mixture model to construct the diversified query result [19]. Dang and Croft introduce a proportionality based approach that takes into account the representation proportion of each aspect in the top-ranked query results [7]. In a recent study, Ozdemiray and Altingovde adapt score- and rank aggregation methods to the result diversification problem and show that they are both effective and efficient in comparison to the earlier methods [15]. Our work in this paper employs representative approaches from both of the implicit and explicit diversification methods to shed light their performance in the context of tweet search.

6   Conclusion

In this paper, we presented an empirical analysis of a variety of search result diversification methods adopted from the text summarization and web search domains for the task of tweet ranking. Our experiments revealed that the implicit diversification methods outperform a popular explicit method, xQuAD, due to the vocabulary gap between the official query sub-topics and tweets. Among the implicit methods, while Sy seems to be the most promising one; there is no clear winner, and different strategies yield the best (or comparable) results for different diversity-aware evaluation metrics.

As a future work, we plan to incorporate additional features such as re-tweet counts, media links, and user popularity. We also aim to investigate the performance of explicit diversification methods with better sub-topic descriptions, and explore how such sub-topic descriptions can be automatically extracted using the clues available in a microblogging platform.

Acknowledgments. This work is partially funded by METU BAP-08-11-2013-055 project. I. S. Altingovde acknowledges the Yahoo! Faculty Research and Engagement Program.

References

1. Agrawal, R., Gollapudi, S., Halverson, A., Ieong, S.: Diversifying search results. In: Proc. of WSDM 2009, pp. 5–14 (2009)
2. Busch, M., Gade, K., Larson, B., Lok, P., Luckenbill, S., Lin, J.: Earlybird: Real-time search at twitter. In: Proc. of ICDE 2012, pp. 1360–1369 (2012)
3. Carbonell, J., Goldstein, J.: The use of mmr, diversity-based reranking for reordering documents and producing summaries. In: Proc.of SIGIR 1998, pp. 335–336 (1998)
4. Carterette, B.: An analysis of NP-completeness in novelty and diversity ranking. In: Azzopardi, L., Kazai, G., Robertson, S., Rüger, S., Shokouhi, M., Song, D., Yilmaz, E. (eds.) ICTIR 2009. LNCS, vol. 5766, pp. 200–211. Springer, Heidelberg (2009)
5. Carterette, B., Chandar, P.: Probabilistic models of ranking novel documents for faceted topic retrieval. In: Proc. of CIKM 2009, pp. 1287–1296 (2009)
6. Clarke, C.L.A., Kolla, M., Cormack, G.V., Vechtomova, O., Ashkan, A., Büttcher, S., MacKinnon, I.: Novelty and diversity in information retrieval evaluation. In: Proc. of SIGIR 2008, pp. 659–666 (2008)
7. Dang, V., Croft, W.B.: Diversity by proportionality: an election-based approach to search result diversification. In: Proc. of SIGIR 2012, pp. 65–74 (2012)
8. Erkan, G., Radev, D.R.: Lexrank: graph-based lexical centrality as salience in text summarization. J. Artif. Int. Res. 22(1), 457–479 (2004)
9. Feng, W., Wang, J.: Retweet or not?: personalized tweet re-ranking. In: Proc. of WSDM 2013, pp. 577–586 (2013)
10. Gollapudi, S., Sharma, A.: An axiomatic approach for result diversification. In: Proc. of WWW 2009, pp. 381–390 (2009)
11. He, J., Meij, E., de Rijke, M.: Result diversification based on query-specific cluster ranking. JASIST 62(3), 550–571 (2011)
12. Jabeur, L.B., Tamine, L., Boughanem, M.: Uprising microblogs: a bayesian network retrieval model for tweet search. In: Proceedings of the 27th Annual ACM Symposium on Applied Computing, SAC 2012, pp. 943–948. ACM (2012)
13. McCreadie, R., Macdonald, C.: Relevance in microblogs: Enhancing tweet retrieval using hyperlinked documents. In: Proc. of the 10th Conference on Open Research Areas in Information Retrieval, OAIR 2013, pp. 189–196 (2013)

14. Otterbacher, J., Erkan, G., Radev, D.R.: Biased lexrank: Passage retrieval using random walks with question-based priors. Inf. Process. Manage. 45(1), 42–54 (2009)
15. Ozdemiray, A.M., Altingovde, I.S.: Explicit search result diversification using score and rank aggregation methods. In: JASIST (in press)
16. Radlinski, F., Dumais, S.T.: Improving personalized web search using result diversification. In: Proc. of SIGIR 2006, pp. 691–692 (2006)
17. Rafiei, D., Bharat, K., Shukla, A.: Diversifying web search results. In: Proc. of WWW 2010, pp. 781–790 (2010)
18. Rodriguez Perez, J.A., Moshfeghi, Y., Jose, J.M.: On using inter-document relations in microblog retrieval. In: Proc. of WWW 2013, pp. 75–76 (2013)
19. Santos, R.L., Macdonald, C., Ounis, I.: Exploiting query reformulations for web search result diversification. In: Proc. of WWW 2010, pp. 881–890 (2010)
20. Sharifi, B., Inouye, D., Kalita, J.K.: Summarization of twitter microblogs. Comput. J. 57(3), 378–402 (2014)
21. Tao, K., Abel, F., Hauff, C., Houben, G.-J., Gadiraju, U.: Groundhog day: Near-duplicate detection on twitter. In: Proc. of WWW 2013, pp. 1273–1284 (2013)
22. Tao, K., Hauff, C., Houben, G.-J.: Building a microblog corpus for search result diversification. In: Banchs, R.E., Silvestri, F., Liu, T.-Y., Zhang, M., Gao, S., Lang, J. (eds.) AIRS 2013. LNCS, vol. 8281, pp. 251–262. Springer, Heidelberg (2013)
23. Teevan, J., Ramage, D., Morris, M.R.: #twittersearch: a comparison of microblog search and web search. In: Proc. of WSDM 2011, pp. 35–44 (2011)
24. Vieira, M.R., Razente, H.L., Barioni, M.C.N., Hadjieleftheriou, M., Srivastava, D., C. T. Jr., Tsotras, V.J.: On query result diversification. In: Proc. of ICDE 2011, pp. 1163–1174 (2011)
25. Vosecky, J., Leung, K.W.-T., Ng, W.: Collaborative personalized twitter search with topic-language models. In: Proc. of SIGIR 2014, pp. 53–62 (2014)
26. Wang, J., Zhu, J.: Portfolio theory of information retrieval. In: Proc. of SIGIR 2009, pp. 115–122 (2009)
27. Zhai, C., Cohen, W.W., Lafferty, J.D.: Beyond independent relevance: methods and evaluation metrics for subtopic retrieval. In: Proc. of SIGIR 2003, pp. 10–17 (2003)
28. Zhai, C., Lafferty, J.D.: A risk minimization framework for information retrieval. Inf. Process. Manage. 42(1), 31–55 (2006)
29. Zhang, X., He, B., Luo, T., Li, B.: Query-biased learning to rank for real-time twitter search. In: Proc. of CIKM 2012, pp. 1915–1919 (2012)
30. Zuccon, G., Azzopardi, L., Zhang, D., Wang, J.: Top-k retrieval using facility location analysis. In: Baeza-Yates, R., de Vries, A.P., Zaragoza, H., Cambazoglu, B.B., Murdock, V., Lempel, R., Silvestri, F. (eds.) ECIR 2012. LNCS, vol. 7224, pp. 305–316. Springer, Heidelberg (2012)

WikipEvent: Leveraging Wikipedia Edit History for Event Detection

Tuan Tran, Andrea Ceroni, Mihai Georgescu, Kaweh Djafari Naini,
and Marco Fisichella

L3S Research Center, Appelstr. 9a, Hannover, Germany
{ttran,ceroni,georgescu,naini,fisichella}@L3S.de

Abstract. Much of existing work in information extraction assumes the static nature of relationships in fixed knowledge bases. However, in collaborative environments such as Wikipedia, information and structures are highly dynamic over time. In this work, we introduce a new method to extract complex event structures from Wikipedia. We propose a new model to represent events by engaging multiple entities, generalizable to an arbitrary language. The evolution of an event is captured effectively based on analyzing the user edits history in Wikipedia. Our work provides a foundation for a novel class of evolution-aware entity-based enrichment algorithms, and considerably increases the quality of entity accessibility and temporal retrieval for Wikipedia. We formalize this problem and introduce an efficient end-to-end platform as a solution. We conduct comprehensive experiments on a real dataset of 1.8 *million* Wikipedia articles to show the effectiveness of our proposed solution. Our results demonstrate that we are able to achieve a precision of 70% when evaluated using manually annotated data. Finally, we make a comparative analysis of our work with the well established Current Event Portal of Wikipedia and find that our system *WikipEvent* using *Co-References* method can be used in a complementary way to deliver new and more information about events.

Keywords: Event Detection, Temporal Retrieval, Wikipedia, Clustering.

1 Introduction

Wikipedia is one of the largest online encyclopedias available in multiple languages. The enormous volume and the fairly reliable quality of information makes Wikipedia a popular source in several research topics. Research utilizing Wikipedia has attracted a large spectrum of interest over the past decade, including knowledge discovery and management, natural language processing, social behaviour study, information retrieval, etc. Much of existing work considers Wikipedia as a static collection, i.e. information once stored is stable or rarely changed over time. However, in practice, Wikipedia grows very rapidly (from 17 millions articles in 2011 to 30 millions articles in 2013 [1]), with new

---

[1] http://en.wikipedia.org/wiki/Wikipedia

B. Benatallah et al. (Eds.): WISE 2014, Part II, LNCS 8787, pp. 90–108, 2014.

articles published and edited everyday by a large community of active contributors worldwide. This calls for an effective way to analyze and extract information from Wikipedia, with the awareness of temporal dynamics.

In this work, we address the problem of extracting from Wikipedia complex event structures, consisting of a set of entities that are connected at a given time period. We exploit the edit history in Wikipedia, which covers a full evolution of articles' content over a long time period. Our method is agnostic to any language constraints, i.e. it can be applied to an arbitrary language, and it does not depend on the number of entities to be known a priori. In principle, our method can detect events from simple schema (such as the release of *new movies*) to complex ones (such as a *revolution*). Our method works naturally with the dynamics of information in Wikipedia; thus, it is able to detect several events pertaining to a number of articles, as the articles' contents change over time. In contrast to previous work in detecting events from Wikipedia [12], our model requires no training data, nor does it need the prior information about the entities establishing an event.

Detecting such dynamic relationships and associated events poses multiple interesting technical challenges. First, these relationships do not conform to any pre-existing schema and therefore can not be discovered by leveraging language patterns as in previous works on static relationship extraction. Second, the underlying events often have a flexible timeline that is difficult to know a priori, e.g., one event can last for a short time (e.g. over a day or week), while others could last over several weeks or months. Third, the entities display a great deal of flexibility in their participation in the underlying events, mainly reflected in the number of participants (some events can involve two entities while others are among several entities [8]). Fourth, as a real-life event happens, the Web community mobilizes itself to report that. Some information generated in a particular time period will no longer be available in a future version of the articles of the entities involved in the event. Thus, it is important to provide users the possibility to access historical information, giving a comprehensive evolution-aware entity-based view.

In this work, we make the following contributions:

1. Presentation of a general model which is agnostic to linguistic constraints, thus it can be applied to Social Media in different languages.
2. Establishment of new methodology for detecting events based on explicit relationships identification.
3. Introduction of the temporal aspect as a fundamental dimension to enrich content with semantic information via historical user edits.

The rest of this paper is organized as follows. We discuss related work in Section 2. In Section 3, we introduce and formalize the problem of dynamic relationships and event discovery, and present a pipeline framework for our approach. Sections 4 and 5 describe details of our approach. In Section 6, we evaluate our approach and demonstrate its effectiveness. Additionally, in this Section we conduct an extensive comparative analysis of our work with the well established

Current Event Portal of Wikipedia [2] by analyzing the events described manually by Wikipedia users versus the events detected by our model. Finally, we provide our conclusions in Section 7.

2   Related Work

2.1   Temporal Information Retrieval and Event Detection

In the area of temporal information retrieval, previous works (e.g. [2,9,24]) leverage time dimension in different ranking models to improve the retrieval effectiveness for temporal queries, often with temporal information preprocessing. In our work, we circumvent the need of indexing or extracting temporal information from Web Archives by using Wikipedia. The link structures in Wikipedia has been shown to be a good indicator for the historical influence of people [23]. Here we propose to exploit the time dimension in articles' revision to identify historical events.

The problem of identifying events has been examined using web articles as part of a broader initiative named topic detection and tracking [1]. A rich body of work has been devoted to identifying events as a landscape view of a web collection, which answers questions such as "What Happened?", "What is New?". Related to our work in this area are two approaches, article-based [1,11] and entity-based [14]. In the article-based approach, events are detected by clustering articles based on semantics and timestamps. In the entity-based approach, temporal distributions of entities in articles are used to model events. In our work, we identify events by dynamic connections among entities that are coupled at a given time period. We formalize this problem and introduce an end-to-end pipeline as a solution.

The approach of using dynamic connections to model events has been recently proposed in [8]. The authors exploited proprietary query logs to measure the temporal interest towards an entity, using word matching as major techniques. The connections between two entities are cast into whether their query histories exhibit peaks at the same time, arguing that this is the proxy of the concurrent interests to both entities. The empirical experiments, while showing relative improvement in event discovery, were obviously affected by the performance of the word matching techniques, as well as the burst correlation algorithms. In our work, we provide a systematic framework to investigate the different types of dynamic connections of entities in many directions. In addition, the choice of Wikipedia eliminates the effects of ambiguity onto event detection performance. We also utilize high-quality ontologies such as YAGO2 [16] to support different entity classes and event domains.

2.2   Mining from Wikipedia

There has been a large amount of research done on Wikipedia. A survey [19] categorizes and presents the different areas of research to which Wikipedia is relevant.

---

[2] http://en.wikipedia.org/wiki/Portal:Current_events

Related to our work is a rich body of work on measuring semantic relatedness of words and entities, from making use of expert-curated taxonomy such as Wordnet [4] exploiting the structure of Wikipedia to compute at larger scale [13]. Wikipedia has been also used as a rich source to measure the semantic relatedness between entities (such as people, songs, organizations, etc.), based on the inter-linking structures of Wikipedia articles [22,20], or on the phrasal overlaps extracted from Web articles [15]. Our work distinguishes itself from existing work in the sense that it emphasizes the temporal dynamics of entities on Wikipedia, i.e. two entities with low correlation can be very relevant to one another within a particular time span, driven by an underlying event (for example, *Barack Obama* and *Iraq*). We adapt the existing measurements to our own metrics accordingly.

There are also recent attempts in extracting and summarizing historical events from Wikipedia articles' text [18]. Analyzing the trends in article view statistics, instead of article edits, the authors of [7] identify concepts with increased popularity for a given time period, and [21] proposes a system to visualize and explore the temporal correlations between different entities. A machine learning based framework for identifying and presenting event-related information from the Wikipedia edits is proposed [12], but only for individual entities. Other work proposed building a set of related entities to build the events of one entity as reference [25]. Contrast with previous work, we exploit the increased Wikipedia editing activity in the proximity of news events, and we use the edit history to identify events and entities showing similar behaviour that might be affected by the same event, in order to extract relationships. To the best of our knowledge, we are the first to propose detecting complex events of multiple entities from Wikipedia edit history in an unsupervised fashion.

3   Approach

3.1   Overview

In this work, we aim to detect events from Wikipedia users' edit history. Existing works model events by actions, for instance, through an RDF triple of subject-predicate-object [18]. In our work, we model an event indirectly through its participating entities: one event consists of a set of entities that are connected at a given time period. For example, the event "83rd Academy Best Actor Awards" on January 25, 2011, can be described by its nominees and winners "Colin Firth", "Jeff Bridges", "James Franco", etc. This way of representing events has a benefit of being agnostic to linguistic constraints of a certain language. On the other hand, it is crucial to define the notion of entity relationships to govern an event. Such relationships must capture well the temporal dynamics of entities in Wikipedia, where information are constantly added or updated over time.

Entity Relationships. We adopt two strategies to identify entity relationships. The *Explicit Relationship Identification* uses links between Wikipedia articles to establish the relationship between their corresponding entities. The intuition behind is that each link newly added or updated in each article revision indicates explicitly a tie between the source and destination entities. For example, during the Egypt Revolution 2011, the Wikipedia article "Hosni_Mubarak" admits many revisions published. In many of them, the link to the article "Tahrir_Square" is added or refined several times. This reveals a strong relationship between the two corresponding entities with respect to the revolution. We detail methods of this strategy in Sections 4.1 and 5.1. In the *Implicit Relationships Identification* we adapt the approach presented in [8] to our domain, in order to define the entity relationships through burst patterns. This is also in line with existing work, which suggests that Wikipedia article view or article edit statistics follow bursty patterns, with spikes driven by real-world events of the entities [7,12]. However, to avoid the coincidence of two independent entities which burst around the same time period by chance, we further impose that the entities must share sufficient textual or structure similarities during time period of study. Besides the pointwise mutual information (PMI), which was used in [8], we propose other classes of similarity measures to estimate the confidence of each implicit relationship. This is discussed in more detail in Section 4.2.

Event Detection. Having defined entity relationships, we detect events by building groups of highly related entities, each representing an event. We cast this problem to the connected components extraction from the graph, with the nodes corresponding to Wikipedia entities and edges corresponding to entity relationships. One subtle problem in identifying such components is that the graph is highly dynamic, i.e. edges change as entity relationships evolve over time, new relationships can be established in a given time and dissipate later, when the tie between two entities gets weaker within its respective revisions. For instance, two entities "Barack Obama" and "Mitt Romney" are highly related during the US presidential election 2012, but they rarely correlate long before and after the event. In this work, we propose an adaptive algorithm that handles this temporal dimension, which will be detailed in Section 5.

3.2   Problem Formalization

Data Model. In our model, time is represented as a sequence of discrete points, each corresponding to a day and indexed by $i = 1, 2, 3, \ldots$ Let E denote the entity collection derived from Wikipedia, where each entity e is associated to a Wikipedia article, and let τ be the set of time points that we consider. At a given time point i, e is represented as a textual document $d_e^{(i)}$, which is the revision of the article at a latest timestamp t that was before i. Given such assumptions, we further define the *edit* of e at the time point i as $m_e^{(i)} := d_e^{(i)} - d_e^{(i-1)}$ and the *edit volume* $v_e^{(i)}$ as the number of revisions between two time points $i - 1$ and i.

Dynamic Relationships. A dynamic relationship is a tuple $r := (e_1, e_2, i)$, where $e_1, e_2 \in E$ are the entities for which r holds, and $i \in \tau$ is the time when r is valid. Dynamic relationships can be of two types: *explicit* and *implicit*. They are identified according to different strategies (Section 4).

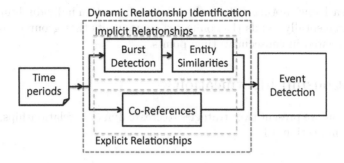

Fig. 1. Architecture for identifying events and relationships between entities

Events. We define an *event* v as a tuple $v := (E_v, \tau_v)$, where $E_v \subset E$ is the *representative entity set*, i.e. entities that participated to v and contributed to its interpretation, and $\tau_v := \{i : i \in \tau, i_{start} \leq i \leq i_{end}\}$ is the *time period* when v occurred.

Problem Statement. Given an entity set E, a time window $W \subseteq \tau$, detect all events $v = (E_v, \tau_v)$ such that $E_v \subset E$ and $\tau_v \subset \tau$.

3.3 Workflow

The detailed workflow of our system *WikipEvent* is described in Figure 1. In short, our system consists of two steps: *Dynamic Relationship Identification* and *Event Detection*. Givena time period as input, our system detects a set of events in which the entities were involved, together with the relationships between them. Such relationships, together with the specified time period, will enable users to fully interpret the detected events (e.g., causes and effects)[3]. The first phase computes the dynamic relationships between entities, using one of the two strategies discussed in Section 4.1 (*Explicit Relationships Identification*) and in Section 4.2 (*Implicit Relationships Identification*). The *Explicit Relationships Identification* strategy uses explicit links in the articles to establish the relationship between two entities. Its intuition is that each link added or modified in one revision encodes a binding between the source and destination articles, hence the entities. The adapted and extended *Implicit Relationships Identification* strategy is based on two steps. First, we use *Burst Detection* to detect salient bursts of activity in the edit history as described in Section 4.2. The output is a set of pairs of entities that have bursts during the same time point. We employ a variety of methods to measure their similarity in the *Entity Similarity* step, described in Section 4.2. The pairs of entities are aggregated using the previously computed similarities to build the co-burst graph for each individual time point.

The second phase, *Event Detection*, generates events described by representative entities and time intervals of involvement. It first builds a sequence

---

[3] The semantic extraction between entities is not the focus of this work.

of graphs, each one capturing the entity relationships at an individual time point. It then incrementally builds the connected components that group entities that are highly related in consecutive time points.

4 Relationship Identification

In this section, we present two strategies to create dynamic relationships, already introduced in Section 3.1.

4.1 Explicit Relationships Identification

In our established strategy, hereafter called *Co-References*, we define entity relationships as follows. For each entity e and an edit $m_e^{(i)}$ at time i, if $e_2 \in m_e^{(i)}$ then it implies that there exists a link to the Wikipedia article of the entity e_2 in the content of $m_e^{(i)}$. A relationship between e_1 and e_2 is established if we have links in both directions (from e_1's edit to e_2 and vice versa).

Definition 1. *Given two entities e_1, e_2 and a time interval $I = [i, i + \delta]$, an explicit dynamic relationship between e_1 and e_2 at time point i is a tuple $r_{exp} \in E \times E \times \tau$ such that $r_{exp}(e_1, e_2, i)$ iff $\exists j, k \in [i, i + \delta], e_1 \in m_{e_2}^{(j)}$ and $e_2 \in m_{e_1}^{(k)}$.*

Intuitively, an explicit dynamic relationship captures the mutual references between two entity edits that are made at close time points. The parameter δ accounts for the possible time delay when adding links between two entities. As an example, while the entities "Cairo" and "Hosni Mubarak" are explicitly related during the Egypt revolution in 2011, the two mutual references can be added at different (close) time points; for instance the link from the article of "Hosni Mubarak" to "Cairo" can be added first, and the inverse link can be added one day after.

4.2 Implicit Relationships Identification

We adapted to the Wikipedia domain the strategy proposed in [8]; in this adapted approach, we identify a relationship between two entities when their edit histories exhibit bursts in the same or overlapped time intervals (co-burst). A burst of an entity is the time interval in which the edit volume of the entity is significantly higher than the preceding and following volumes. More formally, given an entity e and a time window W, we construct a time series $v_e := [v_e^{(i_0)}, \ldots, v_e^{(i_f)}]$ containing the edit volume of entity e at every time point $i \in \tau$. A burst b_e is a sequence of time points $b_e := [i, i+1, \ldots, i+k]$ for which the edit volumes of e are *significantly* higher than the edit volumes observed in neighbouring time points: $v_e^{(i-1)} \ll v_e^{(j)}$ and $v_e^{(j)} \gg v_e^{(i+k+1)} \forall j = i, i+1, \ldots, i+k$. We say two entities e_1 and e_2 *co-burst* at time i iff $i \in b_{e_1}$ and $i \in b_{e_2}$.

A *co-burst* merely identifies entities that admit high volume of edits at the same time. In practice, two entities can have bursts in the same time interval, even if they are of little relevance. To remedy this, we further assume that the edits of two entities must share sufficient resemblance, which we assess through 4 different similarities. Let us denote the similarity of two entities e_1 and e_2 at time i as $S_{method}(e_1, e_2, i)$, then we have the following similarities:

1. *Textual*: it measures how close two entities are in a given time by comparing the content of their corresponding edits. We construct the bags of words $\mathbf{bw}_{e_1}^{(i)}$ and $\mathbf{bw}_{e_2}^{(i)}$, and use Jaccard index to measure the similarity: $S_s(e_1, e_2, i) := J(\mathbf{bw}_{e_1}^{(i)}, \mathbf{bw}_{e_2}^{(i)})$.

2. *Entity*: This is similar to textual similarity, but with the *bag of entities* $\mathbf{be}_e^{(i)}$ (entities that are linked from the edit): $S_e(e_1, e_2, i) := J(\mathbf{be}_{e_1}^{(i)}, \mathbf{be}_{e_2}^{(i)})$.

3. *Ancestor*: it measures how close two entities are in terms of their semantic types. For each entity, we use an ontological knowledge base where the entity is registered, and extract all its ancestors (entities that are connected to through a *subsumption* relation). Given $\mathbf{be}_e^{(i)}$, a *bag of ancestors* $\mathbf{ba}_e^{(i)}$ is filled with the ancestors of every entity in $\mathbf{be}_e^{(i)}$. We then measure the similarity $S_a(e_1, e_2, i)$ by Jaccard index accordingly.

4. *PMI*: This measures how likely two entities co-occur in the edits in all other entities. Given $i \in \tau$ and $e_1, e_2 \in E$, we construct the graph involving all entities linking to e_1 and e_2 from all edits at i. Let $IN(e)^{(i)}$ denote the number of incoming links for e in this graph, we estimate the probability of generating e by $p(e) = \frac{IN(e)^{(i)}}{N^{(i)}}$, with $N^{(i)}$ being the total number of incoming links in the graph at time i. We then computed the link similarity as $S_{PMI}(e_1, e_2, i) := \log \frac{p(e_1, e_2)}{p(e_1)p(e_2)}$.

5 Event Detection

Having defined the dynamic entity relationships, we now aim to detect events by identifying their representative entity sets and the corresponding time period. The event detection is done via an incremental approach as follows. At each individual time point $i \in \tau$, we consolidate different relationships into one unified group, called *temporal graph*. The temporal graph can be thought of as one event's snapshot at the specified time point (Section 5.2). To accomodate the development of events from time points to time points, as well as to factor the event time period, we compare entity clusters of two adjacent temporal graphs, and incrementally merge two clusters if a certain criterion is met. The resulting merged set of entities represents the event that evolves across several continuous days, from temporal graph to temporal graph. Our event detection is detailed below. We first start with the formal definition of the temporal graph and entity clustering.

5.1 Temporal Graph and Entity Clustering

Definition 2. *A temporal graph $G(i)$ at time $i \in \tau$ is an undirected graph (E, P), where E is an entity set and $P = \{(e_1, e_2) | r(e_1, e_2, j)\}$ is the set of edges defined by dynamic relationships at a time point $j \in \mathbf{I} = [i, i + \delta]$.*

In the above definition, the value δ reflects the lag of edit activities between different Wikipedia articles in response to one real-world event. Note that depending on the type of the dynamic relationships, we have two different types of *explicit* and *implicit* temporal graph respectively.

Explicit Temporal Graph Clustering. In an explicit temporal graph, an edge is defined by the relationship $r_{exp}(e_1, e_2, j)$ and it reflects the mutual linking structure of two Wikipedia entities within interval I. From the temporal graph, we identify the set of maximum cliques C to form clusters of entities that are mutually co-mentioned from i to $i+\delta$. Each maximum clique $c \in C$ represents an event that occurs at i. The choice of cliques in favor of connected components in this case ensures the high coherence of the underlying events encoded in the group of entities. For example, considering three entities "Anne Hathaway", "James Franco" and and "Minute To Win It" during January 2011. The first two entities are connected by the fact that the two actors co-hosted the ceremony of the 83rd Academy Awards, while the second and third entities are connected because James Franco was at that time a co-performer in the show. This forms a connected component, but putting the three entities together reveals no obvious event.

Implicit Temporal Graph Clustering. In an implicit temporal graph, a candidate edge will be established from two entities which co-burst at a time point $j \in I = [i, i + \delta]$. To mitigate the "co-burst by chance" (Section 4.2), we define a *maximum* similarity function:

$$S_{max}(e_1, e_2, I) = \max_{j \in [i, i+\delta]} \{S(e_1, e_2, j)\}$$

and create an edge (e_1, e_2) iff $S_{max}(e_1, e_2, I) \geq \theta$. Intuitively θ is the threshold value used to perform a selective pruning preserving only entity pairs with maximum similarities exceeding it. Unlike in an explicit temporal graph, here we relax the entity clustering requirements by representing the events occuring at i as the connected components. This is due to the nature of the implicit dynamic relationships, where two entities e_1 and e_2 that are not directly connected can still co-burst, through an intermediate entity e' during the interval I, by following one path in the graph.

5.2   Event Identification

To identify an event, we need to form a representative entity set from a number of temporal graphs, as well as to specify the time interval in which the entity set lies in. This entails aggregating entity clusters of temporal graphs at consecutive time points. The algorithm named *Local Temporal Constraint* (LTC), proposed in [8], detects events in the dynamic programming fashion. At each time point i, LTC maintains a set of merged clusters from the beginning until i, and merges these clusters with those in the temporal graph at time $i + 1$. Two clusters are merged if they share at least one edge; i.e. one dynamic relationship. However, we observe that this can end up merging a lot of clusters where entities across clusters are very loosely related. We adapt the algorithm as follows. For the

---

Algorithm 1. Entity Cluster Aggregation Algorithm

Input : E, W, γ -cluster merging threshold, *strategy*
Output: C_{ret} as entity sets representing events
Set $C_{ret} = \emptyset$
for each $i \in |W|$ **do**
 Construct temporal graph $G(i)$
 C_{I_i} = clusters in $G(i)$
 for each $c_k \in C_{I_i}$ **do**
  merged = False
  for each $c_j \in C_{I_{i-1}}$ **do**
   if $(strategy = explicit$ & $\frac{|c_k \cap c_j|}{|c_k \cup c_j|} \geq \gamma)$ **or**
   $(strategy = implicit$ & $|c_k \cap c_j| \geq 1)$ **then**
    $c_k = c_k \cup c_j$ (merge the two events)
    merged = True
   end
  end
  if not merged **then**
   $C_{ret} = C_{ret} \cup c_k$
  end
 end
end
return C_{ret}

---

implicit temporal graph, we keep merging two connected components if they share one edge (i.e. original LTC algorithm), while for explicit temporal graph, we only merge two clusters c_1 and c_2 if their Jaccard similarity is greater than a threshold γ. The pseudocode of our algorithm is detailed in Algorithm 1.

Complexity. Let M be the maximum number of events within each event set C_I. Let n be the maximum number of relationships which can be found inspecting each event within each C_I. Let $\mathcal{T}$ be the number of intervals, then the computational cost of LTC is $O((\mathcal{T} - 1)nM^2)$.

6 Experiments and Evaluations

In order to analyze the performances of the proposed methods, we ran experiments on the quantitative (Section 6.3) and qualitative (Section 6.4) characteristics of our extracted events. For the specific task of detecting event structures in Wikipedia, to the best of our knowledge, a comprehensive list of real-world events to be used as universal ground truth does not exist: existing resources (e.g. Wikipedia Current Event Portal, YAGO2 [16]) are limited in terms of number, complexity, and granularity of events. Moreover, since a comprehensive event repository does not exist, fairly computing recall for event detection methods is infeasible. Thus, we performed a manual evaluation as follows. Detected events were manually assessed by five evaluators who had to decide if they were corresponding to real events. In detail, for each detected event, the annotators were asked to check all involved entities and identify a real-world event by examining web-based sources (Wikipedia, official home pages, search engines, etc.), that best explained the co-occurrence of these entities in the event during the specified time period. For each set of entities, a label was assigned in order to represent a *true* or *false* event. These assessments contributed to measure the performances of our methods.

Finally, we conducted an extensive comparative analysis of our work with the well established WikiPortal by analyzing the events described manually by Wikipedia users versus the events detected by our best performing method (Section 6.5).

6.1   Dataset

To build our dataset, we used the English Wikipedia. Since Wikipedia also contains articles that do not describe entities (e.g., "List of mathematicians"), we selected Wikipedia articles corresponding to entities registered in YAGO2 and belonging to one of the following classes: person, location, artifact, or group. In total we got $1,843,665$ articles, each corresponding to one entity. We chose a time period ranging from the 18th January 2011 until the 9th February 2011, because it covers important real-world events such as the Egypt Revolution, the Academy Awards, the Australian Open, etc. The choice of a relatively short time span simplifies the manual evaluation of the detected events. Since using days as time units has been shown to effectively capture the news-related events in both social media and newswire platforms [3], we used the day granularity when sampling time. We name the whole dataset, containing all the articles, as *Dataset A*. Furthermore, we created a sample set, called *Dataset B*, by selecting entities that were actively edited (more than 50 times) in our time period. The intuition behind this selection is that a large number of edits is more likely to be caused by an event. Consequently, this sample contains just $3,837$ Wikipedia articles.

6.2   Implementation Details

Entity Edits Indexing. To store the whole Wikipedia edit history dump and to identify the edits, we made use of the JWPL Wikipedia Revision Toolkit [10]. JWPL solves the problem of storing the entire edit history of Wikipedia by computing and storing differences between two revisions.

Similarity. To resolve the ancestors of a given entity, we employed the YAGO2 knowledge base [16], an ontology that was built from Wikipedia infoboxes and combined with Wordnet and GeoNames to obtain 10 million entities and 120 million facts. We followed facts with *subClassOf* and *typeOf* predicates to extract ancestors of entities. We limited the extraction to three levels, since we observed that going to a higher level included several extremely abstract classes (such as "Living people"). This lowered the discriminating performance of the similarity measurement.

Burst Detection and Event Detection. We implemented Kleinberg's algorithm using the modified version of CShell toolkit [4]. We set the density scaling to 1.5, the transition cost to 1.0, and the default number of burst states to 3 (for more details, refer to [17]). We observed that changing parameters of the burst detection did not affect the order of performance between different event detection methods. For the dynamic relationships, we set the time lag parameter δ to 7 days and γ to 0.8, as these values yielded the most intuitive results in our experiments.

6.3 Quantitative Analysis

The goal of this section is to numerically evaluate our approach under different metrics: (i) total number of detected events, and (ii) the precision, i.e. the percentage of *true* events. For the parameter selection, note that the graph created based on the explicit strategy does not have any weights on its edges. On the contrary, the implicit strategy creates a weighted graph based on the similarities, and the temporal graph clustering depends on the threshold θ to filter out entity pairs of low maximum similarity. We varied the value of θ and noticed that lowering it resulted in a larger number of entity pairs that coalesced into a low number of large events. These events containing a large number of various entities could not have been identified as real events. Therefore, for the following experiments we used $\theta = 1$.

We evaluate approaches for the implicit relationship identification strategies as defined in Section 4.2, referred to as the following *methods*: *Textual*, *Entities*, *Ancestors*, and *PMI*, as well as for the explicit strategy as defined in Section 4.1, referred to as the *Co-References* method. The results are presented in Table 1 and Table 2. The number of events detected for the different similarity setups is presented in the third column of the tables. As expected, we detect more events in *Dataset A* as in *Dataset B*, due to the higher number of entities taken into consideration. The biggest number of detected events is provided by *Co-references* in both datasets. This is attributed to the parameter-free nature of the explicit strategy, while for the implicit strategy, a portion of events are removed by a threshold. Comparing the methods used by the implicit strategy, *PMI* detects more events than any other method. This is caused by the difference in computing the entity similarity $S(e_1, e_2, t)$. *PMI* considers the sets of incoming links, that account for relevant feedback to our e_1 and e_2 from all the other entities in Wikipedia. This results in more entity pairs, and more clearly defined and coherent events, while the other implicit strategy methods tend to conglomerate most of the entities in larger but fewer events. *Textual*, *Entities*, and *Ancestors* compute $S(e_1, e_2, t)$ starting from the edited contents of two entities at a given time. A large amount of content concerning entities that are not explicitly referring to e_1 and e_2 will be taken in consideration as well, making the value of $S(e_1, e_2, t)$ lower. Therefore, using the same value for θ as for the *PMI*, produces a lower number of entity pairs, and consequently of detected events.

---

[4] http://wiki.cns.iu.edu/display/CISHELL/Burst+Detection

Table 1. Performance on Dataset A

Strategy	Method	Events	Precision
Explicit	Co-References	186	70%
Implicit	PMI	124	39%
	Ancestors	33	51%
	Entities	21	62%
	Textual	78	1%

Table 2. Performance on Dataset B

Strategy	Method	Events	Precision
Explicit	Co-References	120	80%
Implicit	PMI	80	69%
	Ancestors	18	50%
	Entities	12	60%
	Textual	15	7%

The precision of every setup, i.e. the percentage of true detected events, is summarized in the fourth column of Table 1 and Table 2. Among the implicit strategy methods, we notice a clear benefit of using similarities that take semantics into account (*Entities, Ancestors,* and *PMI*) over the string similarity (*Textual*). *Ancestors* performs worse than *Entities* in both datasets, showing that the addition of the ancestor entities introduces more noise instead of clarifying the relationships between the edited entities. *Entities* achieves similar performances on both datasets. *PMI* achieves better performance in *Dataset B* than the other implicit similarities since it is leveraging the structure of incoming-outgoing links between Wikipedia articles. However, *PMI* performs worse on *Dataset A* due to the higher number of inactive entities considered, introducing noisy links.

Finally, *Co-References* outperforms all the implicit strategy methods on both datasets, showing that a clear and direct reciprocal mention is stronger than similarities inferred from the text of the edit. Generally, all methods performed better or comparable on *Dataset B* in comparison to *Dataset A*. This shows that selecting only the entities that are edited more often improves the quality of our methods. Although less events are detected in *Dataset B*, more of them correspond to real life events.

6.4   Qualitative Analysis

In this section, we do a qualitative evaluation of the events identified in *Dataset A*. First, we focus on and describe some of the events detected by our best method *Co-References* (highlighted in Section 6.3). Second, we analyze some cases where our methods failed, proposing the causes.

In Table 4 we present and discuss some events identified by our best method *Co-References* matching real-world events. For each detected event, we report the entities involved, the time when the event occurred, and a human description of the event extracted from web-based sources.

Moreover, we show the graph structure of two good examples of identified events:

Example 1. We depict the relationships associated to the real-world event known as the Friday of Anger, in the context of the Egyptian Revolution in Figure 2a. On January 28, 2011, tens of thousands filled the streets across Egypt, to protest against the government. One of the major demonstrations took place in *Cairo*. The organization of the protests was done with the help of internet

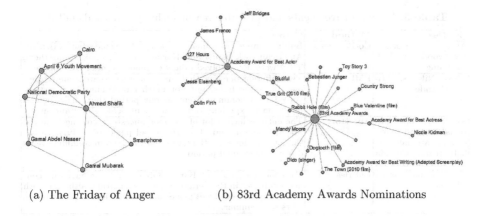

(a) The Friday of Anger   (b) 83rd Academy Awards Nominations

Fig. 2. Example detected events

and *smartphones*, and some of the organizers were the *April 6 Youth Movement* and the *National Democratic Party*. Protesters held portraits of former president *Gamal Abdel Nasser*. The aviation minister and former chief of Air Staff *Ahmed Shafiq*, as well as *Gamal Mubarak*, were seen by the government as probable successors of Hosni Mubarak.

Example 2. The graph structure of another outstanding detected real world: the announcement of the nominees for the 83rd Academy Awards, on January 25, 2011, is shown in Figure 2b. We can see as a the biggest central node the *83rd Academy Awards*, and as a secondary node the *Academy Award for Best Actor*, having as connecting nodes *True Grit* and *Biutiful*, which were nominated for more categories.

In Table 3, we report some failures of our methods to identify real-world events, together with the causes that lead to such erroneous output. Depending on the method, we can notice different patterns that cause false positives. The *entity-based similarity* usually fails because of updates containing a large number of common entities that are not involved in any common event. Using the *ancestor-based similarity* can provide false events because some entities that are very similar, and share a large number of ancestors, have coincidentally concurrent edit peaks in the same period. The *PMI* fails because of similar causes: entities that share a lot of common incoming links to the entities contained in the edits done on the same day. Finally, the *Co-References* method seems to fail when the reciprocal mentions originate in relationships that are independent of any event.

6.5   Comparative Analysis

We compare events detected by our approach with events present in WikiPortal in the same time period. Users in WikiPortal publish a short description of an event in response to the occurrence of a real-world happening and annotate the entity mentions with the corresponding Wikipedia articles. The event

Table 3. False positive events and probable reasons why our methods failed

Entities	Method	Cause
Alexis Korner, Fleetwood Mac, Bob Brunning	Co-References	Bob Brunning was a member of the Fleetwood Mac (a British-American rock band) and Korner wrote a book about them. They were not involved in any common events in our period.
Alexa Nikolas, Ariana Grande	PMI	Two different persons, that just look alike, share some of their own incoming links, but not much more. They were not involved in a common event in our period, but their articles might have experienced unrelated edit peaks simultaneously.
Saudi Arabia national football team, Ghana national football team, Canada national soccer team	Ancestors	The entities have a lot of common ancestors, coming from the Sports domain, and all of them had peaks of activity in the same time. However, they were not involved in common events during the studied period.
Tura Satana, Barack Obama	Entities	Tura Satana died, but Barack Obama did not have any connection to her in the period under investigation, although both entities experienced edit peaks and the entities contained in the edits were similar.

Table 4. Co-References based extracted events and the matching real-life events, along with their date and a human description

Entities	Date(2011)	Human Description	Category Explanation
Category: Green			
Brian Cowen, Michel Martin, Mary Hanafin, Mary Coughlan(politician), Fianna Fáil	From January 18 to January 26	The Irish PM Brian Cowen announced his stepping down as leader of the ruling Fianna Fail party, and different candidacies for the leadership follow his decision.	The event is globally reported in WikiPortal through different daily events.
Saad Hariri, Najib Mikati	January 25	Supporters of Lebanese caretaker Saad Hariri call for a day of protests following Hezbollah's support for Najib Mikati as Prime Minister	The entities participating to the event are all mentioned in an event within WikiPortal.
Category: Yellow			
Gamal Abdel Nasser, Ahmed Shafik, Smartphone, Cairo, April 6 Youth Movement, Gamal Mubarak, National Democratic Party	January 28	In the context of the Egyptian Revolution, the Friday of Anger takes place: tens of thousands filled the streets across Egypt, to protest against the government.	Most of the entities appear in WikiPortal within different events. The entities Gamal Mubarak and Gamal Abdel Nasser do not appear.
Li Na, Kim Clijsters	January 19	Australian Open 2011 women's final: Li Na vs Kim Clijsters.	The Australian Open is mentioned two times, but always focusing on men's matches. The women's final is not reported.
James Franco, Colin Firth, Biutiful, True Grit, 83rd Academy Awards, ...(Figure 2b)	January 25	Announcement of the nominees for the 83rd Academy Awards	The event is reported in WikiPortal, but few participating entities are mentioned.
Category: Red			
Vickie Guerrero, Hornswoggle, Layla El, Dolph Ziggler, Booker T, Professional wrestling, Kane, Santino Marella	January 30	The 2011 Royal Rumble organized by WWE takes place, involving a lot of wrestlers.	The event and no one of its entities are reported in WikiPortal.
Silent Witness, Bruce Forsyth, Loose Women	January 26	In the context of the 16th National Television Awards, presented by Bruce Forsyth, Loose Women and Silent witness are nominated.	The event and no one of its entities are reported in WikiPortal.
Catwoman, The Dark Knight, Bane	January 19	Warner Bros. Pictures announced that Anne Hathaway has been cast as Catwoman and Tom Hardy as Bane in "The Dark Knight Rises".	The event and no one of its entities are reported in WikiPortal.

descriptions can also be grouped into bigger "stories" (such as Egypt revolution), or can be organized in different categories such as sports, disasters, etc.

We conducted the comparison by considering the 130 (70% of 186) true events detected with the *Co-References* method on *Dataset A*, since this is the setup that gave the largest set of true events.

Then, we collected 561 events from WikiPortal within the period of interest by considering all event descriptions inside the same story as representing a single event. We further considered only those event descriptions annotated with at least an entity contained within *Dataset A*, getting in total 505 events. In principle, these events can be detected by our method.

In order to assess the overlap between the two event sets, we classify events according to the following categories:

1. **Green:** The event in one set, with all its participating entities, is present in the other event set either as a single event or as multiple events.
2. **Yellow:** The event is partially present in the other event set, i.e. only a subset of its participating entities appears in one or more events in the other set.
3. **Red:** The event is not reported in the other event set.

We provide explanatory examples for each category in Table 4, along with explanations of each classification choice. We can observe 2 patterns for the events in our set belonging to the green category: (i) one event in *Co-References* is spread over different events in WikiPortal; for instance, the event regarding the candidacies for the Fianna Fail party is reported in WikiPortal through different events, each one focusing on a single candidacy; (ii) one event in *Co-References* corresponds to one event in WikiPortal. The yellow category generally covers the case where an event represents a non mentioned aspect of an event in WikiPortal. For instance, for the Australian Open tennis tournament only the men's semi-final and final matches are reported in WikiPortal, without mentioning the other matches, which are reported in *Co-References*. Similarly, the Friday of Anger (in the context of the Egyptian revolution) and the Academy Awards nominations are present in WikiPortal, but our detected events are endowed with more entities that do not appear in the portal. Finally, the red category collects those events that are not reported in WikiPortal at all, like the Royal Rumble wrestling match.

We noticed that 60% of the events detected by *Co-References* are present fully or partially in the WikiPortal. For the sake of clarity, in Table 5 we present some of the events that are present in WikiPortal but our method was unable to detect, along with an explanation. The main patterns are: (i) the events involve just one entity; (ii) the events involve entities that are highly unlikely to reference each other because of their different roles in the common events.

In conclusion, WikiPortal and *Co-References* can be seen as complementary methods for event detection. While WikiPortal is user-contributed and requires human effort to curate events, our method is fully unsupervised, and can detect additional events without the human intervention.

Table 5. Examples of events from WikiPortal that were not detected by our method, Co-References

Event Description	Date(2011)	Explanation
Apple records record profits of $6 billion as consumers consumed more of its products than was thought (BBC)	January 18	The event involves just one entity
Chinese President Hu Jintao begins a four-day state visit to the United States.	January 18	It is highly unlikely that the prominent entities have mentioned each other
Exotic birds are found to have been driven into Britain's back gardens by the extreme cold, as more than half a million people participate in the largest wildlife survey in the world	January 29	It is highly unlikely that the event attracted the attention of the Wikipedia community
Researchers report that fishing rates in the Arctic are 75 times higher than those reported by the U.N., suggesting future increased exploitation is less possible than previously thought.	February 4	It is highly unlikely that the prominent entities have mentioned each other

7   Conclusions and Future Work

In this work, we propose incorporating temporal aspect with semantic information to capture dynamic event structures. Focusing on Wikipedia, we are able to find historical information, events. Because of the specific task we consider, no annotated collections were available, thus we manually assessed the performance of our methods using a data set of 1.8 million articles. Over an extensive set of experiments we established the effectiveness of out proposed approach and investigated different strategies and methods. We have shown that an explicit relationship identification strategy performs better than an implicit one, achieving a maximum precision of 70%. We observed a further increase to 80% in precision when using only actively edited articles. We further conducted a comparison between events detected by WikipEvent, using the *Co-References* approach, with events present in WikiPortal in the same time period, highlighting that they can be seen as complementary sources of events. Future work in this direction includes using Web as a complementary resource for validating news events (e.g. [5]).

For the future work, we are investigating how a current approach using cross-references between two entities can be combined with an observed low correlation in previous time window of the same pair to improve the quality of event detection. Another direction includes adding more semantics to the event detected via text analysis, or summarizing events on Wikipedia, taking into account the evolution of engaging entities over time [6].

Acknowledgements. This work was partially funded by the European Commission for the FP7 projects CUbRIK and ForgetIT (under grants No. 287704 and 600826 respectively), the ERC Advanced Grant ALEXANDRIA (under grant No. 339233), and the L3S-run project WikipEvent [5]. We thank the anonymous reviewers for constructive advices about our paper and suggestions for direction of future work.

---

[5] https://www.l3s.de/en/projects/iai/~/wikipevent

References

1. Allan, J., Papka, R., Lavrenko, V.: On-line new event detection and tracking. In: ACM SIGIR, pp. 37–45 (1998)
2. Alonso, O., Strötgen, J., Baeza-Yates, R., Gertz, M.: Temporal Information Retrieval: Challenges and Opportunities. In: WWW (2011)
3. Bandari, R., Asur, S., Huberman, B.A.: The pulse of news in social media: Forecasting popularity. In: ICWSM (2012)
4. Budanitsky, A., Hirst, G.: Evaluating wordnet-based measures of lexical semantic relatedness. Comput. Linguist. 32(1) (2006)
5. Ceroni, A., Fisichella, M.: Towards an entity–based automatic event validation. In: de Rijke, M., Kenter, T., de Vries, A.P., Zhai, C., de Jong, F., Radinsky, K., Hofmann, K. (eds.) ECIR 2014. LNCS, vol. 8416, pp. 605–611. Springer, Heidelberg (2014)
6. Ceroni, A., Georgescu, M., Gadiraju, U., Djafari Naini, K., Fisichella, M.: Information evolution in wikipedia. In: Proceedings of the 10th International Symposium on Open Collaboration, OpenSym 2014. ACM (2014)
7. Ciglan, M., Nørvåg, K.: Wikipop: personalized event detection system based on Wikipedia page view statistics. In: CIKM, pp. 1931–1932 (2010)
8. Das Sarma, A., Jain, A., Yu, C.: Dynamic relationship and event discovery. In: WSDM, pp. 1931–1932 (2011)
9. Efron, M., Golovchinsky, G.: Estimation methods for ranking recent information. In: ACM SIGIR, pp. 495–504 (2011)
10. Ferschke, O., Zesch, T., Gurevych, I.: Wikipedia revision toolkit: Efficiently accessing wikipedia's edit history. In: HLT, pp. 97–102 (2011)
11. Fisichella, M., Stewart, A., Denecke, K., Nejdl, W.: Unsupervised public health event detection for epidemic intelligence. In: CIKM, pp. 1881–1884 (2010)
12. Georgescu, M., Kanhabua, N., Krause, D., Nejdl, W., Siersdorfer, S.: Extracting event-related information from article updates in wikipedia. In: Serdyukov, P., Braslavski, P., Kuznetsov, S.O., Kamps, J., Rüger, S., Agichtein, E., Segalovich, I., Yilmaz, E. (eds.) ECIR 2013. LNCS, vol. 7814, pp. 254–266. Springer, Heidelberg (2013)
13. Hassan, S., Mihalcea, R.: Semantic relatedness using salient semantic analysis. In: AAAI (2011)
14. He, Q., Chang, K., Lim, E.-P.: Analyzing feature trajectories for event detection. In: ACM SIGIR, pp. 207–214 (2007)
15. Hoffart, J., Seufert, S., Nguyen, D.B., Theobald, M., Weikum, G.: KORE: keyphrase overlap relatedness for entity disambiguation. In: CIKM, pp. 545–554 (2012)
16. Hoffart, J., Suchanek, F., Berberich, K., Weikum, G.: YAGO2: A spatially and temporally enhanced knowledge base from Wikipedia. Artif. Int. J. (2012)
17. Kleinberg, J.: Bursty and hierarchical structure in streams. In: KDD, pp. 91–101 (2002)
18. Kuzey, E., Weikum, G.: Extraction of temporal facts and events from wikipedia. In: Proceedings of the 2nd Temporal Web Analytics Workshop, pp. 25–32. ACM (2012)
19. Medelyan, O., Milne, D., Legg, C., Witten, I.H.: Mining meaning from Wikipedia. Int. J. Hum.-Comput. Stud. 67 (2009)

20. Milne, D., Witten, I.H.: An effective, low-cost measure of semantic relatedness obtained from wikipedia links. In: AAAI (2008)
21. Peetz, M.-H., Meij, E., de Rijke, M.: Opengeist: Insight in the stream of page views on wikipedia. In: SIGIR Workshop on Time-aware Information Access (2012)
22. Ponzetto, S.P., Strube, M.: Knowledge derived from wikipedia for computing semantic relatedness. J. Artif. Int. Res. 30(1) (2007)
23. Takahashi, Y., Ohshima, H., Yamamoto, M., Iwasaki, H., Oyama, S., Tanaka, K.: Evaluating significance of historical entities based on tempo-spatial impacts analysis using wikipedia link structure. In: ACM HyperText (2011)
24. Tran, T.: Exploiting temporal topic models in social media retrieval. In: SIGIR (2012)
25. Tran, T., Elbassuoni, S., Preda, N., Weikum, G.: CATE: context-aware timeline for entity illustration. In: WWW, pp. 269–272. ACM (2011)

Feature Based Sentiment Analysis of Tweets in Multiple Languages

Maike Erdmann, Kazushi Ikeda, Hiromi Ishizaki,
Gen Hattori, and Yasuhiro Takishima

KDDI R&D Laboratories,
Fujimino Ohara 2-1-15, 356-8502 Saitama, Japan
{ma-erdmann,kz-ikeda,ishizaki,gen,takisima}@kddilabs.jp

Abstract. Feature based sentiment analysis is normally conducted using review Web sites, since it is difficult to extract accurate product features from tweets. However, Twitter users express sentiment towards a large variety of products in many different languages. Besides, sentiment expressed on Twitter is more up to date and represents the sentiment of a larger population than review articles. Therefore, we propose a method that identifies product features using review articles and then conduct sentiment analysis on tweets containing those features. In that way, we can increase the precision of feature extraction by up to 40% compared to features extracted directly from tweets. Moreover, our method translates and matches the features extracted for multiple languages and ranks them based on how frequently the features are mentioned in the tweets of each language. By doing this, we can highlight the features that are the most relevant for multilingual analysis.

Keywords: Feature based sentiment analysis, Twitter, multilingual.

1 Introduction

Nowadays, many people express their satisfaction or dissatisfaction with purchased products on the Internet. This information is invaluable for consumers, product developers, marketing analysts and many others. Since it is impossible to analyze the enormous amount of data manually, sentiment analysis, also known as opinion mining, has become a very popular research area. Especially feature based sentiment analysis is very promising, since it estimates not only the overall sentiment towards a product, but assigns a separate sentiment score for each of the product's features. For instance, "battery" is a typical feature of a smartphone, whereas "engine" is a typical feature of a car.

Traditionally, sentiment analysis is conducted for product review Web sites, which are comparatively easy to analyze, since the reviews are structured well and written in relatively formal language. Since only a small minority of consumers write review articles, some attempts have been made to perform basic sentiment analysis using messages posted on social networking services such as Twitter. An analysis of Twitter messages revealed that a product or brand is

B. Benatallah et al. (Eds.): WISE 2014, Part II, LNCS 8787, pp. 109–124, 2014.

mentioned in about 19% of all tweets [6]. The advantage of sentiment analysis using Twitter is that those messages are up to date and represent the sentiment of a large population on a huge variety of products in many different languages. When a product is sold in multiple countries, it is also interesting to understand the difference of customer satisfaction in those countries. This can be achieved by translating the product names and features of each product and comparing the sentiment analysis results of all languages.

In this research, we conduct feature based sentiment analysis of Twitter messages written in multiple languages. Since the extraction of product features from short and informally written tweets is very difficult, we extract features from online review articles first, and then collect tweets containing both the product name and one of the identified features. In that way, we combine the accurate extraction of features using formally written review texts with up to date sentiment analysis using social networking content.

Besides, we match the features extracted for each language by translating them and arranging them into groups of synonym features. After that, we rank the feature groups to highlight interesting features. Frequently mentioned features are ranked higher than infrequently mentioned features. Features that are mentioned much more frequently in one language than in other languages are emphasized as well.

2   Related Work

A lot of research on sentiment analysis has been described in the last decade. Liu and Zhang [7] as well as Pang and Lee [12] give a comprehensive overview of related work in that area.

The topic of feature based sentiment analysis has also received a lot of attention. Recently, Eirinaki et al. [3] have proposed a method for identifying features by extracting all nouns in the review texts and ranking them by the numbers of adjectives surrounding them. Naveed et al. [9] identify product features from topic keywords created through topic classification with LDA (Latent Dirichlet Allocation) [1] and then estimate sentiment for each product feature separately. Sentiment analysis using dedicated product review Web sites has the disadvantage that those reviews are not necessarily up to date, since most reviews are written shortly after the purchase of a product. Furthermore, only a small minority of consumers review their purchases using review Web sites, so the reviews do not represent all customers' opinions.

Therefore, sentiment analysis using social networking systems is also increasing in popularity. Sentiment Analysis in Twitter was one of the tasks of the International Workshop on Semantic Evaluation (SemEval) [11] in 2013. For instance, Mohammad et al. [13] proposed the usage of hashtags containing opinion bearing terms and emoticons. Günther and Furrer [4] preprocessed the tweets to facilitate their analysis. However, when conducting sentiment analysis using Twitter or other networking services, it is very difficult to obtain accurate results due to the short length, informal writing style and lack of structure of these texts. Moreover, Twitter messages are not composed with the purpose of reviewing a product, but only to share emotions with others in the social network.

Another problem of sentiment analysis using Twitter is that product features are not frequently mentioned explicitly in tweets and are therefore difficult to extract automatically. Therefore, as far as we know, only basic sentiment analysis, not feature based sentiment analysis, has been described for social networking services.

Not much research has been conducted in the area of cross-language sentiment analysis, and most research in that area focuses on creating a language independent sentiment analysis system [2] or applying machine translation in order to reuse a sentiment analysis system created for one language to analyze sentiment in a different language [14].

Nagasaki et al. [10] extract characteristic terms of texts on controversial topics (e.g. the term "extinction" for the topic "whaling") in multiple languages, which is similar to extracting product features. In addition to calculating the frequency rate of each n-gram in the respective language, they translate the n-grams and calculate a cross-lingual frequency rate, i.e. they compare the frequency of the n-gram to the frequency of its translation. However, the authors do not analyze sentiment, but only rank Weblogs according to their relevance to the topic.

Guo et al. [5] are the first researchers to combine cross-language sentiment analysis with feature based sentiment analysis. The authors introduce a method called Cross-lingual Latent Semantic Association which groups semantically similar product features across languages. However, they conduct sentiment analysis on well structured online review articles, which is much easier than analyzing Twitter messages. Besides, they do not attempt to analyze the differences in sentiment in each language beyond listing the sentiment scores side by side.

3 Feature Based Sentiment Analysis

In this section, we describe how we conduct feature based sentiment analysis on Twitter in multiple languages. The system structure is visualized in Figure 1. First, the product name is translated or transliterated[1] from the source language (L1) into the target language (L2) using bilingual dictionaries or machine translation. After that, feature based sentiment analysis is conducted for each language independently, as described in the following subsections. Since the extraction of product features from short and informally written tweets is very difficult, we extract features from online review articles first, and then collect tweets containing both the product name and one of the identified features. Finally, the sentiment analysis results of each language are matched and ranked as described in Section 4.

---

[1] Transliteration is the process of replacing the letters of a word with corresponding letters in a different alphabet.

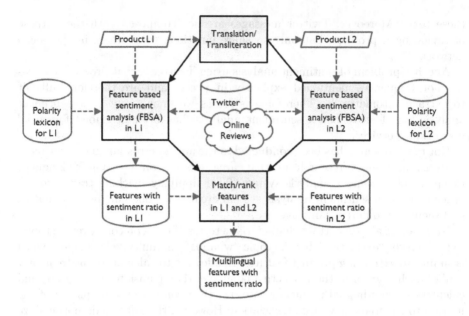

Fig. 1. Overview of the System Structure

3.1 Extraction of Product Features

The feature extraction process is visualized in Step 1a of Figure 2. First, a list of relevant review articles is collected using a Web search engine with a search query consisting of the product name in combination with one or more manually selected keywords, such as "review", "features" or "specification" in the respective language. The plain text in the collected Web sites is crawled using software to remove all markup language tags as well as text that is not part of the review article (e.g. navigation, advertisement).

For the set of review articles texts, product features are extracted using LDA (Latent Dirichlet Allocation) [1]. In the second step, the top ranked topic keywords for each topic are ranked based on the co-occurrence of sentiment bearing terminology. The more often a topic keyword is accompanied by a sentiment bearing term, i.e. a term identified as having either positive or negative meaning, the more likely the topic keyword can be used as a product feature.

A polarity lexicon is used to decide which terms are sentiment bearing terms. For each occurrence of a topic keyword tk with a term in the polarity lexicon plt, the feature probability score fps increases as follows:

$$fps = fps + \left(\frac{1}{\text{distance of } tk \text{ and } plt} \right) \tag{1}$$

The top ranked topic keywords of each product in each language are then selected as product features.

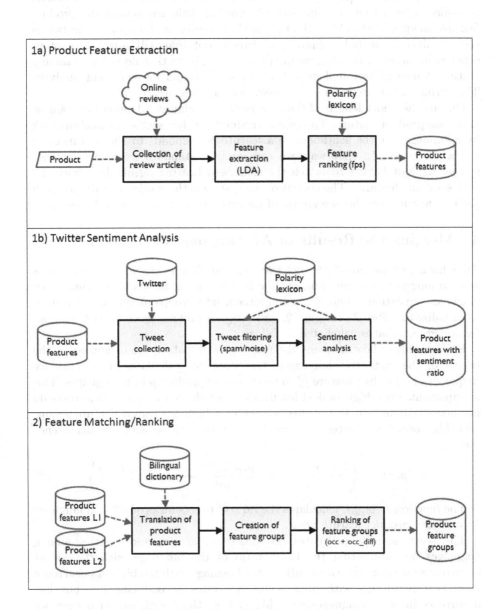

Fig. 2. Flowchart of the Proposed System

3.2 Performing Sentiment Analysis on Tweets

For the top ranked topic keywords estimated to be suitable product features, we collect tweets that contain both the product title and one of the product features as visualized in Step 1b of Figure 2. However, we filter out spam tweets such as advertisements by removing all tweets containing URLs to Web content other than pictures. Besides, we filter out noisy tweets that do not contain any sentiment bearing terminology and thus are not useful for sentiment analysis. The output of this step is a set of tweets for each feature.

Due to the short length of Twitter messages, almost all tweets contain at most one product feature. Therefore, we simplified the sentiment analysis task by assuming that the sentiment of a feature corresponds to the sentiment of the whole tweet. In that way, we can apply a standard sentiment analysis system without having to match the sentiment bearing terminology with the corresponding features. The output of the system is the sentiment ratio for each product feature, i.e. the percentage of positive, negative and neutral tweets.

4 Merging the Results of All Languages

After having extracted all product features and obtained their sentiment ratios in each language, we can translate the features using bilingual dictionaries or machine translation. Then, we arrange them into groups of synonym features, as visualized in Step 2 of Figure 2, and assign a ranking score to each product feature group, that is calculated as follows.

For estimating how frequent the features in a feature group appear in the tweets of the respective languages, the score occ calculates the occurrence frequency of a product feature pf in the tweets of product p in n languages. This is important, since high ranked features extracted from review article texts do not necessarily appear in the tweets in equally high frequency. Feature groups with high occurrence rates are scored higher than those with low occurrence rates.

$$occ(p, pf) = \left(\frac{tr(p, pf, l_1) + tr(p, pf, l_2) + \ldots + tr(p, pf, l_n)}{|n|} \right) \qquad (2)$$

The function $tr(p,pf,l)$ calculates the ratio of tweets for product p in language l containing product feature pf.

The score occ_diff calculates the difference in occurrence ratios among languages by subtracting the tweet ratio of the language with the lowest occurrence rate from the tweet ratio of the language with the highest occurrence rate. Feature groups with high occurrence rates in one language but low occurrence in other languages score higher than those with similar occurrence in all languages.

$$occ\_diff(p, pf) = max\_tr(p, pf) - min\_tr(p, pf) \qquad (3)$$

The function $max\_tr(p,pf)$ calculates the tweet ratio $tr(p,pf,l)$ for the language l that has the largest tweet ratio within the feature group. Correspondingly, the

Table 1. Products Used in Experiment

Category	Products
Smartphones	iPhone 5S, iPhone 5, iPhone 4S, Nexus 5, XPeria Z1, Galaxy S4
Cars	Prius, Lexus, Corolla, Nissan GT-R, Infiniti, Lancer Evolution, Impreza

function min_tr(p,pf) calculates the tweet ratio $tr(p,pf,l)$ for the language l with the smallest tweet ratio within the feature group.

Finally, the overall score is calculated as the weighted sum of the two scores *occ* and *occ_diff* and the highest ranked feature groups are displayed as results.

5 Experimental Results

In this section, we explain and discuss an experiment in which we analyzed the sentiment for 13 products (6 smartphones and 7 cars) expressed in English and Japanese tweets.

5.1 Preparation of the Experiment

In order to extract features from tweets, we collected all tweets containing one of the 13 product titles in the time between December 2013 and March 2014. The products are listed in Table 1. For both categories, we selected products that are popular in the USA as well as in Japan. This limited us to only Japanese car brands, since the market share of foreign car brands in Japan is marginal. The diversity of foreign smartphones sold in Japan is also small, thus half of the smartphones in the experiment are Apple products.

Of course, it would also be interesting to select products that are not popular in one country, in order to find out the reasons why the product is not sold well. However, in order to evaluate our feature extraction method accurately, we had to make sure that we can collect enough data in both languages for each product.

Before starting the experiment, we created a large polarity lexicon by combining the data of several existing polarity lexicons[2,3]. Since polarity labeling is very subjective, i.e. the polarity of terms often depends on the context, terms were sometimes labeled differently in one lexicon than in other lexicons. We resolved these conflicts by choosing the most frequent label for each term. In that way, we were able to not only create a large scale polarity lexicon, but also improved the labeling quality.

In the first part of the experiment, we compared features extracted from online review articles to features extracted directly from tweets to show that features extracted from review articles are significantly more accurate than features extracted from tweets.

---

[2] English polarity lexicons: MICRO-WNOP, SentiWordNet, MPQA, AFINN, DeRose, McDonald, Avaya.
[3] Japanese polarity lexicons: Inui-lab, Avaya.

Table 2. Examples of Removed Tweets

Product	URL	No Sentiment	Example Tweet	Filter
iPhone 5S	no	no	Should I get an IPhone 5S?? Because there is just so many nice iPhone 5S covers.	Keep
	no	yes	I don't know if I should get an iPhone 5S or a Galaxy Note 3...#thestruggle	Remove
	yes	no	Best Free Games for the iPhone 5S: http://t.co/cioZSBHaLB via @youtube	Remove
	yes	yes	Here's YOUR CHANCE to WIN the NEW iPhone 5S! http://t.co/lCs9hGjQeT	Remove
Prius	no	no	Prius' sound like spaceships but that's pretty much the only cool thing about them.	Keep
	no	yes	I think @aaabbbiiii is the only person in our school with a Prius	Remove
	yes	no	Great deals, everday low price on Prius #Prius Share this http://t.co/BSjvJpf1YQ	Remove
	yes	yes	Used 2010 #Toyota #Prius, 82,617 miles, listed for $16,000 under used cars http://...	Remove

We noticed that about 80% of the collected tweets were either spam tweets such as advertisement or noise, i.e. tweets containing no sentiment, that interfere with the identification of accurate product features. Therefore, we decided to remove all tweets containing URLs to Web content other than pictures (spam tweets) and all tweets not containing any sentiment bearing terminology (noisy tweets). A few example tweets filtered out by these two rules are shown in Table 2. After filtering, approximately 1,250,000 English tweets and 250,000 Japanese tweets remained for the smartphone category. For the car category, about 260,000 English and 290,000 Japanese tweets remained.

In order to extract features from review article texts, we collected the top 10 search engine results for each of the product names in combination with the keyword "review". Then, we applied the boilerpipe[4] software to remove all markup language tags and parts of the Web site that are not part of the review article (e.g. navigation, advertisement).

5.2   Extraction of Product Features

We extracted topic keywords from both tweets (baseline) and reviews articles (proposed method) using tf-idf, df-idf and LDA. We decided to extract one set of features for the smartphone category and one set of features for the car category, since the accuracy is slightly higher than for keywords extracted for each product separately. We applied the feature extraction method described in Section 3.1. The top 100 topic keywords extracted of each method were ranked again according to their feature probability score *fps*. For keyword extraction

---

[4] http://code.google.com/p/boilerpipe/

Table 3. Precision of Top 20 Features

	Smartphones		Cars		
	English	Japanese	English	Japanese	Average
tweets	0.15	0.3	0	0.05	0.125
reviews	0.55	0.45	0.65	0.5	0.538

Table 4. Precision of Top 50 Features

	Smartphones		Cars		
	English	Japanese	English	Japanese	Average
tweets	0.24	0.14	0.04	0.08	0.125
reviews	0.5	0.42	0.54	0.38	0.46

using LDA, we set the number of topics to 10 in the same way as described in related work [5], and extracted the top 10 keywords for each topic.

All extracted topic features were manually evaluated. Since this is a very subjective decision, each topic keyword was evaluated by three judges and the evaluation of the majority of them was used to decide whether a topic keyword is a suitable product feature. Generally, a keyword is considered to be a suitable feature if it meets two criteria. First, the keyword needs to be a part of the product, thus it must be possible to describe the relationship in a "has a" expression. For instance, the "iPhone 5S" has a "camera", but the "iPhone 5S" does not have an "Apple". Second, a feature must directly impact the product quality. The quality of the "Prius" car, for example, is affected by the quality of the "engine", but it is not affected by the quality of the "road".

LDA ranked with the feature probability score *fps* achieved the best overall performance. Therefore, we show only the results of that method. Table 3 shows the precision of the top 20 features and Table 4 shows the results of the top 50 features. As the results show, the precision of our proposed method (features extracted from reviews) performed significantly better than the baseline method (features extracted from tweets). While the performance of the proposed method is certainly not optimal, the precision increased by about 40% for the top 20 and by about 30% for the top 50 features. Besides, even if we replace the feature extraction method by a different method, such as the method proposed by Eirinaki et al. [3] or Naveed et al. [9], the proposed method is likely going to perform better than the baseline method.

The top ranked features for each category are shown in Table 5 (tweets) and Table 6 (reviews). The Japanese keywords are translated into English. The plus and minus signs next to the keywords indicate whether the keyword is a suitable product feature. Notable is that the features extracted from English tweets contain many common terms such as "love", "day" or "people". We believe that this is because only a small percentage of tweets explicitly mention features of a product. The features extracted from Japanese tweets often contained terms

Table 5. Examples of Features Extracted from Tweets

$(+)$ = suitable feature $(-)$ = unsuitable feature

Smartphones		Cars	
English	Japanese	English	Japanese
(+) charger	(+) battery	(−) car	(−) chan
(−) Samsung	(+) cover	(−) love	(−) follow
(+) case	(−) Nokia	(−) day	(+) wheel
(−) love	(+) case	(−) people	(−) fugue
(−) day	(+) ios	(−) LOL	(−) Super
(−) shit	(−) SIM	(−) Mom	(−) crown
(−) year	(−) Docomo	(−) shit	(−) work
(−) time	(−) MNP	(−) girl	(−) Gran Turismo
(−) people	(−) running	(−) bitch	(−) legacy
(−) Christmas	(−) debut	(−) gt	(−) taxi

that are somehow related but not suitable as product features. (e.g. "Nokia", "SIM", "Super").

In the next step, we confirmed that the product features extracted from review articles correspond to the terms that are frequently used in tweets describing the products. While we did not conduct a formal evaluation, we identified only a small number of mismatches. The product feature "display" was rarely found in smartphone related tweets and the product features "interior" and "seat" were rarely found in car related tweets, although these terms frequently appear in review article texts. On the other hand, the terms "charger" and "price" for smartphones and the terms "wheel" and "exhaust" for cars frequently appear in tweets but not in review articles, thus they were not identified as product features. However, the majority of product features seems to match the terms that are commonly used in tweets.

Table 6. Examples of Features Extracted from Reviews

(+) = suitable feature (−) = unsuitable feature

Smartphones		Cars	
English	Japanese	English	Japanese
(−) phone	(−) phone	(−) car	(−) car
(−) Apple	(+) camera	(−) drive	(+) engine
(+) camera	(+) app	(+) front	(−) model
(+) display	(+) Android	(+) interior	(+) hybrid
(+) screen	(+) LTE	(−) ride	(−) corner
(+) app	(−) user	(+) engine	(+) power
(−) feature	(+) display	(+) seat	(+) tire
(+) ios	(+) battery	(+) rear	(+) brake
(+) design	(+) mail	(+) handling	(−) gasoline
(−) Sony	(+) size	(−) model	(+) seat

5.3 Sentiment Analysis Results

Before conducting sentiment analysis for the extracted product features, we calculated the number of features per tweet. Almost all tweets did not contain any features. Of the tweets in which features were detected, about 89.6% contained only one feature, 9.2% contained two features, and only 1.2% contained three or more features. Therefore, we simplified the process of sentiment analysis of tweets by regarding the sentiment of a feature as equivalent to the sentiment of the whole tweet.

After that, we ranked all feature groups (pairs of English and Japanese features) according the sum of their *occ* and *occ_diff* scores (see Section 4), to prioritize the ones that are most interesting for the user. Again, we did not formally evaluate the results, but they corresponded well to our intuitive ranking of the features. The top ranked feature groups created from the top 100 English and top 100 Japanese feature are shown in Table 7 (smartphones) and 8 (cars). Missing tweet ratio scores indicate that the feature was not extracted in the corresponding language. The plus and minus signs next to the features indicate whether they are suitable product features. In the smartphone category, 10 feature groups (59%) are suitable features, whereas 9 feature groups (45%) in the car category are suitable features.

Only for the top ranked feature groups that were considered useful features, we analyzed the sentiment using a simple lexicon based sentiment analysis system [8]. For each product feature, we extracted the newest 1,000 tweets containing both the product title and the product feature. The results of two example products, "iPhone 5S" and "Prius", are shown in Figure 3. The left bar of each feature group represents the English and the right bar the Japanese sentiment.

For the "iPhone 5S", the sentiment expressed in the English tweets is generally more negative than the sentiment expressed in the Japanese tweets. This is not surprising, given that the market share of the "iPhone 5S" is significantly higher in Japan than in the USA and other English speaking countries. The highest

Table 7. Top 20 Ranked Feature Groups (Smartphones)

(+) = suitable feature (−) = unsuitable feature

feature	tweet ratio (en)	tweet ratio (jp)	occ	occ_diff	score
(−) Docomo	−	0.347	0.174	0.347	0.521
(−) phone	0.283	−	0.146	0.283	0.429
(+) Android	0.018	0.100	0.059	0.082	0.141
(−) day	0.083	−	0.043	0.083	0.126
(−) data	−	0.079	0.039	0.079	0.118
(−) year	0.076	−	0.039	0.076	0.115
(+) case	0.067	−	0.034	0.067	0.101
(−) people	0.064	−	0.033	0.064	0.097
(−) user	−	0.060	0.030	0.060	0.091
(+) app	0.016	0.063	0.040	0.047	0.087
(+) LTE	−	0.051	0.026	0.051	0.077
(+) battery	0.022	0.054	0.038	0.033	0.071
(+) size	0.005	0.048	0.027	0.044	0.070
(+) mail	−	0.042	0.021	0.042	0.063
(−) Apple	0.038	−	0.020	0.038	0.058
(−) upgrade	0.031	−	0.016	0.031	0.047
(+) screen	0.031	−	0.016	0.031	0.047
(+) camera	0.035	0.030	0.033	0.005	0.039
(−) game		0.021	0.011	0.021	0.032
(+) picture	0.021	−	0.011	0.021	0.032

difference can be observed for the features "app" and "battery". For "app", only 29% of the English tweets are positive whereas 71% of the Japanese tweets are positive. When analyzing the corresponding tweets manually, we discovered that many English speaking users state that applications crash frequently. Japanese users also state problems with the new operating system, but also praise specific applications, such as the new fingerprint recognition application and the new camera application. For "battery", 40.6

For some features of the "iPhone 5S", sentiment in only one of the languages could be obtained. The feature "LTE", for instance, appears only in Japanese tweets. The reason for that might be that many Japanese people use smartphones while commuting in the train or even replace their home computer with a smartphone. Therefore, they rely on fast Internet access technology more than users in other countries and mention it more frequently in their tweets.

For the car "Prius", the sentiment of English and Japanese tweets is slightly more positive than the sentiment of Japanese tweets. The most notable difference is the sentiment for the feature "engine", for which English tweets express a much more positive sentiment than Japanese tweets (62.4% positive tweets for English, 25.6% positive tweets for Japanese). Deeper analysis of the tweets revealed that while users in both languages complain about the rather small engine power of the Prius, only Japanese tweets declare the small engine sound of the "Prius" as a safety hazard. Streets in Japan are very narrow and often lack sideways, thus pedestrians are afraid of colliding with a car that they cannot hear approaching.

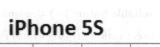

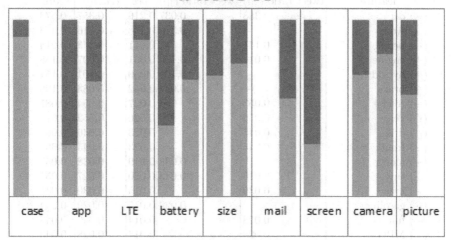

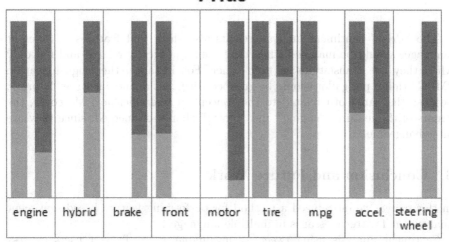

left bar = English sentiment, right bar = Japanese sentiment

Fig. 3. Examples of Sentiment Analysis Results

Table 8. Top 20 Ranked Feature Groups (Cars)

$(+)$ = suitable feature $(-)$ = unsuitable feature

feature	tweet ratio (en)	tweet ratio (jp)	occ	occ_diff	score
$(-)$ car	0.388	0.037	0.216	0.357	0.573
$(-)$ Aqua	–	0.316	0.158	0.316	0.473
$(-)$ people	0.141	–	0.071	0.143	0.214
$(+)$ engine	0.012	0.109	0.061	0.097	0.158
$(+)$ hybrid	–	0.101	0.050	0.101	0.151
$(+)$ brake	–	0.084	0.042	0.084	0.126
$(-)$ road	0.053	–	0.027	0.054	0.081
$(-)$ year	0.052	–	0.026	0.052	0.079
$(-)$ drive	0.045	–	0.023	0.046	0.068
$(+)$ front	0.045	–	0.023	0.045	0.068
$(+)$ motor	–	0.038	0.019	0.038	0.057
$(+)$ tire	0.016	0.043	0.030	0.027	0.057
$(-)$ mile	0.030	–	0.015	0.030	0.045
$(-)$ gasoline	–	0.029	0.015	0.029	0.044
$(-)$ fun	0.028	–	0.014	0.029	0.043
$(+)$ MPG	0.028	–	0.014	0.028	0.043
$(-)$ lot	0.027	–	0.014	0.028	0.042
$(+)$ acceleration	0.001	0.028	0.015	0.027	0.041
$(+)$ steering wheel	–	0.024	0.012	0.024	0.036
$(-)$ silver	–	0.022	0.011	0.022	0.032

The "Prius" sentiment analysis results also show that features in different languages should be matched when they are semantically related and not only when they are translations of each other. For instance, the English feature "MPG" (miles per gallon) could be grouped with the Japanese feature "hybrid", because they are both related to the concept of fuel efficiency. Moreover, the results of the features "engine" and "motor" should be merged, since they are interchangeable.

6 Conclusion and Future Work

In this paper, we introduced a method for performing feature based sentiment analysis on Twitter messages in multiple languages.

Sentiment analysis using tweets is invaluable, since Twitter users express sentiment towards a huge variety of products in many different languages, and because sentiment expressed on Twitter is more up to date and represents the sentiment of a larger population than review articles.

Since the extraction of product features from short and informally written tweets is very difficult, we extracted features from online review articles first and then collected tweets containing both the product name and one of the identified features. In an experiment with English and Japanese Twitter messages on 6 smartphones and 7 cars, feature extraction from review articles increased the precision of the extracted features by about 40% for the top 20 and by about 30% for the top 50 features.

Moreover, we proposed a system to highlight the features that are most relevant for multilingual sentiment analysis. We translated the English and Japanese features and arranged them into groups of synonym features. After that, we ranked the feature groups according to how frequent the features in a feature group appeared in the tweets of the respective languages and according to the difference in occurrence ratios among languages.

Our proposed system allows consumers, product developers and marketing specialists of internationally operating companies to track the sentiment of their products world-wide.

In the future, we want to improve the precision of feature extraction, since only about half of the extracted features in our experiment were useful for sentiment analysis, and combine sentiment analysis of semantically similar features (e.g. "screen" and "display"). Furthermore, we are planning to perform a deeper analysis of the sentiment results in each language, since it is very important to understand not only how sentiment differs among languages but also why.

Apart from that, we are interested in performing feature based sentiment analysis not only on products but on e.g. news, services or events. Nagasaki et al. [10] extracted characteristic terms of English and Japanese texts on controversial topics such as "whaling" or "organ donations" and detected very interesting differences in the terminology used in the texts of each language. Combining the extraction of characteristic terms with sentiment analysis can help us why people from different countries have different opinions on many topics.

References

1. Blei, D.M., Ng, A.Y., Jordan, M.I.: Latent dirichlet allocation. The Journal of Machine Learning Research 3, 993–1022 (2003)
2. Cui, A., Zhang, M., Liu, Y., Ma, S.: Emotion tokens: Bridging the gap among multilingual twitter sentiment analysis. In: Salem, M.V.M., Shaalan, K., Oroumchian, F., Shakery, A., Khelalfa, H. (eds.) AIRS 2011. LNCS, vol. 7097, pp. 238–249. Springer, Heidelberg (2011)
3. Eirinaki, M., Pisal, S., Singh, J.: Feature-based opinion mining and ranking. Journal of Computer and System Sciences 78(4), 1175–1184 (2012)
4. Günther, T., Furrer, L.: Gu-mlt-lt: Sentiment analysis of short messages using linguistic features and stochastic gradient descent. In: Proceedings of the International Workshop on Semantic Evaluation (SemEval), pp. 328–332 (2013)
5. Guo, H., Zhu, H., Guo, Z., Zhang, X., Su, Z.: Opinionit: A text mining system for cross-lingual opinion analysis. In: Proceedings of the ACM International Conference on Information and Knowledge Management (CIKM), pp. 1199–1208 (2010)
6. Jansen, B.J., Zhang, M., Sobel, K., Chowdury, A.: Twitter power: Tweets as electronic word of mouth. Journal of the American Society for Information Science and Technology (JASIST) 60(11), 2169–2188 (2009)
7. Liu, B., Zhang, L.: A survey of opinion mining and sentiment analysis. In: Mining Text Data, pp. 415–463 (2012)

8. Tofiloski, M., Voll, K., Taboada, M., Brooke, J., Stede, M.: Lexicon-based methods for sentiment analysis. Computational Linguistics 37(2), 267–307 (2011)
9. Naveed, S.S.N., Gottron, T.: Feature sentiment diversification of user generated reviews: The freud approach. In: Proceedings of the International AAAI Conference on Weblogs and Social Media (ICWSM), pp. 429–438 (2013)
10. Nakasaki, H., Kawaba, M., Utsuro, T., Fukuhara, T.: Mining cross-lingual/cross-cultural differences in concerns and opinions in blogs. In: Proceedings of the International Conference on Computer Processing of Oriental Languages. Language Technology for the Knowledge-based Economy (ICCPOL), pp. 213–224 (2009)
11. Nakov, P., Kozareva, Z., Ritter, A., Rosenthal, S., Stoyanov, V., Wilson, T.: Semeval-2013 task 2: Sentiment analysis in twitter. In: Proceedings of the International Workshop on Semantic Evaluation (SemEval), pp. 312–320 (2013)
12. Pang, B., Lee, L.: Opinion mining and sentiment analysis. Foundations and Trends in Information Retrieval 2(1-2), 1–135 (2008)
13. Saif, X.Z., Mohammad, M.: Svetlana Kiritchenko. Nrc-canada: Building the state-of-the-art in sentiment analysis of tweets. In: Proceedings of the International Workshop on Semantic Evaluation (SemEval), pp. 321–327 (2013)
14. Wan, X.: Co-training for cross-lingual sentiment classification. In: Joint conference of the Annual Meeting of the ACL and the International Joint Conference on Natural Language Processing of the AFNLP (ACL-IJCNLP), pp. 235–243 (2009)

Incorporating the Position
of Sharing Action in Predicting Popular Videos
in Online Social Networks

Yi Long, Victor O.K. Li, and Guolin Niu

Department of Electrical and Electronic Engineering,
The University of Hong Kong,
Pokfulam Road, Hong Kong, China
{yilong,vli,glniu}@eee.hku.hk

Abstract. Predicting popular videos in online social networks (OSNs) is important for network traffic engineering and video recommendation. In order to avoid the difficulty of acquiring all OSN users' activities, recent studies try to predict popular media contents in OSNs only based on a very small number of users, referred to as experts. However, these studies simply treat all users' diffusion actions as the same. Based on large-scale video diffusion traces collected from a popular OSN, we analyze the positions of users' video sharing actions in the propagation graph, and classify users' video sharing actions into three different types, i.e., initiator actions, spreader actions and follower actions. Surprisingly, while existing studies mainly focus on the initiators, our empirical studies suggest that the spreaders actually play a more important role in the diffusion process of popular videos. Motivated by this finding, we account for the position information of sharing actions to select initiator experts, spreader experts and follower experts, based on corresponding sharing actions. We conduct experiments on the collected dataset to evaluate the performance of these three types of experts in predicting popular videos. The evaluation results demonstrate that the spreader experts can not only make more accurate predictions than initiator experts and follower experts, but also outperform the general experts selected by existing studies.

Keywords: Online Social Networks, Information Diffusion, Video, Prediction.

1 Introduction

Online social networks (OSNs), such as Facebook and Twitter, have now become an indispensable platform for users to access and share videos which are originally hosted by video sharing sites (VSSes), such as YouTube, generating enormous traffic to these VSSes [15]. As is revealed by existing studies, the distribution of video popularity is highly skewed, with the top 0.31% most popular videos accounting for nearly 80% view requests [16]. Therefore, it is important to

B. Benatallah et al. (Eds.): WISE 2014, Part II, LNCS 8787, pp. 125–140, 2014.
© Springer International Publishing Switzerland 2014

identify these popular videos based on their diffusion progress in the early stage, so as to optimize the network traffic engineering system, such as the caching strategy, and to improve the video recommender system in OSNs.

Although considerable studies have been conducted to predict the popularity of videos, most of these proposed methodologies are based on all users' actions in the early stage [17] [19] [22]. Considering that it is rather expensive to acquire all users' diffusion actions in popular OSNs, which usually have millions of users, recent studies try to predict media content popularities based on only a small nunber of users, which are referred to as "experts" [11] [14] or "influencers" [2]. However, these studies mainly focus on other types of media contents, e.g., Wikipedia edits [14] or tweets [2] [11]. Moreover, while these studies simply treat all users' diffusion actions as the same, researchers working on the information diffusion process argue that diffusion actions of users are actually different and depend on their positions in propagation graphs [20] [21] [23]. Therefore, it is meaningful to account for the position information of diffusion actions when selecting experts, and evaluate the performance of selected experts in predicting popular videos.

To fill the above research gaps, we select Renren, the most popular Facebook-like OSN in China, as our research platform, and collect substantial diffusion traces from it. Based on the collected traces, we firstly classify users' video sharing actions of an individual video into three different groups, i.e., initiator actions, spreader actions and follower actions, according to their positions in diffusion process. Then we conduct comparison studies on the impact of these different sharing actions on video diffusion. Motivated by the importance of spreader in distinguishing popular videos, we further incorporate sharing actions' position information into the selection of experts, enabling us to identify three different groups of experts, i.e., initiator experts, spreader experts and follower experts. Experiments based on real dataset have been conducted to evaluate the performance of these experts in predicting popular videos. The main contributions of this paper are summarized as follows:

- Large-scale video diffusion traces, covering the video sharing actions of 2.8 million users for more than 3 years, have been collected from a popular OSN. According to their positions in the diffusion process, users' sharing actions have been classified into three different types, i.e., initiator actions, spreader actions and follower actions.
- When attempting to the identify the initiators of information diffusion, existing studies adopt two different definitions of sharing actions, namely, actions which share media contents independently [20] [24], and those that import media contents from external sites [24] [25]. Based on our collected traces, we investigate these two types of initiator actions, and propose a more comprehensive definition of initiator actions.
- Different from previous empirical studies which mainly focus on characterizing initiator actions [20] [21], we conduct comprehensive studies on all the three types of sharing actions, i.e., initiator actions, spreader actions and follower actions. Surprisingly, while initiator actions are regarded to be

pivotal in theoretical diffusion models [13], our observations show that spreader actions play a more important role in driving video popularity.

- By incorporating positions of sharing actions into the selection of experts, we can identify three different groups of experts, i.e., initiator experts, spreader experts and follower experts. The experiments conducted on the collected real traces demonstrate that spreader experts outperform not only the initiator experts and follower experts, but also the general experts [11], selected without regard to the positions of their sharing actions.

The rest of this paper is organized as follows. Section 2 introduces the video diffusion process of Renren as well as the dataset collected from Renren. Section 3 defines the three types of sharing actions and investigates their roles in video diffusion. Section 4 first introduces how to select the three different types of experts, and then evaluates the popular video prediction performance of these experts. Finally, in Section 5 we summarize our studies and discuss future work.

2   Data Collection and Basic Properties

First, we will give a brief introduction of our research platform, Renren, as well as our video diffusion traces collected from Renren.

2.1   Background of the Renren Social Network

Renren is one of the most popular OSNs in China, and has more than 190 million users. It adopts almost the same interface as well as functionalities of Facebook, and hence is called "the Twin of Facebook in China". In the past few years, an increasing number of users in Renren tend to watch and share/import videos which are originally hosted by external VSSes [16]. In general, a video in Renren is propagated along the friendship links. Initially, if a user finds an interesting video in a VSS, he/she can import this video to Renren by posting the URL of the video in Renren. Then this video will appear in the news feeds of all his/her friends. If any of these friends also think this video is interesting after watching it, he/she may further share this video with his/her friends by simply clicking a button, which will also forward this video to the news feeds of his/her friends, thus generating a cascade in Renren.

While such video propagation process is similar to many other popular OSNs, such as Twitter [5] and Weibo [25], Renren has two unique features making it an ideal platform for studying the video diffusion process: 1) Both Renren users' friendship links and the complete list of all their video sharing actions are publicly accessible by any registered user of Renren [1]; 2) Renren maintains a separate website to store the information of each individual video sharing action, including the original URL of the video in external sites and identity of the Renren user who imports this video into Renren. Therefore, in this paper, we choose Renren as our research platform. The basic statistics as well as properties of video diffusion traces we collected from Renren are presented as follows.

---

[1] The friend lists of Renren users are protected from strangers since April, 2011.

Table 1. Basic statistics of the XJTU dataset

Time period	March 2008 - July 2011
Number of users	2,808,681
Number of video sharing actions	209,409,778
Number of distinct videos	6,416,745

2.2   Data Collection

Due to the large number of users in Renren, it is computationally expensive and nearly impossible for us to acquire the friend list and sharing actions of all Renren users. Fortunately, just like Facebook, Renren evolves from a university-based social network and hence is divided into regional networks with institution affiliation information. Therefore, following the sampling strategy in [12], we perform affiliation-oriented crawls of the Xi'an Jiaotong University (XJTU) network, which is composed of more than 52,685 XJTU students and is a famous online social network community in Renren. To retrieve the position information of these XJTU users' video sharing action, we also identified all of their friends, obtaining more than 2 million users. For both the collected XJTU users and their friends, we further crawled all of their video sharing actions performed in the past 3 years. The statistics of the XJTU dataset are shown in Table 1. Such a substantial dataset from Renren not only captures the information diffusion process better due to its affiliation-based sampling strategy [4], but is also more instructive for optimizing network traffic engineering (e.g., the caching strategies of content delivery network) since most of these users share the same locations and access the Internet via the campus network.

In our collected diffusion traces, each video sharing action can be represented as a tuple (u, v, t, s), which means that at time t, user u shares video v, and this shared video is imported from an external VSS by user s. After collecting video sharing actions of all XJTU users, we can further explicitly track the diffusion process of a video in XJTU community since each video in our dataset is associated with a unique external URL.

2.3   Video Popularity in Renren

Characterizing the distribution of video popularity is essential for many applications, e.g., the skewness of video popularity distribution will determine the caching performance of a video service provider [8]. The degree skewness for popularity distributions are usually described with the Pareto Principle or the 80/20 rule. In this paper, we define the popularity of a video as the number of sharing actions of this video, and calculate the popularities of the 6.4 million videos accounted by our 2.8 million users. To gain insights into the Pareto Principle of these videos, we investigate the cumulative percentage of video popularity for the i most popular videos (shown in Fig. 1). The horizontal axis sorts

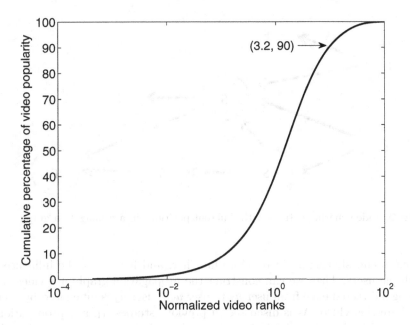

Fig. 1. Test of Pareto principle in video popularity

the videos from the most popular to the least popular, with the rank normalized between 0 and 100. As the coordinate (3.2,90) shows, the top 3.2% most popular videos account for 90% of sharing actions, while the remaining 96.8% of the videos account for only 10% sharing actions. Such extreme skewness of video popularity in OSN is higher than in VSSes (e.g., YouTube), where the top 20% most popular videos account for 90% [3]. This is because while the videos in VSSes are mainly accessed via the search engine or recommendation system, popularity of videos in OSNs are mainly driven by the propagation along inter-personal links. Therefore, it is quite desirable to predict these most popular videos in OSN at an early stage. In our following studies, we will select the top 3% most popular videos as *top videos*, and then conduct comparison studies on users' role in the diffusion process of these *top videos* and all common videos, trying to find which group of users make those *top videos* popular.

3   Sharing Actions in Different Positions of Video Diffusion

Following an approach used in a study of information propagation on Twitter [2], we can track the propagation of a video v among our studied users based on their friendship links as well as sharing actions of video v: if user u_1 and user u_2 are friends, and user u_1 shared video v earlier than user u_2 , we say user u_2 is influenced by user u_1 to share video v. Note that a user may have more

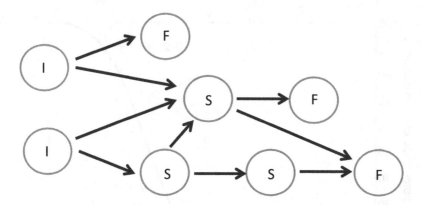

Fig. 2. Video sharing actions with different positions in a propagation graph

than one friend sharing a video before him/her, and hence may be influenced by multiple users. Then we can construct the propagation graph of a video by recording a directed edge from user u_1 to user u_2 if user u_2 is influenced by user u_1 to share this video. As is discussed in previous studies [7], any propagation graph constructed as above is always a directed acyclic graph: it is directed, each node can have zero or more parents, and cycles are impossible due to the time constraint. Based on the propagation graph of a video, we classify sharing actions of this video into three types according to their positions in the propagation graph. Figure 2 describes a simple propagation graph of a video as an example to illustrate our definition of the three types of sharing action. We define the video sharing actions without parents (shown as I) to be **initiator actions**, video sharing actions which have both parents and children (shown as S) to be **spreader actions**, and video sharing actions without children (shown as F) to be **follower actions**.

3.1   Definition of Video Initiators

In fact, in addition to the criteria discussed above, various other criteria [20] [21] [23] [24] [25] have been proposed to classify users' diffusion actions into different types according to their positions in the propagation graph. Although they all investigated initiator actions which start a chain in the propagation graph, they adopted different definition of initiator actions. In particular, while studies working on the diffusion process of Facebook page [21], tweets [20] and email [24] define the initiator actions to be the actions which share media contents independently, i.e., they share media contents when none of their friends have shared them before, recent work focusing on the diffusion process of videos [24] [25] believe the root nodes of video propagation graphs are sharing actions which import videos from external VSSes. Therefore, it is necessary to conduct a comparison studies on these two different definitions of video initiators. Fortunately, based on our collected traces, we can not only obtain the sharing actions which share video independent of friends following the method discussed above, but also identify those sharing actions which import videos from external VSSes by utilizing the last entry of each sharing action (u, v, t, s), i.e., who imports this video v into Renren. Hence, for any video v, we build a set of sharing actions which share video v independently (denoted as action set A), as well as a set of sharing actions which import video v from external VSSes (denoted as action set B)). To explore the differences and similarities of action set A and action set B for each video, we further divide all the actions in $(A \cup B)$ into three disjoint action sets, i.e., $(A \cap B)$, $(A - B)$ and $(B - A)$. The physical meanings of these three sets are discussed as follows.

- Set $(A \cap B)$ includes each sharing action which imports a video from VSSes into Renren when none of the friends have shared this video before. Such actions usually happen in the following way: a user finds an interesting video in VSSes, and it has not been shared by any of his/her Renren friends, then he/she imports this video to Renren.
- Set $(A - B)$ includes each sharing action of a video which has been imported by stranger into Renren when none of the friends of a user have shared this video before. Note that besides forwarding the video shared by each user's friends to him/her, Renren also recommends hundreds of site-wide popular videos which have been imported by any Renren users to all Renren users every day.
- Set $(B - A)$ includes each sharing action which imports a video from VSSes even though one or more of the friends have already shared this video before. Note that such actions are still important as Renren users may miss the news feeds from their friends due to the "visibility and divided attention" [10].

We investigate the average fraction of the above three sets in set $(A \cup B)$ for all videos, and *top videos*, respectively (shown in Fig. 3). The obvious difference between all videos (shown in Fig. 3a) and *top videos* (shown in Fig. 3b) is that while $(A \cap B)$ dominates in all videos, $(A - B)$ becomes the majority in *top videos*. This shows that by broadcasting a video to users who are not socially connected, the recommender system of Renren may enhance the diffusion of the video efficiently.

On the other hand, both Fig. 3a and Fig. 3b demonstrate that the difference of sets A and B, i.e., $(A - B)$ and $(B - A)$, is non-trivial. Hence, simply defining the initiator action to be actions in either set A or B would underestimate the existence of initiators of a video. In our following studies, we define the initiator actions to be $A \cup B$. This comprehensive definition is reasonable because no matter a user shares a video independently, or directly from VSSes, under both situations, he/she is activated by external influence [18], rather than by the diffusion over the friendship of the social network.

3.2   Characteristics of Sharing Actions with Different Position

Based on the positions of sharing actions in the corresponding propagation graphs as well as the definitions discussed earlier, we then classify our collected

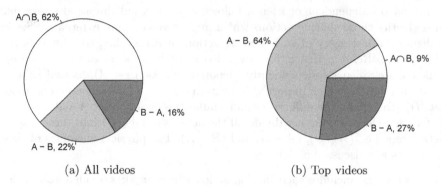

(a) All videos   (b) Top videos

Fig. 3. Breakdown of initiator actions for all videos and for top videos

sharing actions into three types: initiator actions, spreader actions and follower actions. While initiator actions are believed to play a key role in the diffusion process [13] and hence have been intensely investigated in existing empirical studies [20] [21] , hardly any work has been done to characterize spreader actions and follower actions in real traces. In the following, we will investigate the distribution of the elapsed time as well as the proportion of the size for all the three types of sharing actions. In particular, in order to understand which group of actions make these popular videos outstanding, we will compare the observation obtained from all videos and *top videos*.

The Elapsed Time for Different Sharing Actions. The elapsed time of sharing action (u, v, t, s) is defined to be the difference of the time t and the *birth time* of video v. Here, the *birth time* of video v is the time that video v is firstly shared by any user in our studied community. According to the definition of "influence" as well as the three types of sharing actions, we would expect that elapsed time of initiator actions, on average, will be smaller than the elapsed time of spreader actions, and the elapsed time of spreader action will be, on average, smaller than the elapsed time of follower. To verify this intuition, we will investigate the elapsed time distribution of initiator actions, spreader actions and follower actions for all videos and *top videos* respectively (as shown in Fig. 4). While the elapsed times of initiator actions of all videos (as shown in Fig. 4a) and *top videos* (as shown in Fig. 4b) are, on average, smaller than spreader actions and follower actions, which accords our intuition, there is no distinct difference between the CDF of elapsed time for all videos and *top videos*. In contrast, for *top videos*, the elapsed time of spreader actions is, on average, smaller than follower actions. Therefore, "spreader" propagation is an important factor to make a video popular.

The Fraction of Different Sharing Actions. Considering that the above studies on the distribution of elapsed time within each type of sharing actions cannot reveal the size of each type of sharing actions, here we further investigate

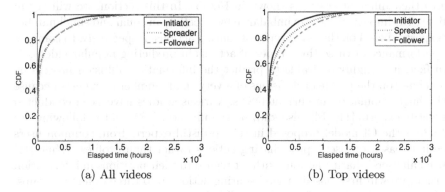

Fig. 4. CDF of elapsed time for initiator actions, spreader actions and follower actions

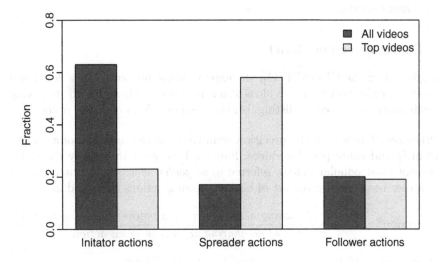

Fig. 5. Fraction of different sharing actions

the fraction of each type of sharing actions for all videos and *top videos*, respectively (as shown in Fig. 5). We observe that, while all videos are mainly driven by initiator actions, *top videos* are driven by spreader actions.

In summary, both the studies on elapsed time and on the fraction of sharing actions suggest that spreader actions play an essential role in making popular videos outstanding.

4   Predicting Popular Videos Based on Specified Experts

Although the above empirical studies demonstrate that spreader actions are important for the diffusion process of popular videos, it is difficult to make a rigorous comparison on the importance of the three different types of actions due to their different sizes (as shown in Fig. 5). In this section, we will try to mitigate the impact of the size imbalance by selecting an equal and small number of "experts" based on the three types of sharing actions, respectively, to compare the performance of these three types of actions in predicting popular videos.

In fact, it is highly desirable to predict the information diffusion process by only relying on the profile or decisions of a very small number of experts because of the huge population in current OSNs. Various studies have been conducted to identify experts [11] [14], also known as "influencers" [2]. In the following, we will adopt the CI model proposed in [11] to distill experts from common users since it is based on users' past sharing actions rather than profiles. While [11] utilized all users' sharing actions without regard to their positions in information diffusion, we will first classify users' sharing actions into three different groups, i.e., initiator actions, spreader actions and follower actions, and then select three different groups of experts accordingly, i.e., initiator experts, spreader experts and follower experts.

4.1   Expert Selection Model

Basically, under the CI model [11], the experts are supposed to be able to find and then share/import *popular* videos at an *early* stage. Therefore, the following two criteria are proposed to distinguish these experts from common users.

1) Precision: Here we use the precision term to measure the ability of a user to identify and share popular videos. Supposed we want to identify the top r percent most popular videos, referred to as *golden videos* here, the precision of a user based on a given set of his/her sharing actions is defined as

$$precision = \frac{number\ of\ sharing\ actions\ of\ golden\ videos\ in\ given\ action\ set}{number\ of\ all\ sharing\ actions\ in\ given\ action\ set}$$

(1)

2) Promptness: As is discussed above, besides the ability to make the right recommendation, i.e., find and share the popular videos, experts are also expected to make the decision promptly. Otherwise, common users could trick our selection criteria by sharing a lot of popular but outdated videos. Hence, we will set a promptness threshold, and only consider sharing actions with elapsed time smaller than the promptness threshold.

In our following experiments, we will use the *precision with promptness threshold (referred to as precision in the following for simplicity)* to measure a user's ability to find and share popular videos promptly. Based on this metric, we can then rank users' precisions, ordered from high to low, and then select the top k percent users as experts. However, such a methodology only focuses on the percentage of desirable video sharing actions, i.e., sharing actions of golden videos, while ignoring the actual number of desirable sharing actions. In other words, if a user u_1 shares one video promptly and one video only, and this only video happens to be a golden video, while another user u_2 shares 100 videos promptly and all 100 videos are golden videos, then user u_1 and user u_2 will be regarded to have the same ability to find and share popular videos promptly in our selection model. Obviously, this result is not reasonable because while user u_1 is likely to choose the single shared video by luck, we have more confidence to regard user u_2 as an expert. Therefore, instead of ranking user according to the precision proposed above, we need to further take the uncertainty into consideration. It is known that confidence level is an intuitive metric to measure the uncertainty of a value. In particular, considering that precision calculated in Equation (1) is binomial, we employ the adjusted Wald confidence interval [1] to measure the uncertainty of the precision. Then in practice, we will use the lower bound of the precision under the 95% confidence interval to rank users, and choose the top $k\%$ users with the highest lower bound to be experts. Next we will conduct experiments to evaluate the performance of selected experts on predicting popular videos at an early stage.

4.2   Experiments on Predicting Popular Videos

Note that given different sharing actions of a user, his/her precision as well as the lower bound will be different according to Equation (1). Thus, based on the three different types of users' video sharing actions discussed in Section 3, we can obtain three different sets of experts, i.e., initiator experts, spreader experts and follower experts, by selecting the top $k\%$ users with the highest lower bounds calculated from the corresponding set of sharing actions. Based on the large-scale video diffusion traces collected from Renren, we conduct experiments to evaluate and compare the performance of the three groups of experts in predicting popular videos. Next, we introduce the dataset used in our experiments.

Selection Dataset and Evaluation Dataset. Our selection of experts is based on all users' past video sharing actions, and we refer to these video sharing actions as *selection dataset*. Then we can evaluate the performance of selected experts in predicting popular videos based on another set of video sharing actions, referred to as evaluation dataset. Following previous work [11] and also considering that our studied users are most active (i.e., the number of sharing actions is largest) from March 1st to June 30th, 2011, we choose videos whose birth time is within these 4 months. In particular, the selection dataset contains sharing actions of videos with birth times between March 1st to April 30th, and the evaluation dataset contains sharing actions of videos with birth times between May 1st to June 30th.

Parameter Settings and Evaluation Method. Given a set of experts, either initiator experts, spreader experts or follower experts, we can then, based on the evaluation dataset, count how many times each video has been shared by these experts within the promptness window, and regard the counted times as the number of votes it has received from experts. Then we further rank all the videos in the evaluation dataset by the number of votes they have received in descending order, and choose the top n videos as the predicted popular videos. Based on the ordered videos as well as the real golden videos, we then employ the precision-recall curve [6] to evaluate the precision as well as recall score of our predicted popular videos under all possible values of n. In fact, the precision-recall curve has been widely used in the performance evaluation of information retrieval algorithms, especially for applications with highly skewed datasets, such as our dataset which has much less golden videos than common videos. In general, the optimization goal of the precision-recall curve is to be in the upper-right-hand corner, which means that in reality, a precision-recall which is closer to the upper-right-hand corner will have a better performance.

Next, we introduce the detailed experimental parameter settings. Following previous studies on predicting video [19] [22], we will try to predict the popular videos at the end of 30 days based on their sharing actions occurring within their first 7 days. This means that the promptness threshold is set to 7 days, and video popularity is set to be the popularity at 30 days. The golden videos are the top 3% because such videos account for nearly 90% sharing actions. As to the ratio of experts, we have tried the values of 0.01, 0.02, 0.05, and 0.1. It is observed that the following conclusions can hold on all these values. Here we will only present the results with the ratio of experts being set to 0.02 due to space limitations.

Performance Evaluation for Different Experts. Recall that after identifying the experts based on all users' video sharing actions in the selection dataset, our objective is to predict the popular videos only based on the sharing actions of selected experts since this can avoid the difficulty of acquiring all users' actions over a long period. As a result, we cannot classify the video sharing actions of selected experts into three types of actions due to the fact that we are not supposed to be aware of all users' sharing actions except for experts. Therefore, when we predict popular videos based on selected experts, no matter they are initiator experts, spreader experts or follower experts, we will make use of all of their video sharing actions. The performance evaluation results of the above three types experts are presented in Fig. 6, and for comparison, we also present the performance of the general experts [11] , which is selected without distinguishing users' sharing actions by their positions.

As mentioned earlier, a precision-recall which is closer to the upper-right-hand corner will have a better performance. While it is intuitive that a follower expert achieves the worst performance due to its position in the diffusion process, two unexpected and important conclusions can be drawn from the precision-recall curves of different experts' prediction results (as shown in in Fig. 6). Spreader experts outperform both the initiator experts and follower experts. This observation contradicts existing work [13] [20] [21] but accords with our earlier empirical observations in Section 3.2; 2) spreader experts even outperform the general experts proposed in existing work, which demonstrates the necessity of distinguishing sharing actions by their positions.

Motivated by the fact that spreader experts would outperform the general experts in existing work, we will further verify the long-lasting debate "Do experts

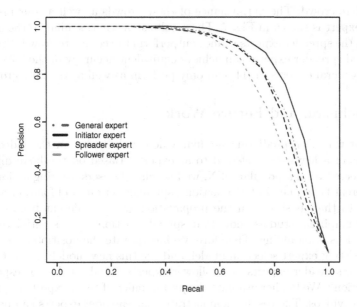

Fig. 6. Precision-recall curve of the prediction performance for different experts

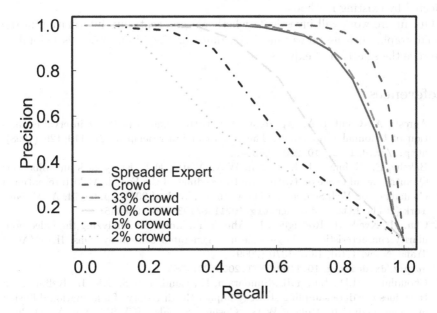

Fig. 7. Precision-recall curve for spreaders experts and crowds

make better decisions than crowds?" [9]. In our context, the "crowd" corresponds to all of the studied users, except the selected experts. We also conduct random sampling on crowds to construct smaller crowds with (2%, 5%, 10%, 33%) of the complete crowd. The performance of these crowds as well as our proposed spreader expert is shown in Fig. 7. The results show that, similar to the conclusion in [9], the spreader experts cannot outperform the complete crowd. However, our selected spreader experts can achieve equivalent accuracy of the 33% crowd while the general experts in [11] can only perform as well as the 20% crowd.

5   Conclusion and Future Work

This paper works on predicting popular videos only based on the video sharing actions of a few users, referred to as experts. Based on the huge diffusion traces collected from a popular OSN, we first classify users' sharing actions into three different types, i.e., initiator actions, spreader actions and follower actions, according to their positions in the propagation graph. Different from existing work, our empirical studies show that spreader actions are essential to make popular videos outstanding. Therefore, we incorporate the position information into the existing expert selection model, and use this new model to select initiator experts, spreader experts and follower experts based on the corresponding sharing actions. We further evaluate the performance of these experts in predicting popular videos. The results demonstrate that spreader experts can not only outperform initiator experts and follower experts, but also the general experts selected by existing methods.

Our future work will investigate whether users' centrality information in the social graph, such as betweenness centrality or PageRank score, is helpful to improve the selection of experts.

References

1. Agresti, A., Coull, B.A.: Approximate is better than "exact" for interval estimation of binomial proportions. The American Statistician 52(2), 119–126 (1998), http://dx.doi.org/10.2307/2685469
2. Bakshy, E., Hofman, J.M., Mason, W.A., Watts, D.J.: Everyone's an influencer: Quantifying influence on twitter. In: Proceedings of the Fourth ACM International Conference on Web Search and Data Mining, WSDM 2011, pp. 65–74. ACM, New York (2011), http://doi.acm.org/10.1145/1935826.1935845
3. Cha, M., Kwak, H., Rodriguez, P., Ahn, Y.Y., Moon, S.: Analyzing the video popularity characteristics of large-scale user generated content systems. IEEE/ACM Trans. Netw. 17(5), 1357–1370 (2009), http://dx.doi.org/10.1109/TNET.2008.2011358
4. Choudhury, M.D., Lin, Y.R., Sundaram, H., Candan, K.S., Xie, L., Kelliher, A.: How does the data sampling strategy impact the discovery of information diffusion in social media? In: Cohen, W.W., Gosling, S. (eds.) ICWSM. The AAAI Press (2010), http://dblp.uni-trier.de/db/conf/icwsm/icwsm2010.html#ChoudhuryLSCXK10
5. Christodoulou, G., Georgiou, C., Pallis, G.: The role of twitter in youtube videos diffusion. In: Wang, X.S., Cruz, I., Delis, A., Huang, G. (eds.) WISE 2012. LNCS, vol. 7651, pp. 426–439. Springer, Heidelberg (2012), http://dx.doi.org/10.1007/978-3-642-35063-4_31

6. Davis, J., Goadrich, M.: The relationship between precision-recall and roc curves. In: Proceedings of the 23rd International Conference on Machine Learning, ICML 2006, pp. 233–240. ACM, New York (2006), http://doi.acm.org/10.1145/1143844.1143874

7. Goyal, A., Bonchi, F., Lakshmanan, L.V.S.: A data-based approach to social influence maximization. Proc. VLDB Endow. 5(1), 73–84 (2011), http://dl.acm.org/citation.cfm?id=2047485.2047492

8. Hefeeda, M., Saleh, O.: Traffic modeling and proportional partial caching for peer-to-peer systems. IEEE/ACM Trans. Netw. 16(6), 1447–1460 (2008), http://dx.doi.org/10.1109/TNET.2008.918081

9. Hill, G.W.: Group versus individual performance: Are n+1 heads better than one? Psychological Bulletin 91(3), 517 (1982)

10. Hodas, N., Lerman, K.: How visibility and divided attention constrain social contagion. In: Privacy, Security, Risk and Trust (PASSAT), 2012 International Conference on and 2012 International Confernece on Social Computing (SocialCom), pp. 249–257 (2012)

11. Hsieh, C.C., Moghbel, C., Fang, J., Choo, J.: Experts vs the crowd: Examining popular news prediction performance on twitter. In: Proceedings of the 22Nd International Conference on World Wide Web, WWW 2013, International World Wide Web Conferences Steering Committee, Republic and Canton of Geneva, Switzerland (2013)

12. Jiang, J., Wilson, C., Wang, X., Huang, P., Sha, W., Dai, Y., Zhao, B.Y.: Understanding latent interactions in online social networks. In: Proceedings of the 10th ACM SIGCOMM Conference on Internet Measurement, IMC 2010, pp. 369–382. ACM, New York (2010), http://doi.acm.org/10.1145/1879141.1879190

13. Kempe, D., Kleinberg, J., Tardos, E.: Maximizing the spread of influence through a social network. In: Proceedings of the ninth ACM SIGKDD International Conference on Knowledge Discovery and Data Mining, KDD 2003, pp. 137–146. ACM, New York (2003), http://doi.acm.org/10.1145/956750.956769

14. Kittur, A., Pendleton, B.A., Suh, B., Mytkowicz, T.: Power of the few vs. wisdom of the crowd: Wikipedia and the rise of the bourgeoisie. In: Proceedings of the 25th Annual ACM Conference on Human Factors in Computing Systems, CHI 2007. ACM (April 2007), http://www.viktoria.se/altchi/submissions/submission_edchi_1.pdf

15. Lai, K., Wang, D.: Understanding the external links of video sharing sites: Measurement and analysis. IEEE Transactions on Multimedia 15(1), 224–235 (2013)

16. Li, H., Liu, J., Xu, K., Wen, S.: Understanding video propagation in online social networks. In: Proceedings of the 2012 IEEE 20th International Workshop on Quality of Service, IWQoS 2012, pp. 21:1–21:9. IEEE Press, Piscataway (2012), http://dl.acm.org/citation.cfm?id=2330748.2330769

17. Li, H., Ma, X., Wang, F., Liu, J., Xu, K.: On popularity prediction of videos shared in online social networks. In: Proceedings of the 22Nd ACM International Conference on Conference on Information and Knowledge Management, CIKM 2013, pp. 169–178. ACM, New York (2013), http://doi.acm.org/10.1145/2505515.2505523

18. Myers, S.A., Zhu, C., Leskovec, J.: Information diffusion and external influence in networks. In: Proceedings of the 18th ACM SIGKDD International Conference on Knowledge Discovery and Data Mining, KDD 2012, pp. 33–41. ACM, New York (2012), http://doi.acm.org/10.1145/2339530.2339540

19. Pinto, H., Almeida, J.M., Gonçalves, M.A.: Using early view patterns to predict the popularity of youtube videos. In: Proceedings of the Sixth ACM International Conference on Web Search and Data Mining, WSDM 2013, pp. 365–374. ACM, New York (2013), http://doi.acm.org/10.1145/2433396.2433443

20. Rodrigues, T., Benevenuto, F., Cha, M., Gummadi, K., Almeida, V.: On word-of-mouth based discovery of the web. In: Proceedings of the 2011 ACM SIGCOMM Conference on Internet Measurement Conference, IMC 2011, pp. 381–396. ACM, New York (2011), http://doi.acm.org/10.1145/2068816.2068852

21. Sun, E., Rosenn, I., Marlow, C., Lento, T.: Gesundheit! modeling contagion through facebook news feed. In: International AAAI Conference on Weblogs and Social Media (2009), http://www.stanford.edu/~{}esun/ICWSM09_ESun.pdf

22. Szabo, G., Huberman, B.A.: Predicting the popularity of online content. Commun. ACM 53(8), 80–88 (2010), http://doi.acm.org/10.1145/1787234.1787254

23. Wang, D., Wen, Z., Tong, H., Lin, C.Y., Song, C., Barabási, A.L.: Information spreading in context. In: Proceedings of the 20th International Conference on World Wide Web, WWW 2011, pp. 735–744. ACM, New York (2011), http://doi.acm.org/10.1145/1963405.1963508

24. Wang, Z., Sun, L., Chen, X., Zhu, W., Liu, J., Chen, M., Yang, S.: Propagation-based social-aware replication for social video contents. In: Proceedings of the 20th ACM International Conference on Multimedia, MM 2012, pp. 29–38. ACM, New York (2012), http://doi.acm.org/10.1145/2393347.2393359

25. Wang, Z., Sun, L., Zhu, W., Yang, S., Li, H., Wu, D.: Joint social and content recommendation for user-generated videos in online social network. IEEE Transactions on Multimedia 15(3), 698–709 (2013)

An Evolution-Based Robust Social Influence Evaluation Method in Online Social Networks

Feng Zhu[1], Guanfeng Liu[1], An Liu[1], Lei Zhao[1], and Xiaofang Zhou[1,2]

[1] School of Computer Science and Technology,
Jiangsu Provincial Key Laboratory for Computer Information Processing Technology,
Soochow University, Suzhou, China, 215006
[2] School of Information Technology and Electrical Engineering,
University of Queensland, Brisbane, Australia, 4072
{gfliu,zhaol,zxf}@suda.edu.cn

Abstract. Online Social Networks (OSNs) are becoming popular and attracting lots of participants. In OSN based e-commerce platforms, a buyer's review of a product is one of the most important factors for other buyers' decision makings. A buyer who provides high quality reviews thus has strong social influence, and can impact a large number of participants' purchase behaviours in OSNs. However, the dishonest participants can cheat the existing social influence evaluation models by using some typical attacks, like *Constant* and *Camouflage*, to obtain fake strong social influence. Therefore, it is significant to accurately evaluate such social influence to recommend the participants who have strong social influences and provide high quality product reviews. In this paper, we propose an Evolutionary-Based Robust Social Influence (EB-RSI) method based on the trust evolutionary models. In our EB-RSI, we propose four influence impact factors in social influence evaluation, i.e., Total Trustworthiness (TT), Fluctuant Trend of Being Advisor (FTBA), Fluctuant Trend of Trustworthiness (FTT) and Trustworthiness Area (TA). They are all significant in the influence evaluation. We conduct experiments onto a real social network dataset Epinions, and validate the effectiveness and robustness of our EB-RSI by comparing with state-of-the-art method, SoCap. The experimental results demonstrate that our EB-RSI can more accurately evaluate participants' social influence than SoCap.

Keywords: Social influence, trust, influence evaluation, social network.

1 Introduction

1.1 Background

In recent years, Online Social Networks (OSNs), like Facebook and Twitter have attracted lots of participants, where they can make new friends and share their experience. In a social network based e-commence platform, like Epinions (epinions.com), each participant can be a buyer or a seller. After a transaction, a buyer can write a product review and give 1 to 5 stars as the ratings for different aspects of the product, such as *Ease of Use*, *Customer Service*, *On-Time*

B. Benatallah et al. (Eds.): WISE 2014, Part II, LNCS 8787, pp. 141–157, 2014.

Delivery, etc. Then, when other buyers want to buy the same product from the same seller, they can view that product review and make the decision based on the review. Based on their own transaction experiences, these buyers can rate the reviews as *Not Helpful, Somewhat Helpful, Helpful,* and *Very Helpful* [14]. If a buyer usually provides *Very Helpful* product reviews, he/she can be trusted by other buyers. As indicated in *Social Psychology* [4,10,24] and *Computer Science* [3], a buyer prefers the recommendation from his/her trusted buyers over those from others. It means that a buyer is more likely to make a purchase decision based on the product reviews given by the trustworthy buyers. Then these trustworthy buyers have strong social influence that can affect others purchase behaviours in OSNs. These buyers are called *advisors* of those participants who trust their product reviews.

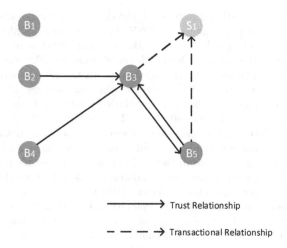

Fig. 1. A social network based e-commence platform

Example 1: Fig.1 depicts a social network based e-commerce platform which contains a seller (i.e., S_1) and five buyers (i.e., B_1 to B_5). B_1 is a new registered buyer, so there is no trust relationship between B_1 and any other buyers. B_3 is an advisor of B_2, B_4 and B_5, so there exist three trust relationships (displayed as a *solid line* in Fig.1) between B_3 and these three buyers. B_3 and B_5 have transactions with S_1, so there exist two transactional relationships (displayed as a *dashed line* in Fig.1) between S_1 and the two buyers. If B_2 wants to buy the product from S_1 and B_3 provides the reviews of that product, then B_2 will make the decision based on B_3's reviews.

1.2   The Problem

Social network based e-commerce platform is in an open environment, where anybody can sign up to become a buyer and write reviews. Moreover, anybody can give the *Helpfulness* of reviews making this product review scheme highly vulnerable to be attacked [14]. A dishonest advisor can cheat the product review system and obtain strong social influence via some typic attacks, like *Constant*[1], *Camouflage*[2], *Whitewashing*[3] and *Sybil*[4] [13]. These dishonest advisors who have strong social influence can harm the benefits of both buyers and sellers. Therefore, it is necessary and significant to develop a robust social influence evaluation method which can accurately evaluate the social influence of participants and defend against these typical attacks.

Example 2: Here we take *Camouflage* attack as a example. Recall Fig.1, suppose B_5 is a dishonest advisor. At first, B_5 provides high quality reviews to accumulate strong social influence. Then, B_3 conspires with S_1 and writes a very good review for the products that are sold by S_1 and have low quality. Then B_2 and B_4 will be cheated to buy the low quality products from S_1.

In the literature, various social influence methods [2,5,8,9,23,25] have been proposed to compute the value of participants' social influences in OSNs, these methods study influence maximization under the popular independent cascade (IC) model [16] and evaluate social influence through the process of information diffusion [18]. These methods mainly focus on the current network status and ignore the trend of participants' historical influence. However, as indicated in [20], an accurate social influence evaluation method requires more influence information that not only the current influence level, but also the influence prediction relevant to forthcoming transactions. Thus, although a participant computed by the existing methods with strong social influence in the current OSN, if the trend of his/her influence is downward, the influence of that participant is more likely to decrease in the near future. In addition, the existing methods do not consider the occurrence of the above mentioned typical attacks from dishonest participants. Therefore, they cannot defend against these attacks to deliver accurate social influence evaluation results.

1.3 Contributions

In order to deliver accurate social influence evaluation results, in this paper, we first propose four influence impact factors, i.e., Total Trustworthiness (TT), Fluctuant Trend of Being Advisor (FTBA), Fluctuant Trend of Trustworthiness (FTT) and Trustworthiness Area (TA). These four factors are significant in social influence evaluation. We then propose a trust evolutionary model based on the Multiagent Evolutionary Trust (MET) model [13], and propose a novel Evolutionary-Based Robust Social Influence (EB-RSI) method based on the trust

---

[1] Dishonest advisors constantly provide unfairly positive/negative ratings to sellers.

[2] Dishonest advisors camouflage themselves as honest advisors by providing fair ratings to build up their trustworthiness first and then gives unfair ratings.

[3] A dishonest advisor is able to whitewash its low trustworthiness by starting a new account with the initial trustworthiness value.

[4] A dishonest buyer creates several accounts to constantly provide unfair ratings to sellers.

evolutionary models and our proposed four impact factors. We conduct experiments onto a real social network dataset, Epinions (epinions.com), and compare our EB-RSI method with the state-of-the-art social influence method, called SoCap [23]. The experimental results illustrate that our EB-RSI method can defend against the typical attacks and deliver more accurate social influence evaluation results than SoCap.

This paper is organised as follows. *Section 2* discusses the related work. *Section 3* introduces the preliminaries. *Section 4* proposes the four influence impact factors. *Section 5* proposes our EB-RSI method. In *Section 6*, we verify the effectiveness and robustness of our EB-RSI by comparing with the state-of-the-art method, SoCap. *Section 7* is the conclusions.

2   Related Work

In the literature, many social influence evaluation approaches [2, 5–7, 11, 15–17, 19, 23] have been proposed to compute the value of participants' social influence in OSNs. Most of these approaches [5–7, 15–17, 19] attempt to model social influence through the process of information diffusion [18]. In addition, the problem of finding influencers is often studied as an influence maximization problem which is to find the *Top-K* (K seed) nodes such that the value of influence is maximized. Kempe et al. [16] consider that the problem of finding a subset of influential nodes is an absolute optimization problem and indicate that influence maximization problem is NP-hard. They propose a greedy algorithm which guarantees $(1 - 1/e)$ approximation ratio. However, this algorithm has low efficiency in practice and thus it is not scalable with the network size. So, the followers devote themselves to renovate the algorithm to spend up the process of computing the influence value or improve the influence propagation model to adapt to the network proliferation. In order to improve the scalability, Chen et al. [6] propose an algorithm, which has a simple tunable parameter, for users to control the balance between the running time and the influence spread of the algorithm. Nevertheless, a single influence evaluation itself is #P-hard, which is also hard to be solved in polynomial time. Jung et al. [15] propose a novel algorithm IRIE that integrates the advantages of influence ranking (IR) and influence estimation (IE) methods for influence maximization. Kim et al. [17] provide a scalable influence approximation algorithm, Independent Path Algorithm (IPA) for IC model. In the model, they study IPA efficiently approximates influence by considering an independent influence path as an influence evaluation unit and it is also easily parallelized by adding a few lines expressions. Furthermore, in order to spend up the evaluation algorithm, Leskovec et al. [19] develop the CELF algorithm, which exploits submodularity to find near-optimal influencer selections, namely the obtained solutions are guaranteed to achieve at least a fraction of $\frac{1}{2}(1 - 1/e)$ of the optimal solution. However, the above methods do not consider the historical data of influence, so they cannot provide influence trend prediction about the forthcoming transactions. In addition, social network is in an open environment, there might be some unreliable reviews given by dishonest participants. But the above methods do not adopt any strategies to identify those unreliable reviews and dishonest participants.

In addition, since the above methods capture only the process of information diffusion and not the actual social value of collaborations in the network. Subbian et al. [23] propose an approach, called SoCap, to find influencers in OSN by using the social capital value. They model the problem of finding influencers in OSN as a value-allocation problem, where the allocated value denotes the individual social capital. However, as ignoring the trend of influence, this method always cannot find high quality influencers who can keep their influences for a long term. In addition, this method is also vulnerable to the unfair rating attacks from dishonest participants. For the problem of unfair rating attacks, there are some models [1, 22] have been proposed to detect the fraudsters and fake reviews in OSNs. These models mainly focus on finding the fraudsters rather than defending against the attacks. Moreover, they did not consider the social influence evaluation methods by fraud detecting.

Furthermore, Yeung et al. [2] have studied the relations between trust and product ratings in online consumer review sites. And they propose a method to estimate the strengths of trust relations so as to estimate the true influence among the trusted users. This kind of strength cannot reflects the whole social influence of this user, because it represents just the local influence among the trusted users and excludes the whole influence effected in all users. In addition, Franks et al. [11] propose a method to identify influential agents in open Multi-Agent systems without centralised control and individuals have equal authority. They find out four single metrics are robustly indicative of influence, and they study which single metric or combined measure is the best predictor of influence in a given network. However, these single metrics need a massive computing time, and extra computing will be used to find out the best predictor of influence. So, this method is also not scalable with the network size.

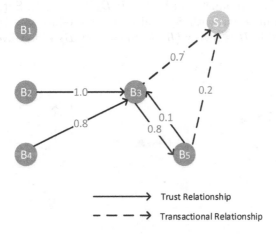

Fig. 2. A trust network with ratings

3   Preliminary

3.1   Social Network

In this paper, a social network is modeled as a directed graph $G = (V, E)$, where V is the set of vertices and E is the set of edges. Each vertex in V represents a buyer or a seller, and each edge in E represents a trust relationship between buyers or a transaction relationship between a buyer and a seller.

3.2   Trust Relationship and Transaction Relationship

We use T and R to denote trust relationships and transaction relationships respectively, where $T(B_i, B_j) \in [0, 1]$ represents buyer B_i's trust value towards advisor B_j, and $R(B_i, S_j) \in [0, 1]$ represents a rating value provided by buyer B_i to seller S_j. In addition, if B_i has no experience with S_j, it is usually set the missing rating value $R(B_i, S_j)$ to 0.5 as a neutral value [13].

Example 3: Fig.2 depicts an OSN which contains four trust relationships, they are $T(B_2, B_3) = 1.0$, $T(B_4, B_3) = 0.8$, $T(B_5, B_3) = 0.1$ and $T(B_3, B_5) = 0.8$, and two transaction relationships, they are $R(B_3, S_1) = 0.7$ and $R(B_5, S_1) = 0.2$.

3.3   Evolutionary Trust Model

The Evolutionary Trust Model [13] is usually used to cope with possible unfair attacks from dishonest advisors. If a buyer has some dissimilar advisors, whose reviews are very different with the buyer's purchase experience. The buyer can evolve its trust relationships to absorb the advisors whose reviews are match the buyer's purchase experience and remove the dissimilar advisors. The evolutionary process have been detailed discussed in [13].

Example 4: In Fig.3, there is a trust relationship $T(B_2, B_3)$, and B_3 provides a good review of the product sold by S_1 ($R(B_3, S_1) = 0.7$). Suppose B_2 experienced a new transaction with S_1, and find the product sold by S_1 is not so good as described in B_3's review ($R(B_2, S_1) = 0.2$), B_2 will evolve his/her trust

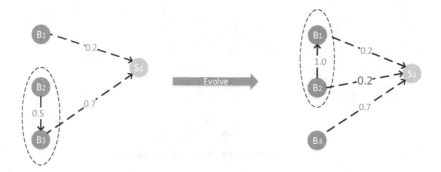

Fig. 3. Evolutionary process

relationships. As B_1 writes the review that contents match the experience of B_2, B_2 builds a new trust relationship with B_1 (i.e., $T(B_2, B_1) = 1.0$) and remove the trust relationship with B_3. Namely B_1 is added as a new advisor into the advisor team of B_2.

4 Influence Impact Factors

As indicated in [20], it is significant to take the current influence level and influence prediction into social influence evaluation. In our method, we propose four important influence impact factors, which are significant in delivering accurate social influence evaluation results.

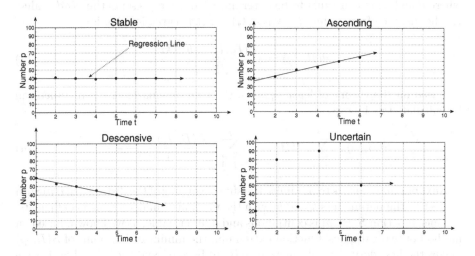

Fig. 4. Typical cases of FTBA

4.1 Total Trustworthiness (TT)

It shows the total trust value of an advisor given by other buyers. We use $TT(B_i, t_j)$ to denote B_i's total trust value at time t_j, and $TT(B_i, t_j)$ is calculated as

$$TT(B_i, t_j) = \sum_{k=1}^{p} T(B_k, B_i) \qquad (1)$$

where p is the number of B_i's in-degree.

4.2 Fluctuant Trend of Being Advisor (FTBA)

FTBA is used to illustrate the fluctuant trend of being advisor in a certain period. Some typical cases of FTBA, that are "stable", "ascending", "descensive" and "uncertain", are depicted in Fig. 4, where X axis is time and the Y axis is the number of being advisor. We use a regression line to model FTBA, and the regression line's gradient (denoted as $grad_1$) and mean distance (denoted as md_1) can measure FTBA well [20]. The regression line is based on the least squares fit.

We denote the number of B_i's in-degree at time t_j as $N_i(B_i, t_j)$, and use $(t_s, N_i(B_i, t_s)), (t_2, N_i(B_i, t_2)), (t_3, N_i(B_i, t_3)), ..., (t_e, N_i(B_i, t_e))$ denote the given data points of B_i's in-degree number from t_s to t_e. Here and in the following of this paper, t_s is the start time of the historical transactions and t_e is the end time of that transactions. Then the regression line can be represented as follows:

$$y = kt + b \qquad (2)$$

where k and b are constants to be determined, and k represents the $grad_1$ value. As the distance from point $(t_j, N_i(B_i, t_j))$ to the regression line is

$$d(B_i, t_j) = \frac{N_i(B_i, t_j) - b - kt_j}{\sqrt{1 + k^2}}. \qquad (3)$$

Based on the theory of least squares, the sum of squares of the distance can be calculated as follows:

$$S(B_i, p) = \sum_{j=1}^{p} d^2(B_i, t_j) = \sum_{j=1}^{p} \frac{(N_i(B_i, t_j) - b - kt_j)^2}{1 + k^2}. \qquad (4)$$

Next our main task is to minimise the sum of squares of the distance $S(i, p)$ with respect to the parameters k and b, with the method of undetermined coefficients.

Since function $S(B_i, p)$ is continuous and differentiable, as we known, based on method of two variables' function extremum, the minimization point of $S(B_i, p)$ makes the first derivative of function $S(B_i, p)$ be zero, and the second derivative positive, which could be easily proved by Taylor formula for function of two variables [21]. For this, we differentiate $S(B_i, p)$ with respect to k and b, and set the results to zero, then we can get following results:

$$k = grad_1(B_i) = (-u - \sqrt{u^2 + 4})/2. \qquad (5)$$

and

$$b = \frac{S_f - kS_t}{n}, \qquad (6)$$

where $S_{f2} = \sum_{j=1}^{p} N_i^2(B_i, t_j)$, $S_f = \sum_{j=1}^{p} N_i(B_i, t_j)$, $S_t = \sum_{j=1}^{p} t_j$, $S_{t2} = \sum_{j=1}^{p} t_j^2$ and $S_{ft} = \sum_{j=1}^{p} N_i(B_i, t_j)t_j$. To indicate the results clearly, we define $u = \frac{pS_{f2} - S_f^2 + S_t^2 - pS_{t2}}{S_f S_t - pS_{ft}}$.

By now, we have worked out the $grad_1$ value (i.e. k). According to above results, the equation of mean distance as follows:

$$md_1(B_i) = \frac{\sum_{j=1}^{p} |N_i(B_i, t_j) - b - kt_j|}{p\sqrt{1 + k^2}}. \qquad (7)$$

4.3 Fluctuant Trend of Total Trustworthiness (FTT)

FTT is used to illustrate the fluctuant trend of total trustworthiness from t_s to t_e. FTT is quite similar with FTBA, we also use a regression line to indicate the FTT, and the regression line is based on the least squares fit. The gradient and mean distance of regression line are denoted as $grad_2$ and md_2 respectively. Let $((t_s, TT(B_i, t_s)), (t_2, TT(B_i, t_2)), (t_3, TT(B_i, t_3)), ..., (t_e, TT(B_i, t_e)))$ denote the given data points of B_i's total trust value. So, we can obtain following equations by above FTBA' methods:

$$k' = grad_2(B_i) = (-u' - \sqrt{u'^2 + 4})/2 \tag{8}$$

$$b' = \frac{S_{Tt} - k'S_t}{n}, \tag{9}$$

where $u' = \frac{pS_{Tt2} - S_{Tt}^2 + S_t^2 - pS_{t2}}{S_{Tt}S_t - pS_{Ttt}}$, $S_{Tt2} = \sum_{j=1}^{p} TT^2(B_i, t_j)$, $S_{Tt} = \sum_{j=1}^{p} TT(B_i, t_j)$, $S_t = \sum_{j=1}^{p} t_j$, $S_{t2} = \sum_{j=1}^{p} t_j^2$ and $S_{Ttt} = \sum_{j=1}^{p} TT(B_i, t_j)t_j$.

$$md_2(B_i) = \frac{\sum_{j=1}^{p} |TT(B_i, t_j) - b' - k't_j|}{p\sqrt{1 + k'^2}}. \tag{10}$$

4.4 Trustworthiness Area (TA)

TA is a measurement method to calculate the quality of trustworthiness from t_s to t_e. We use the trustworthiness areas to represent a buyer's trusted level. In this way, the quality of buyer's trustworthiness can be expressed visibly and measurably. The trustworthiness areas are categorized into the positive area and the negative area by a division line (trust value is 0.5), and we depict some sample areas in Fig.5, where X axis is the times of being advisor and the Y axis

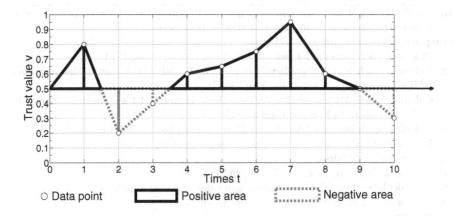

Fig. 5. Trustworthiness area in social network based e-commence platform

is the trust value given by other buyer based on a transaction. Finally, we define TA as:

$$TA(B_i, t_s, t_e) = PA(B_i, t_s, t_e) - NA(B_i, t_s, t_e) \qquad (11)$$

where $PA(B_i, t_s, t_e)$ is positive area value and $NA(B_i, t_s, t_e)$ is negative area value from t_s to t_e.

5 The Evalution-Based Robust Social Influence Method

In this section, we propose an Evalution-Based Robust Social Influence (EB-RSI) method in OSNs. The pseudo-code summaries the process of our EB-RSI method which is given in Algorithm 1. In our method, we adopt the above mentioned four influence impact factors that contain six parameters, they are TT, FTBA's gradient, FTBA's mean distance, FTT's gradient, FTT's mean distance and TA. By means of setting up different weights for these parameters, different inclined results will be generated. The weight set can be provided by individual buyers or domain experts. We normalized the six parameters to help prevent parameters with initially large ranges (e.g., TT and TA) from outweighing parameters with initially smaller ranges (e.g., FTBA' gradient and FTT' gradient). We use the popular normalization methods, Z-Score and Min-max [12]. The details of our method are indicated as follows:

Let $P_i = \{P(B_i, k) | k = 1, ..., 6\}$ be a B_i's parameter set, it represents TT, FTBA's gradient, FTBA's mean distance, FTT's gradient, FTT's mean distance and TA respectively (Line 1 to 4 in Algorithm 1 is to calculate these six parameter values of each buyer). And we use $W = \{W(B_i, k) | k = 1, ..., 6\}$ to indicate corresponding weights of such parameters. Thus, the *Z-score* and *Min-max* methods are calculated as follows:

Algorithm 1. EB-RSI Algorithm

Input: Buyer set B, all trust relationships T, all transaction relationship R, the start time of transactions t_s and the end time of transactions t_e;

Output: All buyers' influence value set M;

1: Calculate each buyer's TT value $P(B_i, 1)$ at time t_e using Eq.1;
2: Calculate each FTBA's gradient value $P(B_i, 2)$ and FTBA's mean distance value $P(B_i, 3)$ from t_s to t_e using Eq.5 and Eq.7;
3: Calculate each FTT's gradient value $P(B_i, 4)$ and FTT's mean distance value $P(B_i, 5)$ from t_s to t_e using Eq.8 and Eq.10;
4: Calculate each buyer's TA value $P(B_i, 6)$ from t_s to t_e using Eq.11;
5: Integrate above six parameters into parameter set P;
6: **for** *each P_i in P* **do**
7:     Normalize P_i to Z_i using *Z-score* method;
8:     Calculate buyer B_i's social influence value V_i using Eq.15;
9:     Normalize V_i to M_i using *Min-max* method;
10: **end for**
11: Return M;

Z-score method is used to normalize the six parameters (Line 7 in Algorithm 1), the range is $[-1.0, 1.0]$, it is calculated as:

$$Z(B_i, k) = \frac{P(B_i, k) - M_k}{S_k}, \tag{12}$$

where M_k is the mean value of P_k ($P_k = \{P(B_i, k)|i = 1, ..., n\}$, where n is the number of buyers) and

$$S_k = \frac{\sum_{k=1}^{6} |P(B_i, k) - M_k|}{6}. \tag{13}$$

Min-max method is used to normalize the social influence value (Line 9 in Algorithm 1), the range is $[0.0, 1.0]$, it is calculated as:

$$M_i = \frac{V_i - V_{min}}{V_{max} - V_{min}}, \tag{14}$$

where V_i is B_i' social influence value (Line 8 in Algorithm 1), and it is calculated as

$$V_i = \sum_{k=1}^{6} W(B_i, k) \cdot Z(B_i, k), \tag{15}$$

V_{min} and V_{max} are that minimum and maximum social influence values, respectively, for the given OSN.

6   Experiments

In this section, we compare our proposed EB-RSI method with SoCap method and conduct experiments based on the following two aspects: (1) In order to investigate the effectiveness of our EB-RSI method, we analyse the *Influence Ranking Trend* (*IRT*), that is the ranking of social influences. This trend can be used to illustrate the stability of influence evaluated by the two methods. (2) In order to investigate the robustness of our EB-RSI, we compare the performances of EB-RSI and SoCap in social influence evaluation when facing some typical attacks.

6.1   Experimental Setting

Experimental Datasets: In our experiments, we adopt a real social network dataset, Epinions (epinions.com), where each node represents a buyer or a seller, and each link corresponds to a trust relationship between a buyer and his/her advisor. Our work focus on studying the effectiveness and robustness of social influence evaluation methods. In order to clearly observe the computation process of participants' social influence, we extract a sub-network that has 362 nodes (200 buyers and 162 sellers) and 5453 links (5055 trust relationships and 398 transaction relationships). The extracting method selects 200 buyers and their corresponding sellers and relationships from original dataset randomly.

Table 1. Four weight sets

Parameter Names	Weight-set-1	Weight-set-2	Weight-set-3	Weight-set-4
TT	0.4	0.2	0.1	0.2
FTBA's gradient	0.1	0.15	0.2	0.1
FTBA's average distance	0.1	0.15	0.2	0.1
FTT's gradient	0.1	0.15	0.2	0.1
FTT's average distance	0.1	0.15	0.2	0.1
TA	0.2	0.2	0.1	0.4

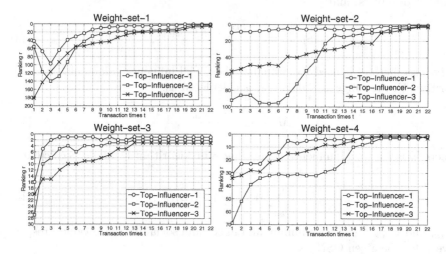

Fig. 6. The *Influence Ranking Trend* delivered by our EB-RSI

Parameters Setting: As introduced in *Section 5*, a buyer could specify different scales of weights for social influence evaluation. In our experiments, we set 4 groups of weights which have different combinations listed in Table 1. To investigate the effects of TT and TA to the influence evaluation, we set TT to 0.4 in Weight-set-1 and TA to 0.4 in Weight-set-4. All the impact factors in the cases of Weight-set-2 and Weight-set-3 have quite similar values. We investigate the *Top-3* influence values after a certain number of transactions with 4 groups of weights, and use "Top-Influencer-1", "Top-Influencer-2" and "Top-Influencer-3" to denote the *Top-3* influencers respectively.

Experimental Environments: All experiments were run on a machine powered by two Intel(R) Core(TM) i5-3470 CPU 3.20 GHz processors with 8GB of memory, using Windows 7. The code was implemented using Java 8 and the experimental data was managed by MySQL Server 5.6.

6.2   Experimental Results

Exp-1: Effectiveness: In order to investigate the effectiveness of our proposed EB-RSI method, we observe the *IRT*s of the *Top-3* influencers in a certain number of transactions. The experimental results are depicted in Fig.6 and Fig.7,

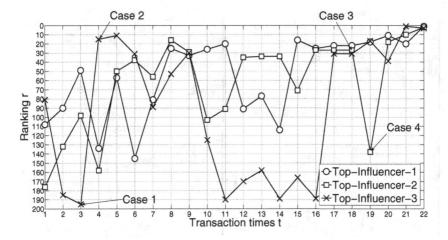

Fig. 7. The *Influence Ranking Trend* delivered by SoCap

where the X axis is transactions times and the Y axis is the influence ranking of buyers. The three curves in each graph represent the *IRT*s of the participants who have the *Top-3* influences after 22 transactions. From Fig.6, the *IRT* of each graph becomes stable after 22 transactions with different weights. From Fig.7, we can find that the *IRT*s of these influencers are fluctuant and unstable. For example, in *Case 1* and *Case 2*, although *Top-Influencer-3* has only one transaction, the influence ranking of *Top-Influencer-3* increases from 195 to 15 (see Table 2). Moreover, in *Case 3* and *Case 4*, *Top-Influencer-2* also has only one transaction, but the influence ranking of *Top-Influencer-2* decreases from 27 to 138 (see Table 2).

Table 2. The data of the cases in Fig.7

ID	Transaction Times	Buyer ID	EB-RSI Value	SoCap Value	EB-RSI Ranking	SoCap Ranking
Case 1	3	Top-Influencer-3	0.548	1.515	58	195
Case 2	4	Top-Influencer-3	0.567	1549.334	85	15
Case 3	18	Top-Influencer-2	0.629	51.991	50	27
Case 4	19	Top-Influencer-2	0.624	1.744	62	138

Analysis: This experimental result illustrates that the influencers identified by SoCap method are unstable with the increase of transactions. Because the SoCap method is complete based on current static OSN and ignore the influence trends. As a result, it cannot take the results of prediction into the social influence evaluation. Thus, SoCap may recommend a low quality influencer with weak influence to other buyers in the near future, e.g., the above mentioned *Top-Influencer-2* in *Case 3* and *Case 4* in Fig. 7. By contrast, our method can recommend the influencers who have stable influence and make other buyers obtain reliable product reviews. Therefore, our EB-RSI outperforms SoCap in the effectiveness.

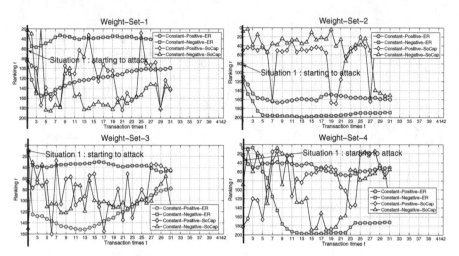

Fig. 8. The *IRT*s of buyers (Constant Attack)

Exp-2: Robustness: This experiment is to investigate if our method can defend against the typical attacks and recommend the honest *Top-K* influencers. As introduced in *Section 1*, it is necessary and significant to maintain the robustness of the social influence evaluation methods. We have introduced four typical attacks in *Section 1*, e.g., Constant[1], Camouflage[2], Whitewashing[3] and Sybil[4]. Since a new participant has only a few trust relationships and transitional relationships, and this new participant has very weak influence computed by our EB-RSI method, we skip the tests of Whitewashing attack and Sybil attack (these two attacks are all based on the new participant). The performances of the two methods when facing Constant attack and Camouflage attack are described as follows:

- *Constant*: We first select two buyers randomly and reset the rating values of all sellers given by the two buyers as 0 and 1 respectively, and give them 2 tags "Constant-Negative" and "Constant-Positive". After 31 transactions, the experimental results of *IRT*s are depicted in Fig.8, where we can see that the two *IRT*s by EB-RSI method (i.e., "Constant-Positive-ER" and "Constant-Negative-ER") decreases dramatically after starting transactions. This kind of downtrend contrast with the *IRT*s by SoCap method (i.e., "Constant-Positive-SoCap" and "Constant-Negative-SoCap"). 0 and 1 are two extreme values in the range of rating, which always reflects nonobjective or dishonest opinions. Therefore, many buyers who have built trust relationships with the two buyers (i.e., "Constant-Positive-ER" and "Constant-Negative-ER") either removed the relationships or reduced the trust values. Then a robust social influence evaluation method should reduce the influences of such two types of buyers. Thus, our proposed EB-RSI method is more robust than SoCap method under the Constant attack.
- *Camouflage*: After 23 transactions, we reset the rating values of all sellers given by the *Top-2* influencers as 0 and 1 respectively, and give them 2

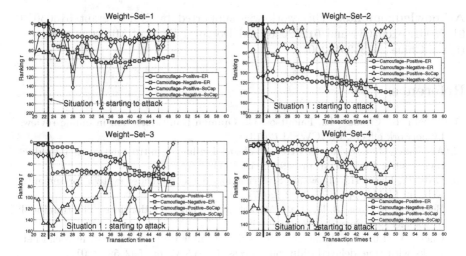

Fig. 9. The *IRT*s of buyers (Camouflage Attack)

tags, i.e., "Camouflage-Negative" and "Camouflage-Positive". The next 26
transactions (from 23 to 49) , The *IRT*s of *Top-2* influencers are depicted
in Fig.9. From the figure, we can see that the two buyers' *Influence Ranking* (i.e., "Camouflage-Positive-ER" and "Camouflage-Negative-ER") by our
EB-RSI method decreases dramatically after starting the attack (see Situation 1 in Fig.9). At the same time, the *Influence Ranking* of the two buyers
(i.e., "Camouflage-Positive-SoCap" and "Camouflage-Negative-SoCap") by
SoCap method are still unstable. As described in above *Constant* attack,
the buyers who provide non-objective or dishonest reviews should be rep-
rimanded (i.e., reduce the attackers' social influence). Therefore, whether
reduce the influences of dishonest attackers becomes a key indicator of the
robustness of influence evaluation method. Therefore, our EB-RSI method
outperforms SoCap method in robustness.

7   Conclusion and Future Work

In this paper, we have proposed a novel EB-RSI method to accurately evaluate
the social influence of participants in OSNs. In our model, we have taken four
influence impact factors into consideration. In addition, to defend against the
typical attacks from dishonest participants, we have adopted an evolutionary
model to evolve trust relationships and transaction relationships. Furthermore,
the effectiveness and robustness of our EB-RSI method have been validated by
comparing with state-of-the-art method SoCap in a real OSN dataset.

In our future work, we plan to improve the efficiency of EB-RSI and apply
it into a real social network based e-commerce platform to accurately evaluate
the social influence of participants and then recommend the buyers with strong
social influence to other buyers or retailers.

Acknowledgements. This work was supported by NSFC grant 61303019, 61073061, 61003044 and 61232006, and Doctoral Fund of Ministry of Education of China 20133201120012.

References

1. Akoglu, L., Chandy, R., Faloutsos, C.: Opinion fraud detection in online reviews by network effects. In: ICWSM (2013)
2. Au Yeung, C.M., Iwata, T.: Strength of social influence in trust networks in product review sites. In: WSDM, pp. 495–504 (2011)
3. Bedi, P., Kaur, H., Marwaha, S.: Trust based recommender system for semantic web. In: IJCAI, pp. 2677–2682 (2007)
4. Berscheid, E., Reis, H.T., et al.: Attraction and close relationships. The Handbook of Social Psychology 2, 193–281 (1998)
5. Chen, W., Lu, W., Zhang, N.: Time-critical influence maximization in social networks with time-delayed diffusion process. In: AAAI, pp. 592–598 (2012)
6. Chen, W., Wang, C., Wang, Y.: Scalable influence maximization for prevalent viral marketing in large-scale social networks. In: KDD, pp. 1029–1038 (2010)
7. Chen, W., Wang, Y., Yang, S.: Efficient influence maximization in social networks. In: KDD, pp. 199–208 (2009)
8. Chen, W., Paik, I., Wang, J., Kumara, B.T., Tanaka, T.: Awareness of social influence on linked social service. In: 2013 IEEE International Conference on Cybernetics (CYBCONF), pp. 32–39 (2013)
9. Cho, Y.S., Ver Steeg, G., Galstyan, A.: Co-evolution of selection and influence in social networks. In: AAAI (2011)
10. Fiske, S.T.: Social beings: Core motives in social psychology. John Wiley & Sons (2009)
11. Franks, H., Griffiths, N., Anand, S.S.: Learning influence in complex social networks. In: AAMAS, pp. 447–454 (2013)
12. Han, J., Kamber, M.: Data Mining Concepts and Techniques: Data Preprocessing. Diane Cerra (2006)
13. Jiang, S., Zhang, J., Ong, Y.S.: An evolutionary model for constructing robust trust networks. In: AAMAS, pp. 813–820 (2013)
14. Jøsang, A., Ismail, R., Boyd, C.: A survey of trust and reputation systems for online service provision. Decision Support Systems 43(2), 618–644 (2007)
15. Jung, K., Heo, W., Chen, W.: Irie: Scalable and robust influence maximization in social networks. In: ICDM, pp. 918–923 (2012)
16. Kempe, D., Kleinberg, J., Tardos, É.: Maximizing the spread of influence through a social network. In: KDD, pp. 137–146 (2003)
17. Kim, J., Kim, S.K., Yu, H.: Scalable and parallelizable processing of influence maximization for large-scale social networks? In: ICDE, pp. 266–277 (2013)
18. Kimura, M., Saito, K.: Tractable models for information diffusion in social networks. In: PKDD, pp. 259–271 (2006)
19. Leskovec, J., Krause, A., Guestrin, C., Faloutsos, C., VanBriesen, J., Glance, N.: Cost-effective outbreak detection in networks. In: KDD, pp. 420–429 (2007)
20. Li, L., Wang, Y.: A trust vector approach to service-oriented applications. In: ICWS, pp. 270–277 (2008)
21. Okelo, B., Boston, S., Minchev, D.: Advanced Mathematics: The Differential Calculus for Multi-variable Functions. LAP Lambert Academic (2012)

22. Pandit, S., Chau, D.H., Wang, S., Faloutsos, C.: Netprobe: a fast and scalable system for fraud detection in online auction networks. In: Proceedings of the 16th International Conference on World Wide Web, pp. 201–210 (2007)
23. Subbian, K., Sharma, D., Wen, Z., Srivastava, J.: Finding influencers in networks using social capital. In: ASONAM, pp. 592–599 (2013)
24. Yaniv, I.: Receiving other people' s advice: Influence and benefit. Organizational Behavior and Human Decision Processes 93(1), 1–13 (2004)
25. Zhang, J., Liu, B., Tang, J., Chen, T., Li, J.: Social influence locality for modeling retweeting behaviors. In: AAAI, pp. 2761–2767 (2013)

A Framework for Linking Educational Medical Objects: Connecting Web2.0 and Traditional Education

Reem Qadan Al Fayez and Mike Joy

Computer Science, University of Warwick, Coventry, CV4 7AL, UK
{r.qadan-al-fayez,m.s.joy}@warwick.ac.uk

Abstract. With the emergence of Linked Data principles for achieving web-scale interoperability, and the increasing uptake of open educational content across institutions, Linked Data (LD) is playing an important role in exposing and sharing open educational content on the web. The growing use of the internet has modified quickly our learning habits. Learning in the medical field is unique in its nature. The educational objects are of various types and should be published by trustworthy organizations. Therefore, medical students and educators face difficulties locating educational objects across the web. To address this problem, we propose a data model for describing educational medical objects harvested from the World Wide Web (WWW) published in Linked Data format. To reduce the burden of navigating through the overwhelming amount of information on the web, we provide a harvesting engine for collecting metadata objects from specified repositories. Then, the harvested educational objects' metadata are represented in our proposed data model named Linked Educational Medical Objects (LEMO). Further enrichment is applied on the metadata records by annotating the textual elements of the records using biomedical ontologies in order to build dynamic connections between the objects. In this paper, we present the framework proposing the data model LEMO along with its implementation and experiments conducted.

Keywords: Linked Data, Web Data Mode, Medical Ontologies.

1 Introduction

The volume of what can be considered an educational material is rapidly increasing on the web which leads to an increased attention to the concerns about managing such online objects. The factors behind the growing number of educational content on the web can be narrowed down to two basic factors: firstly, modern techniques applied in education which encourage the move towards concept-based learning instead of content-based learning. Secondly, the incorporation of web2.0 technologies in education such as wikis, blogs, and social networks [1].

As any area of education, educational content on the web plays a significant role in Medical Education. Multimedia objects such as videos, virtual patients

B. Benatallah et al. (Eds.): WISE 2014, Part II, LNCS 8787, pp. 158–167, 2014.

and pictures have been heavily used in medical e-curricula [2] in the past years. Also, researchers in the field of medical education have been studying the potential use of web2.0 technologies in enhancing the learning and teaching experience and they explained the potential impact which web 2.0 technologies can have on medical education for both students and educators [3].

Searching the educational content needed by a user depends on both the search terms he/she uses when searching for online content, and on the description provided by the educational object's publisher which is the Metadata of an object. Therefore, Metadata is considered the key component for publishing and managing online content. It has been defined as "data describing the context, content, and structure of records and their management through time" [4]. The definition can accommodates different interpretations of what metadata can be. Hence, in e-learning, there is no perfect metadata standard which fits the requirements of different organizations and can be called an ideal standard and different standards exist for describing educational materials [5].

In this paper we propose a framework for harvesting and exposing different types of what can be considered a medical educational object on the web. based on a comparative study conducted on existing standards and conventions for publishing educational objects, we present a comprehensive data model which aims for accommodating different types of educational objects' metadata. The proposed data model is developed in Linked Data structure in order to manage the textual content of the records by annotating keywords which helps in building dynamic relations between the objects. We incorporate annotation of textual content of an object using a well recognized medical ontologies in order to annotate that object with keywords and categories it might belong to. This paper illustrates a work in progress which requires iterations of experiments and enhancements. In this paper, we present the data model proposed along with results of conducted experiments for harvesting and modelling educational objects using the proposed model.

The rest of the paper is organized as follows. Section 2 presents the related background of existing standards and a comparative study conducted. Section 3 introduces the framework of the proposed data model. Section 4 describes the development of the data model in details. Further techniques incorporated in designing and testing the data model are explained in section 5. Experiments conducted on harvesting, mapping and annotating educational content are detailed in section 6. Finally, section 7 concludes and discusses future plans.

2   Background

In the context of e-learning, standards are used to govern the design, description, and publishing of educational materials for the purpose of ensuring reusability, and interoperability between different e-learning systems. It is possible that a standard does not satisfy specific systems needs. Hence, developing an Application Profile (AP) is a flexible way to adopt a standard and satisfy the requirements of a system. Any system or organization can implement a modified version of any standard, either by adding or removing characteristics from the standards set of attributes, and still guarantee its compatibility with the original standard, and the interoperability of its content [6].

Two of the most popular metadata standards used for e-learning are IEEE Learning Objects Metadata (LOM) and Dublin Core Metadata Initiative (DCMI). Several Application Profile where developed in the medical field having one of these standards as their base schema. IEEE LTSC started to develop **Learning Object Metadata (LOM)** standard in 1997 and it has been accredited by IEEE [7]. As defined in LOM standard [8], a LOM instance is designed to record the characteristics of the learning object it describes. LOM categorizes these characteristics into 9 groups: *general, life cycle, meta-metadata, educational, technical, rights, relation, annotation, and classification*. The purpose of such detailed description for a learning object as stated by the working group is to facilitate sharing and exchange of learning objects since the metadata have a high degree of semantic interoperability. **Dublin Core Metadata Initiative (DCMI)** is an open public organization that is non profitable. It supports metadata design and implementation across broad range of purposes. The initiative work resulted in a simple cross domain metadata statement known as Dublin Core Metadata Element Set (DCMES) which has been standardized as ISO standard [9]. Dublin Core standard is used to describe a wide range of resources, where a resource is defined by DCMI as "anything which might be identified" [10]. The simple Dublin Core metadata element set consists of these fifteen elements: *contributor, coverage, creator, date, description, format, identifier, language, publisher, relation, rights, source, subject, title, and type*. All of these elements are optional and maybe repeated if required for a single resource.

Metadata schemas or Application Profiles (APs) were developed by organizations responsible of libraries and repositories for publishing medical educational materials. A brief overview of four popular AP and a comparative study results is presented in this section. **HealthCare LOM**[1] extends the IEEE Learning Object Metadata (LOM). In this schema, one category named HealthCare Education was added to the original 9 categories composing LOM, in which fields to describe health care related metadata were added such as continuing education credits, patient and professional resources, and others [11]. **mEducator**[2] was developed as part of a European funded project called mEducator. The objective of mEducator schema was to enable ease of sharing, discovery, and reuse of medical educational content across EU higher academic institutions [12]. **The Health Education Assets Library (HEAL)**[3] is another metadata schema which extends Educause IMS that is based on IEEE LOM and builds upon it specific fields needed in health science education such as clinical history field and disease process field [13]. Finally, **National Library of Medicine (NLM)**[4] implemented by the world's largest biomedical library. NLM schema is based on

---

[1] http://ns.medbiq.org/lom/extend/v1

[2] www.meducator.net/mdc/schema.rdf

[3] http://www.healcentral.org

[4] http://www.nlm.nih.gov/tsd/cataloging/metafilenew.html

Dublin Core Metadata Initiative and incorporates additional elements identified as requirements by NLM for publishing its content [14].

A comparative study is conducted in order to identify common elements used in the different AP schemas described in the background section. The results of the comparative study is outlined in table 1. The table lists the common elements between the four schemas and how they are represented in each schema. An element in any metadata schema might be represented as a Parent Node (PN) which states that it is a composite element of other leaf nodes, or it might be Leaf Node (LF) which states that this element is a child node of another PN, or it might be a Single Node (SN) which states that the element is represented as a node directly descendent from the root and does not have children.

Table 1. Results of comparative analysis conducted on four medical metadata schemas

	HealthCareLOM		mEducator		HEAL		NLM	
1	General	PN	General	PN	General	PN	General	-
2	Identifier	LN	Identifier	LN	Resource URN	LN	Identifier	PN
3	Title	LN	Title	LN	Title	LN	Title	SN
4	Description	LN	Description	LN	Description	PN	Description	PN
5	Lifecycle	PN	Lifecycle	PN	Creation Date	LN	Date	PN
6	Rights	PN	Rights	PN	IPR	LN	Rights	SN
7	Resource Type	LN	Resource Type	LN	Resource Type	LN	Resource Type	SN
8	Keywords	LN	Keywords	LN	Keywords	LN	Keywords	LN
9	Classification	PN	Classification	PN	Classification	PN	Subject	PN
10	Educational	PN	Educational	PN	Educational	PN	Educational	-
11	Relation	PN	Relation	PN	Companion	-	Relation	SN
12	Technical	PN	Technical	PN	Technical	PN	Technical	-

From the analysis of the four metadata schemas, some elements have been found common between these four schemas as shown in table 1. Identifier, Title, and Description are main elements in all of the four schemas and in most cases they are LN or PN. other common attributes are the Date, Intellectual Property Rights (IPR), the type of the resource, keywords, and classification. The last three attributes in table 1 Educational, Relation, and Technical are not implemented in all four schemas. The elements used in each schema represent the need of the organization developing the AP. From the analysis above, we can say that the goal of the application or the organization for which an AP is developed plays vital role in the elements chosen and their structure in the schema.

3 Overview of System Framework for Developing LEMO Data Model

Several hypothesis were derived after conducting the comparative study. The main aim for proposing this data model is having a metadata in which we can describe different types of educational objects, in addition to the ability of building dynamic connections between the educational objects. The hypothesis focused on the set of elements necessary to have in the proposed data model to achieve its goal which are: 1. General content descriptors of the objects content, 2. People responsible for the development of the object, 3. The category in which the object might be classified into, 4. The objects type, and 5. The objects access rights. The framework shown in figure 1 illustrates the main aims of proposing the data model and the conceptual model design on this model.

The proposed data model focuses on discoverability since the main goal of the project is to expose and link harvested medical educational objects from various sources having different metadata standards or schemas.

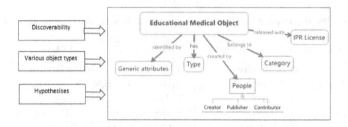

Fig. 1. An Overview of Our Framework for the proposed Data Model Design

4   LEMO Proposed Data Model

Our proposed data model is based on DCMI standards. DCMI was chosen to be the base schema for what we named Linked Educational Medical Objects (LEMO) Application Profile because of its wide usage in different repositories. DCMI is a simple schema with flat structure of elements which is beneficial for introducing further refined elements to the original set of elements. The LEMO AP adopts most of DCMI elements in its schema though changing the structure of some elements based on the conceptual model, shown in the framework presented in the previous section. The changes on DCMI standards are only refinements on its elements without adding new elements. The refinements are applied to specific elements when possible to add extra data in order to build dynamic connections between the educational objects based on the content of their metadata records. Figure 2 represent the schema of the proposed model with DCMI labels to specify the original elements derived from DCMI standard.

5   Harvesting, Mapping and Enriching Educational Objects

The term Linked Data is a new method for publishing data on the web using URIs and RDF. The data published in Linked Data is machine readable and can

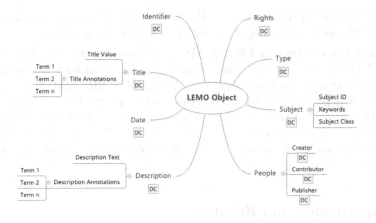

Fig. 2. Proposed LEMO Data Model with DCMI elements labeled

be linked to external data sets in a simple way [15]. In this data model, every educational medical object's metadata is represented using URIs and RDF which make the whole data set a Linked Open Data set of medical educational objects. LEMO AP is implemented as Linked Data using RDF/XML and enriched with existing ontologies such as SNOMED ontology.

Sample of a Record Presented in LEMO AP

```
<rdf:RDF xmlns:rdf="http://www.w3.org/1999/02/22-rdf-syntax-ns#"
         xmlns:dc="http://purl.org/dc/elements/1.1/"
         xmlns:lemo="http://www.warwick.ac.uk/ias/lemo#"
         xmlns:rdfs="http://www.w3.org/2000/01/rdf-schema#"
  <rdf:Description rdf:about="http://www.warwick.ac.uk/ias/lemo#1234">
        <dc:identifier>a</dc:identifier>
        <dc:creator>
            </dc:creator>b<dc:title>
            <rdf:Description>
                <rdf:value>title text</rdf:value>
                <lemo:textAnnotation rdf:parseType="Resource">
                    <lemo:term rdf:parseType="Resource">
                        <lemo:ID>term id</lemo:ID>
                        <lemo:from>1</lemo:from>
                        <lemo:to>2</lemo:to>
                        <lemo:text>word</lemo:text>
                        <lemo:class>class</lemo:class>
                    </lemo:term>
                </lemo:textAnnotation>
            </rdf:Description>
        </dc:title>
    </rdf:Description>
</rdf:RDF>
```

Harvesting and Mapping EMOs: We implemented two harvesting endpoints to collect Educational Medical Objects from repositories. OAI-PMH harvesting protocol is one of the protocols supported in LEMO in order to harvest content from repositories supporting this protocol. This protocol supports harvesting content in different formats based on the library or repository support of this protocol. The harvested records are mapped into LEMO AP using XSL Transformation (XSLT) where XSL stands for Extensible Stylesheet Language and it is a style sheet for XML documents. Harvesting other types of content such as blogs and videos is implemented in LEMO using RSS feeds reader.

Enriching the content of LEMO AP: After mapping the records harvested into LEMO AP, further enrichments are applied to textual elements of the records. The enrichment are basically implemented on two fields *title*, and *description* elements by annotating the raw text using medical ontologies and adding the annotated terms to the refined elements of the AP in *dc:title* and *dc:description* elements. Annotating textual content is done using SNOMED ontology. It is one of the most widely used ontologies for categorizing medical libraries content.

6   Experiments and Results

The experiments were conducted on three different repositories for medical educational content. The repositories chosen varies in their content types in order to validate the data model against the goals determined in the framework. The three repositories participating in the experiments were chosen based on the results of a survey conducted with Warwick Medical School educators. The results showed that most of the educators use popular search engines to find educational resources. They also mentioned using PubMed Central Library and medical journals when searching for educational content to help them in their teaching process. Based on that we have chosen *PubMed Central library*[5] and two other Journals hosting blogs and videos. The video collection is hosted by *The New England Journal of Medicine* (NEJM) on their YouTube channel[6], and the blogs are provided by *The American Journal of Medicine* (AMJMED)[7].

At the first stage of the experiments we harvested medical educational objects from the three repositories. The educational objects were harvested from PubMed Central using OAI-PMH harvesting protocol. The library provides OAI-PMH service which allows us to access and download metadata records of the PubMed Central library content. We implemented OAI-PMH service in our application for harvesting recent records published in PubMed Central library and retrieved more than 10,000 records representing educational materials from different subjects. The data is retrieved in Dublin Core format and stored as XML files. Using RSS feeds for reading videos and blogs, we implemented RSS feeds reader and were able to collect 208 videos posted by NEJM channel on YouTube, and a sample of 10 blogs using the same RSS Reader from AMJMED blogs.

The second stage of the experiments was concerned with mapping the harvested files into the proposed data model. All the content harvested was stored as XML files representing educational objects in different metadata formats. The first step was to build XSL files for transforming the content harvested to the proposed data model format. The benefits of using XSL are: converting the XML

---

[5] http://www.ncbi.nlm.nih.gov/pmc/

[6] https://www.youtube.com/user/NEJMvideo

[7] http://amjmed.org/

files into RDF/XML files, transforming the existing data format of the XML files to the desired RDF/XML LEMO AP proposed in section 4, and enabling the addition of new refined elements needed for data enrichment. Figure 3 is a snippet of a metadata record for an item retrieved from NEJM video collection in its XML format retrieved by the RSS Reader. All metadata records are mapped to LEMO AP using XSL Transformation services developed in the application. Figure 4 shows snippet of the same metadata record in RDF/XML format after the mapping process.

```
<item>
    <title>Leg Venographic Images Obtained before and after Thrombolysis</title>
    <link>http://www.youtube.com/watch?v=a5iuJQOXFxA&amp;feature=youtube_gdata</link>
    <atom:updated xmlns:atom="http://www.w3.org/2005/Atom">2014-05-06T08:44:24.000Z</atom:updated>
    <gd:comments>
    <media:group xmlns:media="http://search.yahoo.com/mrss/">
        <media:category label="Science &amp; Technology" scheme="http://gdata.youtube.com/schemas/2007/ca
        <media:content url="http://www.youtube.com/v/a5iuJQOXFxA?version=3&amp;f=videos&amp;app=youtube_g
        <media:description type="plain">In the first video, a cine loop of contrast-material injection be
        <media:keywords />
        <media:title type="plain">Leg Venographic Images Obtained before and after Thrombolysis</media:ti
        <yt:duration xmlns:yt="http://gdata.youtube.com/schemas/2007" seconds="28" />
    </media:group>
    <gd:rating xmlns:gd="http://schemas.google.com/g/2005" average="5.0" max="5" min="1" numRaters="2" ra
    <yt:statistics xmlns:yt="http://gdata.youtube.com/schemas/2007" favoriteCount="0" viewCount="105" />
    <description>In the first video, a cine loop of contrast-material injection before thrombolysis shows
    <category domain="http://schemas.google.com/g/2005#kind">http://gdata.youtube.com/schemas/2007#video<
    <category domain="http://gdata.youtube.com/schemas/2007/categories.cat">Tech</category>
    <pubDate>Mon, 05 May 2014 14:17:15 GMT</pubDate>
    <guid isPermaLink="false">http://gdata.youtube.com/feeds/api/videos/a5iuJQOXFxA</guid>
    <dc:creator>NEJMvideo</dc:creator>
    <dc:date>2014-05-05T14:17:15Z</dc:date>
</item>
```

Fig. 3. XML Snippet for RSS feeds object from NEJM YouTube Channel

As shown in figure 4, the metadata record is represented in RDF format where the record metadata are represented as triples (subject, predicate, object) i.e. the subject is the video resource identified by *rdf:about*, and has predicate such as *dc:title* and the object is another resource representing the title as seen in the picture. The flexible nature of RDF/XML format make it easier for us to dynamically add elements to the records content for enriching it.

```
<rdf:Description rdf:about="http://gdata.youtube.com/feeds/api/videos/a5iuJQOXFxA">
    <dc:identifier>http://www.youtube.com/watch?v=a5iuJQOXFxA&amp;feature=youtube_gdata</dc:identifier>
    <dc:title rdf:resource="http://gdata.youtube.com/feeds/api/videos/a5iuJQOXFxA:title"/>
    <dc:description rdf:resource="http://gdata.youtube.com/feeds/api/videos/a5iuJQOXFxA:desc"/>
    <dc:date>2014-05-05T14:17:15Z</dc:date>
    <dc:creator>NEJMvideo</dc:creator>
    <dc:type rdfs:type="Video">28 seconds</dc:type>
</rdf:Description>
<rdf:Description rdf:about="http://gdata.youtube.com/feeds/api/videos/a5iuJQOXFxA:title">
    <rdf:value>Leg Venographic Images Obtained before and after Thrombolysis</rdf:value>
</rdf:Description>
<rdf:Description rdf:about="http://gdata.youtube.com/feeds/api/videos/a5iuJQOXFxA:desc">
    <rdf:value>In the first video, a cine loop of contrast-material injection before thrombolysis shows
</rdf:Description>
```

Fig. 4. Snippet for the Same Record in LEMO AP RDF/XML style

The third stage of the experiments was applied for annotating textual elements in LEMP AP records which are the *Title* and *Description* fields. For the annotation service we used a popular and widely used ontology in the medical field called SNOMED ontology. The annotation service was implemented in our application using BioOntology annotation API[8]. The results of annotating the textual fields of a specific record were dynamically added to the records RDF/XML representation. The *Title* resource of any object after mapping contains only the textual value of a title without further enrichments. Figure 5 represents a sample of the *Title* resource after annotating its textual value. Predicates and their objects were dynamically added to the *Title* resource.

```
<rdf:Description rdf:about="oai:pubmedcentral.nih.gov:2606822:title">
    <rdf:value>Dipeptidyl Peptidase-4 Inhibition by Vildagliptin and the Effect on
        Insulin Secretion and Action in Response to Meal Ingestion in Type 2 Diabetes </rdf:value>
    <lemo:lemoTitleAnnotation rdf:resource="oai:pubmedcentral.nih.gov:2606822:title:term1"/>
    <lemo:lemoTitleAnnotation rdf:resource="oai:pubmedcentral.nih.gov:2606822:title:term2"/>
    <lemo:lemoTitleAnnotation rdf:resource="oai:pubmedcentral.nih.gov:2606822:title:term3"/>
    <lemo:lemoTitleAnnotation rdf:resource="oai:pubmedcentral.nih.gov:2606822:title:term4"/>
    <lemo:lemoTitleAnnotation rdf:resource="oai:pubmedcentral.nih.gov:2606822:title:term5"/>
    <lemo
    <lemo   <rdf:Description rdf:about="oai:pubmedcentral.nih.gov:2606822:title:term1">
    <lemo       <lemo:lemoClassLabel>Hydrolase</lemo:lemoClassLabel>
    <lemo       <lemo:lemoTermClass>http://purl.bioontology.org/ontology/SNOMEDCT/74628008</lemo:lemoTermClass>
    <lemo       <lemo:lemoTermText>PEPTIDASE</lemo:lemoTermText>
    <lemo       <lemo:lemoTo>20</lemo:lemoTo>
    <lemo       <lemo:lemoFrom>12</lemo:lemoFrom>
    <lemo       <lemo:lemoTermID>http://purl.bioontology.org/ontology/SNOMEDCT/130202003</lemo:lemoTermID>
    <lemo   </rdf:Description>
    <lemo   <rdf:Description rdf:about="oai:pubmedcentral.nih.gov:2606822:title:term3">
    <lemo       <lemo:lemoClassLabel>General metabolic function</lemo:lemoClassLabel>
    <lemo       <lemo:lemoTermClass>http://purl.bioontology.org/ontology/SNOMEDCT/47722004</lemo:lemoTermClass>
    <lemo       <lemo:lemoTermText>INHIBITION</lemo:lemoTermText>
    <lemo       <lemo:lemoTo>33</lemo:lemoTo>
    <lemo       <lemo:lemoFrom>24</lemo:lemoFrom>
    <lemo       <lemo:lemoTermID>http://purl.bioontology.org/ontology/SNOMEDCT/61511001</lemo:lemoTermID>
    <lemo   </rdf:Description>
    <lemo   <rdf:Description rdf:about="oai:pubmedcentral.nih.gov:2606822:title:term5">
    <lemo       <lemo:lemoClassLabel>Dipeptidyl peptidase IV inhibitor</lemo:lemoClassLabel>
    <lemo       <lemo:lemoTermClass>http://purl.bioontology.org/ontology/SNOMEDCT/422403005</lemo:lemoTermClass>
    <lemo       <lemo:lemoTermText>VILDAGLIPTIN</lemo:lemoTermText>
            <lemo:lemoTo>46</lemo:lemoTo>
            <lemo:lemoFrom>35</lemo:lemoFrom>
            <lemo:lemoTermID>http://purl.bioontology.org/ontology/SNOMEDCT/428807006</lemo:lemoTermID>
        </rdf:Description>
```

Fig. 5. A sampe of an Object's Title Resource after annotation

7   Conclusions and Future Work

In this paper, we proposed a data model that is part of a work in progress project for harvesting and exposing Medical educational objects for facilitating the process of search and finding of educational objects. We have presented the framework for developing and implementing the proposed data model called LEMO AP and conducted experiments for applying this AP on different repositories. The proposed data model was able to accommodate different metadata schemas harvested from these repositories, and annotations were successfully

---

[8] http://bioportal.bioontology.org/annotator

added to the records harvested. Based on the proposed data model, we are aiming at building a metadata repository for harvesting and querying content from repositories and Web2.0 technologies such as RSS Feeds. In the near future, we will study the possibility of inferring categories for the objects retrieved using the annotation terms discovered from SNOMED ontology. Further work will focus on dynamic linking of objects and evaluating the accuracy of these linkages discovered.

References

1. Caswell, T., Henson, S., Jensen, M., Wiley, D.: Open content and open educational resources: Enabling universal education. The International Review of Research in Open and Distance Learning 9(1) (2008)
2. Fleiszer, D.M., Nancy, H.P., Sean, P.S.: New directions in medical e-curricula and the use of digital repositories. Academic Medicine 79(3), 229–235 (2004)
3. Popoiu, M.C., Grosseck, G., Holotescu, C.: What do we know about the use of social media in medical education? Procedia-Social and Behavioral Sciences 46, 2262–2266 (2012)
4. Franks, P., Kunde, N.: Why METADATA Matters. Information Management Journal 40(5), 55–61 (2006)
5. Devedic, V.: Semantic web and education, vol. 12. Springer (2006)
6. Reba, I., de La Passardie, B., Labat, J.M.: A Formalism to Describe Open Standards in order to Generate Application Profiles. In: Proceedings of the International Conference on Computer in Education, pp. 491–498 (2008)
7. Learning Technology Standards Committee of the IEEE: Final 1484.12.1-2002 LOM Draft Standard (2005), http://ltsc.ieee.org/wg12/20020612-Final-LOM-Draft.html
8. Learning Technology Standards Committee of the IEEE: Draft Standard for Learning (2002), http://ltsc.ieee.org/wg12/files/LOM_1484_12_1_v1_Final_Draft.pdf
9. Information and documentation: The Dublin Core metadata element set (2009), http://www.iso.org/iso/home/store/catalogue_tc/catalogue_detail.htm?csnumbe=r52142
10. Powell, A., Nilsson, M., Naeve, A., Johnston, P., Baker, T.: DCMI abstract model. DCMI Recommendation (2007), http://dublincore.org/documents/abstract-model
11. Healthcare LOM, ANSI/MEDBIQ LO. 10.1-2008, Healthcare Learning Object Metadata (2011), http://txlor.org/docs/medbiquitoushealthcarelomspecification-200607.pdf
12. Bamidis, P.D., Kaldoudi, E., Pattichis, C.: mEducator: a best practice network for repurposing and sharing medical educational multi-type content. In: Leveraging Knowledge for Innovation in Collaborative Networks, pp. 769–776 (2009)
13. Candler, C.S., Uijtdehaage, S.H.J., Dennis, S.E.: Introducing HEAL: The health education assets library. Academic Medicine 78(3), 249–253 (2003)
14. National Library of Medicine, Fact Sheet: Medical Subject Headings(MeSH) (2013), http://www.nlm.nih.gov/pubs/factsheets/mesh.html
15. Bizer, C., Heath, T., Berners-Lee, T.: Linked Data - the story so far. International Journal on Semantic Web and Information Systems 5(3), 1–22 (2009)

An Ensemble Model for Cross-Domain Polarity Classification on Twitter

Adam Tsakalidis, Symeon Papadopoulos, and Ioannis Kompatsiaris

Information Technologies Institute, CERTH,
57001, Thessaloniki, Greece
{atsak,papadop,ikom}@iti.gr

Abstract. Polarity analysis of Social Media content is of significant importance for various applications. Most current approaches treat this task as a classification problem, demanding a labeled corpus for training purposes. However, if the learned model is applied on a different domain, the performance drops significantly and, given that it is impractical to have labeled corpora for every domain, this becomes a challenging task. In the current work, we address this problem, by proposing an ensemble classifier that is trained on a general domain and and adapts, without the need for additional ground truth, on the desired (test) domain before classifying a document. Our experiments are performed on three different datasets and the obtained results are compared with various baselines and state-of-the-art methods; we demonstrate that our model is outperforming all out-of-domain trained baseline algorithms, and that it is even comparable with different in-domain classifiers.

Keywords: Sentiment analysis, polarity detection, ensemble classifier.

1 Introduction

Twitter[1] is a microblogging platform that has seen increasing popularity during the latest years. The content produced in this network reflects its users' thoughts on different topics and has proven beneficial for various applications such as modeling public behavior [1], summarisation of events [6] or predicting election results [8]. Sentiment analysis is of particular importance in Twitter: brands can learn what people think of their products, politicians can learn the users' opinions on them, people can aggregate opinions on topics of their interest and so on. Given the huge amounts of broadcasted content, automating the process of opinion mining becomes a rather crucial task for creating real-time insights.

The two mostly studied sentiment analysis tasks are *subjectivity* (given a set of documents, find the subjective ones) and *polarity* detection (discrimination of positive/negative documents). The most common approach to deal with these tasks is to train a classifier on a labeled corpus and apply the learned model on the desired test set. However, accuracy drops significantly when the applied model is trained on a different type of document, domain or time [14]. In order

---

[1] https://twitter.com/

B. Benatallah et al. (Eds.): WISE 2014, Part II, LNCS 8787, pp. 168–177, 2014.

to overcome this problem, lexicon-based approaches are often employed, using a predefined dictionary of words along with their corresponding prior polarity weight. For a given document, they detect its polarity based on the majority class sum-of-weights of its keywords; however, those fail to perform comparably to in-domain learned models. Furthermore, the task of sentiment analysis becomes more difficult when dealing with short, noisy content and well-known approaches applied in well-formed documents seem ineffective for content of such type.

In the current work, we tackle the problem of the domain-dependent nature of the polarity detection task in Twitter. We train different classifiers on various sets of features and combine them in an ensemble model that achieves an average accuracy boost of 10.22% over our main baseline model (text-based learning) when trained on a different domain. We compare our method with out-of-domain state-of-the-art approaches on public datasets achieving better results; we compare different lexicons and show that our method outperforms them by 12.5% on average; most importantly though, we show that the accuracy of our approach is highly competitive against traditional in-domain training methods, following by only a 3.86% the best in-domain algorithm.

2 Background

Representation Forms: Given a set of documents to classify, the first step is to create a vector space representation of them (usually as n-gram features), often using the $tf \cdot idf$ formula to emphasize characteristic words of a document:

$$tf \cdot idf_{i,j} = \ln(1 + tf_{i,j}) \cdot \ln(|D|/df_i) \tag{1}$$

Here i is a term occurring in document j, tf and df its frequency on j and the total number of documents this term appears in respectively and $|D|$ is the total number of documents in the corpus. Other common preprocessing steps include stop-word removal and stemming; however, these were found to decrease [2] or offer no increase [4] in accuracy in various sentiment analysis tasks.

Part-Of-Speech (POS) tags are also used as features ([2], [3], [7], [10]), along with n-grams, since several tags can reveal the presence of subjectivity in a document or help in word sense disambiguation. Saif et al. [2] demonstrated a boost in accuracy ranging from 0.9% to 8.1% when POS tags were used along with unigrams, whereas a slight decrease was found in Go et al.'s work [11]. The role of several other features has been explored, such as punctuation [3], semantic entities [2] and consecutive letters in a word [15], with results varying.

Learning Methods: Sentiment analysis approaches can be separated into *supervised* and *unsupervised* methods. Supervised approaches require a labeled corpus of documents to learn a model from and apply it to a test set. On the contrary, unsupervised approaches apply a predefined list of rules (usually given by a lexicon) in the test set, overcoming the training step of supervised approaches.

Despite the high accuracy reported by many supervised approaches in microblogs, their algorithms are tested on the same domain that they are trained on. However, one cannot expect to find a labeled corpus for training for all different types of problems. Even worse, classifiers are not only domain-dependent but also topic-, document- and time-dependent ([14], [16]), making it impossible to be applied in real-life problems achieving the same accuracy. This is probably the reason that online sentiment analysis services tend to disagree in their outputs. For example, three different online sentiment analysis services were used in [3], revealing a low kappa statistic ranging from 0.4 to 0.6, whereas the average pair-wise agreement of eight different methods ranged from 48% to 72% in [5].

Using a Naïve Bayes (NB) and Support Vector Machine (SVM), Read revealed that both algorithms perform significantly better when tested on the same dataset that they were trained on, in almost all cases [14], arguing that a more general training set should be constructed. A common approach for this task (e.g. [11], [14]) is to collect a large number of documents (tweets) containing happy/sad emoticons and assign to them the corresponding label (positive/negative), whereas another way to tackle the problem is by applying an unsupervised method.

Most unsupervised methods use a sentiment lexicon, e.g., SentiWordNet [17] ("SWN"; about 150,000 synsets with double values indicating their sentiments), Subjectivity Lexicon [12] or Bing Liu's Opinion Lexicon [18] ("OL"; about 6,800 words marked as "positive"/"negative"). Compared to in-domain supervised methods, these approaches perform worse, but achieve comparable or better results than out-of-domain supervised algorithms (e.g., a lexicon-based method achieved an average accuracy of 60.6% on sentences compared to 67.9% and 57.2% of an SVM algorithm trained in- and out-of-domain respectively in [16]).

In [5] the authors studied different methods and combined them in a unique system that failed to perform better than their best individual model. A boost of about 1% for a 4-class sentiment task is reported in [12] when keywords are combined with their prior polarity, whereas a gain of about 5% for the polarity task is reported in [10] when lexicon features are used along with content ones using an in-domain classifier. A weighted classifier was developed in [16] that combined a supervised and a lexicon-based approach based on their precision on every class and revealed a significant increase in accuracy of 13.65% on average for the polarity task. However, their approach assumes that every algorithm should perform fairly well on one class and vice versa.

In the current research we try to overcome the domain-dependence problem by creating an ensemble classifier. Instead of learning one model to apply to our test data, we combine different algorithms based on different document representation forms and highlight their role in the polarity detection task.

3   Methodology

Using the Twitter API[2] over a two-day period in mid-March 2014, we gathered 250,000 tweets written in English containing happy/sad emoticons (":)", "":("; equally balanced), removed all retweets (7, 469) and used the rest as a training

---

[2] https://dev.twitter.com/

set ("Emoticons Dataset", "ED"). Working on ED, we created four different tweet representations and trained one classifier on each one of them, trying to find the parameters that achieve the highest accuracy[3].

Text-Based Representation (TBR): We used three representations of the tweets using binary, term frequency (tf) and $tf \cdot idf$ n-grams. We set $n = 1, 2, 3$, resulting into nine representations in total. Stop-word removal and stemming were ignored, as suggested by previous works ([2], [4]).

Feature-Based Representation (FBR): We represented every tweet as a set of binary values indicating the presence of several features. These included consecutive dots, exclamation marks, mentions, URLs and negations; hashtags were added by removing the "#" sign; words written in upper-case were lower-cased and added by inserting the word "very" upfront (e.g., "very big" for "BIG"); words containing more than two consecutive letters were also added, by replacing the repetitions with two consecutive ones (e.g., "biig" for "biiig").

Lexicon-Based Representation (LBR): We used two lexicons (SentiWord-Net and Opinion Lexicon). Instead of assigning the majority class label on a tweet, we counted the sum of nouns, verbs, adjectives, adverbs and the overall sum of the words as indicated by SWN and the overall sum of words as indicated by OL. In the case of presence of negation, the polarity score of the term that follows was inverted. We use these six features to learn a model from and we compare our results with the simple counting methods of both lexicons.

Combined Representation (CR): We used a combination of TBR with POS tags, using the Stanford POS Tagger [19] and we tested the same parameters as in the case of TBR. Finally, in TBR, FBR and CR, features that appeared only once in the training set were eliminated to achieve noise reduction.

Ensemble Classifier: The main idea behind our ensemble model is to combine the different algorithms' outputs in a weighted scheme in order to classify a tweet. We have separated our classifiers in two categories, based on their domain-(in)dependent nature: the *hybrid* classifier (HC) and the *lexicon-based* one (LC), which is the algorithm that was tested on LBR achieving the highest accuracy. The HC assigns one value per class to a document based on the outputs of the individual (probabilistic) classifiers that are trained on TBR, FBR and CR:

$$hval_c(i) = \sum_r w_r \cdot p_r(i, c). \tag{2}$$

Here i corresponds to the tweet, r to the representation model, w is the model's weight and $p(i, c)$ the probability assigned by the classifier on the respective tweet and class. The weight was set equal to the difference of every classifier's accuracy compared to the random classifier (50%), based on the ED. Finally, the HC assigns the polarity class with the highest $hval$ to every tweet.

The predictions HC and LC on a test set are combined by the ensemble classifier and the documents for which they agree on are automatically assigned

---

[3] All features along with the learned models can be found at
    https://github.com/socialsensor/sentiment-analysis

the corresponding label, under the assumption that they are most likely to have been classified correctly. Then, those "agreed" documents are considered as a new training set, whereas the remaining ones of the test set comprise the new test set. At the final stage, a model is learned on the "agreed" documents and applied on the remaining ones. This technique alleviates us from the domain-dependence problem; however, the final training stage is highly dependent on the accuracy of the ensemble classifier achieved on the "agreed" documents.

4 Experimental Study

4.1 Twitter Test Datasets

We used three datasets for testing our approach, focusing on the tweets written in the English language. These datasets will be used in section 5, while the current section will focus on the results based on ED.

Stanford Twitter Dataset Test Set (STS): We have used the non-neutral part of two versions of this dataset [11]; the first one (referred here as "STS-1") consists of 177 positive and 184 negative tweets (see [2]); the second one ("STS-2") consists of 108 positive and 75 negative tweets (see [13]).

Obama Healthcare Reform (HCR): This dataset contains tweets related to the healthcare reform introduced by Barrack Obama in 2010 ([13]). This set is split into three parts. We focused on the positive and negative tweets contained in the "dev" and "test" set separately. In the first one we found 135 positive and 328 negative tweets, whereas in the second one 116 and 383 respectively.

Obama-McCain Debate (OMD): This dataset contains 3,269 annotated tweets related to the 2008 Obama vs McCain debate ([2], [13]). We used the "positive" and "negative" tweets that have been annotated by at least three people for which the annotators' agreement was above 50% on one of our examined classes. This resulted in 1,897 tweets (707 positive and 1,190 negative ones).

4.2 Model Building

Some common pre-processing steps on ED include the replacements of all user mentions with "usrmntn", URLs with "urlink" and hashtags with the actual hashtag by removing the "#" sign. Negations and common abbreviations were transformed into their reference form (e.g., "I've" to "I have", "isn't" to "is not"). We expanded a list of some commonly used abbreviations[4] to transform them into their proper form, removed all emoticons from the training set and replaced all emoticons with their latent meaning in the test sets only (e.g., ":)" to "feeling happy"). Finally, words containing more than two consecutive letters (e.g. "hiiii") were shortened so that they contain only two repetitions ("hii").

We used Multinomial Naïve Bayes (MNB) and Support Vector Machines (SVM) on ED, as provided by Weka[5]. We randomized ED and used a 66%/33%

---

[4] http://www.englishclub.com/esl-chat/abbreviations.htm
[5] http://www.cs.waikato.ac.nz/ml/weka/

split for training/evaluation purposes before working on any test set. We chose MNB to be applied on TBR, FBR and CR, as in previous works ([4], [9]) and SVM for LBR due to their capability of dealing with double-valued attributes.

TBR: Table 1 presents the accuracy for all examined TBR forms. Bigrams outperformed unigrams and trigrams. Following previous claims regarding the appropriate weighting scheme [9], we expected to see some differences as moving from binary to $tf \cdot idf$ forms; however, we find no such differences. This may be because we have not yet moved to another domain, in which case words that appear much more frequently than in our training set could affect the results.

FBR: We extracted $29,269$ features (mainly hashtags, repetitions and upper-case words) and achieved a relatively low accuracy (61.95%). One explanation of this can be found in the recall (0.84/0.4 for positive/negative class respectively), revealing a bias towards the positive class. Nevertheless, we decided to apply the learned model on our test sets to test the impact of FBR on a different domain.

LBR: Table 2 presents the results obtained by SVM compared to the "count-ing" methods using both lexicons individually and each POS tag from SWN, revealing the superiority of SVM. Our findings consistently support that adjec-tives carry more sentimental weight [18]. OL achieved better results than SWN, despite that there are far less words documented in this lexicon and are only marked as "positive" or "negative", whereas every synset is carrying a double-valued polarity weight in SWN. What is important from the results presented here though is that a learning algorithm over some lexicon features can boost traditional lexicon-based approaches by an average of 3.9% in accuracy.

CR: The results on the ED support previous findings ([2], [11]) on the use of POS tags along with unigrams, revealing an average boost of 4.73% in accuracy across different weighting models (see Table 1). However, we notice a slight decrease on bigrams and trigrams, most likely because the sparsity of these representations increases along with the increase of the "n" value in n-grams much faster than in TBR, resulting into information loss. This is explained by the number of extracted features: there exist about 33% more features in the case of unigrams for CR than for TBR ($28,491$ vs $37,796$), while there are fewer trigrams ($150,398$ vs $145,178$), because of the tf threshold we applied.

Table 1. Accuracies achieved on the 33% of the ED for TBR and CR

Represent.	binary			tf			tfidf		
n-gram	1	2	3	1	2	3	1	2	3
TBR	74.59	**81.07**	78.07	74.62	80.92	78.07	73.46	81.02	78.88
CR	79.64	80.86	77.42	79.61	80.62	77.15	77.60	**80.95**	78.13

Ensemble Classifier: We created a new development set ("DEV") of $10,000$ equally-balanced tweets aggregated from Twitter in the same way that the ED was created; we removed all retweets and used the rest for tuning the parameters

Table 2. Accuracy using lexicon counting methods and an SVM applied on LBR

	Adv	Noun	Vrb	Adj	SWN	OL	LC
Acc.(%)	49.86	51.23	54.85	57.41	59.88	62.44	65.06

Table 3. Accuracies of all classifiers in DEV. LBR and LC refer to the same classifier

Method	TBR	FBR	LBR	CR	HC	LC	Agreed	Disagr.	Total
Acc.(%)	82.16	65.73	79.96	80.56	82.54	79.96	91.13	69.88	85.72

of the ensemble classifier, while using the full ED for training. The parameters for HC in Equation 2 were set to 31.07, 11.95 and 30.95 for TBR, FBR and CR respectively, as indicated by the best achieved accuracy of every model on ED. Table 3 summarizes the results achieved by all individual classifiers on the DEV. The results are similar to the ones achieved in the ED set, with the exception of LBR, which achieved significantly better results in the DEV because of the replacement of emoticons with their latent meaning.

The column "HC" shows the accuracy of the hybrid classifier, whereas the "LC" is copied from LBR for clarity. HC achieved slightly better results, compared to its individual components. In total the two classifiers agreed on 74.55% of the DEV set and they achieve a very high accuracy on these "agreed" tweets, revealing that these can serve as a slightly noisy new training set. In the final stage, we trained a MNB using TBR based on the "agreed" tweets, since this representation achieved the best results in both ED and DEV sets. We notice that the accuracy falls down to 69.88% on the remaining 25.45% "disagreed" tweets, yielding an overall 85.72%. This reduction is because 8.87% of the training set was wrongly classified upfront and the learned model was partially based on these tweets. Nevertheless, our approach managed to outperform the best individual algorithm (MNB on TBR) by a 3.56%; this difference may be greater when we apply our model to a specific domain, in which case TBRs usually lead to poor results.

5   Results

We compare our results with several baselines and state-of-the-art approaches: the majority class classifier (MC); the four methods that comprise our ensemble classifier; three different lexicon (counting) approaches; and the results obtained by our four models (TBR, FBR, LBR, CR) with 10-fold cross validation. All "averages" presented here are calculated by first averaging the accuracies on every test set (e.g., the average of HCR-dev and HCR-test) and then calculating the overall average.

On average, our model outperformed all out-of-domain models presented in Table 4. The average boost in accuracy is 11.65% compared to the MC, 10.22% compared to the CR trained on ED and 6.2% compared to the best lexicon. Moreover, it manages to compete even with various in-domain models, achieving

Table 4. Comparison of results. The "training" column refers to whether the training was based in the same domain with the test set or not. Results copied from other words are cited. The best in- and out-of-domain classifier is highlighted on every dataset.

Feature	Training	STS-1	STS-2	HCR-dev	HCR-test	OMD	Average
Majority Class	Out	50.14	58.33	**70.75**	**76.65**	62.72	63.55
TBR	Out	77.52	76.40	52.70	50.80	62.52	63.74
FBR	Out	60.52	63.48	49.68	52.81	62.84	58.70
LBR	Out	77.52	76.40	68.68	72.69	70.53	72.73
CR	Out	75.50	71.35	58.10	55.22	64.84	64.98
SentiWordNet	Out	49.86	73.60	51.19	52.10	56.19	56.52
Opinion Lexicon	Out	54.18	76.97	68.90	72.14	70.90	69.00
Subj. Lexicon [13]	Out	-	72.10	54.30	58.10	59.10	62.47
TBR (10-fold)	In	83.29	74.16	**77.75**	**80.36**	79.39	**79.06**
FBR (10-fold)	In	60.23	61.80	76.03	75.15	63.42	66.68
LBR (10-fold)	In	78.10	**83.71**	71.71	79.36	71.32	75.92
CR (10-fold)	In	81.27	70.22	76.46	75.75	**80.92**	77.59
Ensemble	Out	**83.57**	80.90	69.11	69.68	**73.96**	**75.20**

only 3.86% lower accuracy than the best such model. OL is the only lexicon outperforming the MC, while SWN performs poorly. This is probably due to the informal nature of tweets, in which the most common words usually appear and a simplistic approach can achieve better results. Integrating POS tags to TBR provides a small boost, only when trained in a different domain. The LBR model is rather competitive with TBR and CR for in-domain tasks, whereas it confidently outperforms them for cross-domain problems. while the FBR model fails to perform equally well with the other models.

In order to explain our classifier's results, we move on analysing its componenents. HC and LC agree in more than half of the tweets on average (see Table 5, "agreement"), on which the average accuracy is high (81.81%). The higher the agreement level (%), the higher the accuracy that we achieve (correlation = 0.997); if such a claim holds universally, then one could argue about whether the ensemble model could be used effectively or not based on the agreement level. However, with only three datasets in use one cannot draw safe conclusions.

We used the "agreed" tweets to train a MNB classifier on TBR. On average, 59.87% of the originally considered as test data was used for training purposes of our ensemble classifier, whereas the rest 40.13% was used for testing. This means that our ensemble classifier learns a model on a training set in which 19.19% of the tweets are wrongly labeled. However, the same argument may apply on the ED set as well, since a tweet containing a happy emoticon may be ironic instead of positive. As expected, the accuracy is lower in the "disagreed" tweets (65.82%) due to the noisy training set.

Our results indicate that in the HCR set the MC achieves higher accuracy than all out-of-domain methods, probably due to its high imbalance. The HCR-test set is also the only one in which an individual classifier outperforms our ensemble model. This dataset is the one with the lowest agreement level between HC and LC). Taking into account that this agreement level correlates with the accuracy

Table 5. Agreement level and results obtained on the test set by the ensemble classifier

	STS-1	STS-2	HCR-dev	HCR-test	OMD	Average
Agreement(%)	70.03	66.29	52.92	51.00	55.98	59.87
Accuracy (agreed)	88.07	86.44	75.10	77.95	81.64	81.81
Accuracy (disagreed)	73.08	70.00	62.39	61.07	64.19	65.82

on the agreed tweets, there might exist a threshold on the agreed documents, below which the classifier's algorithm should be switched to another model, instead of text-based; we leave this as an open problem for the future.

6   Conclusion

This paper proposed an ensemble algorithm to deal with the domain-dependence problem for the polarity classification task on Twitter. The basic idea is to automatically categorize some tweets of a given domain-specific test set and use them as a new training set. Our results show that combining algorithms trained on different features on a generic training set can prove beneficial, achieving high accuracy (81.81%) on the resulting training set. Using this new dataset for training, we achieve better results than all other out-of-domain approaches tested here, but also to compete against widely-applied in-domain learning methods.

Future work includes the incorporation of the "neutral" class in our problem as well as enhancing syntactic rules in our approach [10]. Finally, we plan to test our approach on different datasets for wider justification and test whether the number of documents for which our two combined classifiers agree on is in fact correlated with the accuracy that will be achieved on the whole test set.

Acknowledgments. This work is supported by the SocialSensor FP7 project, partially funded by the EC under contract number 287975.

References

1. Bollen, J., Mao, H., Pepe, A.: Modeling public mood and emotion: Twitter sentiment and socio-economic phenomena. In: Proceedings of the Fifth International AAAI Conference on Weblogs and Social Media (2011)
2. Saif, H., He, Y., Alani, H.: Semantic sentiment analysis of twitter. In: Cudré-Mauroux, P., Heflin, J., Sirin, E., Tudorache, T., Euzenat, J., Hauswirth, M., Parreira, J.X., Hendler, J., Schreiber, G., Bernstein, A., Blomqvist, E. (eds.) ISWC 2012, Part I. LNCS, vol. ISWC 2012, pp. 508–524. Springer, Heidelberg (2012)
3. Barbosa, L., Feng, J.: Robust sentiment detection on twitter from biased and noisy data. In: Proceedings of the 23rd International Conference on Computational Linguistics: Posters, pp. 36–44. ACL (2010)
4. Bermingham, A., Smeaton, A.F.: Classifying sentiment in microblogs: is brevity an advantage? In: Proceedings of the 19th ACM International Conference on Information and Knowledge Management, pp. 1833–1836. ACM (2010)

5. Gonçalves, P., Araújo, M., Benevenuto, F., Cha, M.: Comparing and combining sentiment analysis methods. In: Proceedings of the First ACM Conference on Online Social Networks, pp. 27–38. ACM (2013)
6. Schinas, E., Papadopoulos, S., Diplaris, S., Kompatsiaris, Y., Mass, Y., Herzig, J., Boudakidis, L.: Eventsense: Capturing the pulse of large-scale events by mining social media streams. In: Proceedings of the 17th Panhellenic Conference on Informatics, pp. 17–24. ACM (2013)
7. Zhang, L., Ghosh, R., Dekhil, M., Hsu, M., Liu, B.: Combining lexicon-based and learning-based methods for twitter sentiment analysis. HP Laboratories, Technical Report HPL-2011 (2011)
8. Tumasjan, A., Sprenger, T.O., Sandner, P.G., Welpe, I.M.: Predicting Elections with Twitter: What 140 Characters Reveal about Political Sentiment. In: ICWSM, vol. 10, pp. 178–185 (2010)
9. Bifet, A., Frank, E.: Sentiment knowledge discovery in twitter streaming data. In: Pfahringer, B., Holmes, G., Hoffmann, A. (eds.) DS 2010. LNCS, vol. 6332, pp. 1–15. Springer, Heidelberg (2010)
10. Jiang, L., Yu, M., Zhou, M., Liu, X., Zhao, T.: Target-dependent twitter sentiment classification. In: Proceedings of the 49th Annual Meeting of the Association for Computational Linguistics: Human Language Technologies, vol. 1, pp. 151–160. ACL (2011)
11. Go, A., Bhayani, R., Huang, L.: Twitter sentiment classification using distant supervision. CS224N Project Report, Stanford, 1–12 (2009)
12. Wilson, T., Wiebe, J., Hoffmann, P.: Recognizing contextual polarity in phrase-level sentiment analysis. In: Proceedings of the Conference on Human Language Technology and Empirical Methods in Natural Language Processing, pp. 347–354. ACL (2005)
13. Speriosu, M., Sudan, N., Upadhyay, S., Baldridge, J.: Twitter polarity classification with label propagation over lexical links and the follower graph. In: Proceedings of the First Workshop on Unsupervised Learning in NLP, pp. 53–63. ACL (2011)
14. Read, J.: Using emoticons to reduce dependency in machine learning techniques for sentiment classification. In: Proceedings of the ACL Student Research Workshop, pp. 43–48. ACL (2005)
15. Brody, S., Diakopoulos, N.: Cooooooooooooooollllllllllllllll!!!!!!!!!!!!!!!: using word lengthening to detect sentiment in microblogs. In: Proceedings of the Conference on Empirical Methods in Natural Language Processing, pp. 562–570. ACL (2011)
16. Andreevskaia, A., Bergler, S.: When Specialists and Generalists Work Together: Overcoming Domain Dependence in Sentiment Tagging. In: Proceedings of ACL 2008, pp. 290–298. ACL (2008)
17. Esuli, A., Sebastiani, F.: Sentiwordnet: A publicly available lexical resource for opinion mining. In: Proceedings of LREC, pp. 417–422 (2006)
18. Hu, M., Liu, B.: Mining and summarizing customer reviews. In: Proceedings of the tenth ACM SIGKDD International Conference on Knowledge Discovery and Data Mining, pp. 168–177. ACM (2004)
19. Toutanova, K., Klein, D., Manning, C.D., Singer, Y.: Feature-rich part-of-speech tagging with a cyclic dependency network. In: Proceedings of the 2003 Conference of the North American Chapter of the Association for Computational Linguistics on Human Language Technology, pp. 173–180 (2003)

A Faceted Crawler for the Twitter Service

George Valkanas, Antonia Saravanou, and Dimitrios Gunopulos

Dept. of Informatics and Telecommunications,
University of Athens, Athens, Greece
{gvalk,antoniasar,dg}@di.uoa.gr

Abstract. Researchers, nowadays, have at their disposal valuable data from social networking applications, of which Twitter and Facebook are the most prominent examples. To retrieve this content, the Twitter service provides 2 distinct *Application Programming Interfaces* (APIs): a probe-based and a streaming one, each of which imposes different limitations on the data collection process. In this paper, we present a general architecture to facilitate *faceted crawling* of the service, which simplifies retrieval. We give implementation details of our system, while providing a simple way to express the crawling process, i.e., the *crawl flow*. We experimentally evaluate it on a variety of faceted crawls, depicting its efficacy for the online medium.

Keywords: Faceted Crawling, Twitter, Architecture, Design, Applications.

1   Introduction

The numerous prototypes built on top of social networks are proof of their importance [19,16,23,22,18,15,14,6,7]. These applications focus on different key properties of the underlying data. **Big data** problems require *voluminous* datasets. **Emergency management** requires access to *locational* information [5,21,12]. **News reporting** must access high-quality information from *high-quality* posters. **Social graphs** are another major asset.

In this paper we focus on Twitter because it is prevalent, and has an open-data policy. Building a system to retrieve the desired information is both time consuming and technically challenging. For example, the service provides two distinct *Application Programming Interfaces* (APIs), with certain limitations: Access is restricted to authenticated (i.e., registered) users, and to public tweets, i.e., tweets visible to anyone. The default *streaming* API returns only 1% of public tweets [1], while the *REST* API limits the number of requests issued within a specific timeframe, imitating the politeness principle [11].

In this paper, we consider the problem of an efficient crawler for the Twitter service, to fetch content with desired properties. Those properties refer to different *facets* of the data (tweets, users, graph, location, etc.), giving rise to the *faceted crawling problem*. We discuss our system's design and implementation details. In summary, we make the following contributions:

---

[1] Elevated access can be granted for a fee.

B. Benatallah et al. (Eds.): WISE 2014, Part II, LNCS 8787, pp. 178–188, 2014.

- We present the architecture and implementation of a *faceted crawler* for the Twitter service.
- We simplify the crawling process through a *crawl flow*, which can multiplex queries in an elegant, yet effective way. We have implemented default crawl flows, which can be used in numerous scenarios.
- We present results on our crawler's performance and discuss lessons learned from our interaction with the service's APIs.

2 Related Work

Harvesting web documents is as old a task as the web itself. Search engines rely on web crawlers [9,10,17] to fetch online documents, which they subsequently index and make available. These are built for the *surface* web, where webpages are reachable through hyperlinks. However, Twitter's multiple information *facets* do not allow a straight-forward modification of existing crawlers, which would also violate the *politeness policy*, given the *real-time* nature of the medium. Similar problems exist for crawlers of the *Hidden Web* [8,13], where information can be accessed through query forms.

Several libraries exist [2], to access Twitter's APIs programmatically, one resource at a time. Therefore, these are not complete solutions, whereas we facilitate the *faceted* crawling process through the *crawl flow*.

The work in [4] also discusses *facets* on Twitter, but in a conceptually differ-ent way. More importantly, our research goals are different: [4] is interested in enriching tweets with "context", whereas we aim at the implementation of an efficient and robust crawler. Therefore, [4] can be thought of as an application, that one can build on top of our proposed infrastructure.

3 Twitter API Background

Twitter provides two main Application Programming Interfaces (APIs) to access publicly available data, i.e., data that anyone can see. The first API is a REST-ful one, and requires probing the service with HTTP requests. The second API is a streaming one, and resembles a publish-subscribe mechanism. In both cases, the user can apply filters, to restrict the information they are looking for. In both cases the user needs to be authenticated through one of the available options. In the next paragraphs, we give a more detailed overview of these two APIs.

3.1 REST API

The REST API uses HTTP requests (i.e., GET, POST) to perform the communi-cation between the end user and the Twitter service. This API supports multiple query types, each of which can be employed by contacting a carefully constructed URL, with all the necessary information.

From the REST API specification [1], we identify four types of restrictions, which we must take into account in our crawler. Table 1 gives additional details.

- **Rate restrictions:** The number of queries of a specific type that the developer can issue within the 15 minute window.
- **Maximum Result Size:** The upper bound on the results of a particular query. For instance, even if a user has posted 5000 tweets, we are only able to access the most recent 3200.
- **Probing Result Size:** The number of results that we can retrieve each time we probe the service with that particular query. For instance, a query for a user's timeline will return *at most* 200 tweets.
- **Maximum Query Size:** The number of objects that we can query simultaneously with a single probe to the service. Typically this is 1, (e.g., 1 tweet each time, using its id), but there are some exceptions (e.g., *lookup* at most 100 users).

3.2 Streaming API

Through this API, one can receive data as a *flow* of tweets. The API returns a 1% sample of all public posts, though not uniformly. Consequently, data received through this API may reflect fluctuations of the actual stream, e.g., increase / decrease of posts, temporal patterns of user interactions, etc. A drawback of this API is that it can not be used for all information facets, e.g., the social graph.

4   Faceted Crawler Architecture

Figure 1 shows the architecture of a classic web crawler [10] on the left, compared against our Faceted crawler architecture, on the right. The two designs appear to be similar for the most part. The contents of the frontier queue, however, in the two cases are different, because surface crawlers need only handle URLs of the next pages to fetch. On the contrary, our crawler needs to handle different query types, each of which takes different parameters.

The components SEEDER, RANKER and STREAMER are also different. The latter exists to harvest data using the *streaming* API. The SEEDER exists to support various applications in a unified way. The RANKER is separate from the scheduler, because one application may combine multiple query types, and to simplify application development.

Table 1. Restrictions for some major query types

Query	Rate	Max Result	Probe Result	API limit
USER LOOKUP	180	∞	100	100
TWEET SHOW	180	1	1	1
FRIENDS	15	∞	5000	1
FOLLOWERS	15	∞	5000	1
TIMELINE	180	3200	200	1
RETWEETS	15	100	100	1

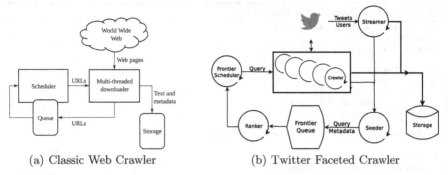

(a) Classic Web Crawler   (b) Twitter Faceted Crawler

Fig. 1. Architectural designs of both classic web crawler and our Twitter faceted crawler

4.1 Streaming API

To obtain information from the **Streaming** API, we employ the STREAMER component, as shown in Figure 1. The component receives the stream, and may forward it for processing, storage and seeding.

4.2 REST-Based Crawling

Aside the STREAMER, the components in Figure 1 largely resemble a classic crawler architecture. However, the actual design is quite different. We have already pointed out the difference in the frontier queue. Moreover, each application may have a different crawling process, therefore we need to efficiently multiplex queries. For this reason, we have decoupled crawling (i.e., accessing the service) from seeding.

4.3 Scheduler

A major component of our system is the SCHEDULER, responsible for queueing crawl tasks. A crawl task contains information about the query type and all of the parameters that accompany it. The SCHEDULER is also responsible for enforcing the rate limits. To achieve this, the component operates in an event-driven manner, shown in Algorithm 1.

The component starts with the query types of the crawling process. It will enqueue these queries to a *timedQueue*, and will trigger the event, leading to the execution of the "EventTrigger()" method. The queries that triggered the alarm are dequeued and passed to the *queue* for crawling. We then reset the timer to trigger for the next query item.

Items in *queue* are processed one by one, by the main scheduler thread. For the current query type, we probe th "Frontier Queue" (Line 3), implemented as a database relation, and pass the result for crawling. The scheduler stores metadata in the database (e.g., statistics) and requeues the query for timely execution, computed through its rate limit.

Algorithm 1. Scheduler Algorithm

> **Input:** Database *db*, Ranker *ranker*
> **Output:** *outQueue*
> **Shared Queue** *queue*, *timedQueue*
>
> //Main Thread
> 1: **while** *!stopped* **do**
> 2:   *qry* ← *queue*.dequeue();
> 3:   *data* ← *ranker*.getNext(*qry*);
> 4:   *outQueue*.enqueue(*qry*, *data*);
> 5:   *db*.store(*qry.qryMeta*);
> 6:   *timedQueue*.enqueue(*qry*, NOW + *qry*.ival);
>
> EventTrigger()
> 7: *nextQuery* ← *timedQueue*.dequeue();
> 8: *top* ← *timedQueue*.top();
> 9: *queue*.enqueue(*nextQuery*);
> 10: resetTimer(*top*.TIME - NOW);

Algorithm 2. Seeder Algorithm

> **Input:** Database *db*, ResultQueue $\mathcal{RQ}$, CrawlFlow $\mathcal{CF}$
>
> 1: **while** *!stopped* **do**
> 2:   (*result*, *qry*) ← $\mathcal{RQ}$.dequeue();
> 3:   storeResult(*result*);
> 4:   update = $\mathcal{CF}$.stepSeeding(*qry*, *result*);
> 5:   **if** (update) **then**
> 6:   *nxtQrs* ← $\mathcal{CF}$.nextQueries(*qryMeta*);
> 7:   **for** (*i* = 0; *i* ¦ *nxtQrs*.size; *i*++) **do**
> 8:   *nxtQrs*(*i*).stepSeeding(*qry*, *result*);
> 9:   *db*.store(*qry.qryMeta*);

4.4 Ranker

As seen in Algorithm 1, the scheduler relies on a **Ranker** object. A **Ranker** implements our **IRanker** interface, shown in Figure 2. The **init()** method is used to properly initialize resources (e.g., database relations). The *getNext()* method returns the next item to submit to a crawler as our next query. The *id* is decided by the scheduler to simplify the architecture. The *query* also contains the **RateLimit** information. This allows for a common interface across queries.

4.5 Seeder

As shown in Algorithm 2, the SEEDER operates in an endless loop, much like the SCHEDULER. It receives information from the result queue $\mathcal{RQ}$ (Line 2), where crawlers write the result of probing Twitter. Results are forwarded for storage (line 3). We then update the frontier in two steps. First, update the

```
public interface IRanker{
  public void init();
  public List getNext( long qid, Query query );
}
```

Fig. 2. The IRanker interface

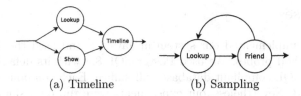

(a) Timeline   (b) Sampling

Fig. 3. Schematic representation of Crawl Flow examples

current query (line 4). The result is a binary variable (TRUE/FALSE), that determines whether we should move to the second step, i.e., update subsequent frontiers (lines 5-8).

4.6   The QUERYLOG Relation

To restart after a (forceful) shutdown, and monitor our system's performance, we store appropriate information, in a relational table, called QUERLOG. A partial view is shown in Table 2. Statistics of the system include rank time, seed time, etc (not shown here). Fields **qid** and **result** are straightforward. Values of the **result** field can be found in [2]. The next three fields ensure that the rate limits are enforced in cases of failure or restarts.

4.7   Crawl Flow

To further simplify the crawling process, we introduce the concept of a CRAWL FLOW. The idea is that on Twitter, a crawl is driven by the underlying application, which can be generally expressed as a sequence of faceted probes (with cycles). The CRAWL FLOW can be thought of as a state automaton, and defines the sequence of the queries to the service. Figure 3 shows a schematic representation of two Crawl Flows that we provide, a user's timeline, and sampling the social graph. Through the CRAWL FLOW, the user specifies:

- The general execution sequence of queries. The sequence may contain loops (including self-loops), depending on the goal.
- An object implementing the IRANK interface.
- An object implementing the ISEED interface.

Table 2. Fields of the QUERYLOG relation

Field	Description
qid	Unique Identifier for the Query
result	Code Signifying how the query
toq	The type of query associated with this tuple
crawler	An identifier for a crawler
tssched	Timestamp when this query was scheduled

5   Use Cases

We have fully implemented our system in Java 1.6 and used the Twitter4j library[2] for method probes. We used PostgreSQL 8.4 with its default configuration, but any SQL-compliant database will suffice. Each component runs on a separate thread. Nevertheless, our experiments were run on a single quad core machine @3.4GHz, with 16Gb of RAM, though half of it was set as Java's heap space, and Ubuntu Linux 64bit.

5.1   Crawling by Location

Location is a very important aspect of tweets. Tweets with location can be retrieved through a geographical filter, specified as 2D bounding box with GPS. Despite its accuracy, GPS is not the sole approach to geocode data. External geocoders [21] can be applied directly to the streaming API. Figure 4 shows the number of tweets (on the left) and users (on the right). Custom geocoding (GEOCODED) can extract an additional 10% to GPS-filtered information (CRAWLED).

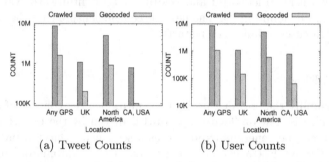

(a) Tweet Counts   (b) User Counts

Fig. 4. Comparing raw counts between crawled and geocoded locational information

Table 3. Jaccard Similarity between the GPS enabled crawls using the Streaming API

	ANY GPS	UK	N AMER	CA, USA
ANY GPS	1.0	0.069	0.249	0.038
UK	0.057	1.0	6×10^{-4}	0.001
N AMER	0.218	0.0	1.0	0.138
CA, USA	0.042	0.0	0.145	1.0

Changing the bounding box of a crawl returns different results, even when the boxes overlap. Table 3 depicts the similarities in terms of received users (upper right, in red) and tweets (lower left, in blue), where each crawler is configured to monitor the corresponding location. Despite some commonly shared users and tweets, a lot of new content is being delivered by each stream.

---

[2] http://twitter4j.org

5.2   Crawling User Timelines

User timelines are useful in several cases, e.g., behavior analysis. To efficiently crawl a user's timeline, it is best to know the number of expected results.

Crawling Basic User Information: We improve user information extraction by 1%, as seen in Figure 5(a), through the combination of two query types. Figure 5(b) depicts this improvement (blue line) as a percentage. The 1 less query out of every 4, shown in Figure 5(a), is due to our best-effort approach for crawling the service, which has tight time constraints.

Crawling the Timeline: Figure 6(a) shows how much time is spent on the *get-Next* method by the RANKER component, which is below 10ms. Both the *getNext* and *stepSeeding* methods do not take, on average, more than 4ms (Figure 6(b)).

The average elapsed time between consecutive schedulings of Timeline queries is shown in Figure 6(c). The interval does not increase, but stabilizes over time. Even with multiple crawlers, our system maintains its performance. With 10 crawlers, we are at more than 98.5% of the optimal case for Timeline queries (Figure 7(a)) perform similarly for Lookup queries (Figure 7(b)). Evidently, more crawlers yield more results faster (Figure 7(c)).

5.3   Sampling

For large networks, sampling is important. We have implemented the Metropolis-Hastings algorithm [20], through the IRANK and ISEED interfaces. Instead of *thinning*, we use reservoir sampling on the collected nodes, which has the same effect. Figure 8a) shows the time required for ranking and seeding. This is the only case where the timings are high, compared to other use cases, and eventually reach an average of ~2.5 seconds. The reason is that sampling, synchronizes on shared resources. Regardless, our implementation is well within the timeframe between subsequent queries of this type (60 seconds).

5.4   Additional Use Cases

Retweet Graph. Retweets are a key concept in Twitter, allowing users to re-post / endorse tweets of others. They can be used to identify cascades of

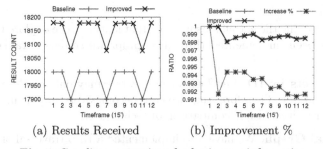

(a) Results Received   (b) Improvement %

Fig. 5. Crawling comparison for basic user information

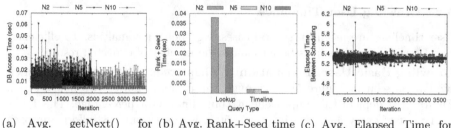

(a) Avg. getNext() for Lookup

(b) Avg. Rank+Seed time

(c) Avg. Elapsed Time for Scheduling Timeline

Fig. 6. Average crawler performance for harvesting (a) user information, (b)-(c)and user timelines

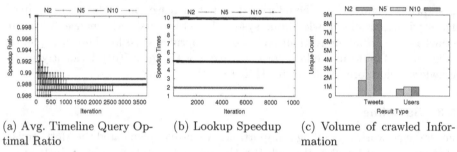

(a) Avg. Timeline Query Optimal Ratio

(b) Lookup Speedup

(c) Volume of crawled Information

Fig. 7. Collective performance of the crawlers used for harvesting user timelines

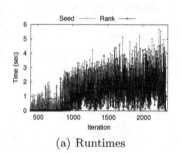

(a) Runtimes

Fig. 8. Sampling scenario statistics

information, leaving the actual reason as a latent feature to be explored. Crawling retweets has been implemented through appropriate seeders / rankers in our framework.

Retrieve tweets by ID. Per Twitter policy, one may only disclose the tweet IDs of their dataset. Therefore, this use case becomes very important for reproducibility of results and fair comparison of techniques.

Crawl Social Graph. We have also implemented a BFS traversal of the social graph of Twitter, through Seeders and Rankers.

6   Conclusion

In this paper we presented a framework for faceted crawling of the Twitter service. Our framework respects rate limits imposed by the service, and interleaves queries to boost performance. We simplify the crawling process through the *Crawl Flow* concept.

Acknowledgements. This work has been co-financed by a Heraclitus II fellowship and EU project INSIGHT.

References

1. https://dev.twitter.com/docs/api/1.1
2. https://dev.twitter.com/docs/twitter-libraries
3. https://dev.twitter.com/docs/twitter-libraries
4. Abel, F., Celik, I., Houben, G.-J., Siehndel, P.: Leveraging the semantics of tweets for adaptive faceted search on twitter. In: ISCW, pp. 1–17 (2011)
5. Ahmed, A., Hong, L., Smola, A.J.: Hierarchical geographical modeling of user locations from social media posts. In: WWW (2013)
6. Bakshy, E., Hofman, J.M., Mason, W.A., Watts, D.J.: Everyone's an influencer: quantifying influence on twitter. In: WSDM, pp. 65–74 (2011)
7. Barbieri, N., Bonchi, F., Manco, G.: Influence-based network-oblivious community detection. In: ICDM (2013)
8. Bergman, M.: The deep web: Surfacing hidden value. Technical report (2001)
9. Brin, S., Page, L.: The anatomy of a large-scale hypertextual web search engine. In: WWW, pp. 107–117 (1998)
10. Castillo, C.: Effective web crawling. SIGIR Forum 39(1), 55–56 (2005)
11. Cho, J., Garcia-Molina, H.: Effective page refresh policies for web crawlers. ACM Trans. Database Syst. 28(4), 390–426 (2003)
12. Eisenstein, J., O'Connor, B., Smith, N.A., Xing, E.P.: A latent variable model for geographic lexical variation. In: EMNLP (2010)
13. Garcia-Molina, H.: Challenges in crawling the web. In: James, A., Younas, M., Lings, B. (eds.) BNCOD 2003. LNCS, vol. 2712, p. 3. Springer, Heidelberg (2003)
14. Ghosh, S., Korlam, G., Ganguly, N.: Spammers' networks within online social networks: a case-study on twitter. In: WWW, pp. 41–42 (2011)
15. Gjoka, M., Kurant, M., Butts, C.T., Markopoulou, A.: Walking in facebook: A case study of unbiased sampling of osns. In: INFOCOM, pp. 2498–2506 (2010)
16. Grier, C., Thomas, K., Paxson, V., Zhang, M.: @spam: The underground on 140 characters or less. In: CCS, pp. 27–37 (2010)
17. Heydon, A., Najork, M.: Mercator: A scalable, extensible web crawler. World Wide Web 2(4), 219–229 (1999)
18. Kamath, K.Y., Caverlee, J.: Content-based crowd retrieval on the real-time web. In: CIKM, pp. 195–204 (2012)
19. Kwak, H., Lee, C., Park, H., Moon, S.: What is twitter, a social network or a news media? In: WWW, pp. 591–600 (2010)

20. Stutzbach, D., Rejaie, R., Duffield, N.G., Sen, S., Willinger, W.: On unbiased sampling for unstructured peer-to-peer networks. In: Internet Measurement Conference, pp. 27–40 (2006)
21. Valkanas, G., Gunopulos, D.: Location extraction from social networks with commodity software and online data. In: ICDM Workshops (SSTDM) (2012)
22. Valkanas, G., Gunopulos, D.: How the live web feels about events. In: CIKM, pp. 639–648 (2013)
23. Weng, J., Lee, B.-S.: Event detection in twitter. In: ICWSM (2011)

Diversifying Microblog Posts

Marios Koniaris[1], Giorgos Giannopoulos[2],
Timos Sellis[3], and Yiannis Vassiliou[1]

[1] School of ECE, NTU Athens, Greece
[2] IMIS Institute, "Athena" Research Center, Greece
[3] RMIT University, Melbourne, Australia

Abstract. Microblogs have become an important source of information, a medium for following and spreading trends, news and ideas all over the world. As a result, microblog search has emerged as a new option for covering user information needs, especially with respect to timely events, news or trends. However users are frequently overloaded by the high rate of produced microblogging posts, which often carry no new information with respect to other similar posts. In this paper we propose a method that helps users effectively harvest information from a microblogging stream, by filtering out redundant data and maximizing diversity among the displayed information. We introduce microblog posts-specific diversification criteria and apply them on heuristic diversification algorithms. We implement the above methods into a prototype system that works with data from Twitter. The experimental evaluation, demonstrates the effectiveness of applying our problem specific diversification criteria, as opposed to applying plain content diversity on microblog posts.

Keywords: Diversification, information retrieval, microblogging services, twitter, data mining.

1 Introduction

Microblog platforms, such as Twitter, host vast quantities of user generated information content. The wealth of information that is produced is, however, lost in its vast numbers and, as a result, cannot be exploited by users (with limited time for reading microblog posts) without first being analysed and refined. In this work, what we consider as refinement is presenting to users a subset of microblog posts, that contains highly relevant and, at the same time, heterogeneous information, representing different aspects of the microblog post, according to user desired diversification criteria.

There has been extensive work on query results diversification [see related work section 2], where the key idea is to select a small set of web results that are sufficiently diverse, according to a (dis-)similarity metric. These similarity metrics are usually based on the content of the items to be diversified (e.g. documents, web pages) or their inclusion to specific classes of a taxonomy describing them. Although these models have proven to be effective on the scenario of search result diversification, we propose that they are insufficient in our setting:

B. Benatallah et al. (Eds.): WISE 2014, Part II, LNCS 8787, pp. 189–198, 2014.

(i) Microblog posts are too short for text-based distance functions to be effective, (ii) Users search microblog posts to find temporally relevant information and information related to people, places, organizations, events (i.e. named entities) which is different from web search [17], (iii) Users repeat microblog queries to monitor the associated search results and (iv) Microblog posts are overwhelmed by opinions and sentiments expressed through them.

To this end, we define multiple dimensions over the information source, such as geographical and temporal attributes, sentiment, contained named entities, readability and social influence and exploit them as diversification criteria to obtain heterogeneous posts covering user information needs, in order to capture the specific characteristics of microblog posts. Also, we define metrics to measure the diversity of posts using the notion of Information Nuggets (initially defined in [4]), which represent minimum quantities of topics, aspects, opinions identified within a text passage. Finally, we compare our proposed diversification variations with the baseline of diversifying on content, demonstrating, thus, the effectiveness of our proposed diversification criteria.

The remaining paper is organized as follows. We firstly review related work in Section 2. In Section 3 we provide background information on diversification algorithms and present our method for diversifying microblog posts. In Section 4, we present an evaluation study that demonstrates the effectiveness of the proposed method. Finally, Section 5 concludes and discusses further work.

2 Related Work

In this section, we present several approaches that deal with the problems of search result diversification and microblog search.

Query results diversification has attracted a lot of attention in the field of text mining. Drosou and Pitoura [6] present a thorough review of fundamental works in diversification. The maximal marginal relevance criterion (MMR), presented in [2], is one of the earliest works on diversification. In [11] a set of diversification axioms is introduced and it is proven that it is not possible for a diversification algorithm to satisfy them all. The authors of [1] propose a greedy diversification algorithm and extend information retrieval evaluation measures, in order to apply them in the context of diversification. Santos et al. [13] diversify explicit sub-queries of the original query. They maximize the semantic coverage with respect to different aspects of the original query.

Several studies on microblog search use different features and employ different algorithms for ranking microblog posts. The authors of [12] integrate a variety of features in a Bayesian model. In [19] learning to rank algorithms are employed and the relevance of a tweet as the re-tweet likelihood of the tweet is defined. Most of the studies on microblog search are concerned only with query relevance and neglect diversity of posts. Diversity of microblog search results has been addressed in [16] with an approach based on MMR and clustering of tweets. However, their methods are based solely on textual similarity of the handled items. A study on event summarization [15], proposes methods for finding the most representative set of tweets about an event from an initial set of tweets. Although event summarization from Twitter and Twitter search share the ultimate goal of finding the representative set of tweets on a topic, we note that a query is more generic than an event. The problem of diversifying microblog posts is also addressed in [3], where the authors compute the smallest subset of posts that cover all other posts with respect to a diversity dimension. Our work focuses on identifying microblogging posts that refer to multiple dimensions, whereas [3] provides a solution only for one dimension. Furthermore, instead on focusing on the coverage of the result set, we aim at maximizing diversity of the result set. Finally, in our previous works in [10,9] we handled the problem of diversifying user comments in news articles. Our current work extends the set of diversification criteria defined in our previous work so that they capture the characteristics of microblog posts.

3   Microblog Diversification

Within this section, we first define the problem addressed in this paper. Then, an overview of microblog posts features that are relevant to our work is provided. Afterwards we present the proposed set of posts-specific diversification criteria which are used to model the specific characteristics of posts on microblog platforms. Then we describe the diversification algorithms and the distance functions, for similarity and diversity, that we utilize in order to apply the proposed diversification criteria.

3.1   Problem Formulation

Given a set of microblog posts and a query, our aim is to find a set of relevant and representative posts, that, on the same time, maximizes diversity w.r.t. an objective function. More specifically, the problem is formalized as follows:

Definition 1 (Microblog posts diversification). *Let q be a user query and N a set of microblog posts relevant to the user query. Find a subset $S \subset N$ of posts that maximize an objective function f that quantifies the diversity of posts in S.*

Typical diversification techniques measure diversity in terms of content. That is, textual similarity between items is used in order to quantify (dis)similarity. In this work, following our previous work on news comments diversification [10], we extend the notion of diversity on additional dimensions, besides textual similarity, in order to capture the characteristics of microblog posts.

Microblog posts have unique attributes, which differentiate our research from previous works in the literature. First, the maximum length of a microblog post is rather short. Likewise the language used differs significantly: since users post messages to microblog services from many different media, including their cell phones, the frequency of misspellings and slang in their posts is much higher than in other domains. Further, location, time and sentiment are significant determinants within microblog posts.

3.2   Diversification Criteria

Based on the aforementioned characteristics of microblog posts we define the
following diversification criteria:

- **Sentiment.** Users, through their posts, express opinions and sentiment
  about various entities. We define three classes of sentiment [positive, negative, neutral], with each class measured in a scale of [0..4]. For each post,
  we construct a three feature vector, with each feature corresponding to a
  different sentiment class. Within each feature we store the strength scale of
  each sentiment class, as expressed in the post.
- **Named Entities.** The microblog post text may reference Persons, Organizations or Locations. Named Entities (NEs) are important in terms of
  diversity: since microblog posts often follow emerging events, they are expected to revolve around e.g. Persons or Places. For the aforementioned NE
  categories, we create a vector, where each feature value corresponds to the
  frequency of each distinct NE identified in the post's text.
- **Location.** Users can have their microblog posts annotated with their exact
  locations, expressed in longitude and latitude. Having the spherical coordinates [1] of users posts ,we make use of the widely adopted haversine formula[2],
  which provides great-circle distances between two points on a sphere from
  their longitudes and latitudes. Values acquired by the haversine formula are
  then normalized in [0..1], divided by the earth radius. Having normalized geographical values associated with users posts we can use a diversity function
  that measures the spatial distance between two microblog posts.
- **Time.** The temporal pattern associated with tweets plays an important role
  in acquiring a diverse set of users' posts since on-line content grows and
  fades over time. Each post has a temporal attribute, a timestamp t_i that is
  measured in milliseconds. Timestamps of users posts are normalized in the
  range [0..1] by applying Min-Max Normalization to the data corpus.
- **Readability.** We propose that post writing quality is a diversification factor,
  since it expresses comprehensibility of the post itself. The most influential
  quantitative measure of text quality is the Flesch Reading Ease Score[3], which
  produces a numerical score, with higher numbers indicating easier texts. [5]
  use a modified version of the Flesch Reading Ease equation to accommodate
  for tweets short format. We also treat each post as having a single sentence.
  For each post in the corpus we apply the modified Flesch Reading Ease Score
  formula and assign a reading score to it.
- **Social Influence.** In microblog services users share a common need; to
  spread their messages to as many users as possible. We propose that the
  influence of microblog users is an important factor to diversify the result
  list of relevant with the user information need posts. In [14], the authors
  proposed a methodology for calculating the importance and the influence of a

---

[1] http://mathworld.wolfram.com/SphericalCoordinates.html
[2] http://en.wikipedia.org/wiki/Haversine_formula
[3] http://en.wikipedia.org/wiki/Readability

Twitter account. We make use of their influence metric, which quantifies the diffusion of information in a microblog service. For each microblog user in the corpus, we acquire their associative Social Influence (SI) score through the use of InfluenceTracker[4]. As before, SI scores of users posts are normalized by applying Min-Max Normalization to the data corpus.

- **Content - Microblog posts similarity.** Finally, we consider the posts content, which is the baseline diversification criterion, used in most works handling diversification. For each post, we construct its term vector, with each feature corresponding to each distinct term found in the whole corpus. Each feature value is computed by normalizing the term's frequency within the post by the total number of terms the post contains.

3.3   Diversification Process

Diversity is closely related to the p-dispersion problem [7]. In the p-dispersion problem the objective is to select p out of n given candidate points, such that the minimum distance between pairs of the selected points is maximized. It has been proven to be NP hard and several heuristics have been proposed. In [11] three diversification objectives where introduced: *Max-Sum*, *Max-Min* and *Mono-Objective*. We adopt these objectives in our framework.

With the previously defined criteria we are able to map each microblog post to the diversification dimensions. For each post we construct distinct vector categories, corresponding to the seven defined diversification dimensions, which represent the post: Sentiment, Named Entities, Location, Time, Readability, Social Influence and Content. We then apply our diversification algorithms on the microblog post, on each of the above dimensions, producing, each time, a separate diversity (dissimilarity) score. Finally, these scores are aggregated into a dissimilarity score that is their weighted sum.

A diversity function, that measures the distance between two items, is needed in order to produce a diversity score. We make use of the widely adopted similarity score and define the diversity score of two items, u, v, for a specific dimension i, as: For diversification dimensions *[Sentiment, Named Entities, Content]i*, expressed in vector forms, $V(u), V(v)$ for items u and v, we apply cosine similarity:

$$d_i(u,v) = 1 - \cos_i(V(u), V(v)) \tag{1}$$

while for **normalized** dimensions *[Location, Time, Readability, Social Influence]* i, expressed in numerical values $N(u), N(v)$, for items u and v, we measure the absolute difference of the two items' values:

$$d_i(u,v) = 1 - |N_i(u) - N_i(v)| \tag{2}$$

The relevance score r of a microblog post u, for a query q is calculated by applying the cosine similarity measure on the query and the posts term vectors:

$$r(u,q) = \cos(V(u), V(q)) \tag{3}$$

Our aim is to gather heterogeneous posts with respect to the described diversification dimensions. However, diversity is not the only objective since these posts ought to be relevant to the initial information need, the user query. The final

---

[4] http://influencetracker.com/

score of each post is a weighted sum of its relevance score to the query and its diversity score to the already selected microblog posts.

Let N be a set of microblog posts relevant to the user query, S be the set of diverse items, $r(u,q)$ the similarity score of the post u to the user query q, $d(u,v)$ the diversity score (distance) between posts u and v and $w \epsilon [0,1]$ a parameter specifying the trade-off between relevance and diversity. The following objectives are considered:

- **Max-Sum:** The first objective aims at maximizing the sum of the relevance to the query and the pairwise dissimilarity between posts in the result set. The score for each candidate post u to be inserted into the set of diverse posts S, for the query q can be expressed as:

$$score_{MAXSUM}(u,v,q) = (1-w) \cdot \frac{r(u,q) + r(v,q)}{2} + w \cdot \sum_{i=1}^{|D|} \lambda_i \cdot d_i(u,v) \quad (4)$$

  where i is the diversification dimension, $|D|$ is the number of diversification dimensions, and $\lambda_i \epsilon [0,1]$ is the weight of each individual diversity score, with $\sum_{i=1}^{|D|} \lambda_i = 1$.

- **Max-Min:** The second objective aims at maximizing the minimum relevance and dissimilarity of the selected set. The score for each candidate microblog post is:

$$score_{MAXMIN}(u,q) = (1-w) \cdot r(u,q) + w \cdot \sum_{i=1}^{|D|} \lambda_i \cdot d_i(u, \min v_{iu}) \quad (5)$$

  where $\min v_{iu}$ is the post from the current diverse set that has the minimum distance to candidate post u.

- **Mono-Objective:** The third objective, combines the relevance and the similarity values into a single value for each document. The score is defined as:

$$score_{MONO}(u,q) = (1-w) \cdot r(u,q) + w \cdot \sum_{i=1}^{|D|} \lambda_i \cdot \frac{1}{|N|-1} \sum_{v \epsilon N} d(u,v) \quad (6)$$

4   Evaluation

4.1   Dataset Used

Our microblog posts dataset consists of data that we collected from Twitter using its streaming API, during the Eurovision song contest. Microblog posts contain novel syntax that makes it challenging for existing information retrieval models to achieve accurate results. Thus, a preprocessing step is necessary. We removed non-English and non-geographically annotated tweets. For each post, we removed uninformative noun phrases (e.g. common words) and specific symbols (e.g. #, @, url). We extracted named entities using the Stanford Named Entity Recognizer[5] [8], sentiment using Sentistrength [18] and social influence using

---

[5] http://nlp.stanford.edu/software/CRF-NER.shtml

InfluenceTracker[6] [14]. Finally, we calculated and normalized the values for the temporal, geographical and readability dimensions for each post.

4.2 Methodology

Our evaluation methodology is based on the concept of Information Nuggets. As presented in [4] Information Nuggets represent properties of information source at one end and components of an information need at the other. Based on our adaptation of the definition of Information Nuggets for user posts, presented in [9], we define the following metrics to quantify diversification:

Nugget Coverage - NC@n. It is a Precision@N-based metric that measures how many Nuggets each result set of posts contains. It is defined as follows:

$$NC@n = \frac{\sum_{k=1}^{n} I_k}{n \cdot |I|} \tag{7}$$

where n is the number of top posts, I_k the number of distinct Information Nuggets contained in post k and $|I|$ the total number of distinct Information Nuggets.

Distinct Nugget Coverage. This measure is complementary to the first one and counts the ratio of the **distinct** Nuggets found in the result set to the total of distinct Information Nuggets:

$$DN@n = \frac{\sum_{i=1}^{|I|} DFI_i}{|I|}, DFI_i = \begin{cases} 1 : FI_i > 0 \\ 0 : FI_i = 0 \end{cases} \tag{8}$$

where FI_i is the frequency of nugget i in the set of top-n comments.

Nugget Uniformity NU@n. With the third measure we try to quantify the variance of Nuggets within posts, demanding that the nuggets are as uniformly distributed as possible. Nugget Uniformity is defined as follows:

$$NU@n = \frac{\sum_{i=1}^{|I|} (FI_i - \bar{I})^2}{|I|}, \bar{I} = \frac{\sum_{i=1}^{|I|} FI_i}{|I|} \tag{9}$$

We applied unsupervised Latent Dirichlet Allocation (LDA) using Mallet[7], to generate 10 topics on the corpus. Each trained topic is a set of keywords with corresponding weights. We keep the top 20 highest-weight keywords for each topic. Then for each topic found by LDA, two researchers in our lab generated a sample user query using an arbitrary number of the highest-weight keywords for each topic. The chosen queries to represent user information needs are: *russia europe, song contest, conchita, eurovision final, watch, poland girls, party time, sing, eurovison winner and tonight.* We evaluated the following scenarios, utilizing the combination of diversification criteria as:

- **Content Diversity - CONTENTDIV.** The baseline that applies diversification using only the criterion of Content diversity.
- **Named Entities Diversity - NEDIV.** The diversification variation that uses only the criterion of Named Entities.

---

[6] http://influencetracker.com/
[7] http://mallet.cs.umass.edu/

- **Sentiment - SENTIDIV.** The variation that uses only the criterion of Sentiment.
- **Readability - READABILITYDIV.** The variation that uses only the criterion of Readability.
- **Time - TIMEDIV.** The variation that uses only the criterion of time.
- **Location - PLACEDIV.** The variation that uses only the criterion of Location.
- **Social Influence - USERINFLUENCEDIV.** The variation that uses only the criterion of Social Influence.
- **HYBRID - HYBRIDDIV.** The variation that uses all the above criteria, with equally weighted factors.
- **SEMIHYBRID - SEMIHYBRIDDIV.** The variation that uses the ContentSim, NamedEntity, Sentiment, Time, Place criteria, with equally weighted factors.

Each of the above variations was run for each of the diversification algorithms presented and for each of the generated user queries. For each of the evaluation scenarios presented previously we return the top-10 result tweets. We note that, when diversification criteria are combined, their scores are weighted equally to produce the final diversification score. Also, for all variations that apply diversity, we set a fixed weight for the diversity score to $w = 0.5$ and, thus, the weight for query-to-post similarity is $(1 - w) = 0.5$.

In terms of satisfying users' information needs, the vast amount of microblog posts poses a barrier for the human user to be involved in the evaluation process. Thus, we designed an automatic approach for extracting Information Nuggets from the diversification results. Specifically, we used OpenCalais Web Service[8], a service which automatically creates rich semantic metadata for user submitted content. We extracted, for each result, it's corresponding Information Nuggets and then calculated the previously defined diversification metrics.

In Figure 1, we present the Nugget Coverage, Distinct Nugget Coverage and Nugget Uniformity values respectively, for post result positions from 1 to 10. We note that Nugget Coverage and Distinct Nugget Coverage are normalized, by definition, in the interval [0..1]. For Nugget Coverage, our first observation is that the baseline (CONTENTDIV) is outperformed by SEMIHYBRID variation, until the sixth result, (and by HYBRID and PLACEDIV until the third result) and after result 6, it accomplishes greater values. In total, the baseline seems to perform slightly better than the rest variations. However, in Distinct Nugget Coverage, the SEMIHYBRID and PLACEDIV variations distinctively outperform all other variations (including the baseline), followed by the variation of combining all criteria (HYBRID). This is a stronger indication of diversity, since it regards distinct "pieces" of diverse information found within posts.

Nugget Uniformity measures, through a variance-like formula, the differences in Nugget frequencies within each method's result post sets. We note that here, in contrast with the other two measures, lower values mean better performance. Here, PLACEDIV and SEMIHYBRID clearly outperform all others variations,

---

[8] http://www.opencalais.com/

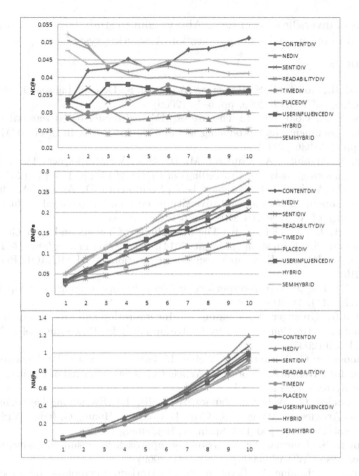

Fig. 1. Average effectiveness of each criterion variation of all different algorithms

signifying that the heterogeneity of nuggets is better distributed with the combination of diversification criteria. Overall it is demonstrated that more refined criteria than plain content similarity can improve the effectiveness of the diversification process.

5   Conclusion

In this paper we proposed a set of diversification criteria on which we adapt state of the art diversification algorithms, in order to present diverse sets of microblog posts on user queries. Our objective is to compute the quintessential subset of posts that contains heterogeneous information, representing different aspects of the microblog post. We conducted an evaluation study that demonstrated the effectiveness of our proposed method, as opposed to applying plain content diversity on microblog posts. In the future, we intend to test whether hybrid algorithm combinations could achieve even better effectiveness by enhancing the definitions of diversification criteria. Also we aim to study how our solution can be adapted in other settings, e.g. Question Answering systems.

References

1. Agrawal, R., Gollapudi, S., Halverson, A., Ieong, S.: Diversifying search results. In: Proceedings of WSDM 2009, pp. 5–14 (2009)
2. Carbonell, J., Goldstein, J.: The use of MMR, diversity-based reranking for reordering documents and producing summaries. In: Proceedings of ACM SIGIR 1998, pp. 335–336 (1998)
3. Cheng, S., Arvanitis, A., Chrobak, M., Hristidis, V.: Multi-Query Diversification in Microblogging Posts. In: Proceedings of EDBT 2014, pp. 133–144 (2014)
4. Clarke, C.L.A., Kolla, M., Cormack, G.V., Vechtomova, O., Ashkan, A., Büttcher, S., MacKinnon, I.: Novelty and diversity in information retrieval evaluation. In: Proceedings of ACM SIGIR 2008, pp. 659–666 (2008)
5. Davenport, J.R.A., DeLine, R.: The Readability of Tweets and their Geographic Correlation with Education (preprint, 2014), http://arxiv.org/abs/1401.6058
6. Drosou, M., Pitoura, E.: Search result diversification. ACM SIGMOD Record 39, 41 (2010)
7. Erkut, E.: The discrete p-dispersion problem. European Journal of Operational Research 46(1), 48–60 (1990)
8. Finkel, J.R., Grenager, T., Manning, C.: Incorporating non-local information into information extraction systems by Gibbs sampling. In: Proceedings of ACL 2005, pp. 363–370 (2005)
9. Giannopoulos, G., Koniaris, M., Weber, I., Jaimes, A., Sellis, T.: Algorithms and Criteria for Diversification of News Article Comments. Journal of Intelligent Information Systems (2014)
10. Giannopoulos, G., Weber, I., Jaimes, A., Sellis, T.: Diversifying User Comments on News Articles. In: Wang, X.S., Cruz, I., Delis, A., Huang, G. (eds.) WISE 2012. LNCS, vol. 7651, pp. 100–113. Springer, Heidelberg (2012)
11. Gollapudi, S., Sharma, A.: An axiomatic approach for result diversification. In: Proceedings of WWW 2009, pp. 381–390 (2009)
12. Jabeur, L.B., Tamine, L., Boughanem, M.: Uprising microblogs. In: Proceedings of SAC 2012, pp. 943–948 (2012)
13. Ounis, I., Santos, L.R.T., Peng, J., Macdonald, C.: Explicit search result diversification through sub-queries. In: Gurrin, C., He, Y., Kazai, G., Kruschwitz, U., Little, S., Roelleke, T., Rüger, S., van Rijsbergen, K. (eds.) ECIR 2010. LNCS, vol. 5993, pp. 87–99. Springer, Heidelberg (2010)
14. Razis, G., Anagnostopoulos, I.: InfluenceTracker: Rating the impact of a Twitter account (preprint, 2014), http://arxiv.org/abs/1404.5239
15. Ren, Z., Liang, S., Meij, E., de Rijke, M.: Personalized time-aware tweets summarization. In: Proceedings of SIGIR 2013, pp. 513–522 (2013)
16. Perez, J.A.R., Moshfeghi, Y., Jose, J.M.: On using inter-document relations in microblog retrieval. In: WWW 2013 Companion Proceedings, pp. 75–76 (2013)
17. Teevan, J., Ramage, D., Morris, M.R.: #TwitterSearch: a comparison of microblog search and web search. In: Proceedings of WSDM 2011, pp. 35–44 (2011)
18. Thelwall, M., Buckley, K., Paltoglou, G., Cai, D., Kappas, A.: Sentiment strength detection in short informal text. Journal of the American Society for Information Science and Technology 61(12), 2544–2558 (2010)
19. Zhang, X., He, B., Luo, T., Li, B.: Query-biased learning to rank for real-time twitter search. In: Proceedings of CIKM 2012, pp. 1915–1919 (2012)

MultiMasher: Providing Architectural Support and Visual Tools for Multi-device Mashups

Maria Husmann, Michael Nebeling, Stefano Pongelli, and Moira C. Norrie

Department of Computer Science, ETH Zurich 8092 Zurich, Switzerland
{husmann,nebeling,norrie}@inf.ethz.ch

Abstract. The vast majority of web applications still assume a single user on a single device and provide fairly limited means for interaction across multiple devices. In particular, developing applications for multi-device environments is a challenging task for which there is little tool support. We present the architecture and tools of MultiMasher, a system for the development of multi-device web applications based on the reuse of existing web sites created for single device usage. Web sites and devices can be mashed up and accessed by multiple users simultaneously, with our tools ensuring a consistent state across all devices. MultiMasher supports the composition of arbitrary elements from any web site, inter-widget communication across devices, and awareness of connected devices. We present both conceptual and technical evaluations of MultiMasher including a study on 50 popular web sites demonstrating high compatibility in terms of browsing, distribution and linking of web site components.

Keywords: Web site mashups, distributed user interfaces, multi-device mashups.

1 Introduction

Despite the widespread use of diverse computing devices in our daily lives, the overwhelming majority of applications are still built to be used on one device at a time. Creating applications that integrate multiple different devices is challenging and there is little tool support to cater for multi-device environments.

While there has been a great deal of research in the area of distributed user interfaces (DUI), most of it is either focused on building new applications from scratch [1,2] or on distributing a single, existing application across multiple devices [3]. The potential of reusing and mixing multiple applications has not been a topic in the DUI community, although the benefits have been recognised in the mashup research community where a range of frameworks and tools for integrating existing components from different web sites into a new application have been proposed [4,5]. However, the resulting applications are usually still developed for a single device.

In [6], we described a first prototype of MultiMasher with an initial set of visual tools designed for creating mashups and distributing them across multiple

B. Benatallah et al. (Eds.): WISE 2014, Part II, LNCS 8787, pp. 199–214, 2014.

connected devices. In this paper, we expand on the concepts of such multi-device mashups and present an updated and extended version of MultiMasher built on a new architecture. In particular, we make the following contributions.

- We introduce and illustrate *concepts* for multi-device mashups in Sect. 3.
- The updated MultiMasher client provides new *visual tools* which are described in Sect. 4. It runs in any modern browser and does not require browser extensions or plugins. It supports the selection of elements from arbitrary web sites and the composition of multi-device mashups in a direct manipulation interface. The selected elements can be arranged in mashups through drag-and-drop operations and new interactions may be defined between them. MultiMasher provides an overview of all connected devices and elements can also be moved easily between devices via drag-and-drop.
- A new *architecture* is presented in Sect. 5 which eliminates several limitations of the first prototype and enables additional functionalities.

In addition, we present both *conceptual and technical evaluations* of MultiMasher in Sect. 6. Conceptually, we have assessed MultiMasher along the dimensions of the logical framework for multi-device user interfaces presented in [7]. We have also taken into account further dimensions covering aspects specific to multi-device mashups, such as the type of components used or inter-component communication. For the technical evaluation, we tested MultiMasher on 50 popular web sites and, despite some limitations mostly affecting highly dynamic, JavaScript-heavy web sites, achieved encouraging results regarding its compatibility and support for browsing, distribution, and linking of web site elements.

2   Background

In [6], we define a multi-device mashup as "a web application that reuses content, presentation, and functionality provided by other web pages and that is distributed among multiple cooperating devices." To focus on the main concepts of a multi-device mashup, we introduce the following simple scenario (Fig. 1), which will be used as the running example in the paper.

> Two friends, Bill and Ted, are on a holiday and want to plan a bike trip to several places nearby. While they each have their smartphone, they also want to make use of the larger, digital TV in their accommodation. As Wikipedia provides good background information, but does not give a good visual impression of a place, they want to simultaneously view Google Image Search results for each place they look up. Bill suggests that they could use MultiMasher to quickly mash up articles from Wikipedia and images from Google for the same place on the large screen, while they could each use their smartphone to enter new locations. However, they want to use a single input field which should update both the article and the images on the large screen rather than searching separately on each page. After collaboratively exploring a couple of places using their

smartphones while sitting on the sofa, they later decide to move to Ted's laptop so that they can meet up with another friend to discuss their plans. Using MultiMasher, the mashup they created across the smartphones and the large screen can easily be migrated while preserving the current state.

Fig. 1. A multi-device mashup composed of two web sites and three devices

This simple scenario illustrates the three areas of related work that our research builds on: mashups, DUIs and collaborative browsing.

The mashup in our scenario is spread across multiple devices and this distribution introduces a set of challenges, such as the migration of interface parts across devices, changing the distribution at run-time, and adaptation to device characteristics. Such problems have been addressed by DUI researchers and a number of frameworks have been built. Similar to our goals, MarcoFlow [8] aims at the composition of distributed mashup-like applications. Due to the high level of complexity, this approach is targeted at skilled developers and is not suitable for non-technical end-users. Other frameworks for DUI development follow a model-based approach, e.g. [9,10]. A framework for the development of distributed interactive applications is introduced in [1]. It is divided into two components, a client side and engine side. Instead of relying on a fixed server, the user can flexibly configure any of the devices as the engine coordinating the distribution. The distributions can be updated at runtime, for example triggered by user interaction. Changing the distribution at runtime requires that interface components migrate from one device to another while preserving their state. This was the focus of their previous work [3], which allows the migration of existing web applications across devices. The system uses a proxy-based approach and uses DOM serialisation to propagate the state from one device to the other. Even though the system supports multiple users, its focus is on sequential interaction. It does not support the simultaneous interaction of multiple users or devices with a web application.

Another aspect in the scenario was the use of one input field to trigger a search in both web sites. The exchange of data and events between widgets is referred to as inter-widget communication (IWC). It is of interest to both the mashup and the DUI community, as it can be used to connect two widgets from different sources, but also to connect widgets across multiple devices. A recent example from the DUI community is DireWolf [11], a framework for web applications based on pre-built widgets. It supports the distribution and migration of widgets at run-time. DireWolf implements both local IWC, as well as IWC across devices. In contrast to our work, a widget can only be present on one device at a time, while we support the replication of a component to multiple devices. Similarly, SmartComposition [12] extends IWC for multi-screen mashups. Both DireWolf and SmartComposition require a specific component model and rely on pre-built widgets, whereas MultiMasher can be used to extract components from arbitrary web sites. In the mashup area, [13] presents a semi-automatic approach targeted at non-technical end-users to extend widgets with IWC capabilities through programming by demonstration. However, this is preliminary work that is still limited to only a few input scenarios. In the current version of MultiMasher, users have to explicitly link widgets from different sources, but their semi-automatic approach could be integrated with ours as they are both based on GUI-level events.

Besides IWC, component extraction is a topic that has been addressed by the mashup research community. In our scenario, components are extracted from existing web sites—Wikipedia and Google Image Search—and connected so that communication between them is possible. Semi-automatic component extraction from existing web sites is, for example, supported by Firecrow [14]. The developer demonstrates the desired behaviour of the component to the tool by performing a series of interactions. The system tries to extract the necessary HTML, JavaScript, CSS and resources. The extracted UI controls may then be embedded into existing web sites. However, the tool does not provide support for linking multiple controls and any inter-widget communication must be implemented manually. On the other hand, mashArt [4] supports this using the approach of universal composition which allows the creation of applications based on the integration of data, application and UI components. The communication between the components is based on events and operations. However, the mashArt tool itself is targeted at advanced web users and the accompanying component library needs to be filled by professional programmers. In contrast, Ghiani et al. [5] present an environment for mashup creation targeted at end-users without programming knowledge. Similar to our work, mashup components can be chosen from arbitrary web sites through direct manipulation of the GUI. In order to connect mashup components, the system generates a list of input and output parameters by intercepting and analysing HTTP requests. The user may then associate input with output parameters across components.

As illustrated in the scenario, our system aims to support multiple simultaneous users: Once web sites are distributed as components and mashed together, multiple users may interact with the same component in collaboration. We achieve this by integrating principles of collaborative browsing [15], or co-browsing, taking care of synchronising the browsers operated by multiple users. The issue of providing co-browsing of dynamic, JavaScript-enabled web pages is tackled in [16]. The presented solution works at the DOM level, can be implemented in JavaScript, and requires no extensions to the browser. The paper describes two mechanisms of JavaScript engine synchronisation. Input synchronisation takes into consideration UI events alone, thus synchronisation happens before the execution of JavaScript. Output synchronisation propagates the changes made on the DOM tree after the JavaScript execution. With MultiMasher, we present a solution which uses similar concepts to those expressed in JavaScript engine output synchronisation. While the authors found input synchronisation to be better for scalability and user experience, it is not clear whether they take into account the fact that propagation of interactions may generate multiple updates to a database, for instance when submitting a form. To address this issue, we opted for an architecture that builds on a remote control metaphor similar to Hightlight [17], where interactions from mobile devices are executed on a proxy browser in the server and changes sent back to the mobile clients.

3   Concepts for Multi-device Mashups

We have developed and integrated concepts for mashups, DUI and co-browsing that play an important role for multi-device mashups. In this section, we give an overview of the concepts in terms of entities and operations, before explaining how they were implemented in MultiMasher in the next sections.

Figure 2 illustrates the three main entities involved in a multi-device mashup: *components*, *mashups* and *devices*: Components extracted from existing web pages are composed into a set of inter-connected mashups which are accessed simultaneously by multiple devices.

We define a component as a subset of web page elements, ranging from a single HTML element, such as a form input field, to the complete body. A mashup contains a specific set of components. All devices accessing the same mashup

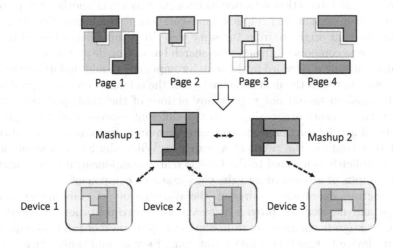

Fig. 2. Overview of a multi-device mashup

will receive the same set of components. In order to obtain a different view for another device, a mashup with a different set of components must be created. Note that there may be an overlap between the set of components used in two mashups. Finally, a multi-device mashup is defined as a set of mashups.

There should be no limitations to the kind of device that can access a multi-device mashup, granted that it is equipped with a modern web browser. The relation between user and device may be one-to-one, many-to-one or one-to-many as, similar to our scenario, a user could use multiple devices at a time, for example a phone and a tablet, while a larger device such as a smart TV may be shared by multiple users.

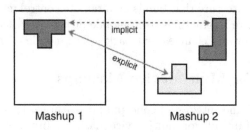

Mashup 1   Mashup 2

Fig. 3. Implicit and explicit inter-component communication

In terms of operations, we distinguish three dimensions: *mashing up, distributing* and *co-browsing*. Mashing up relates to the operations that are also needed in traditional, single-device mashup development. *Component creation* should be supported programmatically or by direct manipulation of the web site. *Component manipulation and adaptation* operations, such as moving, resizing, copying or deleting, can provide a means for quickly building a visually appealing mashup. To obtain rich interactions, *inter-component communication* is required so that interaction with one component may affect another component. We distinguish explicit and implicit inter-component communication (Fig. 3). Components that originate from the same web site communicate implicitly. For example, if one component displays the search bar of Google and another the list of results, entering a keyword in the search component should update the results component. As this is the default behaviour of the original web site, reproducing it in the mashup should not require any actions of the mashup developer. In contrast, components originating from two different sources must be explicitly connected if any interaction between them is required. Introducing a third component that contains the result of a search on Wikipedia into our example, it must be explicitly connected to the Google search component, if it is to update upon the input of a keyword into the Google search component.

Distributing operations addresses the distribution of components across mashups and devices. To provide a means for experimentation, it should be possible to *migrate* components from one mashup to another (and consequently from one device to another), while maintaining its state and configuration, such as connections to other components. *Inter-component communication* should be transparent to the location of the participating component. That is, it should not make a difference whether two components reside in the same or in two different mashups. Note that two components inside the same mashup could also be accessed by multiple devices.

The co-browsing dimension covers the aspect of multiple users accessing a multi-device mashup simultaneously. As a user interacts with a multi-device mashup, updates to another user's view of the mashup are likely to occur. To avoid confusion, a mechanism for *raising awareness for interactions* with the system should be employed.

4   MultiMasher

We have implemented the concepts described in the previous section in MultiMasher which provides visual tools for building and deploying multi-device mashups. It can be loaded into any modern web browser and does not require any extensions or plug-ins. As MultiMasher focuses on direct manipulation and requires no programming knowledge to build a multi-device mashup, our tool was designed to be used by technical and non-technical users alike. MultiMasher does not explicitly distinguish between design and run-time as any changes to a mashup immediately take effect. There are two main views in MultiMasher. The *global view* provides an overview of the complete multi-device mashup and all the mashups of which it is composed. Upon the selection of such a mashup, the *mashup view* is opened where the mashup can be edited and used.

4.1   Global View

The global view shows the state of the whole multi-device mashup and allows users to manipulate it. Initially, the user starts with an empty multi-device mashup. Any number of mashups can be created and added to the multi-device mashup, independent of the devices that are connected to the system. An interactive preview of all mashups is shown (Fig. 4). For each mashup, all the components it contains are colour coded. Components that originate from the same web site have the same background colour. The border colour denotes the set of web site elements that constitute the component. If two mashups contain the exact same component, it will have identical body and border colours in both previews. Components can be migrated between mashups via drag-and-drop and they can be resized directly in the browser. MultiMasher thus supports quick and easy experimentation and adaptation. Since all changes are immediately executed, all connected devices are updated simultaneously. For example, migrating a component from a source to a target mashup, removes the component from the view of all devices connected to the source mashup and introduces it on the target devices. To raise awareness, the global view lists all connected devices and for each mashup displays a colour-coded icon for each device subscribed to that mashup.

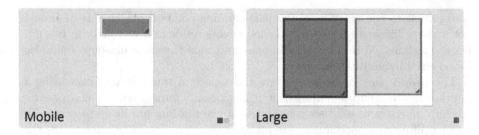

Fig. 4. An interactive preview of a multi-device mashup consisting of two mashups (Mobile and Large). The small coloured squares denote the clients connected to each mashup, two to Mobile, one to Large. The Large mashup contains two components from two different web sites. The Mobile mashup contains one component that originates from the same web site as the component of the left side of the Large mashup, which is indicated by the identical background colours.

4.2   Mashup View

The mashup view displays a single mashup on a device. In this view, the user can interact with the mashup in the role of the end-user, e.g. entering a search query in our example. However, the mashup can also be edited. Components can be created and manipulated. Initially, a freshly created mashup provides an empty canvas. Clicking and dragging the mouse on the canvas specifies the dimensions of a new component. The system then creates a new component and asks the user to provide a source web site by either specifying a URL or choosing one that has been used previously from a list. MultiMasher loads the full web site inside the component boundaries. A component can be in either *execution mode* or *configuration mode*. In execution mode, the user interacts with the source web site of the component. In configuration mode, the user can adapt the component. Once in configuration mode, an *element selection mode* and a *linking mode* are available. During element selection, the user can select any element of the source web site. Subsequently, the component will only display the selected elements and its descendants.

As an example, in Fig. 5, only the main article of Wikipedia has been selected for a component. In the linking mode, users can add explicit inter-component communication. In the current version of MultiMasher, this can be done by attaching tags to change, submit and click events of a web site element. Subsequently, all elements with identical tags for the same event type will be connected and an event triggered on one element will trigger the same event on all connected elements. In our example, we can attach a *search* tag to both the Wikipedia and the Google search button for the click event, so that clicking the Wikipedia search button, also triggers the Google search button and vice versa. In configuration mode, the user can copy a component. This can be useful if the user wants to create another component based on the same web site or to have an exact copy of the component in another mashup. After copying the component in the mashup view, it can be migrated to another mashup in the global

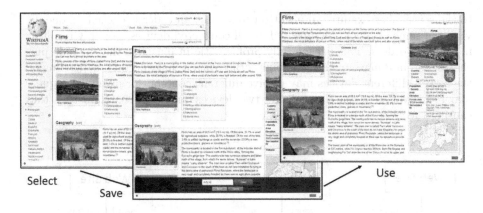

Fig. 5. Element selection mode

view. Also in the mashup view, components can be freely resized and moved to any location in the mashup.

4.3   Co-browsing Feedback

Since MultiMasher is targeted at multi-user scenarios, we added some features for co-browsing feedback to raise awareness for the interactions happening on all connected devices. In the global view, every connected client device is assigned a colour. Whenever that client interacts with a mashup or edits a mashup, all other clients are notified. For example, when a client moves a component inside a mashup, the component will be highlighted with that client's colour while it is being moved on all other devices. Furthermore, a notification message is displayed at the bottom corner of MultiMasher. When a user interacts with the web site inside a component, e.g. entering some text in an input field, the element will again be highlighted with the corresponding colour on all devices.

5   Architecture and Implementation

Our experiments with an early MultiMasher prototype [6] have shown that there is a need for an advanced architecture to support complex scenarios. A major challenge in an infrastructure for multi-device mashups is maintaining a consistent state across all involved devices. In our first prototype, we implemented a solution based on JavaScript engine input synchronisation as described in Sect. 2. DOM events originating from one device are replayed on all other devices. For example, when a user clicks on a button on one device, the click event is intercepted and sent to the server, which forwards it to all connected devices. On the devices, the click event is triggered on the button, thus, the remote click is replayed locally, which should result in the same state on all devices. There are two major drawbacks with this method. First, it can result in multiple updates to the server. For example, if a button to submit an order is clicked on one device, the same click will be repeated on all connected devices, possibly resulting in multiple orders. This is especially problematic in mashups that include purchases, e.g. in a mashup that combines eBay with a map. Second, depending on the server, different devices may receive different content, for example, a query may return different search results for two devices due to the search history, thus causing the system to run out of sync.

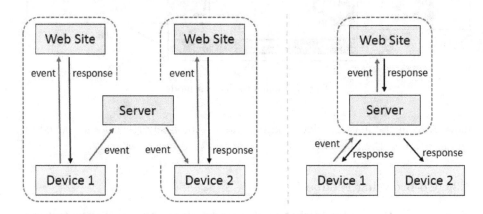

Fig. 6. Session synchronisation in the first prototype (left) and the current version (right) of MultiMasher

The root of the problem stems from having to maintain a consistent global state which is, however, fragmented and distributed across devices, each with its own session. Since we handle arbitrary external web sites which are outside our domain, a possible approach to this issue is to centralise the system: Instead of trying to synchronise the state of multiple sessions that are distributed among devices, we propose to use a single session which is in turn shared across devices. To illustrate the difference, Fig. 6 compares our first prototype to our current approach. The former uses a server for broadcasting events to the devices, which have different sessions with the same web site, while the latter detaches the session from the devices moving it to the server, which becomes a proxy between the devices and the web site.

We propose an approach similar to the remote control metaphor used in Highlight [17]. On the server-side, a headless browser (i.e. a web browser without a GUI) maintains the state of each web site used in the multi-device mashup. Each source web site is loaded in a tab in the headless browser. All requests from client devices are proxied through the MultiMasher server, instead of directly going to the original web servers. For example, if a user clicks a button, the client device sends the click to the server which replays the click in the headless browser. This may lead to an update of the web site, e.g. by loading new content via Ajax or some local JavaScript manipulation. Subsequently, the MultiMasher server serialises the new DOM and sends it to all involved clients. To improve performance, resource requests, for example for images, are sent to the original web servers directly rather than being proxied.

On the client-side, each component representing part of a web site is contained in an iframe. By default, components are displayed in the size that they are received in the headless browser. This may not be suitable for small screens, but MultiMasher supports resizing of components which may be used to adjust the content for the devices in use. Inside the iframe, we disable the JavaScript of the original web site in order to gain full control over the component. All JavaScript is run in the headless browser on the server and the results are sent back to all clients. Thus we ensure a consistent state across all devices. For the selection of components in the client, we use a similar approach to [5]. However, instead of the element ID, we use its path in the DOM tree, thus avoiding the explicit need for IDs. Yet, neither approach solves the problem of evolving web sites where both the DOM structure and element IDs may change.

We implemented MultiMasher using Node.js[1] on the server-side. As a headless browser, PhantomJS[2] is used, which runs in a separate process and communicates with Node.js via a plug-in.

6   Evaluation

In this section, we first present the results of a conceptual evaluation of Multi-Masher along an established framework, followed by a technical evaluation based on 50 top web sites.

6.1   Conceptual Evaluation

In a conceptual evaluation, we assessed MultiMasher along the dimensions of a logical framework for multi-device UIs [7]. As this framework does not take into account aspects specific to mashups, we added the following set of additional dimensions that were derived during the process of building MultiMasher and from related work.

Inter-Component Communication describes how distributed UI elements can be set up to communicate with each other. This might be *automatic*, if no configuration by the user is required; *manual*, if the connection has to be manually established; and *mixed* if both cases are possible. MultiMasher provides a mixed approach. Automatic communication is only possible for components that originate from the same source, otherwise the communication has to be set up manually via tagging.

Synchronisation Consistency defines how complex it is to maintain synchronisation across devices and web sites. In a *consistent* system, resynchronisation is easy to achieve; while in an *inconsistent* system, resynchronisation is hard to achieve. An example of similar observations may be found in [16].

---

[1] http://nodejs.org
[2] http://phantomjs.org

As MultiMasher stores the state of a mashup centrally on the server, a device that is out of sync simply needs to reconnect to the server in order to synchronise.

Type of Components analyses the type of components used in the distribution. These can be *widgets*, i.e. small, pre-built, self-contained web applications; or, as in MultiMasher, arbitrary *HTML elements.*

Component Creation defines the mechanisms that can be supported for creating components. Components may be *pre-built*, i.e. widgets; *scripted* at design-time, i.e. by developers working on the cross-device-mashup; or generated by *direct-manipulation* of web sites by the user as in MultiMasher.

Flexibility to Changes analyses flexibility to changes in the source web sites, e.g. because of evolving web site structure of the re-authored pages [18]. It ranges from *high* if components can adapt well to the new source to *low* if no adaptation is provided. In MultiMasher, an evolving web site can interrupt inter-widget communication. If the DOM elements for a component can no longer be found, it only affects that component. The rest of the mashup remains stable and the user may adjust the component to any changes in the source web page.

Table 1. Conceptual evaluation for multi-device UI dimensions (left) and mashup dimensions (right)

Multi-Device Dimensions	MultiMasher	Mashup Dimensions	MultiMasher
Distribution	Dynamic	Inter-Comp. Comm.	Mixed
Migration	UI elements	Sync. Consistency	High
Granularity	Entire UI to components of UI elements	Type of Comp.	HTML element
Trigger	User	Creation	Direct manipulation
Sharing	Moving information, sharing by interacting	Flexibility	Medium
Timing	Mixed		
Modalities	Multi-modal		
Generation	Run-time		
Adaptation	Resizing		
Architecture	Client/Server		

We applied the original framework and the new dimensions to MultiMasher. A summary of the results can be found in Table 1. Compared to the other frameworks assessed in [7], MultiMasher is the only tool that supports multi-modal interaction. It can in principle be used with any interaction modality that is supported by the browser (e.g. touch). MultiMasher provides the most essential support for UI adaptation in that components can be moved and resized to accommodate different device characteristics. Advanced adaptation operations similar to CrowdAdapt [19] could be added in the future. Overall, our evaluation shows that MultiMasher can be well described in terms of the given dimensions and addresses some of the issues listed as future work in [7] such as preservation of state in UI migration.

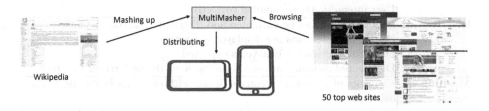

Fig. 7. Evaluation scenario

6.2   Technical Evaluation

Based on the methodology used in [19], we evaluated MultiMasher in terms of its technical compatibility with 50 top web sites, ranked according to popularity and traffic by Alexa[3]. We selected the first 5 web sites from 10 categories: Arts, Business, Games, Health, Home, News, Science, Shopping, Society, and Sports. In order to assess MultiMasher with respect to these web sites, we developed a simple but representative cross-device mashup scenario (Fig. 7), which can be constructed using almost any type of web site. The scenario is composed of two devices, each using a different mashup, and two different inter-connected web sites. These are separated into components, which are then distributed and tested. In particular, we differentiate three categories of components: menu, search bar and main content. Only one of the two web sites is changed at each iteration of the scenario, while the other is fixed. The purpose of the fixed web site is exclusively to test the mashup capabilities (i.e. inter-page communication) of the cross-device mashup and it should not impact the evaluation. We selected Wikipedia, as it has proved to work very well in MultiMasher.

We assessed MultiMasher based on the criteria in Table 2 which can be grouped into three categories: browsing, distribution and mashing up. For each criterion, we assigned a value between *1 = poor* for web sites with major issues to *5 = excellent* for full support.

Overall the evaluation shows that MultiMasher offers good compatibility with web sites from many different domains. 43 of the 50 tested web sites (86%) obtained an average rating of a 4 or higher. Note that the browsing criteria have a higher priority than the other two categories. A low score in terms of browsing implies a low compatibility with MultiMasher overall, despite possibly higher scores in distributing and mashing up. Considering the browsing criteria in isolation, 31 web sites (60%) scored a 4 or higher. Thus, they either had no issues at all or smaller issues affecting non-critical elements, such as ads in separate iframes. 48 sites (96%) scored a 3 or higher, which implies that the majority of critical elements were working as expected. MultiMasher showed

---

[3] http://www.alexa.com

Table 2. Criteria of the technical evaluation

Browsing	Page elements are loaded and displayed correctly. Page elements behave as expected. User interactions are handled correctly.
Distribution	Page elements can be extracted and distributed as components. Distributed elements are displayed correctly. Distributed elements behave as expected. Distributed elements are synchronised with the underlying web page.
Mashing up	Elements of a web page can be linked with elements belonging to another page. Click, change, and submit events are correctly replayed in linked elements.

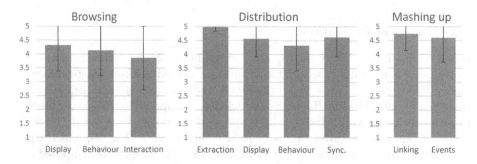

Fig. 8. Results of the technical evaluation (error bars showing standard deviation)

very good compaitiblity with respect to the distribution and the mashing up categories. For the distribution criteria, 48 sites (96%) scored a 4 or higher and 17 sites were rated a 5. Similarly, for the mashing up criteria 44 sites (88%) reached at least a 4 and 39 sites (78%) obtainted the maximum rating of a 5.

Most issues we observed during the evaluation boil down to the following challenges:

– *Heavy use of JavaScript.* In order to prevent de-synchronisations and duplication of updates to the external web servers, we removed JavaScript from the local copy of the page in the front-end. This does not affect the normal behaviour of the page in most cases since JavaScript is run in the back-end, however, it limits cases where dynamic operations are expected to be available directly in the client, e.g. for drag-and-drop or panning a map. Such interactions are triggered by mouse-move events, which are not synchronised, as the performance of the system would suffer from the hundreds of events that may be produced by a single interaction.

– *CSS extraction.* We encountered some instances where the extracted element was not positioned correctly inside the component or where the page background was shown out of place. We manipulate the CSS of the extracted element, so that it is positioned at the top left corner of its component. In some cases this can result in conflicts with existing CSS rules. In the future,

ShadowDom[4], an emerging standard, may alleviate these issues by providing encapsulation of style within elements.
- *DOM evolution.* Certain web sites evolve their DOM, for instance by randomly changing form IDs to prevent spam. However, DOM evolution interferes with the path-based approach that MultiMasher uses to identify elements. DOM re-matching techniques, such as the ones introduced in PageTailor [18], could be used to alleviate this issue.

More generally, the architecture chosen for MultiMasher imposes the following restrictions. Since the state of the system is located centrally on the server, offline use of the system is not possible. Any user interaction that changes the state of the mashup requires the device to be connected to the server. Furthermore, for applications that require user credentials, the central session on the server implies that only one user can log into a service, e.g. an email account, at a time. This is often beneficial as a user may want to access a service from multiple devices concurrently. However, in some cases, this may introduce a security risk, for example, when a mashup is shared with untrusted users.

7   Conclusion

We have presented concepts for multi-device mashups and an implementation in MultiMasher based on an architecture with centralised state in a headless browser. Even though mashups and DUI could be viewed as two orthogonal concepts, we believe that the development of multi-device mashups benefits from an integrated solution, especially when considering an iterative approach to design and development. Future work could explore the potential of MultiMasher to support multi-device development ranging from prototyping to testing and debugging of complex applications. While detailed user evaluations of MultiMasher remain as future work, we have paid particular attention to the design of its visual interface with the goal of allowing the quick and easy creation of user interfaces involving multiple devices.

References

1. Frosini, L., Manca, M., Paternò, F.: A Framework for the Development of Distributed Interactive Applications. In: Proc. EICS, pp. 249–254 (2013)
2. Melchior, J., Grolaux, D., Vanderdonckt, J., Roy, P.V.: A Toolkit for Peer-to-Peer Distributed User Interfaces: Concepts, Implementation, and Applications. In: Proc. EICS (2009)
3. Ghiani, G., Paternò, F., Santoro, C.: Push and Pull of Web User Interfaces in Multi-Device Environments. In: Proc. AVI (2012)
4. Daniel, F., Casati, F., Benatallah, B., Shan, M.-C.: Hosted Universal Composition: Models, Languages and Infrastructure in mashArt. In: Laender, A.H.F., Castano, S., Dayal, U., Casati, F., de Oliveira, J.P.M. (eds.) ER 2009. LNCS, vol. 5829, pp. 428–443. Springer, Heidelberg (2009)

---

[4] http://www.w3.org/TR/2014/WD-shadow-dom-20140617/

5. Thalen, J.P., van der Voort, M.C.: Creating Mashups by Direct Manipulation of Existing Web Applications. In: Piccinno, A. (ed.) IS-EUD 2011. LNCS, vol. 6654, pp. 42–52. Springer, Heidelberg (2011)
6. Husmann, M., Nebeling, M., Norrie, M.C.: MultiMasher: A Visual Tool for Multi-Device Mashups. In: Sheng, Q.Z., Kjeldskov, J. (eds.) ICWE Workshops 2013. LNCS, vol. 8295, pp. 27–38. Springer, Heidelberg (2013)
7. Paternò, F., Santoro, C.: A Logical Framework for Multi-Device User Interfaces. In: Proc. EICS (2012)
8. Daniel, F., Soi, S., Tranquillini, S., Casati, F., Heng, C., Yan, L.: Distributed Orchestration of User Interfaces. Inf. Syst. 37(6), 539–556 (2012)
9. Melchior, J., Vanderdonckt, J., Roy, P.: A Model-Based Approach for Distributed User Interfaces. In: Proc. EICS (2011)
10. Paternò, F., Santoro, C., Spano, L.D.: Maria: A Universal, Declarative, Multiple Abstraction-Level Language for Service-Oriented Applications in Ubiquitous Environments. In: TOCHI, vol. 16(4) (2009)
11. Kovachev, D., Renzel, D., Nicolaescu, P., Klamma, R.: DireWolf - Distributing and Migrating User Interfaces for Widget-Based Web Applications. In: Daniel, F., Dolog, P., Li, Q. (eds.) ICWE 2013. LNCS, vol. 7977, Springer, Heidelberg (2013)
12. Krug, M., Wiedemann, F., Gaedke, M.: SmartComposition: A Component-Based Approach for Creating Multi-screen Mashups. In: Casteleyn, S., Rossi, G., Winckler, M. (eds.) ICWE 2014. LNCS, vol. 8541, pp. 236–253. Springer, Heidelberg (2014)
13. Chudnovskyy, O., Fischer, C., Gaedke, M., Pietschmann, S.: Inter-Widget Communication by Demonstration in User Interface Mashups. In: Daniel, F., Dolog, P., Li, Q. (eds.) ICWE 2013. LNCS, vol. 7977, pp. 502–505. Springer, Heidelberg (2013)
14. Maras, J., Štula, M., Carlson, J.: Reusing Web Application User-Interface Controls. In: Auer, S., Díaz, O., Papadopoulos, G.A. (eds.) ICWE 2011. LNCS, vol. 6757, pp. 228–242. Springer, Heidelberg (2011)
15. Greenberg, S., Roseman, M.: GroupWeb: A WWW Browser as Real Time Groupware. In: Proc. CHI (1996)
16. Lowet, D., Goergen, D.: Co-browsing Dynamic Web Pages. In: Proc. WWW (2009)
17. Nichols, J., Hua, Z., Barton, J.: Highlight: A System for Creating and Deploying Mobile Web Applications. In: Proc. UIST (2008)
18. Bila, N., Ronda, T., Mohomed, I., Truong, K.N., de Lara, E.: PageTailor: Reusable End-User Customization for the Mobile Web. In: Proc. MobiSys (2007)
19. Nebeling, M., Speicher, M., Norrie, M.C.: CrowdAdapt: Enabling Crowdsourced Web Page Adaptation for Individual Viewing Conditions and Preferences. In: Proc. EICS (2013)

MindXpres: An Extensible Content-Driven Cross-Media Presentation Platform

Reinout Roels and Beat Signer

Vrije Universiteit Brussel
Pleinlaan 2, 1050 Brussels, Belgium
{rroels,bsigner}@vub.ac.be

Abstract. Existing presentation tools and document formats show a number of shortcomings in terms of the management, visualisation and navigation of rich cross-media content. While slideware was originally designed for the production of physical transparencies, there is an increasing need for richer and more interactive media types. We investigate innovative forms of organising, visualising and navigating presentations. This includes the introduction of a new document format supporting the integration or transclusion of content from different presentations and cross-media sources as well as the non-linear navigation of presentations. We present MindXpres, a web technology-based extensible platform for content-driven cross-media presentations. The modular architecture and plug-in mechanism of MindXpres enable the reuse or integration of new visualisation and interaction components. Our MindXpres prototype forms a platform for the exploration and rapid prototyping of innovative concepts for presentation tools. Its support for multi-device user interfaces further enables an active participation of the audience which should ultimately result in more dynamic, engaging presentations and improved knowledge transfer.

Keywords: MindXpres, slideware, cross-media transclusion, non-linear navigation.

1 Introduction

With millions of digital presentations that are created every day, their importance in modern society cannot be denied. Presentations are supporting the oral transfer of knowledge and play an important role in educational settings. While digital presentation solutions originated as tools for creating physical media such as photographic slides or transparencies for overhead projectors, very little has changed to the underlying concepts and principles of slide-based presentation tools or so-called slideware for digital presentations. We are, for example, still restricted by the rectangular boundaries of a slide or the linear navigation between slides. As argued by Tufte [1], this form of slideware has some negative consequences for the effectiveness of knowledge transfer. Complex ideas are rarely sequential by nature but the presenter is forced to squeeze them into a linear sequence of slides, resulting in a loss of relations, overview and details. One might

B. Benatallah et al. (Eds.): WISE 2014, Part II, LNCS 8787, pp. 215–230, 2014.

argue that these issues can easily be addressed by creating minimalistic presentations or by introducing some structure via a table of contents. Unfortunately, this does not work in the domain of learning, where complex knowledge or other pieces of rich information need to be presented "as is" [2].

It is important to point out the monolithic nature of slideware presentations where content is spread over many self-contained presentation files. In order to "reuse" previous work, the presenter has to switch between files while giving a presentation or duplicate some slides in the new presentation. Note that the issue is not limited to reusing single slides since there is a wealth of resources available, spread over a wide spectrum of distribution channels and formats. The inclusion of content by reference or transclusion [3] might help to cross the boundaries between different types of media and be beneficial in the context of modern cross-media presentation tools.

There also seems to be an imbalance between the functionality for the authoring and visualisation of content. The main authoring views consist of toolbars and buttons to specify how content should be visualised while there is less support for the authoring of the content itself. While we have seen the addition of basic multimedia types such as videos to modern slideware, most content is still rather static. During a presentation we can, for example, not easily change from a bar chart to a pie chart data visualisation or dynamically change some values to see the immediate effect, which could be beneficial for knowledge transfer [4]. Finally, the audience can be more actively involved via audience response and classroom connectivity systems which provide multi-device interfaces for sharing knowledge and results during as well as after a presentation. The evolution of presentations can be compared with the Web 2.0 movements where users have become contributors, content is more dynamic and interactive and where we have a decentralisation of content via service-oriented architectures.

The rapid prototyping and evaluation of new concepts for the representation, visualisation and interaction with content is essential in order to move a step towards the next generation of cross-media presentation tools. After introducing existing slideware solutions, we discuss the requirements for next generation presentation tools. This is followed by a description of the extensible MindXpres architecture and its plug-in mechanism. The web technology-based implementation of MindXpres is validated based on a number of use cases and MindXpres plug-ins and followed by a discussion of future work.

2   Background

The impact, benefits and issues of slideware have been studied ever since digital slideware has been introduced. While some studies acknowledge slideware as a teaching aid [4], Tufte [1] heavily criticises slideware for sticking to outdated concepts. He addresses the many consequences of spatial limitations or linear navigation and relates them to the human mind which works differently. One of Tufte's conclusions, which is also confirmed by Adams [5], is that slide-based presentations are not suitable for all kinds of knowledge transfer and in particular not in scientific settings. Recent work shows that it is important for the learning process that content is well integrated in the greater whole, both structurally and visually [6], which is influenced by the navigation and visualisation. There have been a number of different approaches to offer non-linear navigation. Zoomable User Interfaces (ZUIs) as used by CounterPoint [7], Fly [8] or Prezi, offer virtually unlimited space. Also Microsoft has experimented with zoomable interfaces in pptPlex. While ZUIs are one way to escape the boundaries of the slide, we have seen other approaches such as MultiPresenter [9] or tiling slideshows [10]. PaperPoint [11] and Palette [12] further enable the non-linear navigation of digital presentations based on a slide selection via augmented paper-based interfaces. Finally, there is a category of authoring tools that use hypermedia to enable different paths through a set of slides. NextSlidePlease [13] allows users to create a weighted graph of slides and may suggest navigational paths based on the link weights and the remaining presentation time. Microsoft follows this trend with their HyperSlides [14] project. The potential of PowerPoint as an authoring tool for hypermedia-based presentations has further been investigated by Garcia [15].

Existing presentation tools require content to be duplicated for reuse, resulting in multiple redundant copies that need to be kept up to date. Even though some attempts have been made to address this issue, there is room for improvement. When it comes to document formats for more general educational purposes, there are formats such as the Learning Material Markup Language (LMML) [16], the Connexions Markup Language (CNXML) and the eLesson Markup Language (eLML) [17]. The common factor of these formats is their focus on the reuse of content, but always at a relatively high granularity level. Content is organised in lessons or modules and users are encouraged to use these, as a whole, in their teaching. When we examined the formats in more detail, we noticed that they support outgoing links to external content but not the inclusion of content via references (transclusion). In the context of presentations, Microsoft's Slide Libraries are central repositories for the storage of slides in order to facilitate slide sharing and reuse within a company. However, one needs to set up a SharePoint server which might represent a hurdle for some users. Slides still need to be searched and manually copied into presentations. Furthermore, users are responsible to push back updates to the repository or update slides when they have been modified on the server side. Other commercial tools with similar intentions and functionality for content reuse are SlideRocket or SlideShare. The SliDL [18] research framework provides a service-oriented architecture for storing and tagging slides in a database for reuse. However, it shares some of the shortcomings of Microsoft's Slide Libraries. The ALOCOM [19] framework for flexible content reuse consists of a content ontology and a (de)composition framework for legacy documents including PowerPoint documents, Wikipedia pages and SCORM content packages. While ALOCOM succeeds in the decomposition of legacy documents, it might be too rigid for evolving presentation formats and the tool is furthermore only supporting the authoring phase.

There is not only a similarity in the evolution of the Web and presentation environments, but a number of the issues presented in this section have solutions in the setting of the Web. It is therefore not surprising that more recently we see the use of web technologies for realising presentation solutions. The Simple Standards-based Slide Show System (S5)[1] is an XHTML-based file format for slideshows which enforces the classical slideware model. The Slidy [20] initiative by the W3C introduces another presentation format which is based on the standard slideware model. While these two formats are too limited for our needs, they have some interesting properties. Both formats show a clean separation of content and presentation via CSS themes. The visualisation is resolution independent and the layout and font size are adapted to the available screen real estate. Finally, we would like to mention recent HTML5-based presentation solutions, including projects such as impress.js, deck.js, Shower or reveal.js. A major benefit of applying a widely used open standard such as HTML is the cross-device support. Nevertheless, also these solutions show some restrictions in terms of visualisation, navigation and cross-media support.

Most of the tools and projects presented in this section focus on specific novel ideas for presentations. However, there is no interoperability between the concepts introduced in different tools. While one project might focus on the authoring, another one focusses on novel content types and a third solution introduces radically new navigation mechanisms. Some slideware tools can be extended via third-party plug-ins but the functionality that is exposed to the developers is often limited by the tool's underlying model. For example, PowerPoint allows plug-ins to interact with the presentation model, but the model dictates that a presentation must consist of a sequence of slides. This lack of freedom is also a shortcoming of existing web-based presentation formats. We therefore see a need for an open presentation platform such as MindXpres which supports innovation by providing the necessary modularity and interoperability [21].

3   Requirements

We now introduce a number of requirements to support a broad range of presentation styles and visualisations which have been compiled based on a review of the more recent presentation solutions presented in the previous section.

Non-Linear Navigation. As outlined earlier, the linear traversal of slides is a concept that has been taken over from early photographic slides. Nowadays, users are accustomed to this form of navigation even if it might come with some disadvantages. Any navigation outside of the predefined linear path (e.g. to answer a question from the audience) is rather complicated, since the presenter either needs substantial time to scroll forwards or backwards to the desired slide or has to switch to the slide sorter view. It is further impossible to include a single slide multiple times in the navigational path without any duplication. There are different ways how this lack of flexible navigation can be addressed, including presentation tools that allow the presenter to define non-linear navigation paths [13,14] or zoomable user interfaces (ZUIs) [7,8,22].

---

[1] http://meyerweb.com/eric/tools/s5/

Separation of Content and Presentation. In order to facilitate experimentation with different visualisations, there should be a clear separation between content and presentation. This allows the authors of a presentation to focus on the content while the visualisation is handled by the presentation tool. Note that this approach is similar to the LaTeX typesetting system where content is written in a standardised structured way and the visualisation is automatically handled by the typesetting system. There is also a LaTeX document class for presentations called Beamer and we were inspired by its structured and content-driven approach. However, the content-related functionality and the visualisation are too limited to be considered as a basis for an extensible presentation tool.

Extensibility. In order for a presentation tool to be successful as an experimental platform for new presentation concepts, it should be easy to rapidly prototype new content types and presentation formats as well as innovative navigation and visualisation techniques. It has to be possible to add or replace specific components without requiring changes in the core. In order to be truly extensible, a presentation tool should provide a modular architecture with loosely coupled components. Note that this type of extensibility should not only be offered on the level of content types but also for the visualisation engine or content structures.

Cross-Media Content Reuse. In the introduction we briefly mentioned the lack of content reuse in existing presentation tools. There is a wealth of open education material available but it is rather difficult to use this content in presentations. On the other hand, the concept of transclusion works well for digital documents and parts of the Web (e.g. via the HTML img tag). A modern presentation tool should also support the seamless integration of external cross-media content. This includes various mechanisms for including parts of other presentations (e.g. slides), transcluding content from third-party document formats as well as including content from open learning repositories.

Connectivity. With the rise of social and mobile technologies, connectivity for multi-device input and output becomes more relevant in the context of presentation tools. Support for multi-directional connectivity is required for a number of reasons. First, it is necessary for the previously mentioned cross-media transclusion from external resources. Second, multi-directional connectivity forms the backbone for audience feedback via real-time response or voting systems [23] as well as other forms of multi-device interfaces.

Interactivity. We mentioned that content might be more interactive and the extensibility requirement addresses this issue since the targeted architecture should support dynamic or interactive content and visualisations. Nevertheless, the use of mouse and keyboard might not be sufficient for components offering a high level of interaction. Therefore, a presentation tool should enable the integration of other forms of input such a gesture-based interaction based on Microsoft's Kinect controller or digital pen interaction [24] as offered by the PaperPoint [11] presentation tool.

Post-Presentation Phase. Even if it was never the original goal of slide decks, they often play an important role as study or reference material. While the sharing of traditional slide decks after a presentation is trivial, this changes when the previously mentioned requirements are taken into account. For instance, the non-linear navigation allows presenters to go through their content in a non-obvious order or input from the audience might drive parts of a presentation. Special attention should therefore be paid to the post-presentation phase. It should not only be easy to play back a presentation with the original navigational path, annotations and audience input, but its content should also be made discoverable and reusable. In accordance with the Web 2.0, we see potential for the social aspect in a post-presentation phase via a content discussion mechanism.

4   MindXpres Platform

In this section, we present the general architecture of our MindXpres[2] cross-media presentation platform which is outlined in Fig. 1 and addresses the requirements presented in the previous section.

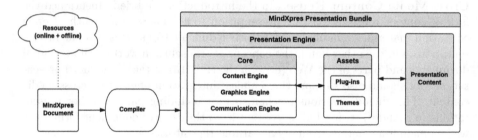

Fig. 1. MindXpres architecture

4.1   Document Format and Authoring Language

Content is stored, structured and referenced in a dedicated MindXpres document format. An individual MindXpres document contains the content itself and may also refer to some external content to be included. A new MindXpres document can be written manually similar to the LATEX approach introduced earlier or in the near future it can also be generated via a graphical authoring tool. In contrast to other presentation formats such as Slidy, S5 or OOXML, the authoring language eliminates unnecessary HTML and XML specifics and focusses on a semantically more meaningful vocabulary. The vocabulary of the authoring language is almost completely defined by plug-ins that provide support for various media types and structures. In order to give users some freedom in the way they present their information, the core MindXpres presentation engine only plays a

---

[2] http://mindxpres.com

supporting role for plug-ins and lets them define the media types (e.g. video or source code) as well as structures (e.g. slides or graph-based content layouts). This is also reflected in the document format as each plug-in extends the vocabulary that can be used. Any visual styling including different fonts, colours or backgrounds is achieved by applying specific themes to the underlying content.

4.2 Compiler

The compiler transforms a MindXpres document into a self-contained portable MindXpres presentation bundle. While a MindXpres document could be directly interpreted at visualisation time, for a number of reasons we decided to have this intermediary step. First, the compiler allows different types of presentations to be created from the same MindXpres document instance. This means that we can not only create dynamic and interactive presentations but also more static output formats such as PDF documents for printing. Similarly, we cannot always expect that there will be an Internet connection while giving a presentation. For this case, the compiler might create an offline version of a presentation with all necessary content pre-downloaded and included in the MindXpres presentation bundle. Last but not least, the compiler might resolve incompatibility issues by, for instance, converting unsupported video formats.

4.3 MindXpres Presentation Bundle

The dynamic MindXpres presentation bundle consists of the compiled content together with a portable cross-platform presentation runtime engine which allows more interactive and networked presentations. Similar to the original document, the compiled presentation content still consists of both, integrated content and references to external resources such as online content that will be retrieved when the presentation is visualised. Note that the content might have been modified by the compiler and, for example, been converted or extracted from other document formats that the runtime engine cannot process. References to external content may have been dereferenced by the compiler for offline viewing.

A presentation bundle's core runtime engine consists of the three modules shown in Fig. 1. The *content engine* is responsible for processing the content and linking it to the corresponding visualisation plug-ins. The *graphics engine* abstracts all rendering-related functionality. For instance, certain presenters prefer a zoomable user interface in order to provide a better overview of their content [25]. This graphical functionality is also exposed to the plug-ins, which can make use of the provided abstractions. The *communication engine* exposes a communication API that can be used by plug-ins. It provides some basic functionality for fetching external content but also offers the possibility to form networks between multiple MindXpres presentation instances as well as to connect to third-party hardware such as digital pens or clicker systems.

In addition to the presentation content and core modules, the presentation bundle contains a set of *themes* and *plug-ins* that are referenced by the content.

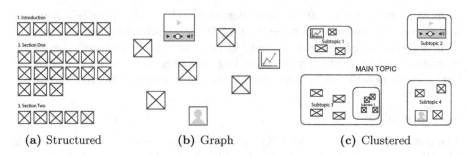

(a) Structured   (b) Graph   (c) Clustered

Fig. 2. Structure plug-in examples

Themes may contain visual styling on a global as well as on a plug-in level. When the content engine encounters different content types, they are handed over to the specific plug-in which uses the graphics engine to visualise the content.

4.4   Plug-in Types

In order to provide the necessary flexibility, all non-core modules are implemented as plug-ins. Even the basic content types such as text, images or bullet lists have been realised via plug-ins and three major categories of plug-ins have to be distinguished:

- *Components* form the basic building blocks of a presentation. They are represented by plug-ins that handle the visualisation for specific content types such as text, images, bullet lists, graphs or videos. The content engine invokes the corresponding plug-ins in order to visualise the content.
- *Containers* are responsible for grouping and organising components of a specific type. An example of such a container is a slide with each slide containing different content but also some reoccurring elements. Every slide of a presentation may for example contain elements such as a title, a slide number and the author's name, which can be abstracted in a higher level container. Another example is an image container that visualises its content as a horizontally scrollable list of images. Note that we are not restricted to the slide format and content can be laid out in alternative ways.
- *Structures* are high-level structures and layouts for components and containers. For example, content can be scattered in a graph-like structure or it can be clearly grouped in sections like in a book. Both are radically different ways of visualising and navigating content but by abstracting them as plug-ins, the user can easily switch between different presentation styles as the ones shown in Fig. 2. Structures differ from containers by the fact that they do not impose restrictions on the media types of their child elements and may also influence the default navigational path through the content.

5   Implementation

We have chosen HTML5 and its related web technologies as the backbone for our MindXpres presentation platform. Other options such as JavaFX, Flash or game engines have also been investigated, but HTML5 seemed to be the best choice. The widely accepted HTML5 standard makes MindXpres presentations highly portable and runnable on any device with a recent web browser, including smartphones and tablets. Furthermore, HTML5 provides rich visualisation functionality out of the box and the combination with Cascading Style Sheets (CSS) and third-party JavaScript libraries forms a potent visualisation platform.

5.1   Document Format and Authoring Language

The MindXpres document format which allows us to easily express a presentation's content, structure and references is based on the eXtensible Markup Language (XML). A simple example of a presentation defined in our XML-based authoring language is shown in List. 1.1. The set of valid tags and their structure, except the **presentation** root tag, is defined by the available plug-ins.

```
1  <presentation>
2    <slide title="Vannevar Bush">
3      <bulletlist>
4        <item>March 11, 1890 - June 28, 1974</item>
5        <item>Amercian Engineer, founder of Raytheon</item>
6      </bulletlist>
7      <image source="bush.jpg"/>
8    </slide>
9  </presentation>
```

Listing 1.1. Authoring a simple MindXpres presentation

5.2   Compiler

The compiler has been realised as a Node.js application. This not only allows the compiler to be used via a web interface or as a web service, but projects such as node-webkit also enable the compiler to run as a local offline desktop application. The choice of using server-side JavaScript was influenced by the fact that Node.js is capable of bridging web and desktop technologies. On the one hand, the framework makes it easy to interact with other web services and to work with HTML, JSON, XML and JavaScript visualisation libraries at compile time. On the other hand, the framework can also perform tasks which are usually not suited for web technologies, including video conversion, legacy document format access, file system access or TCP/IP connectivity.

In order to validate a MindXpres document in the XML format described above, there is an XML Schema which is augmented with additional constraints provided by the plug-ins. After validation, the document is parsed and discovered tags might trigger preprocessor actions by the plug-ins such as the extraction of data from referenced legacy document formats (e.g. PowerPoint or Excel) or the conversion of an unsupported video format. The tag is then converted to HTML5 by simply encoding the information in the attributes of a `div` element. The HTML5 standard allows custom attributes if they start with a `data-` prefix. Listing 1.2 shows parts of the transformed XML document shown in List. 1.1. Note that the transformation does not include visualisation-specific information but merely results in a valid HTML5 document which is bundled into a self-contained package together with the presentation engine.

```
1  <div data-type="presentation">
2    <div data-type="slide" data-title="Vannevar Bush">
3      <div data-type="bulletlist">
4      ...
```

Listing 1.2. Transformed HTML5 presentation content

5.3   Presentation Engine

The presentation engine's task is to turn the compiled HTML content into a visually appealing and interactive presentation. As highlighted in Fig. 1, the presentation engine consists of several smaller components which help plug-ins to implement powerful features with minimal effort. The combination of these components enables the rapid prototyping and evaluation of innovative visualisation ideas. A resulting MindXpres presentation combining various structure, container and component plug-ins is shown in Fig. 3.

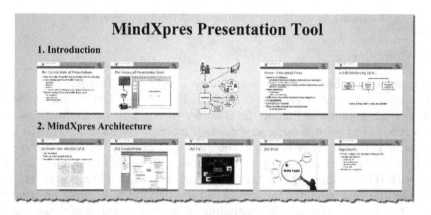

Fig. 3. A MindXpres presentation

Content Engine. When a presentation is loaded, the content engine is the first component that is activated. It processes the content of the HTML presentation by making use of the well-known jQuery JavaScript library. Whenever a `div` element is discovered, the `data-type` attribute is read and the corresponding plug-ins are notified in order to visualise the content.

Graphics Engine. The graphics engine provides support for interesting new visualisation and navigation styles. Next to some basic helper functions, it offers efficient panning, scaling and rotation via CCS3 transformations and supports zoomable user interfaces as well as the more traditional navigation approaches.

Communication Engine. The communication engine implements abstractions that allow plug-ins to retrieve external content at run time. It further provides the architectural foundation to form networks between different MindXpres instances or to integrate third-party hardware [26]. For our MindXpres prototype, we used a small Intel Next Unit of Computing Kit (NUC) with high-end WiFi and Bluetooth modules to act as a central access point and provide the underlying network support. MindXpres instances use WebSockets to communicate with other MindXpres instances via the access point. The access point further acts as a container for data adapters which translate input from third-party input and output devices into a generic representation that can be used by the MindXpres instances in the network. In order to go beyond simple broadcast-based communication, we have implemented a routing mechanism based on the publish-subscribe pattern where plug-ins can subscribe to specific events or publish information. The communication engine provides the basis for audience response systems [26] or even full classroom communication systems where functionality is only limited by the creativity of plug-in developers.

Plug-ins. Plug-ins are implemented as JavaScript bundles which consist of a folder containing JavaScript files and other resources such as CSS files, images or other JavaScript libraries. As a first convention, a plug-in should provide a manifest file with a predefined name. The manifest provides metadata such as the plug-in name and version but also a list of tags to be used in a presentation. The plug-in claims unique ownership for these tags and is in charge for their visualisation if they are encountered by the content engine. As a second convention, a plug-in must provide at least one JavaScript file implementing certain methods, one of them being the `init()` method which is called when the plug-in is loaded by the presentation engine. It is up to the plug-in to load additional JavaScript or CSS via the provided dependency injection functionality. A second method to be implemented is the `visualise()` method which is invoked with a pointer to the corresponding DOM node as a parameter when the content engine encounters a tag to be visualised. A plug-in is free to modify the DOM tree and may also register callbacks to handle future interaction with the content.

Themes. We currently use CSS to provide a basic templating system. These themes offer styling either on a global or on a plug-in level. However, we see this as a temporary solution as it is not well-suited for alternative compiler outputs (e.g. PDF) and a more generic templating scheme is planned for the future.

6   Use Cases

In order to validate the architectural and technological choices, we demonstrate the extensibility and feasibility of MindXpres as a rapid prototyping platform by presenting a number of content- and navigation-specific plug-ins that have been developed so far. Additional plug-ins for audience-driven functionality such as real-time polls, screen mirroring and navigational takeover can be found in [26].

6.1   Structured Overview Plug-in

In Sect. 4 we have explained how structure plug-ins may change the way presentations are visualised and navigated. In order to illustrate this, we have implemented a structure plug-in called *structured layout* which combines a zoomable user interface with the ability to group content into sections. The resulting visualisation of the *structured layout* plug-in is shown in Fig. 3. Whenever a new section is reached, the view is zoomed out to provide an overview of the content within the section and communicate a sense of progress.

6.2   Slide Plug-in

In order to also support the traditional slide concept, we created a slide-like container plug-in. While the benefits and issues of using slides with a fixed size are debatable, we implemented this plug-in as a proof of the framework's versatility. The main function of the slide plug-in is to provide a rectangular styleable component container with a title and some other information. Containers may also offer functionality to layout their content. In this case, the slide plug-in offers a quick and easy layout mechanism which allows the presenter to partition the slide into rows and columns. Content is then assigned to these slots in the order that it is discovered. The use of the slide plug-in together with the resulting visualisation is exemplified in Fig. 4. It also shows the use of the image plug-in

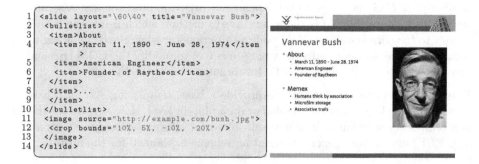

```
1  <slide layout="\60\40" title="Vannevar Bush">
2    <bulletlist>
3      <item>About
4        <item>March 11, 1890 - June 28, 1974</item
           >
5        <item>American Engineer</item>
6        <item>Founder of Raytheon</item>
7      </item>
8      <item>...
9      </item>
10   </bulletlist>
11   <image source="http://example.com/bush.jpg">
12     <crop bounds="10%, 5%, -10%, -20%" />
13   </image>
14 </slide>
```

Fig. 4. Slide plug-in

(a component plug-in) which enables a simple form of cross-media transclusion. A visualised external image can be cropped and filters (e.g. colour correction) may be applied without duplicating or modifying the original source.

6.3 Enhanced Video Plug-in

When videos are used in educational settings, we often need more functionality than what is offered by the average video player [25]. MindXpres provides the enhanced video plug-in shown in Fig. 5 with the possibility to overlay a video with text or arbitrary shapes. This overlay functionality can be used as a basic captioning system as well as to highlight items of interest during playback.

Furthermore, we added the option to trigger certain events at specified times. One can define that a video should automatically pause at a certain point, highlight an object and continue playing after a specified amount of time. Additional features include the bookmarking of certain positions in a video for direct access or the possibility to display multiple videos in a synchronised manner. Our enhanced video plug-in injects the default HTML5 video player and overlays it with a transparent `div` element for augmentation. Currently we make use of the HTML5 video API to synchronise the creation and removal of overlays but a SMIL-based implementation might be used in the future.

```
1   <video source="vid.mp4">
2     <caption start="0:00" duration="1500ms">
3       Lecture 3 - Butterfly Species
4     </caption>
5     <pause start="0:43" duration="5s">
6       <caption>
7         The peacock butterfly (aglais io) ...
8       </caption>
9       <highlight x="30%" y="9%"
10                 width="35%" height="40%" />
11    </pause>
12  </video>
```

Fig. 5. Enhanced video plug-in

6.4 Source Code Visualisation Plug-in

Earlier, we mentioned the difficulty of visualising complex resources such as source code. Our MindXpres source code plug-in exports a `code` tag which allows the presenter to paste their code into a presentation and have MindXpres visualise it nicely by making use of syntax highlighting via the SyntaxHighlighter[3] JavaScript library. Whenever the content engine encounters a `code` tag, it invokes the code plug-in to beautify the code and automatically adds vertical scrollbars for larger pieces of source code as shown in Fig. 6.

---

[3] http://alexgorbatchev.com/SyntaxHighlighter/

```
1  <code>
2    <publications>
3      <publication type="inproceedings">
4        <title>An Architecture for Open Cross-Media
               Annotation Services</title>
5        <author>
6          <surname>Signer</surname>
7          <forename>Beat</forename>
8        </author>
9        <author>
10         <surname>Norrie</surname>
11         <forename>Moira</forename>
12         ...
13  </code>
```

Source Code Visualisation

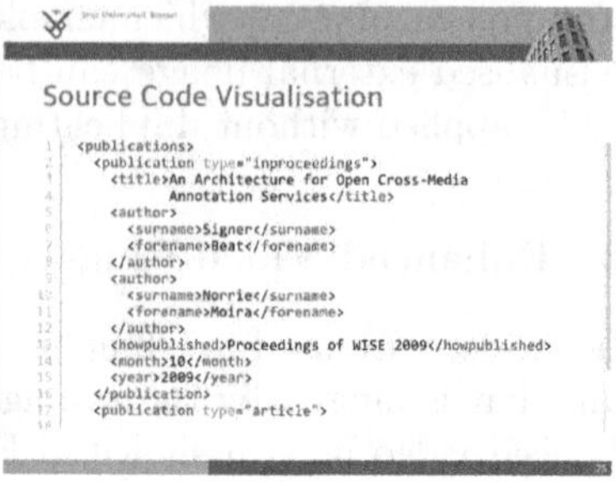

Fig. 6. Source code visualisation

7 Discussion and Future Work

MindXpres currently supports transclusion and cross-media content reuse on the plug-in level. For instance, the image or video plug-in can visualise (and enhance) external resources, a dictionary plug-in might retrieve definitions on demand via a web service or we might create a plug-in that allows us to import content (e.g. PowerPoint slides) from legacy documents at compile time. Nevertheless, we are currently investigating the introduction of generic reuse tags in our document format which would allow the presenter to transclude arbitrary parts of other MindXpres presentations. While our focus has been on the cross-media aspect of resources that can be used in a presentation, we might also investigate the cross-media publishing aspect via alternative compiler output formats.

We are aware that the current authoring of MindXpres presentations has some usability issues. The average presenter cannot be expected to construct an XML document or any CSS themes. In order to tackle this issue and further evaluate MindXpres in real-life settings, we are currently developing a graphical MindX-pres authoring tool. We further intend to provide a central plug-in repository which would make it easy for novice users to find, install and use new plug-ins via the authoring tool. In the long run, we intend to revise the use of monolithic documents and move towards repositories of semantically linked information based on the RSL hypermedia metamodel [27]. This would not only promote content reuse and sharing, but also create opportunities for context-aware as well as semi-automatic presentation authoring where relevant content is recommended by the authoring tool.

8 Conclusion

We have presented MindXpres, an HTML5 and web technology-based presentation platform addressing the lacking extensibility of existing slideware solutions. The extensibility of the MindXpres platform heavily relies on a plug-in mechanism and we have outlined how the tool can be extended on the content, visualisation as well as on the interaction level. By providing different forms of cross-media content transclusion, our solution further avoids the redundant storage and replication of slides. The presented MindXpres document format in combination with a compiler and the corresponding plug-ins offers the opportunity to have compile- or run-time cross-media content transclusion from third-party resources. At the same time, the flexible and extensible document model in combination with a zoomable user interface allows us to escape the boundaries of traditional slide formats. The presented MindXpres solution represents a promising platform for the unification of existing presentation concepts as well as for the rapid prototyping and investigation of new ideas for next generation cross-media presentation solutions.

References

1. Tufte, E.R.: The Cognitive Style of PowerPoint: Pitching Out Corrupts Within. Graphics Press (2003)
2. Farkas, D.K.: Toward a Better Understanding of PowerPoint Deck Design. Information Design Journal + Document Design 14(2) (August 2006)
3. Nelson, T.H.: The Heart of Connection: Hypermedia Unified by Transclusion. Communications of the ACM 38(8) (August 1995)
4. Holzinger, A., Kickmeier-Rust, M.D., Albert, D.: Dynamic Media in Computer Science Education; Content Complexity and Learning Performance: Is Less More? Educational Technology & Society 11(1) (January 2008)
5. Adams, C.: PowerPoint, Habits of Mind, and Classroom Culture. Journal of Curriculum Studies 38(4) (2006)
6. Gross, A., Harmon, J.: The Structure of PowerPoint Presentations: The Art of Grasping Things Whole. IEEE Transactions on Professional Communication 52(2) (December 2009)
7. Good, L., Bederson, B.B.: Zoomable User Interfaces as a Medium for Slide Show Presentations. Information Visualization 1(1) (March 2002)
8. Lichtschlag, L., Karrer, T., Borchers, J.: Fly: A Tool to Author Planar Presentations. In: Proc. of CHI 2009, Boston, USA (April 2009)
9. Lanir, J., Booth, K.S., Tang, A.: MultiPresenter: A Presentation System for (Very) Large Display Surfaces. In: Proc. of MM 2008, Vancouver, Canada (October 2008)
10. Chen, J.C., Chu, W.T., Kuo, J.H., Weng, C.Y., Wu, J.L.: Tiling Slideshow. In: Proc. of Multimedia 2006, Santa Barbara, USA (October 2006)
11. Signer, B., Norrie, M.C.: PaperPoint: A Paper-based Presentation and Interactive Paper Prototyping Tool. In: Proc. of TEI 2007, Baton Rouge, USA (February 2007)
12. Nelson, L., Ichimura, S., Pedersen, E.R., Adams, L.: Palette: A Paper Interface for Giving Presentations. In: Proc. of CHI 1999, Pittsburgh, USA (May 1999)
13. Spicer, R., Lin, Y.R., Kelliher, A., Sundaram, H.: NextSlidePlease: Authoring and Delivering Agile Multimedia Presentations. ACM TOMCCAP 8(4) (November 2012)
14. Edge, D., Savage, J., Yatani, K.: HyperSlides: Dynamic Presentation Prototyping. In: Proc. of CHI 2013, Paris, France (April 2013)
15. Garcia, P.: Retooling PowerPoint for Hypermedia Authoring. In: Proc. of SITE 2004, Atlanta, USA (March 2004)
16. Süß, C., Freitag, B.: LMML - The Learning Material Markup Language Framework. In: Proc. of ICL 2002, Villach, Austria (September 2002)

17. Fisler, J., Bleisch, S.: eLML, the eLesson Markup Language: Developing sustainable e-Learning Content Using an Open Source XML Framework. In: Proc. of WEBIST 2006, Setubal, Portugal (April 2006)
18. Canós, J.H., Marante, M.I., Llavador, M.: SliDL: A Slide Digital Library Supporting Content Reuse in Presentations. In: Lalmas, M., Jose, J., Rauber, A., Sebastiani, F., Frommholz, I. (eds.) ECDL 2010. LNCS, vol. 6273, pp. 453–456. Springer, Heidelberg (2010)
19. Verbert, K., Ochoa, X., Duval, E.: The ALOCOM Framework: Towards Scalable Content Reuse. Journal of Digital Information 9(1) (2008)
20. Raggett, D.: Slidy - A Web Based Alternative to Microsoft PowerPoint. In: Proc. of XTech 2006, Amsterdam, The Netherlands (May 2006)
21. Bush, M.D., Mott, J.D.: The Transformation of Learning with Technology: Learner-Centricity, Content and Tool Malleability, and Network Effects. Educational Technology Magazine 49(2), 3–20 (2009)
22. Haller, H., Abecker, A.: iMapping: A Zooming User Interface Approach for Personal and Semantic Knowledge Management. In: Proc. of Hypertext 2010, Toronto, Canada (June 2010)
23. Dufresne, R.J., Gerace, W.J., Leonard, W.J., Mestre, J.P., Wenk, L.: Classtalk: A Classroom Communication System for Active Learning. Journal of Computing in Higher Education 7(2) (1996)
24. Signer, B., Norrie, M.C.: Interactive Paper: Past, Present and Future. In: Proc. of PaperComp 2010, Copenhagen, Denmark (September 2010)
25. Reuss, E.I., Signer, B., Norrie, M.: PowerPoint Multimedia Presentations in Computer Science Education: What do Users Need? In: Holzinger, A. (ed.) USAB 2008. LNCS, vol. 5298, pp. 281–298. Springer, Heidelberg (2008)
26. Roels, R., Signer, B.: A Unified Communication Platform for Enriching and Enhancing Presentations with Active Learning Components. In: Proc. of ICALT 2014, Athens, Greece (July 2014)
27. Signer, B., Norrie, M.: As We May Link: A General Metamodel for Hypermedia Systems. In: Parent, C., Schewe, K.-D., Storey, V.C., Thalheim, B. (eds.) ER 2007. LNCS, vol. 4801, pp. 359–374. Springer, Heidelberg (2007)

Open Cross-Document Linking and Browsing Based on a Visual Plug-in Architecture

Ahmed A.O. Tayeh and Beat Signer

Web & Information Systems Engineering Lab
Vrije Universiteit Brussel
Pleinlaan 2, 1050 Brussels, Belgium
{atayeh,bsigner}@vub.ac.be

Abstract. Digital documents often do not exist in isolation but are implicitly or explicitly linked to parts of other documents. Nevertheless, most existing document formats only support links to web resources but not to parts of third-party documents. An open cross-document link service should address the multitude of existing document formats and be extensible to support emerging document formats and models. We present an architecture and prototype of an open cross-document link service and browser that is based on the RSL hypermedia metamodel. A main contribution is the specification and development of a visual plug-in solution that enables the integration of new document formats without requiring changes to the cross-document browser's main user interface component. The presented visual plug-in mechanism makes use of the Open Service Gateway initiative (OSGi) specification for modularisation and plug-in extensibility and has been validated by developing data as well as visual plug-ins for a number of existing document formats.

Keywords: Cross-document linking, hyperlinks, open link service.

1 Introduction

As already mentioned by Vannevar Bush in 1945, documents do not exist in isolation but have relationships with other documents [5]. Rather than classifying documents in hierarchical structures, Bush proposed to mimic the working of the human brain by supporting associative links or so-called trails between documents. The trails proposed by Bush were seminal for succeeding hypermedia models and architectures such as Xanadu [18], the Dexter hypertext reference model [10] or the resource-selector-link (RSL) metamodel [22]. The concept of hyperlinks was furthermore instrumental in the success of the World Wide Web by enabling the referencing, annotation and augmentation of content. Nevertheless, existing hypermedia solutions and document formats often only support simple forms of linking. While many document formats offer the possibility to link to entire third-party documents, most of the time it is not possible to address parts of documents. In an HTML document, we can for example create hyperlinks targeting an entire PDF or Word document but it is impossible to link to parts of these documents.

B. Benatallah et al. (Eds.): WISE 2014, Part II, LNCS 8787, pp. 231–245, 2014.
© Springer International Publishing Switzerland 2014

Most of today's document formats offer a simple embedded unidirectional link model. Unidirectional linking implies that a target document is not aware of explicit relationships that have been defined from one or multiple source documents. Furthermore, the use of embedded links means that only the owner of a document can add new links to a document. The growth and acceptance of the Web led to an increasing number of document formats that can be rendered within a browser. More recently, documents of different formats can also be edited and stored in the cloud and we should investigate ways to link parts of documents regardless of the used storage platform.

The advent of the Extensible Markup Language (XML) in combination with its link model (XLink) has been a major step towards advanced linking on the Web. Similar to early hypermedia systems, the combination of XML and XLink can be used to separate the document content from its links. Thereby, XLink provides the typical linking functionality such as bidirectional, multi-directional, multi-source and multi-target hyperlinks. Unfortunately, XLink does not solve the problem of cross-document linking since it only deals with XML documents and does not support other non-XML document models and formats.

A flexible and extensible link model and architecture is not only required to integrate existing document types, but also to deal with new emerging document formats. It should also support more advanced linking features that are currently lacking in most document formats. In this paper, we outline a number of requirements for open cross-document linking solutions. We then present our prototype of a cross-document link service and browser which is based on the open cross-media link service architecture by Signer and Norrie [24]. We paid attention to the aspect of openness as defined by Signer and Norrie and provide a cross-document link solution that does not only provide extensibility on the model layer but also on the application layer by further investigating the concept of visual plug-ins introduced by Signer and Norrie [23].

We begin in Sect. 2 by providing an overview of different link models and mechanisms offered by existing document formats and describe a number of link services and standards. In Sect. 3, we outline the requirements for an open cross-document link service and present a cross-document link service and browser prototype. We discuss a number of data and visual plug-ins that have been developed in order to support text, PDF, HTML and XML documents. A critical discussion of the presented solution is followed by some concluding remarks.

2   Background

Despite the multitude of existing document formats and standards, most document formats adhere to conservative representations of information. As criticised by Nelson [19], the *"What You See Is What You Get"* (WYSIWYG) principle in document processing degraded the computer to a paper simulator, neglecting many features that digital document formats could offer in addition to printed paper. Moreover, many proprietary document formats prevent other documents and applications from accessing and linking to their content. The Extensible

Markup Language was an important step to open the structure of some document formats. The XML Pointer Language (XPointer) [8] and XML Path Language (XPath) [7] can be used to address parts of an XML structure, while the XLink language supports the creation of advanced hyperlinks. However, most XML-based document formats such as DocBook [25], OpenDocument [26] and OOXML [1] have sacrificed the rich linking features and adopted a simple unidirectional link model. Moreover, XML and Semantic Web technologies promote the concept of linked data [14] where data conforming to Semantic Web standards can be linked.

An overview of the support for hyperlinks in a number of popular document formats is provided in Tab. 1. Hyperlinks form a basic building block of the HTML language which offers simple typed and embedded unidirectional links to address arbitrary web resources and link to entire third-party documents (e.g. PDF or OOXML). An HTML link target is rendered via a specific web browser plug-in or opened in a third-party application based on the document's MIME type.

Table 1. Supported link models in existing document formats

Format	Hyperlink Type	Supported Target Resources
HTML	unidirectional	web resources, entire third-party documents
LATEX	unidirectional	web resources, entire third-party documents
PDF	unidirectional	web resources, entire third-party documents, parts of PDF documents
XML	uni-, bi- and multidirectional	web resources, entire third-party documents, parts of XML-based documents
DocBook	unidirectional	web resources, entire third-party documents, parts of other DocBook documents
OOXML	unidirectional	web resources, parts of other OOXML documents

The embedded hyperlinks in HTML web documents and other document formats prevent the management of hyperlinks separately from the underlying documents. A problem of this approach is that new hyperlinks can only be added by the author of a document. This limitation of embedded hyperlinks lead to research in open hypermedia systems where hyperlinks are managed externally in centralised databases or so-called linkbases. Intermedia [11] was an early system managing hyperlinks in an external linkbase. Another example is Microcosm [13] which offered a service for linking within arbitrary desktop applications such as AutoCAD or MS Word. Furthermore, the open hypermedia community tried to enrich the Web with external hyperlinks by considering the Web as another client for open hypermedia systems. Examples of this development include Chimera [2] or Arakne [4]. Open hypermedia systems had a significant impact in enhancing the management of hyperlinks. However, these types of systems had two major shortcomings. First of all, they did not investigate the possibilities for creating hyperlinks between snippets of information in different document formats. Second, it is not clear how to extend these systems to support cross-document linking on the model as well as on the application layer.

While the XML language does not provide a mechanism to create hyperlinks, XLink can be used to create links that go beyond simple embedded unidirectional hyperlinks. The XLink standard has been developed to improve the linking on the Web without relying on open hypermedia systems. Besides the simple unidirectional hyperlinks, XLink also supports so-called *extended hyperlinks*. With extended hyperlinks, bi- and multi-directional hyperlinks can be realised. XLink links can address web resources and parts of XML documents which means that the cross-document linking is limited to XML-based document formats.

Based on XLink and XPointer, various applications have been built to open web documents to third-party annotations and associations to external web documents. Annotea [15] is an RDF-based standard which enhances collaboration via shared web annotations and bookmarks. These annotations can be notes, explanations or comments that are externally attached to a webpage. Annotea uses XPointer to address specific parts of a webpage. A number of tools that implement the Annotea standard have been developed, including the W3C's Amaya[1] web browser or the Annozilla[2] Firefox extension. MADCOW [3] is another tool that enables the "opening" of webpages to arbitrary users. MADCOW offers richer media support than Annotea-based tools by enabling the annotation of parts of images or videos. All these annotation tools do not go beyond the features offered in XLink and hence the addressing of document parts for link sources and targets is still limited to XML-based document formats. Moreover, even if a number of these tools are extensible on the model layer, they lack extensibility on the application and visualisation layer. For example, if a new media type should be supported in MADCOW which is implemented as a monolithic component, a new version of the user interface has to be deployed.

Annotation tools for digital libraries are another trend to open different document formats for linking to external resources. An example of this family of tools is the Flexible Annotation Service Tool (FAST) [17], a standalone annotation system for digital libraries. An interesting feature of FAST is that it separates the core annotation model from the functionality offered by digital libraries. Any digital library information management system can benefit from the features of FAST by creating a new FAST interface (*gate*). Thereby, FAST does not make any assumptions about the structure of the annotated resource but defines a general *handle* concept for the resource to be annotated. Also FAST does not explicitly deal with extensibility on the application layer and is mainly targeting simple annotations rather than richer forms of hyperlinks.

Goate [16] represents another attempt to enhance the HTML link model. It is based on an HTTP proxy architecture for augmenting HTML documents with features of the XLink model such as bidirectional and n-ary hyperlinks.

---

[1] http://www.w3.org/Amaya

[2] http://annozilla.mozdev.org

Similar to Goate, XLinkProxy [6] is using a web proxy to augment HTML documents with XLink features. However, like with other XLink-based solutions the linking in these systems in limited to XML-based documents.

The DocBook and OOXML standards only support simple unidirectional hyperlinks. DocBook hyperlinks allow us to address any web resource, entire external documents or parts of another DocBook document. Since the main goal of OOXML is to facilitate extensibility and interoperability between multiple office applications and on multiple platforms, it does not go beyond the simple unidirectional hyperlinks that were already supported in WYSIWYG office documents. With OOXML hyperlinks it is possible to address any web resource, parts of the document itself and parts of other OOXML documents.

While various approaches try to improve the linking between documents and web resources, there is still no single solution that supports arbitrary document formats (e.g. not only XML-based formats) in combination with easy extensibility. Some approaches can be extended for future emerging document formats but not without some major development efforts and the redeployment of the entire application. We present an architecture for open cross-document linking and browsing that offers rich hypermedia functionality and can be extended via data and visual plug-ins without having to redeploy the entire application.

3   Open Cross-Document Linking

We outline a number of issues that should be taken into account in order to achieve a cross-document linking solution. First of all, there exists a variety of document formats and standards including markup languages and WYSIWYG formats. These document formats have different logical representations such as linear document models, unconstrained tree-like document models or constrained heterogeneous tree-like document models as classified by Furuta [9]. Furthermore, document formats with similar document models can still show differences in terms of the granularity of the lowest level of atomic objects supported by the model. An atomic object in a model might, for example, be a text string representing a paragraph, a sentence or even individual characters. Moreover, many document formats suffer from the fact that they are adding new layers of complexity on top of the existing formats when new features have to be supported. The 5585 pages long specification of OOXML [1] is a testament of this never-ending growth of complexity of some document formats. The integration of cross-document linking functionality into existing document formats is a tedious and complex task since it requires some knowledge about other document formats and their logical structure. Therefore, a cross-document link service should not require any changes to the specification of existing document formats and standards. Further, a multitude of media types such as text, images, sound or video clips are supported in different document formats. These media types are informative and a selector within these media types can form the source or target of a hyperlink. One might, for instance, have a hyperlink from a selection of text in an PDF document to a specific time span of a video clip that is embedded in an HTML document. A cross-document linking solution should therefore not only specify how to address specific nodes of a document's logical structure but also define how to deal with a node's media type in order to select parts of it. All theses issues indicate that it is impossible to make any prior assumptions about the types of linked document formats and their content. This is also one of the main reasons why we think that the XLink standard is not suitable for open cross-document linking since it assumes that the source and target documents have a tree-like document structure.

A promising approach for cross-document linking is the use of an external link service which defines, stores and manages the general hyperlink concepts. One possible definition could be that hyperlinks can have one or more sources and one or multiple targets. The source might be a string representing an XPointer expression for tree-like documents or start and end indices for WYSIWYG formats. However, the definition of sources and targets should be kept abstract and each document format then has to provide a concrete definition of how parts of a resource can be addressed by extending the abstract hyperlink concepts via a plug-in mechanism. The definition of the document addressing part can be achieved by using third-party document APIs, different programming language libraries for specific document formats or existing implementations for standards such as the Document Object Model (DOM). A link service should further be extensible to support existing as well as emerging new document formats.

An interesting idea that can be adopted to realise such a link service is the proposal of the general open cross-media annotation and link architecture by Signer and Norrie [24]. Similar to FAST, its basic idea is the separation of linking and annotation functionality from the annotated media. The annotation and link service knows how to deal with the core link and annotation model and is extensible to support new media types. This cross-media annotation and link architecture by Signer and Norrie shows a number of advantages. First of all, the annotation and link service is based on the RSL hypermedia metamodel [22] which is general and flexible enough to support evolving hypermedia systems. RSL overcomes some limitations of existing hypermedia models mixing technical and conceptual issues. The RSL model is based on the concept of linking arbitrary entities, whereby an entity can either be a resource, a selector or a link. A *resource* defines a media type such as a text, a video or a complete document. The *selector* is always attached to a resource and used to address parts of a resource. Finally, a *link* can be a one-to-one, one-to-many, many-to-one or many-to-many bidirectional association between any entities. In addition, the RSL model offers other features such as user management and support for overlapping hyperlinks via layering. To the best of our knowledge, the cross-media annotation and link architecture by Signer and Norrie [24] was the first to propose the concept of *"extensibility on the visualisation layer"* of a link service. Normally, when a link or annotation service is extended to support a new media type, also the user interface has to be extended to support the visualisation of the new media type. Signer and Norrie [24] recommend to use visual plug-ins in order to avoid a re-implementation and deployment of the entire user interface.

In the remaining part of this paper, we elaborate on how we have developed a cross-document link service based on the ideas of the open cross-media architecture and discuss a number of plug-ins that have been realised so far for our cross-document link service and browser.

3.1   General Open Cross-Document Link Service Architecture

The general architecture of the link service is illustrated in Fig. 1. A central component is the core link service which is based on the RSL metamodel. The core is extensible to support arbitrary document formats by providing a *data plug-in* consisting of a media-specific implementation of the resource and selector concepts. The extensible visualisation component contains the user interface in the form of our link browser that visualises the supported document formats. For each document format to be rendered in the browser, a *visual plug-in* must be implemented. The visual plug-in has two main responsibilities. First, it has to render a specific document format and visualise any selectors that have been defined. Second, it provides a visual handle for the basic create, read, update and delete (CRUD) operations for a given document and its selectors. Taking into account that many document formats come along with their own proprietary third-party applications, visual plug-ins also have to be provided for these applications. These third-party visual plug-ins do not directly support the CRUD operations on the underlying documents but have to communicate with our link browser component in order to exchange information about selectors to be activated or created. Finally, our architecture consists of a data layer that is in charge of storing all the RSL metadata such as resources, selectors or hyperlinks.

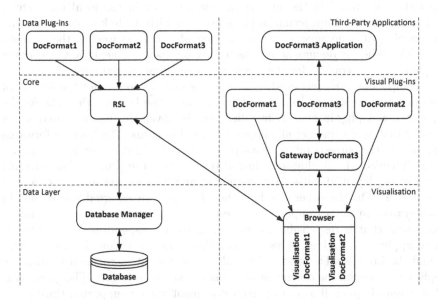

Fig. 1. General open cross-document link service architecture

We decided to make use of the Open Service Gateway initiative (OSGi) [12] for the development of the link service. The OSGi specifications defines a dynamic modular system for Java. Various server applications such as IBM Websphere apply the OSGi framework and also the Eclipse IDE uses OSGi to enable the modularisation of its components as well as to support dynamic extensions via plug-ins. We decided to use OSGi for several reasons. First of all, apart from reducing the application complexity, OSGi offers a decent mechanism for code sharing between different modules. In contrast to Java JAR files, OSGi modules do not share arbitrary code but explicitly define export packages to be shared and import packages to be used. Due to the clear definition of exported functionality, code cannot be "misused" by other installed modules. Second, the link service should be dynamically extensible to support new document formats without a need for redeployment. Dynamic extensibility also allows users to download plug-ins for new document formats on demand and thereby supports emerging document formats. The dynamic extensibility of the link service can be realised based on OSGi and its built-in support for the implementation of dynamic applications. Last but not least, our link service might provide different visual plug-ins (versions) for the same document format. Managing different versions of a module in a pure Java application often causes the so-called "JAR hell" problem, while the OSGi framework offers a specific mechanism for the versioning of modules and dependency resolution.

3.2   Communication with Plug-ins

After launching the link service and its link browser, the user can open any document in a format that is supported by the service. Thereby, documents can either be stored in the link service database or in the local file system. Figure 2 shows the main scenarios for interacting with the link service, including the opening of a document, the navigation of a link as well as the creation of a link. When a document is selected to be opened, the browser retrieves supplementary metadata for the given document from the database via the core RSL component. The retrieved data contains information about the format of the document as well as its associated selectors. The browser then checks the type of visual plug-in that is installed for the given document format via a registry that keeps track of all supported visual plug-ins. The browser forwards a request to the corresponding visual plug-in in order to visualise the document. If the intended visual plug-in is installed locally in the browser (DocFormat1 in our example), it visualises the document and its selectors in a panel within the browser. On the other hand, if the plug-in is an external visual plug-in (DocFormat3 in our example), it receives the request via a special gateway component that is responsible for launching the corresponding external third-party application and manages any communication between the visual plug-in and the browser. The external visual plug-in tells its associated third-party application to open the document and render existing selectors. The question is what should happen if a user opens a document with a supported third-party application rather than with the link browser? In this case, the external visual plug-in notifies its gateway about the document that is currently being visualised. The gateway then communicates with the browser to retrieve potentially stored data about the currently visualised document. If any selectors have been defined for the currently opened document, the gateway returns the list of selectors to the external visual plug-in in order to visualise them.

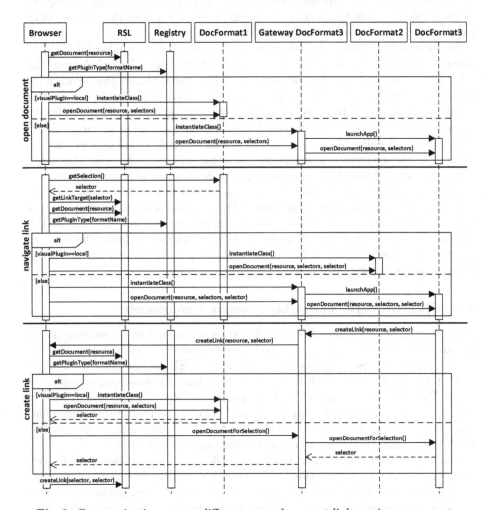

Fig. 2. Communication among different cross-document link service components

When a user clicks on a hyperlink in a document that is visualised in the browser, the browser communicates with the core RSL component to retrieve the target document and its selectors including the target selector of the selected link. The target document is then visualised with its selectors and the target selector of the followed link is highlighted in a different colour than the other selectors. The two documents can, for example, be visualised next to each

other in the browser (`DocFormat1` and `DocFormat2`) or one in the browser while
the other document is rendered in a third-party application (`DocFormat1` and
`DocFormat3`). In the case that a link is selected in a document that is visualised
in a third-party application, the visual plug-in sends a request to its gateway
with the selected link source (selector) as a parameter. The gateway then forwards the request to the browser. From there on the browser handles the request
in the same way as in the previous scenario to either visualise the link target in
the browser or in a third-party application.

In the case of creating a hyperlink in a document that is visualised in a third-
party application, the user selects the option of creating a hyperlink from the
GUI actions supported by the visual-plug-in. The visual plug-in then sends a request to its gateway with the document and the source selector as a parameter.
The gateway forwards the request to the browser which offers the possibility to
open another document in order to define the link's target. The target document is then visualised in the browser or via an external visual plug-in and the
browser or the external visual plug-in listens for any user selection. The user's
selection is retrieved and the browser requests the creation of a hyperlink via the
core RSL component. Note that the scenario shown in Fig. 2 assumes that the
external documents have the same format. On the other hand, if the document is
visualised in the browser, the user selects parts of the document and chooses the
option to create a hyperlink from the supported browser actions. The browser
handles the request similar as in the previous case by allowing the user to choose
another document and listening for any target selector.

3.3   Open Cross-Document Link Service Components

A number of modules have been developed to realise the main components of
the architecture. The core link service has been realised in a standalone module
containing the necessary classes to implement the RSL metamodel. This module
further provides a Java API for CRUD operations on resources, selectors or
links. The core RSL package can be imported by any other module which is
reflected by the `Export-Package: org.rsl.core` metadata in the manifest file.
This allows data plug-ins to extend the RSL resource and selector concepts and
the visualisation component to communicate with the core API.

The visualisation component has been realised as a standalone module. Besides the support for GUI actions for hyperlink CRUD operations and the navigation of hyperlinks between documents, this module is extensible to visualise
arbitrary document formats via the plug-in mechanism mentioned earlier. The
extensibility of the user interface is supported via a specific `DefaultDocument`
class that has to be extended by the visual plug-ins for individual document
formats. The `DefaultDocument` class extends the `JPanel` component and offers the abstract `getSelection()`, `setSelection()` and two different abstract
`openDocument()` methods that are used to open a document with its selectors
to either browse or highlight the target of a hyperlink. In addition to the core
and visualisation module, we realised a separate module for storing documents
and their hyperlinks. This module further offers the flexibility to use different

database management systems for the persistent storage of documents and link data. In our current implementation, we use the db4o[3] object database for storing system objects.

Last but not least, two plug-ins have to be provided for each document format as mentioned earlier. The data plug-ins are standalone OSGi modules that extend the core RSL module, whereas the visual plug-ins are either OSGi modules that are installed locally in the link service or extensions for third-party application user interfaces. Each external visual plug-in needs an extra module (a gateway plug-in) that needs to be installed in the OSGi platform. In order to manage the different data, gateway and visual plug-ins, we have implemented a *module tracker* which maintains a registry of supported document formats and the necessary metadata to instantiate the corresponding classes. Each plug-in has to specify its type via the `Extension-Type` which can be either `data`, `visual` or `gateway`. Furthermore, a plug-in has to specify the supported format via the `Extension-Format` key. Last but not least, local visual plug-ins use the `Extension-Class` to define the classpath of the class implementing the abstract methods of the `DefaultDocument` class. The same metadata is used in the gateway plug-in to define the classpath of the class handling the communication between the external visual plug-in and the browser.

The data plug-in for a specific document format must provide the definition of its logical structure by extending the RSL resource class and further define how to create selectors within this structure by extending the RSL selector class. The definition of the logical structure can vary even for a single document format. For example, if the link service communicates with an HTML visual plug-in that extends the FireFox web browser, a URI string is sufficient to define the HTML document resource. However, if the browser of our link service should support the HTML visualisation, sufficient information about the HTML tree syntax has to be provided.

Any local visual plug-in has to implement the abstract methods of the `DefaultDocument` class which are used by the browser when visualising a document. Furthermore, the browser can easily retrieve or highlight a selector within a document by using the `getSelection()` and `setSelection()` methods. Thereby, third-party visualisation libraries might be used when implementing a visual plug-in. We have, for example, used the ICEpdf library[4] to implement a visual browser plug-in for PDF documents. In doing so, the visual plug-in provides a class extending the `DefaultDocument` class and acts as a proxy between the link browser and a third-party document visualisation library.

On the other hand, an external visual plug-in has to provide some methods to get and set selections in a document that is visualised in a third-party application. This can be achieved since most document applications provide their own SDKs or APIs such as the Foxit Reader Plug-in SDK[5] or Microsoft's Office

---

[3] http://www.db4o.com

[4] http://www.icesoft.org/java/projects/ICEpdf/overview.jsf

[5] http://www.foxitsoftware.com

Developer tools[6]. Furthermore, the external visual plug-in also has to provide the GUI actions for the necessary hyperlink CRUD operation which are offered by the browser for local visual plug-ins.

Last but not least, the gateway plug-in class defined in the `Extension-Class` metadata needs to implement three methods: the `openDocumentForSelection()` method which listens for any selection in external documents as described in the link creation scenario and the two `openDocument()` methods. The gateway also has to implement methods for handling the communication between the browser and the visual plug-in.

4   HTML, PDF, Text and XML Plug-ins

Our cross-document link service currently supports the linking of HTML, PDF, text and XML documents. The browser is able to visualise XML, text and PDF documents via local visual plug-ins while the visualisation of HTML documents is delegated to the Google Chrome web browser. Figure 3 shows a bi-directional hyperlink between a PDF and XML document which are both visualised in our browser prototype as well as a bi-directional hyperlink between a PDF document and an HTML document visualised in the external web browser shown in the lower left part of Fig. 3. Note that we see the support for the XML document format as a first step towards the integration of different Open Office document formats.

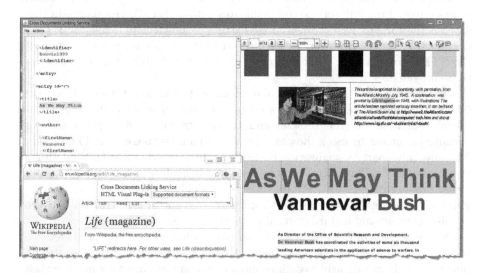

Fig. 3. Links between an XML and PDF as well as a PDF and HTML document

The data plug-in for the HTML document format defines an HTML resource by its URI, while the selector consists of an XPointer-like expression. The HTML

---

[6] http://msdn.microsoft.com/en-us/library/jj620922.aspx

visual plug-in has been implemented as Google Chrome extension which uses the Google Chrome API for accessing documents. Furthermore, the visual plug-in for HTML offers a number of simple GUI actions within the Chrome browser in order to create and navigate hyperlinks. Rangy[7], a cross-browser JavaScript range and selection library, is used to retrieve (in an XPointer-like expression) and highlight selections in HTML documents. To distinguish between the hyperlinks coming from our link service and embedded HTML hyperlinks, the visual plug-in uses a different colour for the visualisation of hyperlinks. The communication between the visual plug-in and its gateway plug-in has been realised by using the WebSocket protocol.

The data plug-ins for PDF, text and XML documents specify their resources via the path and name of the documents in the user's local storage. The selector within a PDF document is defined through a page index and a rectangular area within a page. The local PDF visual plug-in uses the viewer of the ICEpdf library for the rendering of PDF documents. The visual plug-in uses the ICEpdf methods to get and create rectangular selections in a PDF document.

A selector within an XML document is defined via DOM ranges. Note that there are also some libraries that use XPointer such as XInclude, but these libraries are targeting XML inclusion. The local XML visual plug-in extends the StyledEditorKit component for a better visualisation of XML documents. Furthermore, it uses the javax.xml.parsers library for reading XML documents. Last but not least, the XML visual plug-in applies the org.w3c.dom library to retrieve and highlight nodes and ranges within an XML document.

Finally, a selector within a text document is defined by the start and end indices of the selection. The local visual plug-in for text uses a JTextPane for the visualisation of arbitrary text documents.

5 Discussion and Future Work

The open cross-document link service and browser goes beyond the simple annotation concept offered by most annotation tools where only the reading, creation, saving, updating and retrieving of annotations is supported while support for bi-directional hyperlinks between document content is missing. Furthermore, different document formats can be integrated in our link service regardless of their document models. Two features were essential in achieving the presented prototype of an open cross-document link service. First, similar to early open hypermedia systems our cross-document link service uses an external link representation and storage. This implies that there are no changes necessary to the specification of existing document formats in order to integrate them with our link service. Second, through generalisation and the treatment of hyperlinks as first-class objects in the core link service (RSL), each document format to be supported can extend the RSL resource for its own logical document model and specialise the RSL selector with a definition of its selector. Moreover, the link service overcomes to some extent the issue of broken hyperlinks and consistency

---

[7] https://code.google.com/p/rangy/

of hyperlinks when documents evolve. When a hyperlink source has been deleted from a document, the link service automatically removes the link target from the other document. Furthermore, an archive of linked documents is a simple approach to ensure link consistency. To the best of our knowledge, the presented open cross-document link service is the first prototype to introduce flexibility and extensibility on the model as well as on the information visualisation layer as proposed by Signer and Norrie [24]. The extensibility of our open cross-document link service is further supported by the dynamic modular OSGi framework.

We also considered to realise our cross-document link service as an Open Web Platform-based solution. However, for a number of reasons we decided to go for a Java-based system rather than an Open Web Platform solution. There are currently only a limited number of Web-based open source libraries available for different document formats and most of them do not support the editing of documents. Furthermore, most web browsers offer only limited support (e.g. WebSockets) for communicating with third-party applications.

We are currently working on a dynamic plug-in extensibility where plug-ins will be automatically installed on demand. This dynamic extensibility is based on the well-known OSGi extender pattern which listens for the installation of new bundles in the OSGi platform by using a Secure Shell or Telnet protocol. Furthermore, we plan to investigate the extensibility of the link service in a study with developers and foresee to evaluate the usability of the presented solution in an end user study. Last but not least, we are planning to integrate some media plug-ins, for example for video and audio [21,20], into our link service. These media plug-ins could then be used for addressing different media types forming part of specific documents.

6   Conclusion

We have presented a prototype of a cross-document link service and browser for integrating and linking different document formats. Based on ideas from the Open Hypermedia community and by using the RSL hypermedia metamodel, we have realised an extensible cross-media architecture. The presented cross-document link service prototype does not only support the extensibility on the data layer but more importantly also on the application and visualisation layer via visual plug-ins and a modular and extensible architecture that is based on the OSGi standard. While we have introduced various plug-ins for integrating HTML, PDF, XML and text documents with our browser as well as with external third-party applications, the presented cross-document link service and browser presents an ideal platform for investigating future innovative forms of cross-document linking.

References

1. Standard ECMA-376: Office Open XML File Formats, 3rd edn. (June 2011)
2. Anderson, K.M., Taylor, R.N., Whitehead Jr., E.J.: Chimera: Hypermedia for Heterogeneous Software Development Environments. ACM Transactions on Information Systems 18(2) (July 2000)

3. Bottoni, P., Civica, R., Levialdi, S., Orso, L., Panizzi, E., Trinchese, R.: MADCOW: A Multimedia Digital Annotation System. In: Proc. of AVI 2004, Gallipoli, Italy (May 2004)
4. Bouvin, N.O.: Unifying Strategies for Web Augmentation. In: Proc. of Hypertext 1999, Darmstadt, Germany (February 1999)
5. Bush, V.: As We May Think. Atlantic Monthly 176(1) (1945)
6. Ciancarini, P., Folli, F., Rossi, D., Vitali, F.: XLinkProxy: External Linkbases with XLink. In: Proc. of DocEng 2002, McLean, USA (November 2002)
7. Clark, J., DeRose, S.: XML Path Language (XPath) Version 1.0 (November 1999)
8. DeRose, S., Maler, E., Daniel Jr., R.: XML Pointer Language (XPointer) Version 1.0 (January 2001)
9. Furuta, R.: Concepts and Models for Structured Documents. In: Structured Documents. Cambridge University Press (1989)
10. Grønbæk, K., Hem, J.A., Madsen, O.L., Sloth, L.: Designing Dexter-based Cooperative Hypermedia Systems. In: Proc. of Hypertext 1993, Seatle, USA (November 1993)
11. Haan, B.J., Kahn, P., Riley, V.A., Coombs, J.H., Meyrowitz, N.K.: IRIS Hypermedia Services. Comunication of the ACM 35(1) (1992)
12. Hall, R., Pauls, K., McCulloch, S., Savage, D.: OSGi in Action. Manning Publications (2011)
13. Hall, W., Davis, H., Hutchings, G.: Rethinking Hypermedia: The Microcosm Approach. Kluwer Academic Publishers (1996)
14. Heath, T., Bizer, C.: Linked Data: Evolving the Web into a Global Data Space. Morgan and Claypool Publishers (2011)
15. Koivunen, M.-R.: Semantic Authoring by Tagging with Annotea Social Bookmarks and Topics. In: Proc. of SAAW 2006, Athens, Greece (November 2006)
16. Martin, D., Ashman, H.: Goate: An Infrastructure for New Web Linking. In: Proc. of the International Workshop on Open Hypermedia Systems at HT 2002, Maryland, USA (June 2002)
17. Model, M.A.: Architecturesti, and N. Ferro. A System Architecture as a Support to a Flexible Annotation Service. In: Proc. of the 6th Thematic Workshop of the EU Network of Excellence DELOS, Cagliari, Italy (June 2004)
18. Nelson, T.H.: Literary Machines. Mindful Press (1982)
19. Nelson, T.H.: Geeks Bearing Gifts: How the Computer World Got This Way. Mindful Press (2009)
20. Signer, B.: Fundamental Concepts for Interactive Paper and Cross-Media Information Spaces. Books on Demand GmbH (May 2008)
21. Signer, B., Norrie, M.C.: A Framework for Cross-Media Information Mangement. In: Proc. of EuroIMSA 2005, Grindelwald, Switzerland (February 2005)
22. Signer, B., Norrie, M.C.: As We May Link: A General Metamodel for Hypermedia Systems. In: Proc. of ER 2007, Auckland, New Zealand (November 2007)
23. Signer, B., Norrie, M.C.: An Architecture for Open Cross-Media Annotation Services. In: Vossen, G., Long, D.D.E., Yu, J.X. (eds.) WISE 2009. LNCS, vol. 5802, pp. 387–400. Springer, Heidelberg (2009)
24. Signer, B., Norrie, M.C.: A Model and Architecture for Open Cross-Media Annotation and Link Services. Information Systems 6(36) (May 2011)
25. Walsb, N.: DocBook 5 The Definitive Guide. O'Reilly (2010)
26. Weir, R., Brauer, M., Durusau, P.: Open Document Format for Office Applications (OpenDocument) Version 1.2 (March 2011)

Cost-Based Join Algorithm Selection in Hadoop

Jun Gu[1], Shu Peng[1], X. Sean Wang[1], Weixiong Rao[2], Min Yang[1], and Yu Cao[3]

[1] School of Computer Science, Fudan University, Shanghai, China
Shanghai Key Laboratory of Data Science, Fudan University, Shanghai, China
{gujun,pengshu,xywangCS,m_yang}@fudan.edu.cn
[2] School of Software Engineering, Tongji University, Shanghai, China
wxrao@tongji.edu.cn
[3] EMC Labs, Tsinghua Science Park, Beijing, China
yu.cao@emc.com

Abstract. In recent years, MapReduce has become a popular computing framework for big data analysis. Join is a major query type for
data analysis and various algorithms have been designed to process join
queries on top of Hadoop. Since the efficiency of different algorithms differs on the join tasks on hand, to achieve a good performance, users need
to select an appropriate algorithm and use the algorithm with a proper
configuration, which is rather difficult for many end users. This paper
proposes a cost model to estimate the cost of four popular join algorithms. Based on the cost model, the system may automatically choose
the join algorithm with the least cost, and then give the reasonable configuration values for the chosen algorithm. Experimental results with the
TPC-H benchmark verify that the proposed method can correctly choose
the best join algorithm, and the chosen algorithm can achieve a speedup
of around 1.25 times over the default join algorithm.

Keywords: Join algorithm, Cost model, Hadoop, Hive.

1 Introduction

In recent years, MapReduce [1] has become a popular computing framework for
big data analysis. Hadoop [2], an open-source implementation of MapReduce, has
been widely used. One example of the big data analysis is log processing, such as
the analysis of click-streams, application access logs, and phone call records. Log
analysis often requires a join operation between log data and reference data (such
as information about users). Unfortunately, MapReduce was originally designed
for the processing of a single input, and the join operation typically requires two
or more inputs. Consequently, it has been an open issue to improve Hadoop for
the join operation.

Many works have appeared in the literature that tackle the join operation
in Hadoop. Such works roughly fall into the two categories. The first is to design novel join algorithms on top of Hadoop [6][7][8][9][10][11]. The second is to
change the internals of Hadoop or build a new layer on top of Hadoop for the
optimization of traditional join algorithms [3][12][14][15][16][18][19].

B. Benatallah et al. (Eds.): WISE 2014, Part II, LNCS 8787, pp. 246–261, 2014.

Given the many existing join algorithms, it is hard for users to choose the best one for their particular join tasks since different algorithms differ significantly in their performance when used on different tasks. Usually Hadoop users have to define a `map` function and a `reduce` function and to configure their own MapReduce jobs. It's a much harder task for the users to change the internals of Hadoop or build a new layer on top of Hadoop. Hive system [3] is probably the most popular open source implementation to execute join queries on top of Hadoop. However, even with the help of Hive, it is still hard for a Hive user to choose the best join algorithm among all those implemented in Hive. Finally, suppose that a join algorithm is chosen, the users are required to tune some key parameters. Unfortunately, choosing the best join algorithm and tuning the parameters involve non-trivial efforts.

To help the end users select the best join algorithm and tune the associated parameters, in this paper, we propose a general cost model for four widely used join algorithms. Based on our cost model, we adaptively choose one of the join algorithms with the least cost, and then set the reasonable configuration values.

In summary, we make the following contributions:

- First, we design a cost model for the four popular join algorithms in Hive, and propose a tree structure based on which a pruning method is designed for the automated selection of the best join algorithm.
- Second, as a part of the selection process, we provide a method to tune the key parameters needed by the chosen join algorithm.
- Third, based on the TPC-H benchmark [4], we conduct experiments to evaluate our proposed method. The experimental results verify that our method can correctly choose the best join algorithm. Moreover, the chosen algorithm can achieve a speedup of around 1.25, compared with the default join algorithm.

The rest of the paper is organized as follows. In Section 2, we introduce the background of our work. We present the cost model in Section 3, and in Section 4 design an algorithm, along with a tree-structured pruning strategy, to choose the best join algorithm. We report an evaluation of our method in Section 5. In Section 6, we investigate related works, and finally conclude the paper with Section 7.

2   Background

In this section, we first introduce the two popular open source systems: Hadoop and Hive (Section 2.1), and review four join algorithms that are implemented in Hive (Section 2.2).

2.1   Hadoop and Hive

Hadoop, an open-source implementation of MapReduce, has been widely used for big data analysis. The Hadoop system mainly contains two components:

Hadoop Distributed File System (HDFS) [5] and MapReduce computing framework. Hadoop reads input data from HDFS and writes output data to HDFS. The input and output data are maintained on HDFS with data blocks of 64 MB by default. To process a query job, Hadoop starts map tasks (mappers) and reduce tasks (reducers) concurrently on clustered machines. Each mapper reads a chunk of input data, extracts <key, value> pairs, applies the map function and emits intermediate <key', value'> pairs. Those intermediate pairs with the same key are grouped together as <key', list<value'>>. After that, the grouped pairs are then shuffled to reducers. Each reducer, after receiving the grouped pairs, applies the reduce function onto the grouped pairs, and finally writes the outputs back to HDFS.

Hive is a popular open-source data warehousing solution, which facilitates querying and managing large datasets on top of Hadoop. Hive supports queries expressed in an SQL-like declarative language, called HiveQL. Using HiveQL lets users create summarizations of data, perform ad-hoc queries, and analysis of large datasets in the Hadoop cluster. For those users familiar with the traditional SQL language, they can easily use HiveQL to execute SQL queries. HiveQL also allows programmers who are familiar with MapReduce to plug-in their custom mappers and reducers. In this way, Hive can perform complex analysis that may not be supported by the built-in capabilities of the HiveQL language. Based on the queries written in HiveQL, Hive compiles the queries into MapReduce jobs, and submits them to Hadoop for execution. Since Hive can directly use the data in HDFS, operations can be scaled across all the datanodes and Hive can manipulate huge datasets.

2.2   Join Algorithms in Hadoop

In this section, we review four join algorithms that are widely used in MapReduce framework. All of such algorithms are supported in the Hive system. We will design our selection method based on these four join algorithms.

- *Common Join*: We consider that two tables are involved in a join task. In the map function, each row of the two join tables is tagged to identify the table that the row comes from. Next, the rows with the same join keys are shuffled to the same reducer. After that, each reducer joins the rows from the two join tables on the key-equality basis. *Common Join* can always work correctly with any combinations of sizes of the join tables. However, this join algorithm may incur the worst performance efficiency due to a large amount of shuffled data across clustered machines.
- *Map/Broadcast Join*: For two join tables, this algorithm first starts a local MapReduce task to build a hashtable of the smaller table. The task next uploads the hashtable to HDFS and finally broadcasts the hashtable to every node in the cluster (Note that the hashtable is maintained on the local disk of each node in form of a distributed cache). After finishing the local MapReduce task, this algorithm starts a map-only job to process the join query as follows. First, each mapper reads the hashtable (i.e., the smaller

table) from its local disk into main memory. Second, the mapper scans the large table and matches record keys against the hashtable. By combining the matches between the two tables, the mapper finally writes the output onto HDFS. This algorithm does not start any reducer, but requires that the hashtable of the smaller table is small enough to fit into local memory.

– *Bucket Map Join*: Differing from *Map Join*, *Bucket Map Join* considers that the data size of join tables is big. Thus, in order to reduce the memory limitation of *Map Join* from keeping the whole hashtable of the smaller table, this algorithm bucketizes join tables into smaller buckets on the join column. When the number of buckets in one table is a multiple of the number of buckets in the other table, the buckets can be joined with each other. In this way, as *Map Join*, a local MapReduce task is launched to build the hashtable of each bucket of the smaller table, and then broadcast those hashtables to every nodes in the cluster. Now, instead of reading the entire hashtable of smaller table, mappers only read the required hashtable buckets from distributed cache into memory. Thus, *Bucket Map Join* reduces the used memory space.

– *Sort-Merge-Bucket (SMB) Join*: If data to be joined is already sorted and bucketized on the join column with the exactly same number of buckets, the creation of hashtable is unneeded. Each mapper then reads records from the corresponding buckets from HDFS and then merges the sorted buckets. The *SMB* algorithm allows the query processing to be faster than an ordinary map-only join. The *SMB Join* is thus fast for the tables of any size with no limitation of memory though with the requirement that the data should be sorted and bucketized before the query processing.

3   Cost Model

Given the four popular join algorithms, in this section, we design a cost model to estimate the cost of the four algorithms that are used to process a join query. Here, we only consider the fundamental join SQL query without *WHERE* conditions (and other sub-queries such as *GROUP BY*, etc.):

`SELECT C FROM T1 JOIN T2 ON T1.ci = T2.cj;`

where C denotes the projected columns from join tables T_1 and T_2, and c_i and c_j are the query join keys.

Before presenting our cost model, we first make the following assumptions:

– Firstly, we assume that T_2 is the smaller table. Thus, when the *Map/Broadcast Join* or *Bucket Map Join* algorithm is used, the cost of broadcasting and building the hashtable involves only the table T_2.
– Secondly, for *Bucket Map Join* and *SMB Join*, we assume that the big table T_1 and small table T_2 are associated with the same number of buckets, i.e., $N_1 = N_2$. As described in Section 2, *SMB Join* does require $N_1 = N_2$. Instead for *Bucket Map Join*, it requires that $N_1 \% N_2 = 0$ or $N_2 \% N_1 = 0$, where % is the modulus operator. As a special case, the assumption $N_1 = N_2$ still makes sense for *Bucket Map Join*.

– Lastly, we don't distinguish the cost of building sorted buckets and unsorted buckets in our cost model. The assumption is reasonable because in the MapReduce framework, the sorted bucketizing job only takes a slightly more time than the unsorted one, which doesn't influence the correctness of our cost model.

With the above assumptions, we proceed to presenting our cost model. In this model, the cost of each join algorithm consists of the following five parts.

– *Bucketize cost* to bucketize both join tables, if any.
– *Broadcast cost* to build and broadcast hashtables to all mapper nodes, if any.
– *Map cost* for mappers to read input data from HDFS.
– *Shuffle cost* to shuffle mappers' output to reducers' nodes.
– *Join cost* to operate join.

Table 1 defines the symbols and cost functions we will use in our cost model.

Table 1. Cost Model Symbols

Symbol	Description
N_{nodes}	Number of nodes in the Hadoop cluster
B	Block size in HDFS
$N_{mappers}$	Number of mappers Hadoop sets up for the join query
$R(T)$	Number of rows of table T
$S(C(T))$	The total field size of query columns C of table T
$S(T)$	The size of input join table T
$S_c(T)$	The size of table T only with the query columns C
$N_{buckets}$	Number of buckets for join tables
$B_i(T)$	The ith bucket of table T
$S(B_i(T))$	The size of the ith bucket of table T
$R(B_i(T))$	Number of rows of the ith bucket of table T
$S_{Hashtable}(r)$	The size of hashtable of table T / bucket B with r rows
$N_{bucket-query}$	Number of queries with the same join key need bucketizing
Cost Function	**Description**
$T_{Hashtable}(r)$	Cost to build hashtable of table T / bucket B with r rows
$T_{Join}(r1, r2)$	Cost to join two tables or buckets with $r1$ and $r2$ rows
$T_{ReadHDFS}(m)$	Cost to read m GB data from HDFS
$T_{Transfer}(m)$	Cost to transfer m GB data through network
$T_{Bucketize}(t)$	Cost to bucketize join table t

We highlight the cost used by the four join strategies in terms of the five aforementioned parts:

1. *Common Join*
   Bucketize cost = 0
   Broadcast cost = 0

Map cost $= T_{ReadHDFS}(S(T1) + S(T2))$
Shuffle cost $= T_{Transfer}(S_c(T1) + S_c(T2))$
Join cost $= T_{Join}(R(T1), R(T2))$

2. *Map/Broadcast Join*
   Bucketize cost $= 0$
   Broadcast cost $= T_{ReadHDFS}(S(T2)) + T_{Hashtable}(R(T2)) +$
   $\qquad T_{Transfer}(S_{Hashtable}(R(T2))) * N_{nodes}$
   Map cost $= T_{ReadHDFS}(S(T1) + S_{Hashtable}(R(T2)) * N_{mappers})$
   Shuffle cost $= 0$
   Join cost $= T_{Join}(R(T1), R(T2))$

3. *Bucket Map Join*
   Bucketize cost $= (T_{Bucketize}(T1) + T_{Bucketize}(T2))/N_{bucket-query}$
   Broadcast cost $= T_{ReadHDFS}(S(T2)) + T_{Hashtable}(R(T2)) +$
   $\qquad T_{Transfer}(S_{Hashtable}(R(T2))) * N_{nodes}$
   Map cost $= T_{ReadHDFS}(S(B_i(T1)) + S_{Hashtable}(R(B_i(T2)))) * N_{mappers}$
   $\qquad = T_{ReadHDFS}(S_c(T1) + S_{Hashtable}(T2))$
   Shuffle cost $= 0$
   Join cost $= T_{Join}(\bigcup R(B_i(T1)), \bigcup R(B_i(T2)))$
   $\qquad = T_{Join}(R(T1), R(T2))$

4. *SMB Join*
   Bucketize cost $= (T_{Bucketize}(T1) + T_{Bucketize}(T2))/N_{bucket-query}$
   Broadcast cost $= 0$
   Map cost $= T_{ReadHDFS}(S(B_i(T_1)) + S(B_i(T_2))) * N_{mappers}$
   $\qquad = T_{ReadHDFS}(S_c(T1) + S_c(T2))$
   Shuffle cost $= 0$
   Join cost $= T_{Join}(\bigcup R(B_i(T1)), \bigcup R(B_i(T2)))$
   $\qquad = T_{Join}(R(T1), R(T2))$

Before giving the details to compute the cost of each algorithm, we first look at the map tasks lunched by Hadoop:

$$
N_{mappers} = \begin{cases} \dfrac{S(T1)}{B} & \text{\textit{Common Join / Map Join}}, & (1) \\[2mm] N_{buckets} & \text{\textit{Bucket Map Join / SMB Join}} & (2) \end{cases}
$$

In case (1), for the *Common Join* and *Map Join*, the number of map tasks is determined by the number of splits of the big join table. By default, the split's size is equal to the HDFS block size. In case (2), for *Bucket Map Join* and *SMB Join*, because the tables are bucketized, the number of map tasks is determined by the number of buckets.

Now, we compute the cost of the four algorithms one by one. First for *Common Join*, in the map phase, mappers need to read the whole join tables' data $S(T)$ from HDFS. During the map function, unused columns will be filtered, and the records that are relevant to the join query are shuffled to reducers. Thus, the size of shuffle data is $S_c(T) = S(C(T)) * R(T)$. The cost of join is estimated by comparing the entire records $R(T1)$ and $R(T2)$ in two tables from begin to end.

Second for *Map Join*, it is required to build the hashtable of the smaller table $T2$, and the associated cost includes the one used to read $T2$ from HDFS, to build the hashtable, and finally to broadcast the hashtable to the number N_{nodes} of clustered nodes. After that, each mapper reads a split of $T1$ and the whole hashtable of $T2$. Hence, all mappers in total read the entire $T1$ and the number $N_{mappers}$ of times to load the hashtable of $T2$. The size of hashtable, $S_{Hashtable}(r)$, depends on the number of rows to build the hastable, and we compute $S_{Hashtable}(r) = \beta * r$ *bytes*. In Hadoop, each row of the hashtable occupies around 1 byte, for simplicity, we set $\beta = 1$.

Next, for *Bucket Map Join*, we first need to bucketize both tables. We divide the total bucketizing cost by $N_{bucket-query}$, which means the bucketizing cost can be shared by $N_{bucket-query}$ queries and all these queries will benefit from the bucketizing job. Each bucket needs to be broadcasted to all other N_{nodes} nodes, so in total the broadcast cost is the same as *Map Join*. For map cost, $N_{mappers}$ mappers need to read buckets of $T1$ and corresponding hashtable of buckets of $T2$. Since we don't take *WHERE* conditions into account, all buckets of $T1$ and buckets' hashtables of $T2$ will be read. For the join cost, the $\bigcup R(B_i(T))$ represents the required buckets in T. Given the two tables T_1 and T_2, the buckets in T_1 are loaded to compare with the ones in T_2 for the join processing. All buckets will be read and compared with no *WHERE* clause in current model.

Finally, for *SMB Join*, its cost is different from *Bucket Map Join* in two parts: (i) It doesn't have to build and broadcast hashtables, and (ii) the mappers need to read all the buckets of $T1$ and $T2$, instead of hashtable of $T2$.

4   Cost-Based Selector

Based on the aforementioned cost model, in this section, we first design a tree structure (Section 4.1) and next propose a method to select one of the four join algorithms (i.e., *Common Join, Map Join, Bucket Map Join* and *SMB Join*) as the best algorithm to process a join query (Section 4.2).

4.1   Pruned Join Algorithm Candidates Tree

We design a *Join Algorithm Candidates Tree* (*JACTree*), shown in Fig. 1, as a pruning method for the automated selection of best join algorithm.

In this tree structure, the root means the bucket size of the smaller join table. We compare it with the size of the smaller table. If bucket_size is larger than the table_size (i.e., the left brunch), it means it's unnecessary to create any bucket. Otherwise, the tables need to be bucktized (i.e., the right brunch).

Now for the internal node with bucket_size $>=$ table_size, we next need to consider the hashtable size of the bucket, and compare it with the available memory of the task nodes in the cluster. In case that the hashtable size is larger than the memory size, we then reach the leave node ① *Common Join*, and otherwise the leave node ② *Common Join* and *Map Join*. Similarly in the right branch, we can compare the hashtable size with the memory size and reach either

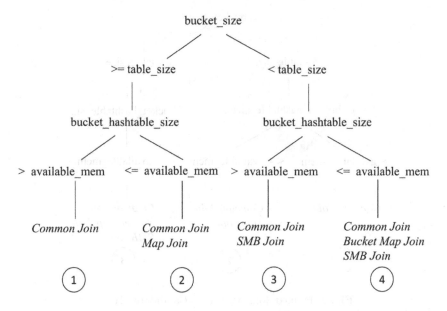

Fig. 1. Join Algorithm Candidates Tree

the leave node ③ or the leave node ④. As we can see, the two input parameters including the table size and available memory may lead to the selection of a different join algorithm.

In the tree structure of Fig. 1, we note that due to limited buckets, the leaf node ③ *SMB Join* always perform worse than the leaf node ④. *Common Join* has already been contained in all the rest leave nodes. Consequently, we further prune the leave node ③, and have a new structure as shown in Fig. 2. In the new tree structure, namely *Pruned Join Algorithm Candidates Tree (Pruned-JACTree)*, now only one leave node contains the *SMB Join*.

4.2 Select Join Algorithm Based on Cost Model

In order to enable the proposed cost model, we need to know the values of key parameters used by the cost model. To this end, we estimate such parameters as follows.

We first estimate the parameters including the size of join tables $S(T1)$, $S(T2)$, the number of rows of them $R(T1)$, $R(T2)$, and the total field size of query columns $S(C(T1))$, $S(C(T2))$. In detail, when a table is uploaded to HDFS, with the help of Hive log, we can find the values of $S(T1)$, $S(T2)$ and $R(T1)$, $R(T2)$. Next, we use MapReduce `RandomSampler` utility to estimate the total field size of query columns. By the `RandomSampler`, we first set the number of input splits that will be sampled. The sampler will then randomly sample the selection columns of two join tables for the estimation of $S(C(T1))$ and $S(C(T2))$.

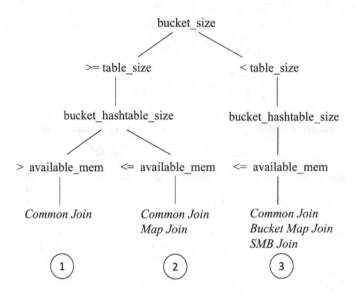

Fig. 2. Pruned Join Algorithm Candidates Tree

Second, we estimate the *available_mem* by the minimum JVM memory in task nodes. In terms of the memory used by the hashtable, we can compute the value by the number of rows in the table.

$$Hashtable\_Mem(T) = R(T) * \alpha \tag{3}$$

In Eq. (3), the parameter α indicates the number of bytes per record in the hashtable needed by the main memory. In our experiment, we empirically set α by 200 bytes.

Finally, we need to decide the suitable bucket number $N_{buckets}$. A small data size per bucket may lead to too many but small size of files in HDFS and slow down the query. Alternatively, a very large data size per bucket will incur very few **map** tasks, and the hashtable of one single bucket is too large to fit into memory for *Bucket Map Join*. Thus, we determine the number of bucket number with the following function:

$$Bucket\_Num(T, available\_mem) = \frac{\beta * Hashtable\_Mem(T)}{available\_mem} \tag{4}$$

In the Eq. (4), a higher β (> 1) means a larger number of buckets, to ensuring that the buckets should be loaded into memory on the node with higher probability. In our experiment environment, we empirically set $\beta = 1.3$.

When the above parameters are ready, we use Fig. 3 to describe the steps that our selection method (namely a selector) chooses the best algorithm among the

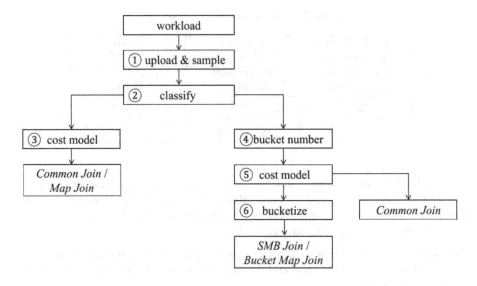

Fig. 3. Cost-based Selector Workflow

four available join algorithms. ① For an input workload, selector first collects join tables' information for cost model. ② Selector next makes a decision to choose an algorithm by following either of the two brunches. ③ If one of the join table in the query is small enough to fit into memory, the selector computes the cost of *Common Join* and *Map Join*, and finds the one with a smaller cost as the chosen join algorithm. ④ Otherwise, our selector uses Eq. (4) to calculate the reasonable bucket number for the join tables. ⑤ Based on the proposed cost model, the selector next decides to use either bucketed join (i.e., *Bucket Map Join* or *SMB Join*) or *Common Join*. ⑥ If the bucketed join is chosen, our selector will only select the columns needed by the queries in the workload to the buckets while bucketizing.

5 Experiment

We present experiments to evaluate our proposed selection method, and study (1) how the number of buckets affects the performance of bucketed joins, and (2) whether the proposed selection method can correctly choose the best join algorithm for a given workload.

Cluster Setup. We evaluate the experiments on a 5-node cluster with 1 namenode and 4 datanodes. Each of the nodes is installed with Ubuntu Linux (kernel version 2.6.38-16) and equipped with 750 GB local disk and 4 GB main memory. We implement our cost-based join algorithm selector on the top of Hadoop 1.2.1 and Hive 0.11.0. The default block size is 64 MB.

Table 2. # Rows and Size of Data Sets

Table Name	# Rows	Size
lineItem0.5x	3,000,000	365 M
lineItem30x	180,000,000	22.0 G
lineItem100x	600,000,000	74.1 G
orders1x	1,500,000	164 M
orders30x	45,000,000	4.9 G
orders100x	150,000,000	16.6 G

Datasets. We generate the synthetic data sets (LineItems and Orders) by the TPC-H benchmark, and generate different scales of join tables to simulate different combinations of input tables. Table 2 shows the size and number of rows of different scales of input join tables.

Workload. We only generate the join query workload with the simple equi-join queries between the LineItems and Orders tables. In our experiment, we design the following queries:

```
Q1: SELECT l.l_shipmode, o.o_orderstatus
FROM lineitem100x l JOIN orders1x o ON l.l_orderkey = o.o_orderkey;

Q2: SELECT l.l_shipmode, o.o_orderstatus
FROM lineitem0.5x l JOIN orders1x o ON l.l_orderkey = o.o_orderkey;

Q3: SELECT l.l_shipmode, o.o_orderstatus
FROM lineitem30x l JOIN orders30x o ON l.l_orderkey = o.o_orderkey;

Q4: SELECT l.l_linenumber, o.o_orderkey
FROM lineitem30x l JOIN orders30x o ON l.l_orderkey = o.o_orderkey;

Q5: SELECT l.l_shipmode, o.o_orderdate
FROM lineitem30x l JOIN orders30x o ON l.l_orderkey = o.o_orderkey;

Q6: SELECT l.l_shipmode, o.o_orderstatus
FROM lineitem100x l JOIN orders100x o ON l.l_linestatus = o.o_orderstatus;
```

Practical Cost Estimation. During our experiment, to use the proposed cost model, we need to estimate the cost of some key operations such as reading HDFS blocks, etc. At first, we denote the cost of transferring 1 GB data through network cost as 1 G, and the CPU cost of comparing 10^6 times as 1 C. Based on the denotation, by around tens of empirical test, we empirically draw the following equations:

(1) $T_{ReadHDFS}(m) = 0.6 * T_{Transfer}(m) = 0.6 * m \quad G$
(2) $T_{Bucketize}(t) = T_{ReadHDFS}(S(t)) + 4 * T_{Transfer}(S_c(t)) \quad G$
(3) $T_{Hashtable}(r) = 30 * r * 10^{-6} \quad C$
(4) $T_{Join}(r1, r2) = (r1 + r2) * 10^{-6} \quad C$

In the above first equation, considering that the data is partially located on local nodes, we empirically compute the cost of reading HDFS by 60% of the cost of transferring data through network. Next, we estimate the bucketizing cost by the cost of reading, transferring and writing data. Since writing data to HDFS is always ineffcient, we empirically set the cost of writing HDFS 3 times as the cost of transferring data. After that, we calculate the cost of building one key-value pair of hashtable as 30 times of CPU comparison. Finally, the cost of join operation is estimated as processing all rows of two join tables.

In our experiments' environment, the network bandwith is about 100 M/s and CPU latency for one comparison is around 0.1 microsecond, we determine 1 C = 10^{-2} G.

5.1   Effect of the Number of Buckets

In order to demonstrate how the number of buckets can affect the performance of *Bucket Map Join* and *SMB Join*, we have conducted an experiment with the query Q_6 on the tables LineItems and Orders of scale 30. We vary the number of buckets from 1 to 2000, and the corresponding query time is shown in Fig. 4.

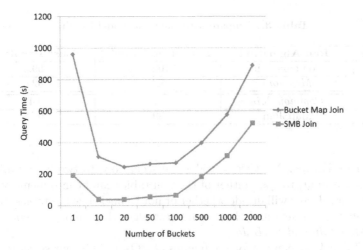

Fig. 4. Effect of the Number of Buckets

As shown in Fig. 4, with a smaller number of buckets, each bucket contains a larger data size, and the mapper has to process a large amount of data. For *Bucket Map Join*, since the hashtable of the buckets may be too large to fit into memory, it may fail to broadcast the hashtable and Hive has to use the default *Common Join*. On the other hand, too many buckets will trigger too many mappers and each mapper only joins a very fewer number of records.

In addition, Fig. 4 indicates that the number 20 of buckets can achieve the best performance, which is consist with our analysis. In our experiment cluster, the maximum JVM memory on each tasknode is around 900 MB and the average usage of it is around 25%. So we assume the minimum JVM memory available on the tasknode is 600 MB. orders30x has 45,000,000 rows. According to Eq. (3), the hashtable of it will need 9,000 MB memory space. With Eq. (4), our selector will set the reasonable bucket number as 20. The result shows when the bucket number is 20, the join performance is the best for both bucketed joins.

5.2 Correctness of Cost-Based Selector

In this section, given multiple scenarios, we evaluate the correctness of our cost model and verify whether or not our selector can pick the best join algorithm.

One Small Table First, we consider the scenario that one of the join table's hashtable is small enough to fit into memory. For the Q_1 and Q_2 in our workload, orders1x's hashtable can be fit into memory. Table 3 illustrates the real execution time and our cost model theoretical result. We can see the order of the cost result is roughly the same as the order of execution time, which means our selector can pick the join algorithm with the lowest query latency.

Table 3. Compare with the Cost-model Result

Query	Join Algorithm	Execution Time(s)	Cost Result(G)
Q1	Common Join	1022	59.5
	Map Join	659	52.0
Q2	Common Join	31	0.41
	Map Join	39	0.78

Two Large Tables. Next we consider the scenario with two large tables. For the queries from Q_3 to Q_6, neither of the join tables can fit into memory. Given this case, our selector will decide whether or not to bucketize the tables, tune the best number of buckets and choose one best algorithm among *Common Join*, *Bucket Map Join* and *SMB Join*.

From Table 4, we can see for the queries Q_3, Q_4 and Q_5, our selector chooses *SMB Join*. It is because all these three queries benefit from bucketizing tables. Therefore the cost of bucketizing is divided by 3. In terms of Q_6, it is the only query to join the tables on l_linestatus and o_orderstatus, and our selector chooses *Common Join* because bucketizing is also a time-consuming MapReduce job.

In addition, the experiments results indicate that our selector does not pick *Bucket Map Join* for any of the six queries. By careful analysis, we find that the

Table 4. Compare with the Cost-model Result

Query	Join Algorithm	Execution Time(s)	Cost Result(G)
Q3	Common Join	367	24.3
	Bucket Map Join	498/3+269=435	32.2/3+24.7=29.1
	SMB Join	498/3+40=206	32.2/3+4.7=15.4
Q4	Common Join	356	23.8
	Bucket Map Join	498/3+249=415	32.2/3+24.7=29.1
	SMB Join	498/3+40=206	32.2/3+4.7=15.4
Q5	Common Join	368	24.7
	Bucket Map Join	498/3+218=484	32.2/3+24.7=29.1
	SMB Join	498/3+43=209	32.2/3+4.7=15.4
Q6	*Common Join*	1237	72.5
	Bucket Map Join	1271+1700=2971	86.4+59.7=146.1
	SMB Join	1399+184=1583	86.4+13.9=100.3

reason why *Bucket Map Join* always performs worst is that it not only needs to bucketize in advance but also needs to build and then broadcast the hashtable for each bucket. These efforts incur expensive cost.

As a summary, for the given workload with six queries from Q_1 to Q_6, the order of our cost model result is roughly the same as the real execution time. With our cost-based selector, the total execution time of the workload is accelerated 24.4%, compared with the default join algorithm (*Common Join*). Compared with the wrong join algorithm, which new users may randomly pick, the total execution time is accelerated 58.7%.

6   Related Work

First, some previous works implement new join algorithms on top of MapReduce. For example, Lin *et al* propose a new scheme called "schimmy" to save the network cost during the *Common Join* [6]. In this scheme, mappers only emit messages and reducers read the data directly from HDFS and do the reduce-side join between the messages and data. Okcan *et al* propose how to efficiently perform θ−join with a single MapReduce job [7]. Their algorithm uses a Reducer-centered cost model that calculates the cost of Cartesian product of mapped output. With this cost model, they assign the mapped output to the reducers that minimizes job completion time. Blanas *et al* propose the process of *Semi-Join* and *Per-Split Semi-Join* in MapReduce framework [8]. Lin *et al* propose the concurrent join that performs a multi-way join in parallel with MapReduce [9]. Afrati *et al* focus on how to minimize the cost of transferring data to reducers for multi-way join [10]. Verica *et al* propose a method to efficiently parallelize set-semilarity joins with MapReduce [11].

Second, there exist some works to hackle the internals of MapReduce. Map-Reduce-Merge is the first that attempts to optimize join operation in the MapReduce framework [12]. Map-Reduce-Merge extends MapReduce model by adding

Merge stage after Reduce Stage. Yang *et al* have proposed an approach for improving Map-Reduce-Merge framework by adding a new primitive called *Traverse* [13]. This primitive can process index file entries recursively, select data partitions based on query conditions and feed only selected partitions to other primitives. Jiang *et al* propose Map-Join-Reduce for one-phase joining in MapReduce Framework. This work introduces a filtering-join-aggregation model as another variant of the standard MapReduce framework [14]. This model adds a Join Stage before the original Join Stage to perform the join. Besides the introduction of more stages, some works columnar to improve the join queries. Llama [15] is a recent system that combines columnar storage and tailored join algorithm. Llama also proposed joining more than two tables at a time using a concurrent join algorithm. Clydesdale [16], a novel system for structured data processing is aimed at star schema, using columnar storage, tailored plans for star schemas and multi-core aware execution plans to accelerate joins.

Finally, some works propose to build new layer on top of Hadoop in order to process a join query. A typical work is Hive, a data warehouse infrastructure built on top of Hadoop. Hive optimizes the chains of map joins with the enhancement for star joins. Olston *et al* have presented a language called *Pig Latin* [17] that takes a middle position between expressing task using the high-level declarative querying model in the spirit of SQL and the low-level/procedural programming model using MapReduce. *Pig Latin* is implemented in the scope of the *Apache Pig* project [18]. Pig Latin programs are compiled into sequences of MapReduce jobs which are executed using the Hadoop MapReduce environment. Pig optimizes join by using several specialized joins, such as fragment replicate joins, skewed joins, and merge joins. The *Tenzing* system [19] has been presented by Google as an SQL query execution engine which is built on top of MapReduce and provides a comprehensive SQL92 implementation with some SQL99 extensions. Tenzing's query optimizer applies various optimizations and generates a query execution plan that consists of one or more MapReduce jobs.

7   Conclusion

In this paper, we proposed a general cost model for the four popular join algorithms in Hive and designed a method to choose the best join algorithm with least cost. Our experiment results verified that our cost-based selector can correctly choose the best join algorithm and tune the key parameters (such as the number of buckets). As the future work, we will further extend our work to generally support more complex join queries such as multiway join.

Acknowledgments. This work was partially funded by EMC, and was a part of a joint research project between Fudan University and EMC Labs China (the Office of CTO, EMC Corporation). This work was also partially funded by the NSFC (Grant No. 61370080).

References

1. Dean, J., Ghemawat, S.: MapReduce: Simplified Data Processing on Large Clusters. In: OSDI, pp. 137–150 (2004)
2. Hadoop, http://hadoop.apache.org/
3. Hive, http://hive.apache.org/
4. TPC-H, Benchmark Specification, http://www.tpc.org/tpch/
5. HDFS architecture, http://hadoop.apache.org/docs/r0.19.1/hdfs_design.html
6. Lin, J., Dyer, C.: Data-intensive text processing with MapReduce. Synthesis Lectures on Human Language Technologies 3(1), 1–177 (2010)
7. Okcan, A., et al.: Processing Theta-Joins using MapReduce. In: Proceedings of the 2011 ACM SIGMOD (2011)
8. Blanas, S., et al.: A comparison of join algorithms for log processing in MapReduce. In: Proceedings of the 2010 ACM SIGMOD, pp. 975–986 (2010)
9. Lin, Y., et al.: Llama: Leveraging Columnar Storage for Scalable Join Processing in the MapReduce Framework. In: Proceedings of the 2011 ACM SIGMOD (2011)
10. Afrati, F.N., Ullman, J.D.: Optimizing joins in a map-reduce environment. In: Proceedings of the 13th EDBT, pp. 99–110 (2010)
11. Vernica, R., et al.: Efficient parallel set-similarity joins using mapreduce. In: Proceedings of the 2010 ACM SIGMOD, pp. 495–506 (2010)
12. Yang, H., et al.: Map-reduce-merge:simplifiedrelational data processing on large clusters. In: Proceedings of the 2007 ACM SIGMOD, pp. 1029–1040 (2007)
13. Yang, H.-c., Parker, D.S.: Traverse: Simplified Indexing on Large Map-Reduce-Merge Clusters. In: Zhou, X., Yokota, H., Deng, K., Liu, Q. (eds.) DASFAA 2009. LNCS, vol. 5463, pp. 308–322. Springer, Heidelberg (2009)
14. Jiang, D., et al.: Map-join-reduce: Towards scalable and efficient data analysis on large clusters. IEEE Transactions on Knowledge and Data Engineering (2010)
15. Lin, Y., Agrawal, D., Chen, C., Ooi, B.C., Wu, S.: Llama: leveraging columnar storage for scalable join processing in the MapReduce framework. In: SIGMOD Conference, pp. 961–972 (2011)
16. Balmin, A., Kaldewey, T., Tata, S.: Clydesdale: structured data processing on hadoop. In: SIGMOD Conference, pp. 705–708 (2012)
17. Olston, C., Reed, B., Srivastava, U., Kumar, R., Tomkins, A.: Pig latin: a not-so-foreign language for data processing. In: SIGMOD, pp. 1099–1110 (2008)
18. Pig, http://pig.apache.org/
19. Chattopadhyay, B., Lin, L., Liu, W., Mittal, S., Aragonda, P., Lychagina, V., Kwon, Y., Wong, M.: Tenzing A SQL Implementation On The MapReduce Framework. PVLDB 4(12), 1318–1327 (2011)

Consistent Freshness-Aware Caching
for Multi-Object Requests

Meena Rajani[1], Uwe Röhm[2], and Akon Dey[2]

School of Information Technologies
The University of Sydney
Australia
meenakrajani@gmail.com,
{uwe.roehm,akon.dey}@sydney.edu.au
http://www.usyd.edu.au

Abstract. Dynamic websites rely on caching and clustering to achieve high performance and scalability. While queries benefit from middle-tier caching, updates introduce a distributed cache consistency problem. One promising approach to solving this problem is *Freshness-Aware Caching (FAC)*: FAC tracks the freshness of cached data and allows clients to *explicitly* trade freshness of data for response times. The original protocol was limited to single-object lookups and could only handle complex requests if all requested objects had been loaded into the cache at the same time. In this paper we describe the *Multi-Object Freshness-Aware Caching (MOFAC)* algorithm, an extension of FAC that provides a consistent snapshot of multiple cached objects even if they are loaded and updated at different points of time. This is done by keeping track of their *group valid interval*, as introduced and defined in this paper. We have implemented *MOFAC* in the JBoss Java EE container so that it can provide freshness and consistency guarantees for cached Java beans. Our evaluation shows that those consistency guarantees come with a reasonable overhead and that *MOFAC* can provide significantly better read performance than cache invalidation in the case of concurrent updates and reads for multi-object requests.

Keywords: Freshness, Distributed cache, Replication, Invalidation, Consistency.

1 Introduction

Large e-business systems are designed as n-tier architectures: Clients access a web-server tier, behind which an application server tier executes the business logic and interacts with a back-end database. Such n-tier architectures scale-out very well, as both the web and the application server tiers can be easily clustered by adding more servers. However, there is a certain limit to the performance and scalability of the entire system due to the single database server in the back-end. It is essential to minimize the number of database calls to alleviate this bottleneck.

B. Benatallah et al. (Eds.): WISE 2014, Part II, LNCS 8787, pp. 262–277, 2014.

A distributed cache layer at the application server level strives to minimize these access costs between the application server tier and the database tier. While this works very well for read-only access, updates induce a distributed cache consistency problem. Keeping the contents of the cache consistent with the backend database is necessary to ensure correctness of the overall system. The well-known techniques for handling this are cache invalidation and cache replication, both of which have certain disadvantages with regard to the best performance of either read- or update-intensive workload. In an n-tier architecture, business logic is processed in the application server making it more efficient by keeping the data closer to it.

The Java application server comes with Enterprise Java Bean (EJB) to manage business logic and persist application state. An entity bean usually represents a row from the database table and is hence costly to to create, presenting the need for a a second level cache. As a result, when an entity bean is updated in one node, either a replication or invalidation message is sent across all the nodes in the cluster. Both synchronous and asynchronous replication of entity beans are costly when data consistency issues are created due to invalidation. Invalidation of entity beans on the other hand causes cache misses and in-turn can cause database bottlenecks.

FAC showed that we can trade-off freshness with high availability but was limited to freshness management on the basis of a single Enterprise Java Bean (EJB). However, in real world applications most transactions access more than one object that exist in binary or ternary relationships with other objects. As a result, when an object is updated in a database, that object and all its associated objects are removed from the cache resulting in a high performance penalty. This must be avoided in order to maintain high performance. In this paper, we make the following contributions:

- We present a Multi-Object Freshness Aware Caching (MOFAC) algorithm which guarantees both the freshness and inter-object consistency of the cached data.
- We implemented this algorithm in the JBoss 6 middle-tier application server cache.
- We present results of a performance evaluation and quantify the impact of the different parameters of MOFAC on its performance.

2   Freshness-Aware Caching

In Freshness-Aware Caching (FAC) [11], each cache node keeps track of how stale its content is and only returns data that is fresher than the freshness limits set by the client. In this paper, we describe algorithms that can support this, and which also ensures that every request that touches multiple objects is given a consistent view of them; that is, there was a time, within the freshness limit, when all the information read was simultaneously up-to-date.

2.1   Freshness Concept

Freshness of data is a measure on how outdated (stale) a cached object is in comparison to the up-to-date master copy in the database [12]. There are several approaches to measuring this: *Time-based* metrics rely on the time duration since the last update of the master copy, while *value-based* metrics rely on the value differences between cached object and its master copy. A time-based staleness metric has the advantage that it is independent of data types and does not need to access the back-end database. On the other hand, a value-based metric needs the up-to-date value to determine the value differences allowing for local freshness decisions without making a trip to database to improve scalability. Hence, MOFAC uses a time-based metric.

Definition 1 (Staleness Metric). *The* staleness *of an object o is the time duration since the object's stale-point, or 0 for freshly cached objects. The stale-point $t_{update}(o)$ of object o is the point in time when the master copy of o was last updated while the cached object itself remained unchanged.*

$$stale(o) := \begin{cases} (t_{now} - t_{update}(o)) & | \; if \; master(o) \; updated \; at \; t_{update}(o) \\ 0 & | \; otherwise \end{cases}$$

Definition 2 (Freshness Intervals). *The cache lifetime of an object o consists of two disjoint intervals, $vi(o)$ and $di(o)$. Its* valid interval, $vi(o)$, *is the half open interval $[t_{load}, t_{update})$ when the object is loaded into the cache. The object is in the valid interval when it retains its state over a period of time until some event occur in the present or future that changes its state in database. The object's deferred interval $di(o)$ is defined as the half open interval $[t_{update}, t_{expire})$ such that the object enters the deferred interval as soon as it is updated in database.*

The timestamp of the object is recorded when it is loaded from the database to the cache node. When an object is updated via the cache node or any other cache nodes in the cluster, the update timestamp is recorded. The expiry time is calculated using the update timestamp and freshness constraint. In other words, the length of the deferred interval is adjusted by the freshness constraint. This is illustrated in Figure 1.

If an object is inside the valid interval, it is considered to be fresh and consistent with the backend database; if its timestamp falls inside the deferred interval, it means that the master copy on the database has been updated, but the staleness of the object

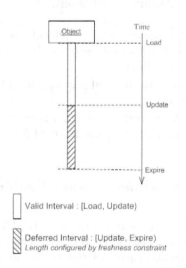

Fig. 1. Cache Objects Freshness Intervals

still meets the freshness constraint. Only objects that have exceeded the expiry timestamp, are considered to be too old with respect to the freshness constraint and are evicted from the cache.

3 Multi-Object Freshness-Aware Caching (MOFAC)

The FAC algorithm, as described, cannot guarantee the consistency of several related objects that are accessed within the same transaction. We introduce a concept of object grouping to address this limitation.

An object group represents a set of logically related objects in which two entities are considered to be part of the same group of objects if they are associated with each other in an explicit or implicit relationship. For example, this can be done by leveraging the foreign-key-relationships in the schema of the underlying database which is often explicitly defined in the object-relational mapping at the middle-tier.

The objects in a group are mutually consistent when all of them have been persisted together in the database. Object grouping enables *snapshot consistency* to the objects in the group by ensuring that modifications to any object in the group results in notifications making all cached copies of the group stale. However, they may continue to reside in the cache as long as the freshness conditions are met using the formula described in Section 3.1. All members of the group are guaranteed to be loaded

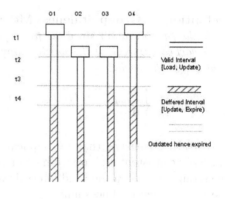

Fig. 2. Multi Object Group Freshness

with the same snapshot of the database reflecting the freshness interval of the entire object group.

Definition 3 (Group Freshness Intervals). *Suppose we have a group of objects* $G = \{o_1, o_2, ..., o_n\}$. *The group's* valid interval, *denoted by* gvi, *is the intersection of all valid intervals of its group members:*

$$gvi(G) := vi(o_1) \cap vi(o_2) \cap ... \cap vi(o_n), o_i \in G$$

In other words, the gvi is the half-open time interval defined as

$$gvi(G) := [MAX(t_{load}(o_i)), MIN(t_{update}(o_i))),$$

$$1 \leq i \leq n, o_i \in G$$

Figure 2 illustrates an example in which o_1, o_2, o_3 and o_4 are a group of objects with the same meaning of freshness interval in Figure 1. The objects have been loaded into the cache at different points in time within the interval (t_1 and t_2).

When an object from this group is requested, the group's valid interval is calculated to verify that all the objects were in their valid interval together at some point of time; in this case, it is $[t_2..t_3)$. If this satisfies the user's freshness constraint, the user can access the cached group of objects. If an object in a group is updated by some other cache node, the stale point is registered (t_3 or t_4 in this example). Note that subsequent updates on already stale objects do not change the staleness point. We call a cache consistent group where *all* members are in their valid interval, *fresh*, and otherwise, *stale*.

Currently we have considered objects which are in one to many and many to one associations. This can be easily extended to many to many associations. If two different transactions require the same group of objects, only one instance of that group exists in the cache.

Definition 4 (Group Staleness Metric). *The* staleness *of a group of objects, G, is the time duration since the first update to any object in G until now, or is 0 for cached group G, if the master copy of all o in G is not updated since G is loaded into the cache.*

$$stale(G) := \begin{cases} (t_{now} - MIN(t_{update}(o_i))) \mid if\ master(o)\ updated\ at\ t_{update}(o) \\ 0 \qquad\qquad\qquad\qquad\qquad\qquad \mid otherwise \end{cases}$$

In order to ensure that an application gets a group of objects which are mutually consistent and are fresh enough to serve a user request, each object in the group must be in the valid interval and the staleness of group G must satisfy the user-defined freshness limit.

Definition 5 (Group Consistency and Freshness). *A group of objects G is fresh enough and consistent if the group is present in cache, the group valid interval is not empty (i.e. at some point in time in the past, all group objects were valid at the same time), and the staleness of G is within the application's freshness limit.*

$$freshcon(G) : G \in Cache\ \wedge\ gvi(G) \neq \emptyset\ \wedge\ stale(G) \leq freshlimit$$

3.1   Example

Let us consider an online bookstore application where a book entity is brought into the application cache together with its authors and reviews as a group G. This could be because all the parts are needed to construct the content of a dynamic web page.

$$Cache := \{(b, \{a\}, \{r\}) \mid b \in Books, a \in Authors, r \in Reviews : \\ a.id \in b.authors\ \wedge\ r.isbn = b.isbn\}$$

Here, each group consists of one Book object, and a set of corresponding Author and Review objects. In this example, we can guarantee a user is accessing a consistent data snapshot, which is within the required freshness constraint, by using Definition 5.

Figure 3 is an example of a book group. It has one book instance which is associated with two collections: one review collection and one author collection. The key of collection is generated from the root object's key, which is 1 here. When any object is updated in one of the cache nodes, the rest of the cache nodes receive the modified timestamp with the key of the corresponding object. They then register that timestamp and convert the group's status from fresh to stale.

4   MOFAC Algorithm

In this section we describe the multi-object freshness-aware caching (MOFAC) algorithm that handles freshness of multi-objects freshness in the application-tier cache. It consists of three sub-algorithms: the handling of MOFAC reads, the handling of user-level updates to a local cache node, and the processing of update-notifications on a remote cache node.

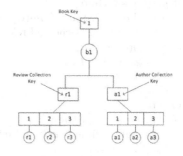

Fig. 3. Example of a book group with two child collections

4.1   Multi Object Freshness Read

A freshness-aware cache tracks the load and stale points of each cached object, group memberships and each groups valid intervals. This meta-data is used to decide whether an object can be returned from the cache or whether it has to be (re-)loaded from the backend database.

In Algorithm 1, it is worth to noting that on an initial cache miss for the whole group, all objects that form the group are loaded into the cache together. This is typically done by the `cacheable` and `association` annotations of an application. In addition, the usual cache granularity is at the level of individual objects and not cache groups. So in most applications, this algorithm is initiated multiple times in a row while the application is traversing the different object links within the same group. Most of these end us as fast in-memory operations because the whole group (complex object) is loaded into the cache due to the earlier cache misses. This cache miss behaviour is handled in detail in the MOFAC read-algorithm in Algorithm 1.

This algorithm assumes that the cache has mechanisms to determine dependencies between associated objects, such as to iterate over all direct child objects of a given object in the cache or to determine whether a complex object is completely cached with all its associated child objects (predicate *isComplete*() in

Algorithm 1. MOFAC Read Algorithm for an arbitrary complex object

---

input : an object key k
input : freshness limit f_{limit}
output: object reference O

$O \leftarrow \text{lookup}(k)$
if $O \notin Cache \lor \neg isComplete(O)$ **then**
    for all $o_i \in O, o_i \notin Cache$: **do**
        retrieve o_i (evtl. with child objects) from database
        $t_{load}(o_i) \leftarrow t_{now}$
        $t_{update}(o_i) \leftarrow t_{max}$
        $Cache \leftarrow Cache \cup o_i$
    end for
end if
if $gvi(O) = \emptyset$ **then**
    for all $o_i \in O, vi(o_i) \notin gvi(O)$: **do**
        evict o_i from Cache and reload (evtl. with child objects) from database
        $t_{load}(o_i) \leftarrow t_{now}$
        $t_{update}(o_i) \leftarrow t_{max}$
        $Cache \leftarrow Cache \cup o_i$
    end for
end if
if $stale(O) > f_{limit}$ **then**
    evict stale O from Cache and reload (evtl. with child objects) from database
    $t_{load}(O) \leftarrow t_{now}$
    $t_{update}(O) \leftarrow t_{max}$
    $Cache \leftarrow Cache \cup O$
end if
return O

---

above's algorithm). This functionality is provided by the Java Persistence API (JPA) layer. The cache can determine, which objects should be cached together and whether a complex object (including (or references) sub-objects is completely cached or not, based on meta-data that is extracted from the annotations in the Java application code.

4.2 Update Handling on Local Cache Node

We have to distinguish between two cases when processing updates on a multi-object freshness aware cache: Firstly, how should updates be processed locally on the cache node that received the user transaction. Secondly, how should the update notifications be processed on the other nodes of the distributed cache.

Algorithm 2 listed in the Appendix describes how updates should be handled at a local cache node: The object to be updated is first persisted to the backend database and then corresponding update-notifications are sent to the other cache nodes in the cluster. These notifications differ slightly based on what kind of object was updated in the cache group (i.e. whether it is the root node, a child object or a child collection). All update notifications are sent within the original user transaction, so that we have a synchronous freshness update to all the nodes. It does not require any expensive 2-phase-commit protocol since the only issue that can arise on a remote node is that it may not have the object in the cache when the update message is received. In this case, the message can safely be ignored on the update-notification. All we need to have is a guarantee that the notifications are delivered so that nodes who indeed do cache the same object get notified.

4.3   Update Handling: Processing Update-Notifications

The second part of the MOFAC update algorithm is the reception of the update-notification on a remote node. Although we conceptually get three different kinds of update-notifications, for either a whole group or just a child object or child collection, the actual handling is the same just differing in the type of target object.

Algorithm 3 listed in the Appendix describes these steps. An update notification not only specifies which object has been modified (the request specifies the unique object identifier, but not the object itself) and the timestamp on when this happened at the original node. If the modified object is also present in the receiving local cache, and no message was received with respect to an earlier update, then the stale point of the cached copy of the updated object is set to the received timestamp.

The algorithm assumes that all nodes in the cluster are time synchronised so that these timestamps between nodes are comparable. This is not difficult assumption for a typical closely-coupled cluster of today's standard. However, when the caches are distributes over a wide-area, we recommend switching to local timestamps of receiving notifications. This might be later than when the original update happened on the remote machine, but would be more consistent with all other timestamps used for MOFAC comparison which also are all locally determined (such as load time or time of an application request).

5   Evaluation

We have implemented the MOFAC algorithm inside the in-memory cache of a Java EE platform version 5.0 server and evaluated the performance characteristics of our proposed method using an exemplified dynamic web application: a simplified online bookstore.

5.1   Benchmark Application

The bookstore benchmark application consists of three components: A client emulator, the clustered bookstore server application, and the backend database.

The client emulator is a multi-threaded Java application that simulates a configurable number of clients that access the bookstore with either browsing (read-only) or buying (read-write) request. For the browsing workload, a method is invoked to find a certain book with all its authors and reviews.

The bookstore server application consists of a session bean, that provides the corresponding browsing and buying calls, as well as the implementation of the three entity beans representing Book, Author and Review entities. It is deployed into a JBoss 6 application server container and configured to run on a variable number of cluster nodes. For the caching side, we configured JBoss to use Infinispan as a distributed caching tier. Note that although Infinispan is a separate product and comes with its own configuration files, it is indeed loaded as part of the JBoss installation into the same JVM when the application server starts. We configured Infinispan so that it tightly integrates with the JBoss container by installing a caching interceptor into the JBoss interceptor chain, so that it gets invoked with any EJB access.

Finally, the bookstore state is stored in a single backend PostgreSQL database that is shared among all JBoss/Infinispan instances. We have used entity bean POJO entity class to persist and load data to and from the database into three kinds of entity beans: Book, Author and Review. Book and Author are in a one-to-many relationship, while Book and Review are also in a one-to-many relationship. The details of these ORM definitions of the three entity beans are shown in the Appendix in Listing 1.1.

5.2   Evaluation Setup

All experiments were conducted on an evaluation system consisting of a small cluster of eight Dell Optiplex servers, each equipped with a quad-code Intel Core2 Q9400 CPU (2.66 GHz), 4 GB RAM, two 500 GB HDDs, and running RedHat Fedora Linux 10 (kernel version 2.6.27.30).

We used a Java-based test client simulator and JBoss version 6.0 application server. The client simulator was running on a dedicated separate computer, and another dedicated server was used as back-end database server, running PostgreSQL Server 9.1. All nodes were interconnected via Fast Ethernet. The communication between the JBoss Server instances in the cluster (partition) is handled by the JGroups group communication library via *channel* for node discovery and reliably exchanging messages among cluster nodes. We have configured the cluster to use a round robin load balancing policy.

Client simulator and application server were Java applications executed under Java version 1.6. The server was executed with a Java heap size of up-to 512 MB (option $-Xmx512m$).

5.3   Evaluation of MOFAC's Overhead

In the first evaluation series, we are interested in measuring the general overhead induced by MOFAC in comparison to the standard cache invalidation techniques of JBoss/Infinispan. We compare the following cache functionalities:

INV synchronous cache invalidation (cache invalidation with synchronous notifications to remote nodes)

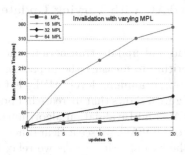
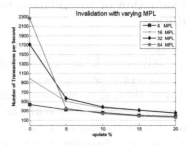

Fig. 4. Mean Response Time and Throughput of Cache Invalidation on Cluster Size 8 with varying Update Ratio

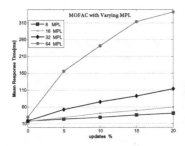
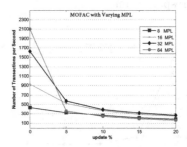

Fig. 5. Mean Response Time and Throughput of MOFAC on Cluster Size 8 with varying Update Ratio and Freshness Limit 0

MOFAC using the synchronous communication mechanism to send freshness notification to remote nodes.

In order to determine the costs of the caching control code, we fixed the freshness limit for all client requests to 0. This ensures that MOFAC caching produces about the same number of cache misses than cache invalidation - any update will trigger the eviction of any of its replica in other cache nodes. We evaluated both the MOFAC cache and the invalidation cache for varying update rates and varying multi programming level (MPL – number of concurrent clients) (Figures 4 and 5). The main difference is that with cache invalidation, this happens eagerly, directly at the end of the original update transaction. While on the other hand, with multi-object freshness-aware caching, it happens 'lazily', only when another transaction with freshness limit 0 tries to access a stale copy in a cache node.

Figures 4 and 5 shows that there is no measurable overhead for multi-object freshness aware caching as compared to the standard cache setting with cache invalidation. The two curves are always within a certain confidence interval of each other, in particular for the higher ratio of updates when a lot of invalidations are triggered within the system.

5.4 Evaluation of Invalidation vs. MOFAC with Varying Update Ratio

In this experiment we have multi-object read and update transaction, and we compare MOFAC with the Invalidation algorithm to measure the impact of the update ratio on the performance. The multi-programming level (MPL) is 64 is kept constant throughout this experiment. Figure 6 shows that increase in update ratio impacts performance of both the MOFAC and Invalidation algorithms.

However, if we increase the freshness level of MOFAC, we see a clear difference in throughput and response time between the two algorithms. As we relax the freshness limit, MOFAC shows reduced response time and increased throughput (even with higher update ratio). Where as in the case of invalidation, even a slight increase in the update ratio results in reduced throughput and increased mean response time.

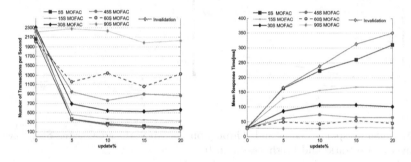

Fig. 6. Mean Response Time and Throughput of MOFAC vs Invalidation on Cluster Size 8 with varying Update Ratio

5.5 Evaluation of Varying Freshness Limit and Varying MPL

In the third experiment, we investigate the effect of varying the freshness limits and MPL. We fix the cluster size to 8 nodes and vary the MPL from 8, 16, 32 through to 64 and use a workload in which read-only transactions concurrently and randomly access a book object with it reviews and authors, while update transactions write to an existing book object. We vary both the amount of update transactions and the freshness limit from 0, 15, 30, 45, 60 to 90 seconds for books, authors and reviews.

As we can see in Figure 7, as the freshness limit increases, we get faster response times. This is exactly what we aim for with multi-object freshness-aware caching – to be able to trade freshness of data for query performance. When the freshness limit is set to 5 seconds, the response time is close to that of cache invalidation. With a freshness limit of 90 seconds, the response time is up-to just half of that of cache invalidation for an update ratio of 20%. With lower update ratios, the saving is proportionally less. We also observe that there is a

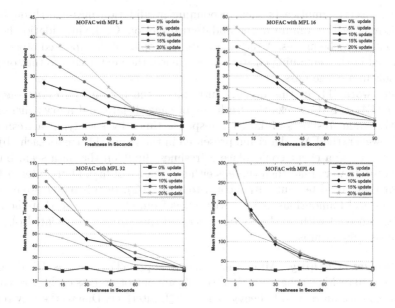

Fig. 7. Performance of MOFAC on Cluster Size 8 with MPL 8, 16, 32 and 64 varying Update Ratio and varying Freshness limit

throughput gain as well as reduced response time when we relax the freshness constraint.

From this experiment we can conclude that with multi-object freshness-aware caching, the more we relax the freshness constraint, the better the throughput and response time of the system becomes.

In the above experiments, we have seen that MOFAC can reduce mean response times and increase throughput for each update level with varying MPL. If we compare MOFAC for MPL 32 in Figure 7 with the invalidation cache algorithm results from Figure 4 we can clearly see that MOFAC performs better. At an update ratio of about 20%, the improvement in response time is about 50% with MOFAC and a (reasonable) freshness limit of 30 seconds. The higher the chosen freshness limit, the more this benefit increases since stale objects can be continue to used in the cache, thus improving throughput.

6   Related Work

Application-Tier Caching. Web caching is an attractive solution for reducing bandwidth demands, improving web server availability, and reducing network latencies. However web caching only supports static content [14]. But the dynamic nature of modern applications requires pages to be generated on the fly.

The state-of-the-art for clustered application servers is an asynchronous cache invalidation approach, as used in, e.g., the BEA WebLogic and JBoss application servers [3,13]: When a cached object is updated in one application server node, that server multicasts a corresponding invalidation message throughout the cluster after the commit of the update transaction. Cache invalidation hence leads to more cache misses. After an update, all copies of the updated object get invalidated in the remaining cluster nodes. Hence, the next access to that object will result in a cache miss.

Earlier work [11] in this domain introduced the notion of *freshness-aware caching (FAC)*. FAC tracks the freshness of cached data and allows clients to explicitly trade freshness-of-data for response times by specifying a freshness limit. However, the FAC algorithm presented in the paper [11] treats each object separately; thus, a client could place a freshness limit on the data seen, but if several objects were read then there is either a chance of high abort rates (Plain FAC) or they could be mutually inconsistent (δ-FAC). In this paper, we extend the theoretical foundations of FAC with ideas from [4] and [8], so that the new MOFAC can deal with *freshness intervals* and the *grouping* of related objects into consistency groups.

Middle-Tier Caching. Midle-tier caching approaches such as IBM's DB-Cache [1,5,10]or MTCache from Microsoft [9] are out-of-process caching research prototypes with an relatively heavy-weight SQL interface. Due to the lazy replication mechanisms, these approaches cannot guarantee distributed cache consistency, although some work around MTCache started at least specifying explicit currency and consistency constraints [8]. Our work differs in that we keep track of the freshness of data of each object separately and only notify about updates to objects that are actually modified; the remaining objects of a group still remain in their valid interval. Furthermore, MTCache works with the relational model while we are working on objects and object relational model.

Data Grid Caching and Replication. In recent years, service infrastructures for sharing large scientific datasets that are geographically distributed have been developed in the form of so-called data grids [6]. A core underlying concept of data grids is caching via adaptive data replication protocols [7]. There are three core differences to the work presented in this paper. Firstly, data grids deal with relatively static, read-mostly datasets, while we are focusing on dynamic web-based applications with frequent updates. Secondly, data grids are optimized for periodically exchanging large datasets, while our focus is on on-demand caching of individual interrelated objects. Thirdly, data grids target the data distribution problem over a wide-area network, while our proposed MOFAC algorithm assumes a closely-coupled caching system inside the same data center.

7   Conclusions and Future Work

This paper proposes a new and promising approach to distributed caching: *Multi-Object Freshness-Aware Caching* (MOFAC). MOFAC tracks the freshness of the cached data and provides clients a consistent snapshot of data with reduced response time if the client agrees to lower its data freshness expectation. MOFAC gives application developers an interesting tuning knob: The more parts of an application can tolerate (slightly) stale data, the better MOFAC can make use of the existing cache content and hence provide a better mean response times compared to cache invalidation. The choice is between performance versus data freshness. The location within the application where this choice has to be made is application specific. In our evaluations, implemented using a MOFAC cache in the JBoss 6.0 application server, we measured savings of up-to 25% on response times, albeit in settings which may not be considered to be very realistic in the context of a real-life bookstore with lots of updates on books objects. However, even with more conservative freshness settings, such as freshness 0 for core book states and more relaxed freshness requirements for reviews, we have observed that a MOFAC cache can improve performance in the of range of 10% to 15%.

When in doubt, a developer has the choice of picking a freshness limit of 0 for requests, in which case, the proposed MOFAC algorithm behaves similar to normal cache invalidation. This makes the proposed multi-object freshness-aware caching a very attractive approach for distributed caching for dynamic web applications delivering significant performance improvements in comparison to cache invalidation, while at the same time providing actual data freshness guarantee within the constraints specified by the application.

We intend to further evaluate and study the characteristics of MOFAC and compare it with FAC and other traditional approached like cache invalidation in more complex and interesting application scenarios. We will use this to develop techniques and tools to enable application developers to choose appropriate settings that best suite the different aspects of the business object hierarchy of the application.

Acknowledgement. The authors would like to thank the members of the Database Research Group at the School of Information Technologies, University of Sydney for their support with this work. We would also like to thank the anonymous reviewers of the paper for their valuable suggestions and comments.

References

1. Altinel, M., Bornhövd, C., Krishnamurthy, S., Mohan, C., Pirahesh, H.: Reinwald. Cache tables: Paving the way for an adaptive database cache. In: VLDB (2003)
2. Amza, C., Soundararajan, G., Cecchet, E.: Transparent caching with strong consistency in dynamic content web sites. In: Proceedings of ICS 2005 (2005)
3. BEA. BEA WebLogic Server 10.0 Documentation (2007), edocs.bea.com
4. Bernstein, P., Fekete, A., Guo, H., Ramakrishnan, R., Tamma, P.: Relaxed-currency serializability for middle-tier caching & replication. In: SIGMOD (2006)
5. Bornhövd, C., Altinel, M., Krishnamurthy, S., Mohan, C., Pirahesh, H., Reinwald, B.: DBCache: Middle-tier database caching for highly scalable e-business architectures. In: Proceedings of ACM SIGMOD 2003, p. 662 (2003)
6. Chervenak, A., Foster, I., Kesselman, C., Salisbury, C., Tuecke, S.: The data grid: Towards an architecture for distributed management and analysis of large scientific datasets. Journal of Network and Computer Applications 23, 187–200 (2001)
7. Chervenak, A., Schuler, R., Kesselman, C., Koranda, S., Moe, B.: Wide area data replication for scientific collaborations. In: Proceedings of 6th IEEE/ACM International Workshop on Grid Computing (Grid 2005) (November 2005)

8. Guo, H., Larson, P.-Å., Ramakrishnan, R., Goldstein, J.: Relaxed currency and consistency: How to say 'good enough' in SQL. In: SIGMOD 2004 (2004)
9. Larson, P.-Å., Goldstein, J., Zhou, J.: MTCache: Transparent mid-tier database caching in SQL Server. In: Proceedings of ICDE 2004, Boston, USA (2004)
10. Luo, Q., Krishnamurthy, S., Mohan, C., Pirahesh, H., Woo, H., Lindsay, B., Naughton, J.: Middle-tier database caching for e-business. In: SIGMOD (2002)
11. Röhm, U., Schmidt, S.: Freshness-aware caching in a cluster of J2EE application servers. In: Benatallah, B., Casati, F., Georgakopoulos, D., Bartolini, C., Sadiq, W., Godart, C. (eds.) WISE 2007. LNCS, vol. 4831, pp. 74–86. Springer, Heidelberg (2007)
12. Röhm, U., Böhm, K., Schek, H.-J., Schuldt, H.: FAS – a freshness-sensitive coordination middleware for a cluster of OLAP components. In: VLDB (2002)
13. Stark, S.: JBoss Administration and Development, 3rd edn. JBoss Group. JBoss Group (2003)
14. Wang, J.: A survey of web caching schemes for the internet. SIGCOMM Comput. Commun. Rev. 29(5), 36–46 (1999)

Appendix

Algorithm 2. Update Handling on Local Cache Node

---

input : object identifier o
input : current transaction context tx
output: object o updated in cache and on backend database
output: update notifications broadcasted to other cache nodes

if object $o \in Cache$ **then**
    update o in local cache node
    persist o in database
    if $isGroup(o) = true$ **then**
        send update-notification to neighbour nodes for group key k
    else if $isCollection(o) = true$ **then**
        send update-notification to neighbour nodes for collection key k
    else
        send update-notification to neighbour nodes for object o
    end if
    commit
end if

Algorithm 3. Processing of Update-Notifications

---

input : object identifier o
input : update-notification timestamp ts
output: object's o stale point is updated if present in cache

if object $o \in Cache$ **then**
    if $stale(o) = 0$ **then**
        $t_{update}(o) \leftarrow ts$ {update meta-data of cache entry o}
    end if
end if

```
 1 @Entity
 2 @Cacheable
 3 @Cache( usage = CacheConcurrencyStrategy.TRANSACTIONAL)
 4 @Table( name = "BOOKENTITY")
 5 public class BookEntity implements Serializable {
 6     private static final long serialVersionUID = 1L;
 7     @Id
 8     private int ISBN;
 9     private String title;
10     private String description;
11
12     @Cache( usage = CacheConcurrencyStrategy.TRANSACTIONAL)
13     @OneToMany( cascade = CascadeType.ALL, fetch = FetchType.LAZY,
14             targetEntity = AuthorEntity.class, mappedBy = "bookEntity")
15     private Collection<AuthorEntity> authors;
16
17
18     @Cache( usage = CacheConcurrencyStrategy.TRANSACTIONAL)
19     @OneToMany( cascade = CascadeType.ALL, fetch = FetchType.EAGER,
20             targetEntity = ReviewEntity.class, mappedBy = "bookEntity")
21     public Collection<ReviewEntity> reviews;
22 }
23
24 @Entity
25 @Cacheable
26 @Cache( usage = CacheConcurrencyStrategy.TRANSACTIONAL)
27 public class ReviewEntity implements Serializable {
28     @Id
29     private int id;
30     private String bookReview;
31
32     @Cache( usage = CacheConcurrencyStrategy.TRANSACTIONAL)
33     @ManyToOne( cascade = CascadeType.ALL, fetch = FetchType.EAGER)
34     @JoinColumn( name = "ISBN")
35     private BookEntity bookEntity;
36 }
37
38 @Id
39     private int author_id;
40     String authorName;
41     String authAddress;
42
43     @Cache( usage = CacheConcurrencyStrategy.TRANSACTIONAL)
44     @ManyToOne( cascade = CascadeType.ALL, fetch = FetchType.LAZY)
45     @JoinColumn( name = "ISBN")
46     private BookEntity bookEntity;
47 }
```

Listing 1.1. ORM definitions of the three entity beans of the Bookstore JEE application including the Java annotations for the cache configuration.

ε-*Controlled-Replicate*: An Improved *Controlled-Replicate* Algorithm for Multi-way Spatial Join Processing on Map-Reduce

Himanshu Gupta and Bhupesh Chawda

IBM Research Laboratory, New Delhi, India
{higupta8,bhupeshchawda}@in.ibm.com

Abstract. Gupta et al. [11] studied the problem of handling multi-way spatial join queries on map-reduce platform and proposed the *Controlled-Replicate* algorithm for the same. In this paper we present ε-*Controlled-Replicate* - an improved *Controlled-Replicate* procedure for processing multi-way spatial join queries on map-reduce. We show that ε-*Controlled-Replicate* algorithm presented in this paper involves a significantly smaller communication cost vis-a-vis *Controlled-Replicate*. We discuss the details of ε-*Controlled-Replicate* algorithm and through an experimental study over synthetic as well as real-life California road datasets, we show the efficacy of the ε-*Controlled-Replicate* algorithm vis-a-vis *Controlled-Replicate*.

Keywords: Map-Reduce, Spatial Data, Multi-way Join Processing.

1 Introduction

In this paper we look at the problem of processing multi-way spatial overlap join processing on map-reduce platform. Given two sets of rectangles R_1 and R_2, a 2-way spatial overlap join finds pairs of rectangles (r_1, r_2) s.t. $r_1 \in R_1$, $r_2 \in R_2$ and rectangle r_1 overlaps with rectangle r_2. A multi-way overlap spatial join query can be written down as a conjunction of pairs of relations $\mathcal{Q} = \{(R_{1,1}, R_{1,2}) \wedge (R_{2,1}, R_{2,2}) \wedge \ldots \wedge (R_{n,1}, R_{n,2})\}$ where $R_{i,1}$ and $R_{i,2}$ denote the two relations being joined on spatial overlap predicate. We consider rectangles instead of a polygon as spatial join is first computed over minimum bounding rectangles (MBR) of objects and only if the two MBRs overlap, the two objects are then checked for overlap. This two step process is more efficient and we refer the reader to [11,14] for further details.

Spatial join queries come up naturally in a number of scenarios and hence the processing of spatial joins has been extensively studied in databases. This includes both optimizing 2-way joins [5–7,10,12,13,17,18] as well as multi-way joins [14,16]. In last few years, map-reduce paradigm has become very popular and as a result, few studies have looked at how we can efficiently process spatial joins on map-reduce platform. The processing of 2-way joins on map-reduce has been looked at by [19–21] while Gupta et al. [11] presented the *Controlled-Replicate* framework - the first algorithms to efficiently process the multi-way

B. Benatallah et al. (Eds.): WISE 2014, Part II, LNCS 8787, pp. 278–293, 2014.

spatial joins on map-reduce. In this paper, we present ϵ-*Controlled-Replicate* which improves on the *Controlled-Replicate* algorithm developed in [11] for processing multi-way spatial joins on map-reduce.

A map-reduce program requires a user to provide the implementation of two functions - *map* and *reduce*. A map function reads the input and converts the input to an intermediate form consisting of a set of key-value pairs. These key-value pairs are collected by the map-reduce engine and are communicated to reducers in a manner that all pairs with identical key are routed to a single reducer. The reducer nodes collect the respective pairs and compute the output. The communication cost is an important contributor to the over-all cost of the map-reduce program. Larger the communication, higher is the cost of the map-reduce program. A good map-reduce implementation of a task hence minimizes the communication among the cluster nodes. The ϵ-*Controlled-Replicate* procedure developed in this paper reduces the communication cost of *Controlled-Replicate* procedure developed in Gupta et al. [11].

Gupta et al. [11] discuss the complexities of handling multi-way spatial joins on map-reduce in detail. A naive way of handling multi-way spatial join is to solve the problem as a series of 2-way joins. However this requires processing big intermediate join results and hence requires huge communication cost. Another naive way is to communicate each rectangle to all reducers however this also entails a very large amount of communication among cluster nodes. Gupta et al. [11] present a novel framework *Controlled-Replicate* which minimizes the communication by exploiting the spatial locations of the rectangles. In this paper we present ϵ-*Controlled-Replicate* which further improves the the communication overhead of *Controlled-Replicate* framework developed in Gupta et al. [11].

Controlled-Replicate requires two map-reduce cycles. We observe that there is a trade-off between the communication costs of two map-reduce cycles in *Controlled-Replicate* procedure. Communication in the second map-reduce cycle can be significantly reduced by increasing the communication in the first cycle by a small amount, thereby leading to a significant reduction in the overall cost. This trade-off is the basic intuition behind ϵ-*Controlled-Replicate* and we discuss this trade-off in detail in this paper.

Contribution: The contributions of this paper are summarized below.

1. We present a new approach ϵ-*Controlled-Replicate* for processing multi-way spatial overlap join queries on map-reduce platform. ϵ-*Controlled-Replicate* significantly improves the performance of *Controlled-Replicate* approach presented in Gupta et al. [11] which is the only study which has investigated multi-way spatial joins on map-reduce. This paper hence improves the state-of-the-art in handling multi-way spatial joins on map-reduce.
2. We discuss the details of ϵ-*Controlled-Replicate* procedure and how it improves *Controlled-Replicate* by significantly reducing the communication among cluster nodes.
3. We carry out an experimental study on both synthetic data as well as real-life California road data-set and present the efficacy of ϵ-*Controlled-Replicate* vis-a-vis *Controlled-Replicate* procedure.

Organization: Section 2 presents the related work. Section 3 presents the basic notation used in this paper. Section 4 presents the basics of designing a map-reduce join program, how 2-way overlap join can be computed on map-reduce and the naive ways of handling multi-way spatial joins. Section 5 summarizes the *Controlled-Replicate* procedure developed in Gupta et al. [11]. Section 6 presents the ϵ-*Controlled-Replicate* procedure and discusses how it improves *Controlled-Replicate* presented in [11]. Section 7 presents the results of experimental evaluation. Section 8 concludes the paper.

2   Related Work

Spatial join processing on RDBMS is a well developed research area. A number of studies have investigated optimization of both 2-way and multi-way spatial joins in an RDBMS [5–7,12–14,16]. In this paper we study the related problem of optimizing multi-way spatial joins on map-reduce platform.

Inspired from the immense popularity of map-reduce paradigm, there have been multiple efforts for optimizing different classes of joins on map-reduce [2,4, 15] etc. 2-way spatial joins on map-reduce have been considered in [19,21]. Gupta et al. [11] is the only prior work to have looked at multi-way spatial joins on map-reduce. In this paper, we present ϵ-*Controlled-Replicate* framework which improves the *Controlled-Replicate* framework presented in Gupta et al. [11] for multi-way overlap spatial joins.

3   Notation

This section presents the basic notation used in this paper. The notation of ϵ-enclosure, ϵ-band, *Enclosure-Split* operation and *Forward* operation is developed in this paper while the rest is imported from Gupta et al [11].

3.1   Start-Point of a Rectangle

We call the top-left vertex of a rectangle its start-point [11]. In Figure 1(a), the start-point of rectangle r_3 is shown in circle.

3.2   Partitioning

Let the complete x-range and y-range of spatial data be $[x_0, x_n)$ and $[y_0, y_n)$ i.e., all rectangles lie within this space. A rectilinear *partitioning* divides the space into a set of disjoint rectangles which taken together cover the whole space [11]. We call these rectangles as partition-cells or simply cells. Figure 1(a) shows an example of partitioning. Here the whole space is partitioned into 16 cells. We represent a partitioning as $\mathcal{C} = (c_1, c_2, \ldots, c_q)$ where the c_j's represent the individual partition-cells.

Distance between a Rectangle and Partition-Cell. The distance between a partition-cell c and a rectangle r is defined as the minimum distance between any point p_1 in partition-cell c and any point p_2 within the rectangle. Mathematically it can be written down as follows:

$$dist(c,r) = \min_{p_1,p_2} dist(p_1,p_2), \forall (p_1,p_2) p_1 \in c, p_2 \in r \tag{1}$$

Cell of a Rectangle. Given a rectangle u, its cell c_u is defined as the partition-cell in which the start-point of rectangle u lies [11]. In Figure 1(a), the cell of rectangle r_1 (i.e., c_{r_1}) is 6 as the start point of rectangle r_1 lies in cell 6. Similarly the cell of rectangle r_2, (i.e., c_{r_2}) is 3.

Cells in the 4th Quadrant wrt. a Rectangle. In this paper we will be heavily using this notion. Consider a rectangle u and its cell c_u. If we divide the whole 2D space by taking the start-point of cell c_u as origin, than the cells lying in the fourth quadrant are said to be the cells in the 4th quadrant wrt. rectangle u and denoted as $\mathcal{C}_4(u)$ [11]. Mathematically it is defined as follows:

$$\mathcal{C}_4(u) = \{c_i\} \text{ s.t. } c_i.x \geq c_u.x \ \& \ c_i.y \leq c_u.y$$

In Figure 1(a), cells 6-8, 10-12 and 14-16 are in the 4th quadrant wrt. rectangle r_1. Cell of rectangle r_1 i.e., c_{r_1} is 6 and the four quadrants (marked by Q_1, Q_2, Q_3 and Q_4) formed by division of 2D space by taking start-point of cell 6 as origin are also shown in Figure 1(a). A cell is also a rectangle and hence we can equivalently define cells in the 4th quadrant wrt. a cell. Cells 6-8, 10-12 and 14-16 are in 4th quadrant wrt. cell 6. We will be simply using the phrase 4th quadrant and from the context it will be clear which cell is being mentioned.

3.3 ϵ-Enclosure and ϵ-Band

In this paper we develop the concepts of ϵ-Enclosure and ϵ-Band which we will be using heavily. Given a cell c, its ϵ-Band, denoted as c_ϵ^b, consists of all points in 4th quadrant which are within a distance of ϵ from cell c and are not within cell c. The bigger region formed by combining cell c and its ϵ-band, denoted as c_ϵ^{en}, is called ϵ-enclosure of cell c. Consider Figure 1(b). The ϵ-Band of cell 6 is shown as shaded. The area within dark lines i.e., cell 6 and the shaded area forms the ϵ-enclosure of cell 6. For $\epsilon=0$, the ϵ-Band of cell c is empty and the ϵ-enclosure identical to cell c.

3.4 Transform Operations

We next define five operations which transform a rectangle wrt. a partitioning in different manners. Three of these *Project, Split, Replicate* are imported from [11] while the two *Forward* and *Enclosure-Split* are developed in this paper. These operations form the building blocks for the approaches developed in this paper.

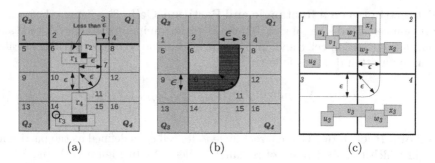

Fig. 1.

Project: Project operation returns the cell of a rectangle i.e., it determines the partition-cell in which the start-point of the rectangle lies. The projection of a rectangle u on a partitioning $\mathcal{C}$ results in the generation of a single key-value pair (c_u, u) where c_u is the partition-cell within which the start-point of the rectangle u lies [11].

Project$(u, \mathcal{C}) \rightarrow (c_u, u)$

Projection of rectangles r_1 and r_2 result in the pair $(6, r_1)$ and $(3, r_2)$ respectively.

Split: Split operation determines all the partition-cells which overlap with the rectangle. Consequently for each such partition-cell, a key-value pair is generated and hence a set of key-value pairs is returned for each rectangle [11].

Split$(u, \mathcal{C}) \rightarrow \{(c_i, u)\}, \forall(i) \text{ s.t. } u \cap c_i \neq \phi$

The split operation on rectangle r_1 returns the pairs $(6, r_1)$ and $(7, r_1)$ while the split operation on rectangle r_2 return the pairs $(3, r_2)$ and $(7, r_2)$.

Enclosure-Split (En-Split): For a rectangle u, en-split returns the set of cells c such that its ϵ-enclosure c_ϵ^{en} overlaps with the rectangle

En-Split$(u, \mathcal{C}) \rightarrow \{(c_i, u)\}, \forall(i) \text{ s.t. } u \cap c_\epsilon^{en} \neq \phi$

The en-split operation on rectangle r_2 returns the pairs $(2, r_2)$, $(3, r_2)$, $(6, r_2)$ and $(7, r_2)$ as cells 2,3,6 and 7 are within distance ϵ of r_2 and r_2 is in fourth quadrant to all these 4 cells. Note that the pair $(4, r_2)$ is not returned even though the distance between rectangle r_2 and cell 4 is less than ϵ as the rectangle r_2 is outside 4th quadrant of cell 4 and hence r_2 does not overlap with 4_ϵ^{en} i.e., ϵ-enclosure of cell 4. For $\epsilon=0$, the en-split operation reduces to split operation.

Forward: For a rectangle u, the *Forward* operation returns those cells which overlap with the ϵ-enclosure of cell c_u (i.e., the cell containing start-point of rectangle u).

$$\text{Forward}(u,\ \mathcal{C}) \rightarrow \{(c_i, u)\},\ \forall(i)\ \text{s.t.}\ c_i \cap c_{u_\epsilon}^{en} \neq \phi$$

The Forward operation on rectangle r_1 returns the pairs $(6, r_1)$, $(7, r_1)$, $(10, r_1)$ and $(11, r_1)$ as the cells 6,7,10 and 11 overlap with the ϵ-enclosure of cell 6.

Replicate: For a rectangle u, the *Replicate* operation returns all cells which are in the fourth quadrant wrt. rectangle u [11].

$$\text{Replicate}(u,\ \mathcal{C}) \rightarrow \{(c_i, u)\},\ \forall(i)\ \text{s.t.}\ c_i \in \mathcal{C}_4(u)$$

The replicate operation on rectangle r_1 returns cells 6-8, 10-12 and 14-16.

Equivalently we define these operations on a relation. The operations *Project(R, C)*, *Split(R, C)*, *ϵ-Split(R, C)*, *Forward(R, C)* and *Replicate(R, C)* represent the output of *Project, Split, ϵ-Split, Forward* and *Replicate* operations on each rectangle u in relation R respectively.

3.5   Query Graph

A multi-way overlap join query can be represented using a graph $\mathcal{G}$. Each relation represents a vertex in this graph. There is an edge between relations R_i and R_j if there exists a join condition R_i and R_j in the query $\mathcal{Q}$ and this edge is represented as $\langle R_i, R_j \rangle$.

Consider the query $\mathcal{Q} = R_1$ *Overlaps* R_2 and R_2 *Overlaps* R_3 and R_3 *Overlaps* R_4. We will be explaining various concepts using this query. The query graph consists of the edges $\langle R_1, R_2 \rangle$, $\langle R_2, R_3 \rangle$ and $\langle R_3, R_4 \rangle$.

3.6   Consistent Rectangle-Sets

A set of rectangles $\mathcal{U}$ is called a consistent set if for each pair of rectangles u and v in $\mathcal{U}$ the following conditions hold:

- Rectangles u and v belong to different relations say R_u and R_v.
- If there is an edge $\langle R_u, R_v \rangle$ in query graph $\mathcal{G}$ than rectangles u and v overlap.

Consider the rectangles in Figure 1(c) and query $\mathcal{Q}$ (Section 3.5). The rectangle-set $\mathcal{U} = (u_1, v_1, w_1)$ is a consistent rectangle-set. All three rectangles (u_1, v_1, w_1) belong to different relations. The query $\mathcal{Q}$ contains the edges $\langle R_1, R_2 \rangle$ and $\langle R_2, R_3 \rangle$; and the rectangle-pairs (u_1, v_1) and (v_1, w_1) overlap. The query graph does not contain the edge $\langle R_1, R_3 \rangle$ and hence the rectangles u_1 and w_1

are not required to overlap. The rectangle-set (u_2, v_1, w_1) is not consistent as the rectangles u_2 and v_1 do not overlap but the overlap condition $\langle R_1, R_2 \rangle$ is present in query $\mathcal{Q}$.

The intuition behind this concept is that each sub-set of an output tuple is a consistent rectangle-set. Consider the output tuple $\mathcal{U}_1 = (u_1, v_1, w_1, x_1)$. Each sub-set of this set is a consistent rectangle-set.

3.7 Crossing Rectangle-Sets

Consider a spatial area $\mathcal{A}$. A set of rectangles $\mathcal{U}$ is said to cross the area $\mathcal{A}$ if the following conditions hold:

- Any two rectangles u and v in $\mathcal{U}$ belong to two different relations.
- Each rectangle u in $\mathcal{U}$ overlaps with spatial area $\mathcal{A}$.
- For each rectangle u in $\mathcal{U}$ the following holds:
  - Say the rectangle u belongs to relation R_u. For each edge of the form $\langle R_u, R \rangle$ in query graph $\mathcal{G}$, if no rectangle exists in the set $\mathcal{U}$ that belongs to relation R then the rectangle u crosses the spatial area $\mathcal{A}$ (in other words, the rectangle u is not wholly contained within the area $\mathcal{A}$).

Again consider the Figure 1(c) and query $\mathcal{Q}$. The rectangle-set $\mathcal{U} = (u_1, v_1, w_1)$ crosses the cell 1. All rectangles belong to different relations and overlap with cell 1. There is an edge $\langle R_3, R_4 \rangle$ in graph $\mathcal{G}$, no rectangle of relation R_4 is present in $\mathcal{U}$ and this requires rectangle w_1 to cross the cell 1; and w_1 crosses the cell 1. However $\mathcal{U}$ does not cross the ϵ-enclosure of cell 1 as the rectangle w_1 does not cross the area 1_ϵ^{en}. The rectangle-set $\mathcal{V} = (u_1, v_1, w_2)$ crosses both the cell 1 area as well as the area 1_ϵ^{en}.

The intuition behind this concept is that a crossing and consistent rectangle-set from one area can combine with a consistent rectangle-set from another area to form an output tuple. The rectangle-set $\mathcal{U}_2 = (u_3, v_3)$ is a consistent set and crosses cell 3 and it can combine with consistent set $\mathcal{U}_3 = (w_3, x_3)$ to form an output tuple (u_3, v_3, w_3, x_3).

4   Map-Reduce Framework and Spatial Joins

In this section we outline a very brief sketch of a map-reduce join algorithm. For want of space we do not describe map-reduce model in detail and we refer the reader to Dean et al. [8]. The basic idea behind designing a map-reduce algorithm is to divide the computation into multiple parts and let each reducer handle an individual part. The map operations read the input and generate a set of intermediate key-value pairs. All pairs with identical keys are routed to a single reducer. Each reducer computes a part of the output.

Consider there are k reducer nodes available. In case of spatial joins, we divide the 2D grid into k partition-cells (i.e., we divide x and y axis in $\sqrt{k}$ partitions each). The map operations process the rectangles and communicate each rectangle to a set of reducers. The basic idea is to get the rectangles forming an output

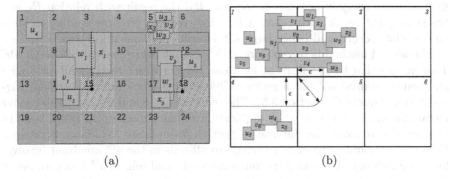

(a)   (b)

Fig. 2.

tuple to at least one reducer. This way, each output tuple can be computed by at least one reducer. Each reducer computes a part of the output. If for an output tuple, all the constituent rectangles are present at more than one reducer, a duplicate avoidance mechanism is employed so as only one reducer computes this output node. Combining the output of all reducers produces the complete join output.

4.1 Handling 2-way Overlap Joins

Let R_1 and R_2 be the two relations being joined. We can process the *overlap* join by splitting the two relations. Let $r_1 \in R_1$ and $r_2 \in R_2$ be two rectangles which satisfy the *Overlap* predicate. Hence there must be at least one reducer which will receive both r_1 and r_2. Such reducers can output that the rectangles r_1 and r_2 overlap. As there can be more than one reducers which receive both r_1 and r_2, we need a duplicate avoidance mechanism. Duplicates can be avoided by letting that reducer compute the output tuple, which contains the start-point of the overlapping rectangular region. Consider the rectangles r_3 ($\in R_1$) and r_4 ($\in R_2$) in Figure 1(a). The overlapped area between rectangles r_3 and r_4 is shown black. Reducers 14 and 15 will receive both the rectangles r_3 and r_4. However the start-point of overlapping area lies in cell 14. The output tuple (r_3,r_4) is hence computed by reducer 14.

4.2 Handling Multi-way Overlap Joins

A multi-way overlap join can not be computed by splitting all relations as for some output tuples, there might not be a single reducer which receives all the constituent rectangles. Consider the Figure 1(c) and query $\mathcal{Q}$ (Section 3.5). Let the rectangles u_1, v_1, w_1 and x_1 belong to relations R_1, R_2, R_3 and R_4 respectively. If all the relations are split, there will not be any reducer which will receive all the four rectangles and hence this output tuple can not be computed.

There are two naive methods to compute a multi-way join. The first is to compute a multi-way join as a cascade of 2-way joins. This method will hence first compute the join of relations R_1 and R_2, the result with relation R_3 and the subsequent result with relation R_4. However this requires reading in large intermediate results and hence large subsequent communication cost.

The second method is to replicate each rectangle which we call *All-Replicate*. This ensures that there will be at-least one reducer which receives all rectangles and hence we need not use a series of 2-way joins. For example, consider Figure 2(a) and query Q (Section 3.5). The reducers 15-18 and 21-24 will all receive rectangles u_1, v_1, w_1 and x_1. A duplicate avoidance mechanism hence need to be employed to avoid the duplicates (Section 5:Duplicate Avoidance Strategy). However communicating each rectangle to all cells in the 4th quadrant involves lot of redundancy. A rectangle is communicated to all cells in 4th quadrant, even if it is not a part of any output tuple (e.g., rectangle u_4). Secondly a rectangle is replicated even if all the output tuples containing this rectangle can be locally computed (e.g., rectangles u_3, v_3, w_3, x_3). As a result, this approach too requires huge communication among cluster nodes. Both the naive approaches are hence not practical.

5   Controlled-Replicate (C-Rep)

5.1   Motivation

2-way Cascade is a naive approach because it solves a multi-way join as a series of 2-way joins with each 2-way join requiring a separate map-reduce cycle. While *All-Replicate* solves a multi-way join in one go, it replicates all the rectangles. Both these approaches hence require a huge communication among cluster nodes.

We need an approach which solves a multi-way join in one go rather than as a cascade of 2-way joins but unlike *All-Replicate* which does not replicate all rectangles. This is what precisely the *Controlled-Replicate* approach achieves. It selectively identifies which rectangles need to be replicated and replicates only such rectangles.

5.2   Outline

The basic idea of *C-Rep* framework is to identify which rectangles need not be replicated. The framework defines a number of conditions which each rectangle must satisfy. Any rectangle which does not satisfy these conditions is not replicated. *C-Rep* runs as a round of two map-reduce cycles. Map operations in the first round split all the relations and hence a reducer c receives all rectangles which overlap with cell c. Reduce tasks figure out which rectangles need to be replicated. Such rectangles satisfy the conditions of *C-Rep* framework. Second round of map operations replicate such rectangles and project the rest. Finally the second round of reduce tasks compute the multi-way join.

As only rectangles marked in the first round are replicated, *C-Rep* replicates much lesser number of rectangles as compared to *All-Replicate* and hence incurs a much smaller communication cost as compared to *All-Replicate*. As *Controlled-Replicate* does not generate any large intermediate joins, it incurs a much smaller reading/writing cost as well as communication-cost vis-a-vis *2-way Cascade* which incurs a huge reading/writing cost due to generation of large intermediate results.

5.3 Conditions of *Controlled-Replicate*

Let U_c be the set of rectangles split on partition-cell c and subsequently received by the reducer c. The reducer c first identifies all rectangle-sets $\mathcal{U}$, $\mathcal{U} \subseteq U_c$ which satisfy the following conditions.

- **C1:**The rectangle-set $\mathcal{U}$ is consistent.
- **C2:**The rectangle-set $\mathcal{U}$ crosses the partition-cell c.
- **C3:**The rectangle-set $\mathcal{U}$ is not an output tuple.

Let US_c be the set of such rectangle-sets. *C-Rep* replicates all rectangles in uS_c [11] where uS_c is defined as the union of all rectangle sets in US_c i.e.,

$$uS_c = \bigcup_{\mathcal{U}} \mathcal{U}, \ s.t. \ \mathcal{U} \in US_c$$

5.4 Duplicate Avoidance Strategy

Consider an output tuple $\mathcal{U}$, let u_r, $u_r \in \mathcal{U}$ be the rightmost rectangle in $\mathcal{U}$ i.e., the rectangle with the largest x-coordinate of the starting-point and let u_l, $u_l \in \mathcal{U}$ be the lowermost rectangle in $\mathcal{U}$ i.e., the rectangle with the smallest y-coordinate of the starting point. Duplicates are avoided by letting only the partition-cell which contains the point $(u_r.x, u_l.y)$ compute the output tuple $\mathcal{U}$.

Consider the query $\mathcal{Q}$, Figure 2(a) and the rectangle-set $\mathcal{U}=(u_1, v_1, w_1, x_1)$. x_1 is the rightmost rectangle in $\mathcal{U}$ and u_1 is the lowermost rectangle. Cell 15 contains the point $(x_1.x, u_1.y)$ (Figure 2(a)) and hence reducer 15 computes the output tuple $\mathcal{U}$. Reducers 16-18, 21-24 will determine that their cells do not contain the point $(x_1.x, u_1.y)$ and hence will not output the tuple $\mathcal{U}$.

5.5 Intuition Behind the Conditions

Unlike *All-Rep*, *C-Rep* replicates only selected rectangles. Specifically *C-Rep* first finds out *consistent* and *crossing* sets in cell c and replicates only rectangles belonging to such sets (Conditions C1 and C2). A set which is not consistent can not be part of an output tuple. Secondly a set which does not cross the cell c, can not combine with any other set from another cell to form an output tuple. Hence such sets need not be included in US_c. The total number of rectangles in uS_c is much smaller as compared to U_c and hence *C-Rep* improves considerably over *All-Rep*. An output tuple is consistent and crosses a cell but is not included in set US_c as an output tuple can not combine with any other set (Condition C3).

5.6   Illustrative Example

Consider Figure 2(b) and cell 1. Cell 1 receives rectangles $u_1, u_2, v_1, v_2, v_3, v_4, v_5$ and v_6 after splitting all relations. There are four consistent and crossing sets (u_1, v_1), (u_1, v_2), (u_1, v_3) and (u_1, v_4) and these rectangle-sets form the set US_1. The set uS_1 is $(u_1, v_1, v_2, v_3, v_4)$ and C-Rep replicates the rectangles in set uS_1.

6 ϵ-Controlled-Replicate (ϵ-C-Rep)

ϵ-C-Rep first carries out en-split operation on all the relations. Let U_c^ϵ be the set of rectangles received by reducer c. Let U_c be the set of rectangles overlapping with cell c. Note that $U_c \subseteq U_c^\epsilon$. We also compute the set uS_c, computed as part of C-Rep procedure as discussed in Section 5. We then find out all rectangle-sets $\mathcal{V}$, $\mathcal{V} \subseteq U_c^\epsilon$ which satisfy the following conditions.

- **C4:** The rectangle-set $\mathcal{V}$ is consistent.
- **C5:** The rectangle-set $\mathcal{V}$ crosses the enlarged partition-cell c^ϵ.
- **C6:** The rectangle-set $\mathcal{V}$ is not an output tuple.

Let VS_c be the set of such rectangle-sets. We define vS_c as the union of all rectangle sets in VS_c i.e.,

$$vS_c = \bigcup_\mathcal{V} \mathcal{V}, \ s.t. \ \mathcal{V} \in VS_c$$

Let S_1 be the set of rectangles which are common to both uS_c and vS_c i.e., $S_1 = uS_c \cap vS_c$. Let S_2 be the set of rectangles which are in uS_c but not in vS_c i.e. $S_2 = uS_c - vS_c$. ϵ-C-Rep forwards rectangles in set S_2 while replicates the rectangles in set S_1.

6.1   Differences With *Controlled-Replicate*

The procedure ϵ-C-Rep differs with C-Rep in two aspects.

1. In the first stage, C-Rep splits all the relations while ϵ-C-Rep en-splits all the relations. Hence ϵ-C-Rep involves more communication in the first stage vis-a-vis C-Rep. In C-Rep, Reducer c receives all rectangles which overlap with cell c while in ϵ-C-Rep, reducer c receives all rectangles which overlap with cell c as well as rectangles overlapping with the ϵ-band of cell c i.e., c_ϵ^b. ϵ-C-Rep has a higher communication cost in this stage.
2. In the second stage, some of the rectangles chosen for replication by C-Rep are instead forwarded by ϵ-C-Rep. ϵ-C-Rep hence has a smaller communication cost than C-Rep

The extra rectangles received by reducer c in the first round, enable reducer c to do a better job in deciding which rectangles to replicate in the second round. The reduced communication cost in the second round offsets the communication cost in the first round leading to an overall reduction in the communication cost.

6.2   Intuition

In *C-Rep*, reducer c computes the crossing sets and replicates them to all reducers in the 4th quadrant. A set crossing cell c is likely to form an output tuple with the rectangles which are close-by. However rectangles in this rectangle-set are still communicated to all reducers in the 4th quadrant. For example consider Figure 2(b). The set $\{u_1,v_1\}$ crosses cell c and forms an output tuple with rectangles w_1 and x_1 which are nearby. In such cases, replicating a rectangle to all reducers in the 4th quadrant involves a lot of redundancy.

This extra communication is exactly what ϵ-*C-Rep* reduces. As each relation is en-splitted, each cell c also receives all rectangles which overlap with ϵ-band of cell c. Hence reducer c can determine the output tuples that are formed by a consistent set crossing cell c and a consistent set in the ϵ-band area. Hence the rectangles which belong to such consistent sets crossing cell c need not be replicated. Such rectangles are only forwarded. As the cost of forwarding a rectangle is much smaller than replicating a rectangle, ϵ-*C-Rep* involves much smaller communication vis-a-vis *C-Rep*.

6.3   Details

$\mathcal{V}S_c$ is the set of rectangle-sets which cross the ϵ-enclosure of cell c i.e., c_ϵ^{en}. vS_c is the set of rectangles which belong to at least one rectangle-set in $\mathcal{V}S_c$. uS_c is the set of rectangles which are replicated by *C-Rep*. The rectangles in uS_c belong to two following cases:

1. The set $uS_c \cap vS_c$ i.e., Rectangles which are common to both uS_c and vS_c. Such rectangles belong to the rectangle-sets which cross both cell c as well as its ϵ-enclosure c_ϵ^{en}. Such rectangles are hence replicated as they can potentially belong to output tuples which may contain rectangles outside c_ϵ^{en}.
2. The set uS_c-vS_c i.e., Rectangles which are in uS_c but not in vS_c. Such rectangles belong to consistent sets which cross cell c but do not cross the ϵ-enclosure of cell c. Such rectangles need not be replicated as they can not form an output tuple with rectangles outside c_ϵ^{en}. Such rectangles can form output tuples only with rectangles overlapping with ϵ-band of cell c. Such rectangles are hence forwarded.

ϵ-*C-Rep* hence forwards many rectangles which were otherwise being replicated by *C-Rep*. As the cost of forwarding a rectangle is much smaller than the cost of replicating a rectangle, ϵ-*C-Rep* involves a significantly smaller communication cost vis-a-vis *C-Rep*. Note that for ϵ=0, ϵ-*C-Rep* reduces to *C-Rep*. The set vS_c is identical to the set uS_c and hence the set uS_c-vS_c is null and no rectangle is forwarded.

6.4   Illustrative Example

Consider Figure 2(b) and cell 1. The set uS_1 is (u_1,v_1,v_2,v_3,v_4) (Section 5:Illustrative Example). ϵ-*C-Rep* en-splits all the relations. Cell 1 hence

Table 1. Performance on Synthetic DataSets

ϵ	Time hh:mm	# Rectangles in ϵ-Band	# Rectangles Forwarded	# Rectangles Replicated	# Intermediate Key-Value Pairs
0	01:27	0	0	658K	16M
100	01:16	41K	36K	622K	15M
500	00:57	208K	171K	487K	12M
1000	00:52	424K	312K	346K	10M
2500	00:47	647K	414K	244K	9M

receives the rectangles overlapping with cell 1 i.e., $\{u_1,u_2,v_1,v_2, v_3,v_4,v_5,v_6\}$ as well as the rectangles overlapping with the ϵ-Band of cell 1 i.e., $\{w_1,x_1\}$. The consistent sets that cross the ϵ-enclosure of cell 1 i.e., 1_ϵ^{en} are $\{(u_1,v_3),(u_1,v_4)\}$ and this forms the set VS_1. These consistent sets crossing 1_ϵ^{en} area can form an output tuple with a consistent rectangle-set outside 1_ϵ^{en} e.g., (u_1,v_3,w_2,x_2). The set vS_1 hence is $\{u_1,v_3,v_4\}$. The set $uS_1 \cap vS_1$ hence is $\{u_1,v_3,v_4\}$ and hence these rectangles are replicated. The set uS_1-vS_1 is $\{v_1,v_2\}$ and these two rectangles are forwarded.

The two extra rectangles received by cell 1 in the first stage in ϵ-C-Rep (i.e., w_1 and x_1) allow the cell 1 to derive the information that the consistent sets crossing set 1 i.e. $\{u_1,v_1\}$ and $\{u_1,v_2\}$ can not form an output tuple with any consistent set outside ϵ-enclosure of cell 1 i.e., 1_ϵ^{en}. None of these two sets are a subset of any set which crosses the ϵ-enclosure of cell 1 i.e., 1_ϵ^{en}. Hence these two sets do not contribute to the set vS_1. The set vS_1 is formed by the rectangles which belong to a rectangle-set which crosses the ϵ-enclosure of cell 1 i.e., 1_ϵ^{en}.

Note that in C-Rep, cell 1 did not receive the rectangles overlapping with the ϵ-band (i.e., w_1 and x_1 here), it could not infer that the sets crossing cell 1 i.e., $\{u_1,v_1\}$ and $\{u_1,v_2\}$ do not form an output tuple with any consistent rectangle-sets outside 1_ϵ^{en}. Hence C-Rep must replicate the rectangle v_1 and v_2.

Note that in Figure 2(b) replication of a rectangle in cell 1 results into 6 intermediate key-value pairs as all six cells are in 4th quadrant of cell 1. While the forward operation results into 4 intermediate key-value pairs as only cells 1,2,4 and 5 overlap with the ϵ-enclosure of cell 1. Hence forwarding rectangles $\{v_1,v_2\}$ instead of replicating them saves 4 extra communications. While in the first stage w_1 and x_1 are sent to cell 1 and that costs two extra communications in ϵ-C-Rep and hence the overall saving is 2 communications. Across large data-sets and more number of partition-cells this adds up and translates into a significant saving in communication cost.

7   Experimental Evaluation

7.1   Experimental SetUp

We next show the efficacy of ϵ-C-Rep over C-Rep through an experimental study. The experiments are run over a 16 core Hadoop cluster built using Blade Servers

Table 2. Performance on Real-life California Road Data

ϵ	Time hh:mm	# Rectangles in ϵ-Band	# Rectangles Forwarded	# Rectangles Replicated	# Intermediate Key-Value Pairs
0	00:21	0	0	83K	8.1M
100	00:18	23K	50K	33K	7.1M
200	00:17	46K	55K	28K	6.9M
300	00:16	70K	57K	26K	6.8M

with four 3 GHz Xeon processors having 8GB memory and 200 GB SATA drives. These machines run Red Hat Linux 5.2. The software stack comprises of Hadoop 0.20.2 with HDFS. All the experiments are executed with 64 reduce processes. The 2D space is hence divided in 8x8 grid.

7.2 Synthetic Data

We first show the results on synthetic data. We generate three sets of rectangles. We execute the query $Q'=R_1$ *Overlaps* R_2 *and* R_2 *Overlaps* R_3. The dimensions of the rectangles are within 0 and 1000. The dimension lengths of the rectangles as well as the start-points are generated using a uniform distribution. Rectangles are generated within a spatial domain of 100K x 100K grid. The number of rectangles in each relation are set to 1M.

We execute ϵ-*C-Rep* for different values of ϵ. We start with $\epsilon=0$ and increase the value of ϵ. Table 1 presents the results. We note down the five metrics: (a) Time Taken to execute the query, (b) Number of Rectangles in ϵ-Band i.e., number of extra rectangles communicated in the first cycle in ϵ-*C-Rep* vis-a-vis *C-Rep*, (c) Number of rectangles forwarded and replicated by ϵ-*C-Rep* by the second round of reducers i.e. the number of rectangles in the set $\sum_c uS_c - vS_c$ and $\sum_c uS_c \cap vS_c$ and (d) Number of Intermediate key-value Pairs i.e., the total number of rectangles communicated (across both cycles of map-reduce).

Note that for $\epsilon=0$, the ϵ-*C-Rep* reduces to *C-Rep*. Hence *C-Rep* takes an hour and 27 minutes to execute the query Q'. *C-Rep* replicates 658K rectangles which results into a total of 16M rectangles being communicated to the reducer nodes. For $\epsilon=100$, 41K extra rectangles are communicated in the first rounds. With this extra information, ϵ-*C-Rep* replicates 622K rectangles and forwards 36K rectangles. With a reduction of 36K replications, the overall number of key-value pairs go down by 1M (From 16M to 15M). This reduction finally translates to a performance gain of 9 minutes in the execution time.

As ϵ increases, the number of rectangles in ϵ-band increases and more rectangles are communicated in the first round which enables the reducers to do a better job of deciding which rectangles to replicate and which to forward. This hence results into a gradual decrease in the total number of rectangles communicated to the rectangles and as a result in the time taken for the query to execute. For $\epsilon=2500$, a significant gain of 40 minutes is observed in the performance.

7.3   California Road DataSet

Census 2000 TIGER/Line shape files consisting of 2092079 roads were used to create the real data set [1]. The ratio of California spatial dimensions i.e., |x-range|/|y-range| was found to be 0.63. The mapping was hence done on 2D space with y-range as [0,100K] and x-range as [0,63K]. Table 2 presents the results. Once again we again get the similar trends. With increasing value of ϵ, a reduction in the number of intermediate key-value pairs is observed and an improvement in the running time of ϵ-C-Rep is observed.

7.4   Choosing the Value of ϵ

Let n be the number of relations in the multi-way join query and let l_{max} be an upper-bound on the rectangle dimensions. Then the value of ϵ should be less than or equal to $n*l_{max}$. Having an epsilon greater than $n*l_{max}$ will not yield any further benefits. As the value of ϵ increases from 0 to $n*l_{max}$, the number of rectangles that need to be replicated will keep decreasing. As ϵ becomes more than $n*l_{max}$, no rectangle will need to be replicated.

8   Conclusion

In this paper we presented ϵ-$Controlled$-$Replicate$, a new procedure for processing multi-way spatial joins on map-reduce. The new algorithm presented in this paper, improves the prior approaches for handling multi-way spatial joins. Using both synthetic and real-life California road data, we provided an experimental proof that the proposed approach works and beats the prior approaches.

As part of future work, we plan to make use of some recent approaches proposed in literature to improve the performance of map-reduce implementations. Two such approaches presented in literature are *CoHadoop* [9] and *HaLoop* [3]. *HaLoop* materializes the input to the reducer thereby eliminating the need of communicating all the rectangles to the reduce phase in both the cycles. *Co-Hadoop* colocates the related data at the same machine. For the case of multi-way spatial join processing, spatially close data from all relations can be colocated. This way we can compute the rectangles which need to be replicated as part of map operations only, thereby altogether eliminating one map-reduce cycle altogether.

References

1. Census 2000 Tiger/Line Data,
   http://www.esri.com/data/download/census2000-tigerline
2. Afrati, F.N., Ullman, J.D.: Optimizing joins in a map-reduce environment. In: EDBT (2011)
3. Bhatolia, P., et al.: HaLoop: Efficient Iterative Data Processing on Large Clusters. In: VLDB (2010)

4. Blanas, S., Patel, J.M., Ercegovac, V., Rao, J., Shekita, E.J., Tian, Y.: A comparison of join algorithms for log processing in map-reduce. In: SIGMOD (2010)
5. Brinkhoff, T., Kriegal, H., Seeger, B.: Parallel processing of spatial joins using R-trees. In: ICDE (1996)
6. Brinkhoff, T., Kriegal, H.P., Schneider, R., Seeger, B.: Multi-step processing of spatial joins. In: SIGMOD (1994)
7. Brinkhoff, T., Kriegal, H.P., Seeger, B.: Efficient processing of spatial joins using R-trees. In: SIGMOD (1993)
8. Dean, J., Ghemawat, S.: MapReduce: Simplified data processing on large clusters. Comm. of ACM 51(1) (2008)
9. Eltabakh, M.: et al. CoHadoop: Flexible Data Placement and its exploitation in Hadoop. In: VLDB (2011)
10. Gunther, O.: Efficient computation of spatial joins. In: ICDE (1993)
11. Gupta, H., Chawda, B., Negi, S., Faruquie, T., Subramaniam, L.V., Mohania, M.: Proceesing multi-way spatial joins on map-reduce. In: EDBT (2013)
12. Lo, M., Ravishankar, C.V.: Spatial hash joins. In: SIGMOD (1996)
13. Lo, M.L., Ravishankar, C.V.: Spatial joins using seeded trees. In: SIGMOD (1994)
14. Mamoulis, N., Papadias, D.: Multiway spatial joins. In: ACM Transaction on Database Systems (2001)
15. Okcan, A., Riedewald, M.: Processing theta-joins using mapreduce. In: SIGMOD (2011)
16. Papadias, D., Arkoumanis, D.: Approx processing of multiway spatial joins in very large databases. In: Jensen, C.S., Jeffery, K., Pokorný, J., Šaltenis, S., Bertino, E., Böhm, K., Jarke, M. (eds.) EDBT 2002. LNCS, vol. 2287, p. 179. Springer, Heidelberg (2002)
17. Patel, J., DeWitt, D.J.: Clone join and shadow join: Two parallel spatial join algorithms. In: ACM-GIS (2000)
18. Patel, J.M., DeWitt, D.J.: Partition based spatial-merge join. In: SIGMOD (1996)
19. Wang, K., Han, J., Tu, B., Dai, J., Zhu, W., Song, X.: Accelerating spatial data processing with map-reduce. In: ICPADS (2010)
20. Zhang, S., Han, J., Liu, Z., Wang, K., Feng, S.: Spatial queries evaluation with mapreduce. In: GCC (2009)
21. Zhang, S., Han, J., Liu, Z., Wang, K., Xu, Z.: Sjmr: Parallelizing spatial join with mapreduce on clusters. In: CLUSTER (2009)

REST as an Alternative to WSRF:
A Comparison Based
on the WS-Agreement Standard

Florian Feigenbutz, Alexander Stanik, and Andreas Kliem

Technische Universität Berlin, Complex and Distributed IT Systems,
Secr. EN 59, Einsteinufer 17, 10587 Berlin, Germany
{florian.feigenbutz}@campus.tu-berlin.de,
{alexander.stanik,andreas.kliem}@tu-berlin.de
http://www.cit.tu-berlin.de

Abstract. WS-Agreement and WS-Agreement Negotiation are specifications that define a protocol and a language to dynamically negoti-
ate, renegotiate, create and monitor bi-lateral service level agreements
in distributed systems. While both specifications are based on the Web
Services Resource Framework standard, that allows using stateful SOAP
services, the WSAG4J reference implementation additionally provides
a RESTful service implementation of the same operations. This paper
evaluates the performance disparity between the standard conformable
and the RESTful implementation of WS-Agreement and WS-Agreement
Negotiation.

Keywords: SLA, WS-Agreement (Negotiation), WSAG4J, REST,
WSRF.

1 Introduction

Nowadays, software architects and developers have the fundamental choice between two major approaches when creating web services: WS-* based or REST-
ful web services [13]. Both acronyms describe popular approaches for distributed
services: the WS-* family describes a large stack of specifications based on the
Simple Object Access Protocol (SOAP) [12] while *Representational State Transfer (REST)* is more an "architectural style" [10] than a standard that strongly
relies on the *Hypertext Transfer Protocol (HTTP)* as the application-level protocol [19]. There is a notable number of well defined WS-* specifications which
are modularly designed in a way that they can be changed, combined, and used
independently of each other [8]. Many of these specifications specify interfaces
which are usually defined by the *Web Service Description Language (WSDL)*
[6]. With the upcoming of *Web Application Description Language (WADL)* [14]
as equivalent to WSDL, such a specification chain can also applied to REST.

This paper investigates to what extent a specific WS-* specification could be
ported to RESTful services and be extended with WADLs. Furthermore it studies both the SOAP and the REST based implementation in terms of the feature

B. Benatallah et al. (Eds.): WISE 2014, Part II, LNCS 8787, pp. 294–303, 2014.

set and the performance. For the comparison we use the *WS-Agreement (WSAG)* [4] and the *WS-Agreement Negotiation (WSAN)* [23] specifications which are built on top of the *Web Services Resource Framework (WSRF)* [1] standard. The intention of WSRF is to provide a stateful WS-* web service which can be used to model, access and manage states in distributed systems [11]. Our comparison is based on an existing open source software framework, named *WS-Agreement for Java (WSAG4J)* [22] [24], that implements the WS-Agreement and the WS-Agreement Negotiation standards. The reason for choosing these specifications and this framework is that a significant effort to design a RESTful approach of both specifications was already investigated by [15] [20] [5]. Any WSAG service acts as a neutral component between a service provider and a service consumer for SLA agreement and contracting. As such it needs to be available to both at any time which implies hard requirements for availability and scalability of such a service. Therefore we expect our performance benchmarks to indicate whether WSRF or REST based WSAG allows to handle more concurrent clients with given hardware. Furthermore, the WSAG4J framework itself provides also both a SOAP and a RESTful service [24] with an appropriate client implementation.

The rest of the paper is structured as follows: In section 2 we present our practical approach for this comparison, where performance benchmarks had been performed and show the differences in terms of scalability, availability, and efficiency. For this evaluation a test scenario was designed that respects not only atomic operations but also workflows for which both WS-Agreement standards were conceived. Moreover, we interpret results, discuss the reasons and analyze their origin. Next we present related work, that compares WS-* based services to RESTful ones in section 3. Finally, section 4 concludes this paper.

2 Evaluation

In order to compare both the RESTful and the WSRF variants it is important to select a real life usage scenario with significant complexity. A typical use case for WSAG is automated SLA negotiation which is already used by research projects in the area of fully automated service-level agreements [22] [7]. Web services handling SLAs via WSAG and WSAN act as neutral notaries which must always be reachable to both agreement parties guaranteeing verification of concluded contracts. For this reason performance, scalability and availability are hard requirements for any production system.

2.1 Test Scenarios

Based on the use case of automated SLA agreement we picked three sample tests that reflect WSAG usage in the field of cloud computing. Terminology is adapted from the WS-Agreement and WS-Agreement Negotiation specifications [4] [23]. Basically every WSAG and WSAN service provides at least one *Agreement Factory* containing one or more *Agreement Templates* that describe the provided services and serve as a sample for incoming *Agreement Offers* the service is willing to accept. *Templates* also hold information for agreement creation such as context or terms and can optionally define creation constraints allowing customization of *Agreement Offers*. *Offers* are created by the *Agreement Initiator* and sent to the *Agreement Responder*. If the latter accepts an *Agreement* is created, otherwise the *responder* replies with a *Fault*. For the given scenarios both WSRF and REST services were configured with one *Agreement Factory* holding three templates and are specified as follow:

– **GetFactories.** The first scenario is a very basic step in which an *Agreement Initiator* requests all *Agreement Factories* served by the *Agreement Responder*. This use case is usually the first step an *initiator* has to go through to discover services of an unknown *responder*.
– **GetTemplates.** The next scenario reflects the follow-up step in service discovery: The *initiator* needs to gain knowledge about available *Agreement Templates* for any *factory* of interest.
– **Negotiation Scenario.** The third scenario runs a complex negotiation process between the *initiator* and the *responder*. In this case the *responder* implements the agreement on behalf of the *Service Provider* while the *initiator* acts on behalf of the *Service Consumer*. The offered service computes resources for certain time frames using negotiable templates. Within the scenario the *initiator* sends a first *Negotiation Counter Offer* to which the *responder* replies with another counter offer for less resources at the same time or an equal amount of resources at a later time. The *initiator* evaluates given options and sends a third counter offer which is finally accepted leading to a *Negotiated Offer* used by the *initiator* to create the offer.

2.2   Test Infrastructure

The load tests used two commodity servers providing four virtual machines as shown in figure 1. Each server was equipped with two Intel Xeon E5430 2.66 GHz CPUs (four cores per CPU) and 32 GB RAM. The nodes were connected via regular Gigabit Ethernet links and ran Linux (kernel version 3.2.0-57). Both nodes ran KVM virtual machines with two cores. Inside the virtual machines we used Ubuntu Linux 12.04 (kernel version 3.2.0-57) and Java 1.6.0.26 (OpenJDK). Tests were coordinated using the Java based load testing framework *The Grinder* in version 3.11 [3].

Host 1 provided *vm1* which ran Apache Tomcat 7.0.50 and served the WSAG4J web apps with a maximum of 2 GB heap space. To allow dedicated usage of available heap space only one of both apps (WSRF and REST) was deployed simultaneously. Host 2 provided *vm2*, *vm3* and *vm4* which executed the test runner component of the Grinder framework named *grinder-agent*. The three agents were coordinated by another host running the Grinder's *grinder-console* component which handled code distribution, test synchronization and collection of measurement results.

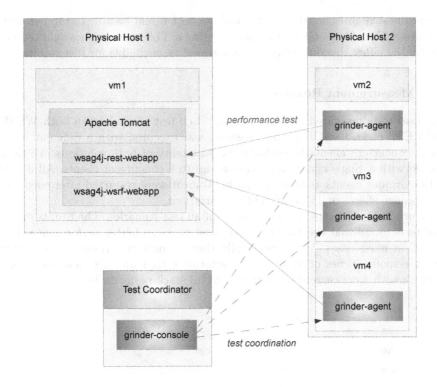

Fig. 1. Physical architecture of test environment

This infrastructure allowed short network paths avoiding biased results due to network issues while still being close enough to real life scenarios in which clients will always be located on different machines than the WSAG service.

2.3 Impact of Security Technology

In the context of SLA negotiation security features such as non-repudiation form the technological foundation for general feasibility and acceptance. We aimed to ensure a comparable level of trust for both approaches: WSRF and REST.

WSAG4J's WSRF based solution utilized WS-Security standards such as *BinarySecurityToken* and *XML signature* [18] [9] by default while the RESTful distribution shipped without adequate replacement. Therefore we chose to run all tests with HTTPS and replaced WSRF's security tokens with *TLS Client Certificates* which verify the identity of request senders. Because message payload was neither encrypted with WSRF nor REST by default we enabled TLS for both variants considering the sensitive nature of SLA agreement to protect communication from any kind of eavesdropping. Nevertheless this setup could not ensure message integrity if any intermediate host would be able to tamper with the message's content. Given that intercepting a TLS connection would require substantial effort this discrepancy was assessed as negligible for test results.

2.4   Measurement Results

Our load tests measured 200 test runs for each test scenario with both WSRF and REST code bases. Each test scenario was executed with a different number of concurrent clients to evaluate the scalability of both solutions. All tests started with a single client and increased up to 8 concurrent clients. All JVMs of the Grinder agents as well as Apache Tomcat were restarted after each run to minimize effects of JVM internal optimizations.

Figure 2, 3 and 4 show response times of all test scenarios. The results of our load tests reveal that the RESTful code base provides better performance than WSRF in most cases. More specifically there is only two results which show better response times of WSRF: *GetFactories* with 1 and 2 concurrent users. Starting with 4 concurrent users the RESTful stack performs better.

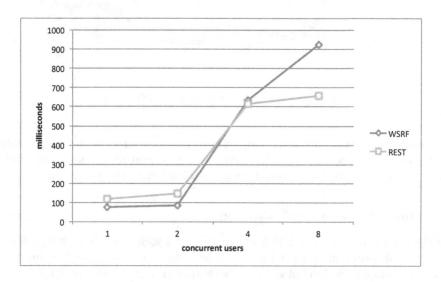

Fig. 2. Response times for GetFactories

For *GetTemplates* and the *Negotiation Scenario* results reveal lower response time of the RESTful approach in all cases. It is important to point out that while running the *Negotiation Scenario* an increasing number of test failures appeared with rising numbers of concurrent users. Using the WSRF stack the first failures appeared with 4 concurrent users and concerned already 80% of all tests while the RESTful stack showed 59% of failures under the same load. This is also the reason why figure 4 only reveals times up to 4 concurrent users. With more than 4

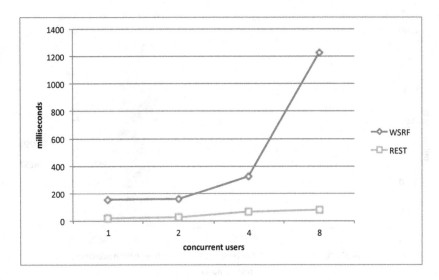

Fig. 3. Response times for GetTemplates

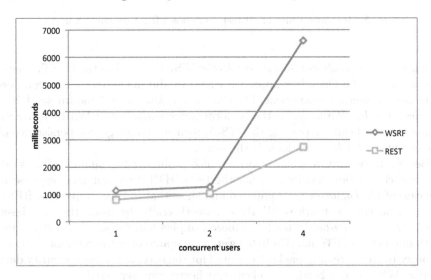

Fig. 4. Response times for NegotiationScenario

concurrent users the number of failures increased rapidly leading to unreliable measurement results.

Last figure 5 compares response times of all test scenarios proving increased complexity of the last scenario in terms of computation time.

Due to the modularity of WSAG4J, the implementation of functionalities for processing agreement offers, creating agreements, monitoring the service quality, and evaluating agreement guarantees is comprised in the SLA *Engine Module*

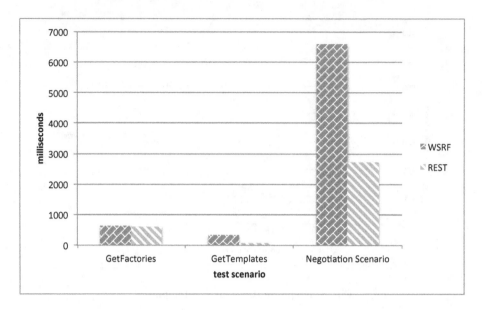

Fig. 5. Response times of all test scenarios with 4 concurrent users

which is used by both web service stacks the WSRF as well as the REST. Therefore we follow the black box approach where we did not separate between performance of internal components like the *Engine Module* and the frontend *Web
Service Modules*. We focus on the performance comparison of WSRF to REST
where the overhead for parsing the WS-Agreement language, for persistence of
agreements or for business logic is the same.

Besides measuring response times we also evaluated the amount of network
traffic each solution required during the tests. HTTP request and response in
the case of *GetFactories* are compared and show that in this sample case REST
required nearly one-tenth of WSRF's network traffic by using the very basic
media type *text/uri-list* instead of a more complex and verbose XML structure.
Both numbers of 3637 and 378 bytes were aggregated over the relevant payload.
In order to compare only payload required for the specific use case, security data
such as WS-Security headers or client certificates was neglected.

3   Related Work

The comparison of WS-* respectively SOAP and RESTful web services has already been performed by several scientists [13] [19] [16] [21] [17]. However, the
comparison of stateful approaches with the intention to include WADL in a WS-*
standard is still an open issue. Thus, the following papers either compared both
in different contexts or migrated applications between WS-* and RESTful approaches.

Pautasso et al. [19] compared both WS-* and RESTful web services from a conceptual and technological perspective and developed advices when to use which approach. They presented a general and comprehensive summary to support architectural decisions. In contrast, the focus of this paper is on a specific standard (WSAG) that requires stateful web services by presenting a performance comparison of both approaches.

Upadhyaya et al. [21] provided a semi-automatic approach to migrate existing SOAP based services into RESTful services and compared performance measurements of both solutions showing slightly better performance of REST based services. Compared to their work, our paper focuses on one single WS-* standard and compares already existing services rather than generating them which allows a more detailed inspection of both solutions.

Mulligan and Gračanin [17] developed a middleware component for data transmission offering both a SOAP and a REST interface. They evaluated their implementations with regard to performance and scalability requirements. Other than their work, this paper compares both approaches using very specific WS-* standards: WSRF and WSAG. We also evaluate the performance of both approaches with real life use cases from SLA contracting.

Kübert et al. [15] analyzed the WSRF based WSAG specification and designed a RESTful service with a feature set close to the standard. Their work proved that porting a WSAG service to REST is possible in theory but their scope ended with the proof of feasibility. We use an existing software framework and gain insights about performance gaps between both solutions. Based on their work as well as the existing RESTful implementation of WSAG4J this paper adds an evaluation of both approaches which has not been shown before.

4 Conclusion

In terms of performance it becomes apparent with an increasing number of concurrent clients that the RESTful stack of WSAG4J scales better than the WSRF based solution. In the given test infrastructure we could test the *GetFactories* case with 24 concurrent REST clients without running into failures while WSRF reported 50% failures with a number of 8 concurrent clients. These results are likely to be influenced by the amount of required network traffic which is significantly larger in the case of WSRF and therefore puts a higher load on the latter's serialization engine.

All tests were executed with HTTPS terminated by Apache Tomcat. To enhance the latter's TLS performance future work could include the *Apache Portable Runtime (APR)* [2] to enhance Apache Tomcat's TLS performance. We expect that enabling APR will reduce RESTful response times as the REST setup relies on Tomcat to verify and handle client certificates while WSRF uses its inbuilt logic to handle WS-Security tokens and would therefore profit less from enabling APR.

Finally we underline that web services providing WS-Agreement and WS-Agreement Negotiation act as neutral notaries which must by definition always be reachable to both agreement parties enabling 24/7 verification of SLAs.

As proved by our measurements the RESTful implementation of the WSAG4J framework scales better than the WSRF based solution and can therefore reduce operation costs and complexity when using WSAG4J for SLA negotiation and monitoring.

The second reason for the REST based solution is its enhanced interoperability compared to the WSRF variant. When providing a public WSAG service it is advisable to support as many different clients as possible. Because REST's technological footprint is lighter than the one of WSRF, it is open to more development environments possibly attracting a larger number of users.

As future work we see a higher investigation into specifying RESTful operations for WS-* specifications, especially for the WS-Agreement and the WS-Agreement Negotiation standards. This is also one of the hot discussed topics of the Grid Resource Allocation and Agreement Protocol Working Group (GRAAP-WG) of the Open Grid Forum (OGF).

Acknowledgment. The research leading to these information and results was partially supported by received funding from the European Commission's Competitveness and Innovation Programme (CIP-ICT-PSP.2012.5.2) under the grant agreement number 325192. The views and conclusions contained herein are those of the authors and should not be interpreted as necessarily representing the official policies or endorsements, either expressed or implied, of the MO-BIZZ project or the European Commission.

References

1. Web services resource 1.2 (ws-resource) (April 2006),
   http://docs.oasis-open.org/wsrf/wsrf-ws_resource-1.2-spec-os.pdf
2. The apache software foundation: Apache portal runtime (2013),
   http://apr.apache.org
3. The grinder, a java load testing framework (2013),
   http://grinder.sourceforge.net
4. Andrieux, A., Czajkowski, K., Dan, A., Keahey, K., Ludwig, H., Nakata, T., Pruyne, J., Rofrano, J., Tuecke, S., Xu, M.: Web services agreement specification (ws-agreement) (March 2007), http://www.ogf.org/documents/GFD.192.pdf (updated version 2011)
5. Blumel, F., Metsch, T., Papaspyrou, A.: A restful approach to service level agreements for cloud environments. In: 2011 IEEE Ninth International Conference on Dependable, Autonomic and Secure Computing (DASC), pp. 650–657 (December 2011)
6. Christensen, E., Curbera, F., Meredith, G., Weerawarana, S., et al.: Web services description language (wsdl) 1.1 (2001), http://www.w3.org/TR/wsdl
7. Comuzzi, M., Spanoudakis, G.: Dynamic set-up of monitoring infrastructures for service based systems. In: Proceedings of the 2010 ACM Symposium on Applied Computing, SAC 2010, pp. 2414–2421. ACM, New York (2010), http://doi.acm.org/10.1145/1774088.1774591
8. Curbera, F., Duftler, M., Khalaf, R., Nagy, W., Mukhi, N., Weerawarana, S.: Unraveling the web services web: an introduction to soap, wsdl, and uddi. IEEE Internet Computing 6(2), 86–93 (2002)

9. Eastlake, D., Reagle, J.: Xml signature (2000), http://www.w3.org/Signature/
10. Fielding, R.T.: Architectural Styles and the Design of Network-based Software Architectures. Ph.D. thesis, University of California (2000), AAI9980887
11. Foster, I., Czajkowski, K., Ferguson, D., Frey, J., Graham, S., Maguire, T., Snelling, D., Tuecke, S.: Modeling and managing state in distributed systems: The role of ogsi and wsrf. Proceedings of the IEEE 93(3), 604–612 (2005)
12. Gudgin, M., Hadley, M., Mendelsohn, N., Moreau, J.J., Nielsen, H.F., Karmarkar, A., Lafon, Y.: Simple object access protocol (soap) 1.2 (2002), http://www.w3.org/TR/soap/
13. Guinard, D., Ion, I., Mayer, S.: In search of an internet of things service architecture: Rest or ws-*? a developers perspective. In: Puiatti, A., Gu, T. (eds.) MobiQuitous 2011. Lecture Notes of the Institute for Computer Sciences, Social Informatics and Telecommunications Engineering, vol. 104, pp. 326–337. Springer, Heidelberg (2012), http://dx.doi.org/10.1007/978-3-642-30973-1_32
14. Hadley, M.J.: Web application description language (wadl) specification (2009), http://www.w3.org/Submission/wadl/
15. Kübert, R., Katsaros, G., Wang, T.: A restful implementation of the ws-agreement specification. In: Proceedings of the Second International Workshop on RESTful Design, WS-REST 2011, pp. 67–72. ACM, New York (2011), http://doi.acm.org/10.1145/1967428.1967444
16. Muehlen, M., Nickerson, J.V., Swenson, K.D.: Developing web services choreography standards the case of {REST} vs. {SOAP}. Decision Support Systems 40(1), 9–29 (2005), http://www.sciencedirect.com/science/article/pii/S0167923604000612 WS-REST 2011
17. Mulligan, G., Gracanin, D.: A comparison of soap and rest implementations of a service based interaction independence middleware framework. In: Proceedings of the 2009 Winter Simulation Conference (WSC), pp. 1423–1432 (2009)
18. Nadalin, A., Kaler, C., Hallam-Baker, P., Monzillo, R.: Web services security: Soap message security 1.0 (2004), http://docs.oasis-open.org/wss/2004/01/oasis-200401-wss-soap-message-security-1.0.pdf
19. Pautasso, C., Zimmermann, O., Leymann, F.: Restful web services vs. "big"' web services: Making the right architectural decision. In: Proceedings of the 17th International Conference on World Wide Web, WWW 2008, pp. 805–814. ACM, New York (2008), http://doi.acm.org/10.1145/1367497.1367606
20. Stamou, K., Aubert, J., Gateau, B., Morin, J.H.: Preliminary requirements on trusted third parties for service transactions in cloud environments. In: 2013 46th Hawaii International Conference on System Sciences (HICSS), pp. 4976–4983 (January 2013)
21. Upadhyaya, B., Zou, Y., Xiao, H., Ng, J., Lau, A.: Migration of soap-based services to restful services. In: 2011 13th IEEE International Symposium on Web Systems Evolution (WSE), pp. 105–114 (2011)
22. Wäldrich, O.: Orchestration of Resources in Distributed, Heterogeneous Grid Environments Using Dynamic Service Level Agreements. Ph.D. thesis, Technische Universität Dortmund, Sankt Augustin (December 2011)
23. Wäldrich, O., Battre, D., Brazier, F., Clark, K., Oey, M., Papaspyrou, A., Wieder, P., Ziegler, W.: Ws-agreement negotiation version 1.0 (March 2011), http://www.ogf.org/documents/GFD.193.pdf
24. Wäldrich, O., et al.: Wsag4j: Web service agreement for java, http://wsag4j.sourceforge.net, version 2.0, Project Website

Enabling Cross-Platform Mobile Application Development: A Context-Aware Middleware

Achilleas P. Achilleos and Georgia M. Kapitsaki

Department of Computer Science, University of Cyprus,
1 University Avenue, Nicosia, Cyprus
http://www.cs.ucy.ac.cy

Abstract. The emergence of mobile computing has changed the rules of web application development. Since context-awareness has become almost a necessity in mobile applications, web applications need to adapt to this new reality. A universal development approach for context-aware applications is inherently complex due to the requirement to manage diverse context information from different sources and at different levels of granularity. A context middleware can be a key enabler in adaptive applications, since it can serve in hiding the complexity of context management functions, promoting reusability and enabling modularity and extensibility in developing context-aware applications. In this paper we present our work on a cross-platform framework that fulfils the above. We elaborate on the need for cross-platform support in context-aware web application development for mobile computing environments identifying gaps in the current state of context support. The paper introduces the architecture of the middleware that fills these gaps and provides examples of its main components. An evaluation based on the development of a prototype, web-based, context-aware application is detailed. The application is compared against an analogous hybrid mobile application showing the evolutionary potential introduced via the middleware in delivering context-aware mobile applications.

Keywords: web applications, context middleware, context-awareness, HTML5, mobile computing.

1 Introduction

The growth of mobile devices capabilities offered by a variety of mobile platforms (e.g., iOS, Android, Windows Phone) increases the interest of users in personalized applications that can be accessed from any place and platform. This is reflected in a mantra coined in 1991: "*giving all the freedom to communicate anywhere, anyplace, anytime, in any form*" [1]. This vision is now more profound due to the explosion of hardware, software and communication technologies provided by an abundance of smart devices. Mobile users are best served if the above goal is fulfilled by keeping users requirements and their need for control in top priority, without implementing applications that overwhelm users with redundant information and futile functionality. Such a personalized technology view

B. Benatallah et al. (Eds.): WISE 2014, Part II, LNCS 8787, pp. 304–318, 2014.

has been adopted by various researchers to improve the user experience of web applications [2], [3]. Currently the personalized perspective has been enriched by the widespread presence of mobile sensors, such as GPS (Global Positioning System) receivers, accelerometers, compasses, microphones and cameras that are integrated in a mobile device [4], as well as the huge volume of social network data available from different locations.

In order to keep up with these advancements, technologies should adapt to the end-user's perspective and aid in tailoring applications to the surrounding context (i.e., location, situation, social data) exploiting the capabilities of smart devices and platforms. Realization of this smart vision demands mechanisms for ubiquitous and reliable acquisition, analysis and sharing of information to improve user requirements, by anticipating user needs while the user remains undisturbed by the underlying technology. Also, application developers that target those environments should have access to a transparent infrastructure and a set of reusable context modules that support cross-platform development of such context-aware applications.

Nowadays mechanisms and technologies that target cross-platform development in the field of web applications are emerging, e.g., HTML5 [5]. Nevertheless, a universal web application framework tailored to the needs of context-awareness is missing. In this work we elaborate on these issues presenting the current state on cross-platform web development and proposing a solution that facilitates the development of such applications providing software engineers with reusable and extensible elements. The contribution of this paper is twofold: on the one hand, it provides an analysis and review on the HTML5 support in mobile web browsers of widely-used mobile devices and, on the other hand, it gives on overview of the proposed HTML5 Context Middleware (H5CM) that serves as a facilitator for cross-platform development of context-aware web applications.

Regarding the former part of our contribution we have performed a study on HTML5 support provided by widely used mobile browsers in different mobile platforms. For the latter part we have exploited the results of this study in order to design and implement a generic context management middleware that relies on the HTML5 specification and the latest developments in mobile platforms and browsers. The H5CM framework provides reusable context sensing and reasoning modules in the form of plugins and offers the ability to extend this pool of plugins through the development of additional modules. Modules are provided in an hierarchical ordering. H5CM supports context-management functions through a common interface to multi-domain context sources for facilitating the development of web-based context-aware mobile applications.

The rest of the paper is organized as follows. Section 2 introduces the notion of context-awareness in mobile platforms settings that is acting as motivation for our work, whereas section 3 analyses the support of HTML5 features relevant to context-awareness on popular web browsers. We then describe in section 4 the main concepts of the H5CM framework. The framework's plugin concept is analysed through examples in section 5 along with the presentation of how the plugins are used in a mobile application. A comparison and evaluation of the context-aware application is performed in section 6, against an analogous mobile application developed using hybrid technology (i.e., PhoneGap). The final section concludes the paper.

2   Context-Awareness in Mobile Platforms

Context-awareness is a key enabling requirement for enhancing the ability of mobile users to exploit efficiently different software applications and services using different mobile devices (e.g., smartphones, tablets, and netbooks) at different locations. In mobile computing context-awareness encompasses various aspects spanning from sensing information at the hardware and network information level, to context-based recommendations at the application level, such as the case of context-based music recommendation presented in a previous work [6].

The popularity gained by different mobile platforms enables users to perform everyday tasks and use diverse devices for various work and leisure activities [7]. In such mobile settings, the user needs to be dynamically assisted, by tailoring the application and providing proactive behaviour. Mobile devices offer clear-cut user benefits, but there are application usability and ease of use limitations (e.g., small screen, low battery) that must be considered. Perhaps the key issue that needs to be addressed is this diversity of platforms even when only one user is considered, since users own and use different devices.

The term of context-awareness as already introduced describes the process of acquiring, managing and distributing different pieces of context to intelligently adapt the application behaviour. We adopt the definition of Dey and Abowd [8] which states that context is *any information that can be used to characterize the situation of an entity, in which the entity can be a person, a place, or a physical or computational object that is considered relevant to the interaction between the entity and the application.* Directly from this definition it is apparent that the requirement arises to be able to support diverse context sources via a modular, reusable and extensible mechanism. In specific, four driving requirements, namely modularity, extensibility, code reusability and cross-platform development, motivated us to define and develop a context-aware, web-based middleware to handle context-awareness support in different environments.

Comparable approaches that handle context data in different platforms can be found in the literature, such as MUSIC [9], [10] for the Java platform and Really Simple Context Middleware (RSCM) [11] for the Android platform. The essential difference of the current work with existing solutions lies on the application type focus, which is on the web development, rather than on native applications that are not universally exploitable in various environments. The former case of MUSIC refers to the results of a European project on Self-Adapting Applications for Mobile Users in Ubiquitous Computing Environments and its proposed middleware runs using Java OSGi. The latter is available in Google code and focuses on Android, allowing the development of context-aware applications using context plugins.

Differentiating from approaches that are tailored to specific platforms the main driving motivation of this work is platform-independence. This point has also been reflected in hybrid technologies that constitute a first attempt to satisfy the assortment of mobile devices and platforms. Examples of such technologies can be found in PhoneGap [12], Apache Cordova, i.e., the open-source engine that runs PhoneGap, and Titanium[1]. The above solutions provide cross-platform mobile development environments via a set of uniform JavaScript libraries that can be invoked by the developer, wrapping (i.e., calling) device-specific native backing code through these JavaScript libraries. This provides access to native device functions, such as the camera or accelerometer from JavaScript. The main criticisms against hybrid development environments is that the developer needs to learn how to use the native libraries for each platform, but most importantly that mobile devices are not fast enough to smoothly run a hybrid application [13]. On the other hand, native applications offer benefits in terms of performance and API coverage. Still native applications lack in terms of instant deployment, since they require manual installation and/or upgrades, and flexibility to combine data from different resources [14].

In contrast to existing solutions we propose to exploit the rapidly developing HTML5 capabilities of mobile browsers to define a middleware that supports cross-platform application development using only web technologies [15]. In this way, performance issues that are apparent in hybrid applications, and complex deployment issues that appear with native applications can be avoided.

3 HTML: Features and Browser Support

The current landscape of mobile application development can benefit from the new features of web technologies that provide interactivity and animation, and support cross-platform development. HTML5 is the W3C specification referring to the new version of HTML [5]. The first public working draft of the specification was made available in January 2008, while the specification is currently at a very mature stage and undergoes continuous development.

The question on whether HTML5 features are supported and can be exploited in miscellaneous platforms remains open and is an issue that is constantly evolving. In this section we present the results of a study performed on different mobile platforms, examining their support for HTML5 features. The comparison was performed with the aid of a simple prototype application that we developed for this purpose. While the survey provides also data on the general support of HTML5 features, the focus of this work is on browser APIs that support features with a context-awareness flavour, such as mobile device geolocation tracking and monitoring, device orientation, motion and acceleration, monitoring battery level, ability to connect to RESTful and SOAP Web Services, etc. The main conclusion drawn is that mobile and tablet browsers support most features including context-relevant ones, since vendors try to keep in line with the evolving HTML5 specification. Moreover, as new sensors are added to mobile devices additional features are continuously added by browser vendors, by implementing new APIs in different browsers.

---

[1] http://www.appcelerator.com/titanium/

Table 1. Browser Support - [Samsung Galaxy Note 3 running Android 4.4.2 - Also tested on S3, Tab 2], [Apple iPad 2 running iOS 7.1.2: Also tested on iPhone 4S]

Name	Default (Android)	Dolphin Browser HD 10.3.1 (Android)	Firefox Mobile 31.0 (Android)	Opera Mini 7.5 (Android)	Opera Mobile 22.0 (Android)	Chrome Dev 36.0.1985.128 (Android)	Safari Browser - Default (iOS)	Dolphin Browser 6.5.1 (iOS)	Opera Mini 8.0.1 (iOS)	Chrome Browser 36 (iOS)	Mercury Browser 8.6.1 (iOS)
Popularity (Max. 5)	-	4.5	4.5	4.4	4.5	4.5	-	4	-	4	4.5
Score (Max. 555)	475	463	483	53	486	492	410	410	410	410	410
Device Orientation	✓	✓	✓	x	✓	✓	✓	✓	x	✓	✓
Device Motion	✓	✓	✓	x	✓	✓	✓	✓	✓	✓	✓
Geolocation	✓(x)	✓(x)	✓	x	✓	✓(x)	✓	✓	✓	✓	✓
Battery Status	x	x	✓	x	x	x	x	x	x	x	x
Device Media	x	✓(x)	x	x	✓	✓(x)	x	x	x	x	x
Device Network	✓	✓	✓	✓(x)	✓	✓	✓	✓	✓	✓	✓
Web Sockets	✓	✓	✓	x	✓	✓	✓	✓	x	✓	✓
RESTful services	✓	✓	✓	✓	✓	✓	✓	✓	✓	✓	✓
SOAP services	✓	✓	✓	✓	✓	✓	✓	✓	✓	✓	✓

Legend - Supported: ✓, Not supported: x

The growing collection of HTML5 features and the ability to display 3D graphics natively in a browser, i.e., via WebGL that is the key standard supported by many browsers [16], make cross-platform development of applications a tangible target. The above table showcases the features considered critical to enable context-awareness. A score is shown as calculated by the website html5test.com in respect to browser support coverage for all HTML5 features. Popularity is also shown based on available user reviews and the total number of downloads, as retrieved for each application from the respective platform store.

In particular, Table 1 presents the support in different browsers for the Android and iOS platforms. Note that the values shown in parenthesis in the table indicate the features that are not supported on the Galaxy Tab 2 tablet, which are though supported on smartphones Note 3 and Galaxy SIII. In overall, smartphone and tablet devices provide support for the HTML5 features (in five out of six main browsers) up to a level of 86.8%, while most of the key features that enable context-awareness are supported. Only the Battery Status and Device Media APIs are currently lacking support in most browsers, with browsers like Firefox Mobile 31.0 and Opera Mobile 22.0 providing support also for these features. In this regard an application that needs to support, e.g., Battery Status, can be executed at least on Firefox Mobile 31.0 for Android.

The iOS platform and the browsers available for iOS offer high support in terms of general HTML5 features, up to the level of 68.8%. At the same time support for context-aware specific features is also at the highest level with 7 out of 9 features being supported through APIs provided by the different browser vendors. In particular, only two features are not supported and these refer again to the Battery Status and Device Media APIs. On the basis though of the continuous and progressive strive to implement new APIs (especially in terms of media support) we strongly believe that support for these features will soon arrive also on the browsers running on the iOS platform [17].

Finally, HTML5 support was also tested for the MS Windows Phone platform on the Nokia Lumia 800 mobile device running Windows Phone 7.1. A detailed presentation is omitted due to the low support encountered in this case in terms of HTML5 features. In specific, the support was exceedingly limited for context-aware features, with mainly the Device Network and Geolocation APIs being supported in some of the five browsers (i.e., Explorer, UC Browser, Explora, MetroUI and Incognito) that were considered in the test. Moreover, the support of generic HTML5 features was approximately 25% for the MetroUI browser, while support in all other browsers was basically non-existent. In the future we aim to test HTML5 support in browsers, as soon as they become available in newer mobile devices that support the Windows Phone 8 platform or higher.

4   HTML5 Context Middleware

Having in mind the necessity of providing cross-platform access to context-aware elements we have designed the HTML5 Context Middleware to allow developers to reuse existing modules making their code more compact. The implementation of the H5CM takes into consideration the four basic requirements of modularity, extensibility, reusability and the "develop once deploy on any platform" approach. H5CM follows a hierarchical structure: at the lower level there are context-sensor modules (s-m) that allow acquiring and distributing low-level context information that is relevant to the mobile device, the end-user and the environment. At the second level of the hierarchy there are context-reasoner modules (r-m) that accept low-level context from one or more sensor modules and apply the appropriate reasoning logic so as to create high-level context information. The context-aware application is at the top of the hierarchy and is

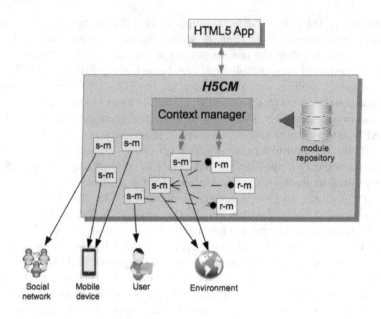

Fig. 1. Main elements of the HTML5 Context Middleware

able to communicate at the software level with sensor and/or reasoner modules to acquire context information that enables the adaptation of the application's logic.

The key architectural elements of H5CM are illustrated in Fig. 1. The *Context manager* constitutes a central point that handles information originating from different sources as gathered by respective plugins and distributes this information accordingly to the context-aware applications that require it. These modules (or plugins) act as enablers of context-awareness, empowering applications to be adapted to end-user preferences and circumstances. The plugins cover four main categories:

- modules that allow context acquisition directly from the mobile device (e.g., battery level)
- modules that provide information retrieved from social networks (e.g., LinkedIn, Delicious)
- modules that obtain input from the end-user, and
- modules that obtain information from the environment connecting to local or remote sensors or servers (e.g., room temperature)

In specific, the *Context manager* orchestrates the management of modules that retrieve information from the environment, the user and the device. The *Context manager* is implemented as a singleton JavaScript web module, while the context plugins are also implemented as JavaScript web modules. Hence, the context-aware application can be also implemented using web technologies (i.e.,

HTML5, CSS, JavaScript). Through the *Context manager*, context modules (i.e., sensors, reasoners) required for an application are registered and loaded. In fact, H5CM and the *Context manager* are based on the popular Q JavaScript framework[2], which offers the Promise object and complies fully with the Promises/A+ Specification[3]. A promise represents the result of an asynchronous operation. Interacting with a Promise object through its *then* method allows registering callbacks and receiving a fulfilled promise return value or the reason for failure. Using this approach it is guaranteed that the context plugins required by the application are loaded successfully.

At this point the application is able to interact and communicate with the loaded context plugins (e.g., GPSCoordinates, BatteryLevel). The loaded plugin(s) is(/are) actually registered with the *ContextManager* singleton, which initializes the associated context property and creates a new *CustomEvent* assigning the name of the context property as the unique identifier for the event. The context-aware application defines an *EventListener* associated with the above context event, which allows monitoring and updating the application with relevant context changes. The context plugin is responsible to update the context property and dispatch a unique event, when new context is available.

The above functionality is accompanied by an extensible and reusable *Context Repository*. The modules define an extensible and reusable repository. On the one hand, they can be reused by developers, since they are generic and can be invoked from any context-aware web application. On the other hand, the module set can be extended by technical users that need additional functionality as features of HTML5 expand and the respective support in mobile browsers is extended. The modules to be loaded are defined as dependencies in the application. Since the values of context data change more or less frequently depending on the type of the data, H5CM is responsible for monitoring all context variables and updating the application when relevant context changes are detected. This functionality offers a context monitoring that allows the adaptation of the application while it is executing.

As aforementioned, in order to provide context information at different levels of granularity a concept introduced in H5CM is the differentiation between sensor- and reasoner-modules. The former provide access to raw context data in the form that these can be collected from context locations (e.g., user geographical coordinates), whereas the latter give access to higher level context information that are derived from the "basic" raw data and are meaningful to the end-user. As a reasoner example consider the use of the raw data provided by: 1) the battery status that actually comprises of the battery level of the device and the information on whether the device is actually charging, 2) the coordinates of the user location and 3) the device motion, to provide the information of whether the user is currently walking on a street or driving. Application developers are able to contribute at this level by developing additional reasoner-modules that aggregate content from different context-sensor modules.

---

[2] Q Promises JavaScript Framework - http://documentup.com/kriskowal/q/.

[3] Promises/A+ Specification - http://promises-aplus.github.io/promises-spec/.

5   Context-Aware Sensor and Reasoner Modules

In this section we present the H5CM in more detail giving information on how to make use of the context-middleware modules to build context-aware applications. The implementation of the H5CM is available at Google Code[4], which includes the context manager web component and the available context sensor and reasoner modules. The presentation of these context modules serves as a guideline for developers on how to develop additional context-aware sensor and reasoner modules, enriching thus the functionality of the middleware, but also how to use these modules for the implementation of context-aware applications. At the time we have developed 8 sensor and 2 reasoner plugins that constitute the current state of the middleware. We present for demonstration purposes in more detail the *BatteryLevel* and the *FacebookConnect* sensor modules along with the *BatteryAnalyzer* reasoner module.

Listing 1. *Battery Level Sensor Module.*

```
1.  function getBatteryLevel() {
2.     var singletonContextManager = ContextManagerSingleton.getInstance("batteryLevel");
3.     var battery = navigator.battery || navigator.mozBattery || navigator.webkitBattery;
4.     if (battery != undefined) {
5.        var bLevel = Math.round(battery.level * 100);
6.        battery.addEventListener("levelchange", updateBatteryStatus);
7.        return bLevel;
8.     }
9.     . . . . .
10. function updateBatteryStatus() {
11.    var bLevel = Math.round(battery.level * 100);
12.    singletonContextManager.batteryLevel.batteryLevel = bLevel;
13. }
```

The *BatteryLevel* sensor-module retrieves and continues monitoring the level of the device battery (e.g., 74%), as required by the context-aware application. This is a hardware specific sensor plugin, while the *FacebookConnect* is a social sensor plugin that provides the means to acquire profile data of the user (i.e., public user data). It also allows asking the user to grant access for retrieving additional restricted data (given in comma separated string values). As a prerequisite the user of the mobile, web application needs to provide his/her Facebook credentials and allow the application to access these restricted data. *FacebookConnect* sensor module provides the connection point to Facebook, but similar functionality can be achieved by implementing modules for additional social networks, such as Google+, LinkedIn or Delicious. Note that security and privacy of user data is supported and respected through the use of advanced mechanisms provided by the different social networks. Moreover, hosting and deployment of the context-aware application on a secure web server (e.g., Apache) provides the capability of layering the Hypertext Transfer Protocol (HTTP) on top of the Secure Sockets Layer (SSL) protocol, thus adding the security capabilities of SSL to standard HTTP communications.

Listing 1 presents part of the implementation of the battery level sensor module that allows at first to register this plugin to the Singleton *ContextManager* instance (i.e., line 2) that in turn attaches this context to the *ContextListener*. Following, the

---

[4] https://code.google.com/p/h5cm/

battery is accessed by using one of the three web engines powering different browsers (i.e., line 3) and the level of the battery is obtained, while an event listener allows monitoring changes to battery level (i.e., lines 4-8). As soon as battery level change event is fired the *updateBatteryStatus* function is invoked (i.e., lines 10-13) that updates and distributes via the singleton *ContextManager* instance the new battery level to the application or reasoner depending on the application scenario.

Listing 2. Invoking Battery Level Sensor and Reasoner modules.

```
1.  var cms = ContextManagerSingleton.getInstance("application");
2.  . . . . .
3.  batteryLevel = getBatteryLevel();
4.  document.getElementById("batteryLevel").innerHTML = batteryLevel;
5.      reasoning = getBatteryAnalyser(parseInt(batteryLevel),30);
6.      . . . . .
7.  var coords = getGPSlocation();
8.  cms.gpsCoordinates.watch("gps", function (id, oldval, newval) {
9.      . . . . .
10. if (reasoning == false){
11.   document.getElementById("gps").innerHTML = "Latitude: " + myLat + " Longitude: " + myLng;
12. }
13. else if (reasoning == true){
14.   document.getElementById("gps").innerHTML = "Latitude: " + myLat + " Longitude: " + myLng;
15.   mapThisGoogle(myLat,myLng);
16. }
17. return newval;
18. });
19. . . . . .
20. var fbData =new Array("name","email","gender","username","birthday");
21. getFBInfo(fbData);
```

Based on the battery level we have defined a module that allows reasoning and adapting the application. The developer is allowed to set a cut off value for the battery level that defines if the application should restrict its functionality to preserve resources. The example application code shown at line 3 (Listing 2), allows invoking firstly the *BatteryLevel* sensor module presented in Listing 1. Then as shown in Listing 2, based on the value detected for the battery level the *BatteryAnalyser* reasoning module (see Listing 3) is invoked, which takes the final decision by comparing the two values and returning the result in the form of a boolean variable. In the example demonstrated this allows to make a decision (lines 7-16), when the location tracking sensor module is called, whether to show the location of the user using a text-based representation or using Google Maps.

Listing 3. Battery Analyser Reasoner Module.

```
1.  function getBatteryAnalyser(batteryLevel, batteryLevelCutoff){
2.      var singletonContextManager = ContextManagerSingleton.getInstance("batteryAnalyser");
3.      if (batteryLevel>=batteryLevelCutoff) {
4.          return true;
5.      }
6.      else if (batteryLevel<batteryLevelCutoff) {
7.          return false;
8.      }
9.  }
```

Listing 4 presents a small part of the sensor module that allows connecting to Facebook. In particular, after user credentials are validated and the permissions to access the profile data are confirmed, the Facebook API is invoked for retrieving these data.

Then via the singleton *ContextManager* instance the application is notified that the profile data have been retrieved and the web application needs to only parse and display these data. This specific module can be invoked from a context-aware web application as shown in lines 20-21 of Listing 2.

Listing 4. Facebook Connect Sensor Module.

```
1.      function getFBInfo(data) {
2.          var singletonContextManager = ContextManagerSingleton.getInstance("facebookInfo");
3.          FB.api('/me', function(response) {
4.          for (i=0;i<data.length;i++){
5.              var data_next = data[i];
6.              if (profile){
7.                  var profile = profile + "@%@" + response[data_next];
8.              }
9.              else{
10.                 var profile = response[data_next];
11.             }
12.         }
13.         singletonContextManager.facebookInfo.facebookInfo = profile;
14.     });
15.     }
```

For the development of these modules we used HTML5 APIs to implement context-sensor plugins that enable access to device-specific hardware information, along with additional APIs that enable acquisition of network-specific information from RESTful and SOAP Web Services. Moreover, we have exploited APIs of social networks to define web-based context-sensor modules that allow retrieving the user profile and other data. Reasoners are also defined that allow acquiring low-level information from context-sensors and applying reasoning logic to generate high-level information.

6   Use Case Evaluation

As a use case for our evaluation we have chosen to implement a typical context-aware application: the restaurant finder application. Usually restaurant finder applications rely on the current user position and propose nearby restaurants allowing users to view the location of the restaurant, browse restaurant reviews and consult opening hours. Our version of *My Restaurant Out There* offers similar features in a context-aware fashion utilizing information collected through a number of context modules. Specifically, the application utilizes the following information:

– The location of the user that is used to return restaurant options in the vicinity of the user,
– The battery level of the device based on which it can either display the returned results in a tabular format (e.g., for battery level < 30%) or on Google maps that offer a resource-demanding alternative (e.g., for battery level >= 30%).

Existing sensor-module plugins from H5CM are used to offer the above functionality to the restaurant finder application. The *BatteryLevel* is used to understand the battery level and status of the device, whereas *Geolocation* gives the current user coordinates or the city the user is located in. Regarding the battery use, instead of using directly the respective sensor-module the *BatteryAnalyser* reasoner presented in the previous section is called, since it offers a finer representation of the context value returning true or false based on the battery level.

Fig. 2. Execution of My Restaurant out There

For evaluation purposes we have implemented two variants of the restaurant finder application: the first variant utilizes the proposed framework, whereas the second is based on one of the most popular framework for hybrid technologies PhoneGap. The latter variant was implemented for Android. Screenshots of the two applications under execution are shown in Fig. 2 with the left side indicating the list of restaurants as returned from the web applicaiton on FireFox (that supports the battery status feature of HTML5) for low battery level levels (i.e., 6%) and the right side depicting a map of restaurants as returned by the hybrid application for higher battery level (i.e., 81%).

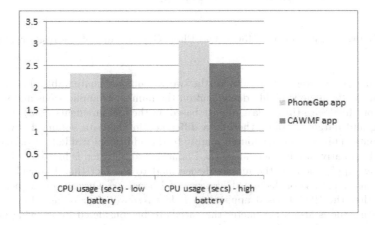

Fig. 3. CPU usage for the H5CM and PhoneGap variants of the restaurant finder application

Since one of the driving forces of the design of the proposed framework was the minimization of resources required from the mobile device we have measured the following parameters for the two variants: (i) CPU usage and (ii) network traffic usage including both outgoing and incoming traffic. For the evaluation a Samsung galaxy S3 device with Android version 4.3 was employed. The measurements were based on the "show CPU usage" feature of Android (available from Developer options) that depicts among other information the CPU usage in the last minute and the Traffic Monitor Android application[5] that provides traffic data information per application for different time intervals including measurements for the current date. We have measured the approximate values for each metric when Firefox or the restaurant finder hybrid application is running on the device, along with a number of Android applications or services (roughly 9). The results are displayed in Figs. 3 and 4 respectively. Note that for network traffic the diagram indicates the traffic measured only for the case when the application is running on a device with low battery level. We have omitted the high battery level case, since the traffic is in that case highly dependent on the interaction of the user with Google maps (zooming etc.) and no concrete conclusions could be drawn.

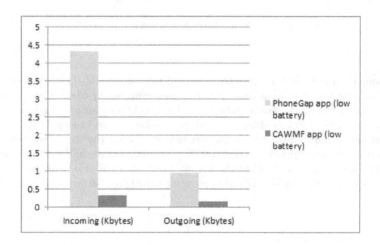

Fig. 4. Network traffic for the H5CM and PhoneGap variants of the restaurant finder application

Regarding CPU usage the results for the two variants are comparable. This is plausible, since CPU usage is highly dependent on the number of applications and services running on the device (in the last minute based on the measurements used). For both incoming and outgoing traffic there is a difference in the data volume between the two variants. This may be attributed mainly to the additional traffic that needs to be handled by a native application, since the PhoneGap application is in essence installed as a native application on the mobile device, and to the fact that PhoneGap-based applications make use of dedicated libraries of PhoneGap. Overall, our experiments indicate that the H5CM-based application is less resource-demanding. However, the application scale is also a parameter that needs to be considered for drawing general conclusions, since the application use case is rather simplistic.

---

[5] https://play.google.com/store/apps/details?id=com.radioopt.tmplus

From the above initial proof of concept and the description of the proposed procedure many benefits can be observed from the developer side. These are mainly found in the elements of modularity, extensibility, code reusability and cross-platform development provided by the H5CM as introduced earlier in the paper. An additional benefit can be found in an increase in code quality, since the application design based on the framework modules produces more compact and comprehensible code, an essential element also for the phase of software maintenance.

7 Conclusions

Cross-platform support in the development of context-aware applications is a desirable feature to simplify application development. In the field of web applications the most popular choice towards this direction is HTML5. Support for HTML5 features on mobile web browsers of popular platforms is necessary in order to take advantage of access to context data as facilitated by HTML5 constructs making thus the user experience adaptive and personalized.

In this paper we have performed a review on the support of HTML5 features on widely used mobile browsers, observing that many features are indeed supported. This tendency will strengthen as the HTML5 specification evolves to embrace the emerging needs of mobile users, while the modular architecture of the HTML5 Context Middleware allows supporting new HTML5 features as soon as they become available, by implementing additional modules. Moreover, we have introduced our context-aware framework, namely the H5CM, that alleviates developers workload and puts application development for different mobile devices in a common perspective by providing a selection of plugins that make the access to context data seamless. Context data are also monitored in order to reflect changes to the actual application. Also, the evaluation served as an initial proof of concept and showcased that HTML5 context-aware applications designed using the H5CM are less resource-demanding than mobile applications developed using hybrid technologies; especially in terms of network load.

As future work we intend to expand the library of plugins available on Google Code. This will enable developers to use additional context-aware features and thus reduce the time in developing context-aware applications. We will investigate also the integration of additional security and privacy guarantees in the use of H5CM, on top of social networks guarantees and the use of an HTTPS-enabled web server, since protection of user-relevant data is vital in context-awareness.

References

1. Weiser, M.: The Computer for the 21st Century. Scientific American (September 1991)
2. Rodden, K., Hutchinson, H., Fu, X.: Measuring the user experience on a large scale: user-centered metrics for web applications. In: Proc. SIGCHI Conference on Human Factors in Computing Systems, pp. 2395–2398 (2010), doi:10.1145/1753326.1753687
3. Rossi, G., Schwabe, D., Guimarães, R.: Designing personalized web applications. In: Proc. 10th Int'l Conf. World Wide Web, pp. 275–284 (2001)
4. Lee, Y., Ju, Y., Min, C., Yu, J., Song, J.: MobiCon: A Mobile Context-Monitoring Platform. Communications of the ACM 55(3), 54–65 (2012), doi:10.1145/2093548.2093567.

5. Berjon, R., Faulkner, S., Leithead, T., Navara, E.D., O'Connor, E., Pfeiffer, S., Hickson, I.: HTML5 - A vocabulary and associated APIs for HTML and XHTML. W3C Candidate Recommendation (February 2014),
   http://www.w3.org/TR/html5/
6. Han, B.-J., Rho, S., Jun, S., Hwang, E.: Music emotion classification and context-based music recommendation. Multimedia Tools and Applications 47(3), 433–460 (2010), doi:10.1007/s11042-009-0332-6.
7. Tarkoma, S., Lagerspetz, E.: Arching over the Mobile Computing Chasm: Platforms and Runtimes. IEEE Computer 44(4), 22–28 (2011)
8. Dey, A.K., Abowd, G.D.: Towards a Better Understanding of Context and Context Awareness, Workshop: What, Who, Where, When, and How of Context Awareness. In: ACM Conf. Human Factors in Computer Systems, The Hague, Netherlands (2000)
9. Paspallis, N., Achilleos, A., Kakousis, K., Papadopoulos, G.A.: Context-aware Media Player (CaMP): Developing context-aware applications with Separation of Concerns. In: IEEE Globecom 2010 Workshop on Ubiquitous Computing and Networks (UbiCoNet), Miami, Florida, USA, December 6, pp. 1741–1746 (2010)
10. Floch, J., Fr, C., Fricke, R., et al.: Playing MUSIC building context–aware and self–adaptive mobile applications. Journal Software: Practice and Experience (2012)
11. Ioannides, F., Kapitsaki, G.M., Paspallis, N.: Demo: Professor2Student – connecting supervisors and students. In: Daniel, F., Papadopoulos, G.A., Thiran, P. (eds.) MobiWIS 2013. LNCS, vol. 8093, pp. 288–291. Springer, Heidelberg (2013)
12. Wargo, J.M.: PhoneGap Essentials: Building Cross-platform Mobile Apps. Addison-Wesley Professional (2012)
13. Gai, D.: Hybrid VS Native Mobile Apps (2013),
    http://www.gajotres.net/hybrid-vs-native-apps/
14. Mikkonen, T., Taivalsaari, A.: Reports of the Web's Death Are Greatly Exaggerated. IEEE Computer 44(5), 30–36 (2011)
15. Mikkonen, T., Taivalsaari, A.: Apps vs. Open Web: The Battle of the Decade. In: Proc. 2nd Annual Wksp. Software Engineering for Mobile Application Development (2011)
16. Khronos Group, WebGL Specification, Editor' s Draft (2011),
    http://www.khronos.org/registry/webgl/specs/latest/
17. Melamed, T., Clayton, B.: A comparative evaluation of HTML5 as a pervasive media platform. In: Mobile Computing, Applications, and Services, pp. 307–325. Springer, Heidelberg (2010)

GEAP: A Generic Approach to Predicting Workload Bursts for Web Hosted Events

Matthew Sladescu[1,2], Alan Fekete[1,2], Kevin Lee[2,3], and Anna Liu[2,3]

[1] The University of Sydney, Australia
[2] National ICT Australia
[3] The University of NSW, Australia
{firstname.lastname}@nicta.com.au

Abstract. A number of recent research contributions in workload forecasting aim to confront the challenge facing many web applications of maintaining QoS in the face of fluctuating workload. Many of these demonstrate good prediction accuracy for periodic and long-term workload trends, but they exhibit poor accuracy when faced with predicting the magnitude, profile, and time of non-periodic bursts. It is such workload bursts that have been known to bring down numerous e-commerce and other web-based systems during events like online sales, as well as product, and result announcements. In this paper, we leverage the implicit link that often exists between such events and workload bursts, and we contribute: a generic approach that can make use of a given event's definition to forecast the time, magnitude and profile of the event's associated workload burst; a burst prediction accuracy metric for evaluating the efficacy of burst prediction methods; and an evaluation to showcase the generic applicability of event aware prediction across multiple domains, using real workload traces from three different domains.

Keywords: Performance, Cloud, Burst Prediction, E-Commerce, Event.

1 Introduction

Many events like product announcements, online sales, sporting events, and elections are often accompanied by workload bursts that have on numerous occasions led to system unavailability and reduced web site QoS. Not only can these effects lead to lost revenue, but they can also negatively impact on a web site's reputation. In these scenarios, cloud computing opens up multiple opportunities for those application providers who can accurately forecast the workload associated with their events, by allowing them to determine: when, how many, and for how long cloud resources should be acquired for an event in order to maintain QoS at minimal cost. Pre-provisioning the right amount of resources in time to handle application workload surges, using accurate resource demand predictions, is especially beneficial in avoiding the detrimental effect [1, 2] that resource initialization lag [3–5] can have on cloud application SLAs. Further, foreknowledge of future demand can help application providers in deciding how to cost-effectively reserve (http://aws.amazon.com/ec2/reserved-instances/) resources. The opening examples of events in this introduction, all point to an important link [6] that associates bursts with events. Earlier work [6] illustrates this implicit event-burst link with

B. Benatallah et al. (Eds.): WISE 2014, Part II, LNCS 8787, pp. 319–335, 2014.

further cases of how events can induce bursts, and is useful in answering "user-directed questions" like "Can the broadcast of a TV show impact search" (in terms of bursts associated with TV show events); but does not provide a direct means to predict the time, magnitude and profile of workload bursts associated with future events. Since events are often triggers for bursts, knowledge of how event characteristics influence associated bursts can be instrumental in providing accurate workload forecasts to address the benefits listed above for a cloud computing context, and can also enable event organizers to engineer their event characteristics to bring them to a desired level of popularity.

This paper forms part of ongoing work [7, 8] on Event Aware Prediction (EAP) and provides a *generic* method to exploit the implicit event-burst link, essentially allowing translation from a given definition (Section 2.1) of an upcoming event to a prediction for that event's associated workload burst. We build on the concepts developed in [7] and enhance them by: (i) Developing a generic EAP approach (GEAP) that is not specific to any domain, but can be re-used to predict event-associated burst profiles for any event type that conforms to the assumptions described in Section 2.5. We aim to show how this re-use is achieved for three different event types, using real workload from auctions, sporting, and product announcement events. (ii) Introducing a new type of forecast correction for GEAP predictions based on past actual/forecast observations. (iii) Showing how similar events can more effectively be identified through partial event feature matching instead of exact feature matching; (iv) as well as how prediction horizon length (short term vs long term prediction) affects GEAP accuracy.

Numerous recent contributions like [9–11] have been put forward to advance the state of workload prediction. Many of these show promising results in predicting periodic and long term trends in workload, however, the sole use of conventional machine learning techniques like neural networks and SARIMA have been shown to be ineffective for workload burst prediction, where many methods treat bursts as outliers during their training phase [7, 12]. Consequently, dedicated workload burst prediction techniques [7, 12] have been introduced, some of which can only predict burst location [12], not magnitude. Unlike all other prediction methods mentioned thus far, [7] did not ignore the implicit link that exists between events and accompanying workload bursts and exhibited superior burst prediction accuracy to the state of the art in [7]; however it's implementation was specific to auction events. We contribute: (a) a generic event aware prediction approach, which can predict the time, magnitude, and profile of event-associated bursts for many different event types; (b) a burst prediction error metric for evaluating burst location *and* magnitude prediction accuracy (Section 3); and (c) an evaluation using real workload traces (Section 4) to assess the performance of GEAP.

2   Generic Event Aware Prediction

GEAP aims to accurately forecast the time, magnitude, and profile of workload bursts associated with future events. Each event is defined as "a scheduled or unscheduled occurrence, (like an online sale or a breaking news event); which can be described by an arbitrary set of features, (like the event starting and ending times); and can induce load for associated resources (like the web pages or database entries associated with a popular online sale)." (adapted from [7]). Unscheduled events refer to un-planned occurrences like breaking news events, which can induce workload bursts for infrastructure hosting information about such breaking news. Earlier work [13, 14] has shown how real-time updates from social media sources like Twitter can be used to detect such unscheduled events. In particular [14] demonstrates how 'sensor' classifiers can be used to detect given target events like the occurrence of earthquakes. In future work, we plan to investigate how such detections can be used to create structured event definitions that can be used with GEAP to predict unscheduled event-associated bursts. A "scheduled" event, which we focus on in this paper, refers to a pre-planned event for which characteristics (eg: event starting time, product being announced at a product announcement event) are known before the event begins. Numerous real-world examples of crashes associated with events hosted by ticketing, result announcement, deal of the day, and auction websites, referenced in [7], illustrate how challenging the management of such *pre-planned* events is for companies who depend on them to conduct their business, as well as for maximizing their revenue and reputation. For application providers hosting future events, GEAP can be used to predict the workload that their application will experience during these events. To achieve this, GEAP learns what the workload looked like for past events with a certain set of characteristics, (described in a repository of past events (Section 2.1)), in order to predict what the workload will look like for current or future events, (described in a repository of upcoming events (Section 2.1)), that have similar characteristics to events that occurred in the past. Information learnt about how workload fluctuates during events with different characteristics is captured by the Workload Burst Model (Section 2.2). Hence, the inputs used by the prediction module (PM), shown in Figure 1, to form a prediction include the repositories of past and upcoming events, and the workload burst model. Conceptually, the PM uses these inputs to *translate* from the *definition* of an upcoming event, to a *prediction* for that event's associated workload. We first discuss the main PM inputs, and then describe how the PM uses these inputs to predict workload.

2.1 Event Repositories

Two of the main GEAP PM inputs shown in Figure 1, are the Past and Upcoming Event Repositories. These repositories each contain a list of events, where each event can be characterized by a set of time features and secondary features, as described in [8]. In summary, the time features represent important times during an event's lifetime, (where these times are usually in close proximity to important workload characteristics eg: the announcement time for a product announcement event is close to the peak of the associated traffic burst of people wanting to view the product; and as observed in the soccer World Cup data set [15], many people check the score for a soccer match close to the ending time of the event). Figure 2 shows an example of the time features associated with product announcement event types. Secondary features, (like the category of an auction for an auction event), represent any other event characteristics that can affect the profile of an event's associated workload burst. Examples of the time and secondary features used to define the events in our evaluation are shown in Table 1. We call events that can be described by the same types of time and secondary features as being events of the same *type*, where each row in Table 1 represents a different event type. To enable a generic EAP approach, we permit the addition of any type of event that can be described

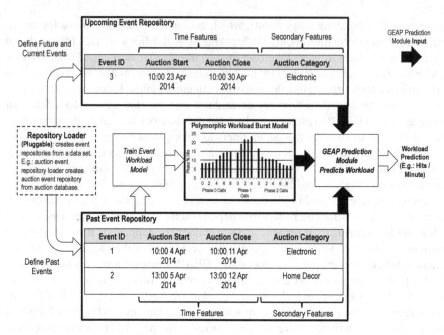

Fig. 1. GEAP Process and Main Inputs

by at least one time feature, and any number of secondary features into the repositories of past and upcoming events. As shown in Figure 1, a repository loader is responsible for determining the types of features that are important to the types of events that they are loading. Examples of repository loaders can include an auction event repository loader, a product announcement event repository loader, and a sporting event repository loader, where such repository loaders can be used to load the types of events shown in Table 1. Notice that we can be flexible in how we define events eg: in Table 1, we define three different versions of a sporting event, where each version corresponds to a different number of soccer matches that occur in a single day. As we will see in Section 2.2, when training a workload burst model, this approach can be useful when bursts for two separate matches that have occurred on the same day overlap and make it difficult to distinguish the traffic attributed to each individual soccer match.

2.2 The Workload Burst Model

As shown in [8], the event associated workload burst profile can vary not only between the *different* event types that are loaded into event repositories, but also between events of the same type. Hence, instead of trying to identify a common workload burst model for all event types, (and attempting to adopt a model like [16, 17]), we use a common *process*, (as indicated by the "Train Event Workload Model" step in Figure 1), that can learn the workload burst profile for each event type based on history from the past event repository. This process will not only learn how the event type affects the workload burst profile, but also how event feature values for events of the same type can influence

Table 1. Examples for Time Features and Secondary Features (adapted from [8])

Event Type	Time Features		Secondary Features
Product Announcement Event	- Announcement time	Derived Time Features:	
Sporting Event (defined as 1 soccer match)	- Soccer match event starting time - Soccer match event ending time	- Event associated burst starting time	- Is weekend match? - Number of concurrent matches
Sporting Event (defined as 2 and 4 soccer matches occurring on the same day)	- First soccer match starting time - First soccer match ending time - Second soccer match starting time - Second soccer match ending time	- Event associated burst ending time	- Game round (eg: quarter final, semi final or final)
Sporting Event (defined as 3 soccer matches occurring on the same day)	- First soccer match starting time - First soccer match ending time - Second soccer match starting time - Second soccer match ending time - Third soccer match starting time - Third soccer match ending time		
Auction Event	- Opening time - Closing time		- Auction category - Closing hour based time period

the burst profile. We adopt the polymorphic workload burst model developed in [8] for this purpose; and create a separate polymorphic model for each event type. In summary, the polymorphic model creates n phases to model the burst profile during different time periods of an event, where each phase is bounded by two adjacent *time features*. Hence an event described by $n + 1$ time features includes n phases. For example, 3 of the time features that can be used to describe a product announcement event are: the announcement time, the burst starting time, and the burst ending time. As shown in Figure 2, these 3 time features are used to bind 2 separate phases. Two of the features that are used to delimit the earliest and latest phases are the burst starting and ending time features, respectively. These are known as *derived* time features, since they do not have well known values before the event, (unlike the product announcement time which is known before the event starts), and must instead be calculated. Derived time features are used to represent burst bounds, and can be calculated for each past event using the gradient descent algorithm described in [12]. The average length of the earliest and latest phases for past events, can be used to estimate the burst bounds for future events. During the training phase shown in Figure 1, the polymorphic model learns the workload burst profile for each of the phases within an event type. Each phase is divided into m equidistant cells, (similarly to [8]), such that each cell represents a certain portion of the phase's lifetime. Each of these cells contains a hypothesis that reports the %hits (%web requests) that are expected to occur during the cell's portion of the phase lifetime relative to the hits that are expected to occur across the entire phase. This hypothesis is trained based on the %hits that have occurred historically in a given cell for events with varying values for secondary features, allowing us to capture variations in

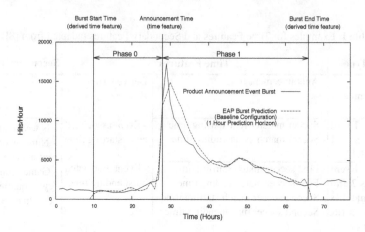

Fig. 2. Product Announcement Event Associated Burst Features

the workload burst profile attributed to such secondary features. Once cell hypotheses are created, the polymorphic model can predict the % hits expected at any time unit using the interpolation techniques described in [8]. Unlike [8], we employ a C4.5 [18] type approach to train and use these cell hypotheses, as described in Section 2.4.

2.3   The Prediction Module

Once burst models have been constructed for required event types, these can be used, together with repositories of past and upcoming events by a prediction module (PM),

---

Algorithm 1. Predicting Workload at Time Unit t_p

input : the desired time of prediction t_p;
         a repository PE of past events;
         a repository UE of upcoming events
output: the hit magnitude expected at time t_p

1 $E_a \leftarrow$ GetActiveEventsAtTime(t_p, UE);
2 $R \leftarrow 0$;
3 **for** $e \in E_a$ **do**
4 $\quad S \leftarrow$ FindSimilarEventsFromPast(e, PE);
5 $\quad P_p \leftarrow$ GetPhaseAtTimeForEvent(e, t_p);
6 $\quad h_m \leftarrow$ GetAverageTotalEventPhaseHits(S, P_p);
7 $\quad$ **if** ActualHitsKnownForPastPhaseForecasts(t_c, e, P_p) **then**
8 $\quad\quad K \leftarrow$ CalculateCorrectionFactor(t_c, e, P_p); $h_m \leftarrow h_m * K$;
9 $\quad B \leftarrow$ CalculateHitBuckets(e, P_p, h_m);
10 $\quad b \leftarrow$ GetHitBucketAtTime(t_p, B);
11 $\quad r \leftarrow$ IsAtBucketHitPoint(t_p, b) ? b.GetHitCount() : 0;
12 $\quad R \leftarrow R + r$;
13 **return** R;

---

to forecast the time, magnitude and profile of event associated bursts, as earlier advertised in Figure 1. The algorithm used by the PM to forecast workload is described in Algorithm 1 which we refer to by line numbers in this section and the next.

The PM first calculates a list of events, E_a, that will be active at the required prediction time t_p, based on the time features of events in the upcoming event repository. The hit magnitude, r, expected for each active event is predicted, and added to a running total, R, that will be reported as the hit magnitude prediction for time t_p once all active events are processed. To predict the hit magnitude expected at t_p for an active event e: we first find locations along the phase lifetimes of e where integer hits are most likely to occur (hit points). This step is particularly important for events, (like auctions), where hits are sparse throughout the event lifetime, so that a prediction can be made as to when these hits occur. We adopt a "hit-bucket" approach (line 9) based on [7] to identify when hits are likely to occur. In summary, the hit bucket approach divides a phase's lifetime into partitions of variable lengths. The first bucket starts at the phase starting time, and spans the minimum length that corresponds to at least 1 integer hit (calculated based on the %hits traversed by the bucket according to the workload model, and the total hits, h_m, expected for the phase at t_p for events similar to e). Subsequent hit buckets are contiguously appended, and also span the minimum time length that contains at least 1 integer hit. The polymorphic model can then be used to determine the hit point locations for each hit bucket based on the time when the highest % hits are reported by this model. If t_p is at a hit point (line 11) for a bucket b, then the expected hit magnitude can be calculated at this point based on the % hits reported by the polymorphic model for the length of the bucket b, and based on h_m. For events that experience a continuous stream of hits (non-sparse) at each time unit (eg: each minute), the hit bucket lengths will all span one time unit, and hits are calculated for each single time-unit length hit bucket. The total hit magnitude h_m for a given event e's phase P_p is calculated by looking at the average total hits that occurred for a set of *similar* events S in the same phase in the past. S is identified based on similarities in secondary features for events of the same type, as will be discussed in Section 2.4. As we will observe in Section 4, these secondary features can have a significant impact on prediction accuracy. As in [11], we also explore the influence of feedback correction on prediction accuracy. We customize a correction factor "K" for our scenario, which needs to provide a correction for the total hits expected to occur in a phase, h_m, since this is the variable used by the hit bucket calculation to determine where and how many hits should be placed along a given event's lifetime. K is applied to h_m as shown in line 8. It is calculated as the ratio between the total hits, T_o, that have been observed to occur up to the current time, t_c, for the current phase, and the corresponding sum of (un-corrected) predictions, T_p, made up to t_c. When the sum of predictions equals 0, K equals 1.

2.4 Identifying Similar Events

The identification approach described in this section aims to find a set of events, S, that are likely to exhibit a similar value for a given variable, v, to a given event e. This approach will be used to identify the group of events, (Line 4), that should be used to provide an initial estimate for h_m (Line 6), as well as to identify the group of events that should be used by the polymorphic model to estimate the %hits expected, g_i, at a given cell i within a given phase of an event type. Earlier work [8] identified the set of events S by first trying to find the events in the repository of past events (PE) that *exactly* matched the values of secondary features for e; and if no such events were found, then S was defined to include all events in PE regardless of secondary feature values. In this paper we aim to improve on the accuracy of the reported value for v by identifying *partial* matches for the secondary feature values of e when no exact match can be found. We follow an approach similar to C4.5 to identify S, and first describe how this works with the use case that needs to estimate h_m (ie: $v = h_m$). In this approach, all events in PE are first placed into a single node labeled c_{parent}. We then identify the secondary feature type, sf, that causes the maximum swing in value for h_m when the secondary feature value changes (we call this the splitting feature). We then split events into a number of child nodes, (we label each child node as c_i), based on the range of values that sf can assume for events in c_{parent}. This process is then recursively repeated for each child node c_i, (where each c_i would play the role of c_{parent}), until all secondary features have been considered. When we need to identify S for a given event e, we navigate down the tree, from the root node based on the secondary feature values that describe e and the splitting feature values encountered along the tree path, until a leaf node is reached or no further edges match the secondary feature values of e. The final node reached by e is labeled as the *matching node*. The deeper e's matching node is, the closer the events in the matching node correspond to e with respect to secondary features. Hence, unlike in [8], we don't need to default to using the entire PE to calculate h_m if we don't find an exact match for the secondary features of e. Instead, we find a partial match for these secondary features that can provide an estimate for h_m that is closer to the value expected for events similar to e. The average value of h_m is taken for all events in the matching node to calculate h_m for e. The same process is used to train and use a hypothesis that can predict the %hits, g_i, expected at cell i for a given phase in the polymorphic workload burst model. We investigate the influence of this partial matching approach on prediction accuracy relative to exact matching in Section 5.

2.5   GEAP Assumptions

The GEAP method is flexible in facilitating workload burst prediction for many different event types, based on a number of known assumptions:
1. Each event includes at least one time feature.
2. Future events will exhibit similar workload bursts to those exhibited by events of the same type that have occurred in the past. Given training and testing sets for events, we test the goodness of fit of a GEAP model as described in Section 4.1, to check if GEAP can be suitable for a given event type. All event types encountered in our evaluation demonstrated repeatable patterns based on tests of significance.

3   Evaluation Metrics

A workload forecasting accuracy metric dedicated to bursts, was recently demonstrated in [12], for evaluating the accuracy of a dedicated workload burst *location* prediction technique. This metric provides an indication of burst *location* prediction accuracy, but cannot assess the accuracy of burst *magnitude* predictions. Burst location is important to understand when compute resources are required, but it is the burst magnitude which will translate into how many compute resources are required to manage a workload burst. Hence, to evaluate the accuracy of GEAP we require a metric that can measure both burst magnitude and location prediction error. We contribute a new metric for this purpose, which we call Burst Prediction Error (BPE). The BPE is a composite metric consisting of 3 error components that are reported using percentage units, and are computed as described in Figure 3. The first 2 components report the average proportion of hits (page requests), across all bursts, that were (1) under-forecasted (we label this under forecasting proportional error component as UFPE) and (2) over-forecasted (we label this over forecasting proportional error component as OFPE) relative to the actual number of hits that occurred during each one of these bursts. These first 2 components are computed based on the Magnitude of Relative Error (MRE) [19] between the bounds of each burst ($MRE = \frac{|Actual-Forecast|}{Actual} \times 100$ %). We set the MRE numerator to the absolute sum of errors having a certain sign, (positive or negative depending on whether we are calculating UFPE or OFPE, respectively); and the MRE denominator to the sum of actual values between burst bounds.

The third error component of the BPE reports the total forecasting proportional error (TFPE), where this proportion includes hits that were both under and over forecasted for predicted bursts. As described in [2], one may not always want to place equal importance on both over-forecasting and under-forecasting errors. Eg: cloud application providers may prefer the over-provisioning of compute resources associated with over-forecasting errors over the reduced QoS associated with under-forecasting the amount

---

To calculate the Burst Prediction Error (BPE) for a given time-series of predictions based on the actual time series (the test set):

1. The lower B_l and upper B_u bounds of each burst in the test time series are identified using the burst identification technique described in [12].
2. For each identified burst, j, the $OFPE_j$, $UFPE_j$ and $TFPE_j$ are calculated as follows:
   (a) Find the predicted time series between B_l and B_u for current actual burst j considered
   (b) Find the difference between each actual, a_{t_i}, and corresponding predicted, p_{t_i}, value at times t_i between B_l and B_u: $(d_{t_i} = a_{t_i} - p_{t_i})$
   (c) Find the sum of all actual values between the burst bounds: $SAV_j = \sum_{i=B_l}^{B_u} a_{t_i}$
   (d) Find the sum of negative errors: $SNE_j = \sum_{i=B_l}^{B_u} |d_{t_i}|$ $(\forall d_{t_i} < 0)$
   (e) Then $OFPE_j = \left(\frac{SNE_j}{SAV_j} \times 100 \right) \%$
   (f) Find the sum of positive errors: $SPE_j = \sum_{i=B_l}^{B_u} d_{t_i}$ $(\forall d_{t_i} > 0)$
   (g) Then $UFPE_j = \left(\frac{SPE_j}{SAV_j} \times 100 \right) \%$
   (h) Then $TFPE_j = a \times UFPE + b \times OFPE$ %
   (i) The BPE for burst j is recorded as: $BPE_j = \{UFPE_j, OFPE_j, TFPE_j\}$
3. Report the burst prediction error, BPE = {UFPE, OFPE, TFPE} as the average of all calculated values of BPE_j for each burst, j, that was considered.

Fig. 3. Calculating the BPE Metric

of resources required during an event. Hence we report both under and over forecasting error components, and allow one to specify weights (a and b in step 2h of Figure 3) to represent the importance of each of these components when calculating the TFPE.

In essence, each burst consists of a total number of web requests (SAV_j); and *for each burst*, the BPE_j identifies the proportion of actual page requests (hits) that are unexpected by a prediction method at certain points in time ($UFPE_j$); and the proportion of hits incorrectly predicted to occur at certain points in time ($OFPE_j$). The $TFPE_j$ reports the total burst prediction error as a proportion of the total number of hits that actually occurred during the burst, in order to reflect the significance of the error. For each burst, we calculate the *total* proportion of requests that were erroneously forecasted, and not the *mean* error proportion (like MAPE [20]) since we are looking for an error metric that is closer to the *total* potential business lost by not provisioning enough compute resources to meet the demand of each burst ($UFPE_j$), and the *total* level of over-provisioning ($OFPE_j$) for each burst.

4   Evaluation

In this section we aim to assess the following GEAP characteristics: (i) BPE for event-associated workload from different domains; (ii) BPE sensitivity to increased prediction horizon length; and (iii) the efficacy of using time and secondary features, partial feature matching, and the correction factor, K, for improving burst prediction accuracy.

To achieve this, we make use of real workload from auction, sporting, and product announcement event domains, as outlined in Table 2. Predictions will be made for bursts present in these data sets using the following GEAP configurations:

1. **Baseline (B):** use of all time and secondary features listed in Table 1 with the standard GEAP configuration as described in Section 2.
2. **Restricted Secondary (-S):** configuration (B), excluding secondary feature usage.
3. **Restricted Secondary, No Correction (-S,NC):** configuration (-S), but also excluding use of a correction factor K. (Evaluates the impact of secondary features on prediction accuracy independently of K.)
4. **Restricted Time (-T):** configuration (B), excluding use of every 2nd time feature, for each of the World Cup soccer match event types shown in Table 1.
5. **Restricted Time, No Correction (-T,NC):** configuration (-T), excluding use of K. (Evaluates the impact of time features on prediction accuracy independently of K.)
6. **Exact Matching (EM):** configuration (B) using exact secondary feature matching instead of the partial C4.5 event matching approach described in Section 2.4.
7. **Exact Matching, No Correction (EM,NC):** configuration (EM), excluding use of K. (Evaluates the impact of using exact instead of partial event feature matching on prediction accuracy independently of K.)
8. **No Correction (NC):** configuration (B) excluding use of K for making predictions.

These configurations are used to predict workload bursts present in the data sets shown in Table 2 using prediction horizons of 1 Hour, 1 Day, 1 Week, and 1 Month. The accuracy of these predictions is measured using the BPE metric, where we set equal weights for UFPE and OFPE in calculating TFPE (ie: $a = b = 1$ in Figure 3).

4.1 Methodology

To assess the burst prediction error (BPE) of each GEAP configuration listed above, for every burst in the data sets shown in Table 2, and for different prediction horizons, we follow the method outlined below:

a) Verify that GEAP assumptions (Section 2.5) hold for all event types considered: (1) The first assumption is validated by looking at the time features available for each event type; (2) We validate the goodness of fit of the un-corrected (NC) GEAP model for the event types being considered using Chi Squared goodness of fit tests [23]. When predictions made by the GEAP (-NC) model (constructed using the training sets shown in Table 2) are NOT statistically different to the actual workload distribution in the test set (ie: test p value > 0.05), we proceed with using GEAP for the event type concerned. In our evaluation all goodness of fit tests yielded p-values in excess of 0.95.

b) Once assumptions are validated, we assess the burst prediction accuracy of GEAP using the BPE metric introduced earlier. Leave One Out Cross Validation (LOOCV)

Table 2. Training and Testing Sets Used for Evaluation

Event Type	Training Set	Testing Set
Product Announcements	Wikipedia load traces (Page Requests/Hour) [21] for the page of a popular tablet device, around the time of product announcements for 4 tablet generations. Sufficient trace was retrieved for each product announcement such that event associated burst boundaries could be identified as described in Section 2.	Workload associated with each tablet announcement event in the product announcement training set is used in turn as a testing set. At no point in time is the same event added to both training and testing sets.
Sporting Event Types	Soccer World Cup 1998 load traces (Page Requests/Minute) [22] for soccer match events occurring in the months of June and July, 1998. Soccer match event types are described in Table 1, and burst boundaries for each event-associated burst are identified as described in Section 2. Starting time for each match is based on information from `http://fifa.com/tournaments/archive/worldcup/france1998/`. Each soccer match is assumed to end 105 minutes after it commences irrespective of any added time.	Workload associated with each soccer match in the training set is used in turn as a testing set. At no point in time is the same event added to both training and testing sets.
Auction Event Types	Auction Set A containing 245 728 auctions ending within the two day period of: 24/9/2013 to 26/9/2013. Both auction Set A and Set B (the testing set) were collected using the same group of auction search keywords from a popular online auction site.	Auction Set B containing 237 347 auctions ending between 28/9/2013 and 30/9/2013. The testing time series is composed of the aggregated bid-history of all auctions in Set B.

[24] is used for the World Cup and Product Announcement data sets where there are fewer events relative to the Auction data set which contains over 200,000 events. This is done in order to allow for greater variety of test events. In summary, LOOCV works in the following way: For a given event type with n sample events, a GEAP prediction model is trained from $n - 1$ of the sample events. This is then used to predict workload for the remaining event that was not seen during the training phase of the GEAP prediction model. This process is repeated in iterations, where each of the n sample events takes turns in acting as the test set. LOOCV experiments are repeated for each GEAP configuration listed earlier in this section; for each applicable data set (see Table 2); and for each prediction horizon. The BPE components yielded by each LOOCV run are recorded for discussion.

5   Results and Discussion

Figures 4, 5, and 6 illustrate how BPE varies with prediction horizon when different GEAP configurations are used to predict workload bursts associated with product announcement, sporting, and auction event types respectively. The TFPE is shown for each prediction horizon when different GEAP configurations are used with different data sets. The TFPE can be read from the y-axis, and the individual UFPE and OFPE components are shown in brackets, next to each data point. The x-axis shows the prediction horizon for each data point.

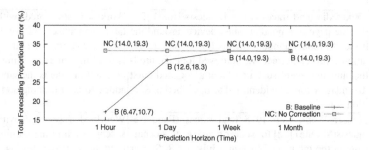

Fig. 4. Announcement BPE Components: (UFPE%,OFPE%) shown in brackets

5.1   Correction Factor and Prediction Horizon Effects

Results for the product announcement data set (Figure 4) and for the sporting event data set (Figure 5) exhibit the most prominent influence of the correction factor, K, (which is used in all GEAP configurations that exclude "NC" from the series name). In all data sets, we notice that the Baseline (B) configuration was able to yield the lowest total forecasting proportional error (TFPE) found for the shortest prediction horizon of 1 hour. This TFPE then increases with the prediction horizon up to a critical point, after which the error remains constant regardless of further increases to the horizon. The critical point represents the moment in time when K can no longer be applied, since no history of predicted/actual workload pairs exists when predicting so far into the future (ie: the

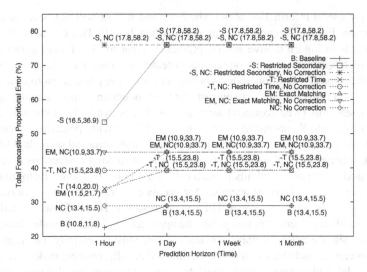

Fig. 5. World Cup BPE Components: (UFPE%,OFPE%) shown in brackets

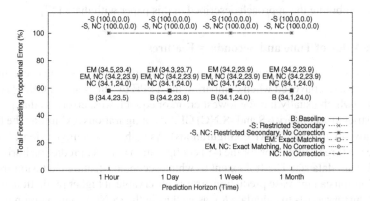

Fig. 6. Auction BPE Components: (UFPE%,OFPE%) shown in brackets

actual workload values in the test set never come into the present for use as history). When K can no longer be used to gather information about prediction error trends, GEAP falls back to *only* using the past event repository and the workload model to calculate predictions, as described in Algorithm 1 of Section 2.3. If the same data in the repository of past events is available for all prediction horizons from the critical point onwards, the prediction will be the same for all of these prediction horizons. We note that auction events (Figure 6) yielded the least benefits from application of K, including at low prediction horizons. As observed in [25], the largest proportion of auctions usually have low total bid counts like 0 and 1, *and* these bids usually occur towards the end of an auction's lifetime. For this large proportion of auctions, which may or may not exhibit a single bid at the *end* of their lifetime, the observed bid history and corresponding bid-predictions could both simply show zero bid count when calculating K. This does not give any useful information about whether a bid will or will not occur at the end of the auction. Further, K cannot be calculated when T_p is 0. This explains why "NC" configurations in Figure 6 have similar prediction error to GEAP configurations that use K. In such scenarios, we rely on secondary features to improve prediction accuracy, and provide some information about whether an auction is popular enough to receive a bid at the end of its lifetime. In contrast, for the product announcement and sporting event data sets, hits (web requests) were almost always available in the workload to be able to calculate K, providing another means of improving prediction accuracy at shorter prediction horizons aside from specifying accurate time and secondary features. Past work specific to auctions [7], shows one way of making corrections even when T_p is 0, based on the observed hits in each cell of a relative frequency histogram. This approach can improve forecast accuracy for events exhibiting sparse hits during their lifetime (especially for events where TFPE in our results is 100%). However, unlike the correction method presented in this paper, the approach in [7] can degrade accuracy for other events since a cell containing a certain % hits must be entirely traversed before corrections can be made based on observed hits in that cell, resulting in delayed corrections. Such delayed corrections can translate to missing corrections for spike peaks. In future work we aim to investigate a means of improving the influence of a correction factor by combining the approach described in this paper with that in [7].

5.2   The Value of Time and Secondary Features

The benefits of including secondary feature information for the sporting event and auction data sets are conveyed when comparing the TFPE of: the baseline GEAP configuration (B), (which usually has the lowest error, or equal lowest error), to that produced by the restricted secondary (-S and -S,NC) GEAP configurations (exhibiting the highest error relative to all other GEAP configurations). As with other configurations, K is able to reduce the forecasting error for the (-S) configuration at lower prediction horizons for the World Cup data set. The full extent to which secondary feature exclusion from the event definition can increase prediction error is observable at higher prediction horizons (when no history exists to calculate K), as well as in the (S,NC) configuration in which K is not used.

The past event repository may not always contain a set of events that exactly match the secondary features of the upcoming event for which workload is being predicted. This is the case for some of the events that are present within the World Cup data set. In such scenarios, instead of estimating the expected % hits and hits per percent based on every event in the past event repository to make forecasts; predictions can be more accurate if they are made based on information about events that are determined to be similar based on a partial match for secondary features. This can be seen in the World Cup data set where exact feature matching approaches (EM and EM,NC) consistently yield higher error than the baseline GEAP configuration that uses partial matching. For auctions, the exclusion of secondary features leads to constant predictions of 0 bids/minute (100% UFPE), since the majority of auctions have 0 bids, and there are no secondary features to differentiate the few auctions that do have bids; meaning no integer hits can be predicted. Also in the auction data set, the exact matching GEAP configuration does not produce significantly different results to the baseline (partial matching) configuration. Since there are more than $200,000$ events in the auction training set, each described by category and closing hour based time period, it is much more likely that exact matches will be found for events considered in this training set. This explains why there is no increase in the TFPE of the EM configuration (Figure 6) relative to the baseline configuration. This is in contrast to the World Cup data set, where there are only 27 one day events in total (events being defined as described in Table 1), making it more likely that exact matches will not be found.

Apart from secondary features, events are also described using time features. As discussed in Sections 2.1 and 2.2, these characteristics provide an indication of where spikes in the event-associated workload profile are likely to be found. The World Cup data set allows us to vary the time features used, and as expected, removal of time features from the sporting event definition increases prediction error. Similarly to our analysis for secondary features, the full extent of this error increase is observed when no K is used (-T,NC), and at higher prediction horizons where no workload history for the event being considered exists to calculate K.

The majority of results thus far imply that the description of an event, given in terms of time and secondary features, can influence the event-associated workload profile predicted, as well as the prediction accuracy yielded by GEAP. This highlights an implicit link between the description of an event, and the event's associated workload profile. Accurate event definitions become particularly important at higher prediction horizons for reducing prediction error when K can no longer be used, but can also be beneficial at lower prediction horizons, as witnessed by our results. They are also differentiating factors for prediction accuracy for event types like auctions where hits are sparse along the event lifetime, and use of K is ineffective. In future work we aim to experiment with techniques for automatically identifying features that can best describe events, and hence allow for improved workload predictions.

6 Conclusion

In this paper we have presented a generic event aware prediction approach that can inform web-application providers about the demand they can expect for their upcoming events. Such information enables application providers to pre-provision resources early to avoid cloud resource initialization lag, consequently having the right capacity available on time to serve event-associated demand spikes. Furthermore, it enables them to cost-effectively reserve cloud resources based on demand forecasts for future events. In our evaluation we have shown the importance of the implicit event-burst link, (which other generic workload prediction methods [9–11] ignore), by illustrating the detrimental effect on prediction accuracy of ignoring secondary and time feature information about upcoming events. We have also shown how partial matching can be used to more effectively find a set of related past events, for an upcoming event, based on secondary features; how a correction factor can be applied to the GEAP approach; and when these techniques can improve prediction accuracy. The understanding gained from how event characteristics, within an event definition, can be translated into resource demand predictions by GEAP lends itself to many applications. Specifically, these include: providing event organizers with the ability to engineer their event characteristics to a desired level of event popularity; assisting web application providers to efficiently maintain QoS during significant events; and providing accurate forecasts to numerous other domains where event-associated burst prediction is important.

Acknowledgement. NICTA is funded by the Australian Government through the Department of Communications and the Australian Research Council through the ICT Centre of Excellence Program.

References

1. Suleiman, B., Sakr, S., Venugopal, S., Sadiq, W.: Trade-off analysis of elasticity approaches for cloud-based business applications. In: Wang, X.S., Cruz, I., Delis, A., Huang, G. (eds.) WISE 2012. LNCS, vol. 7651, pp. 468–482. Springer, Heidelberg (2012)
2. Islam, S., Lee, K., Fekete, A., Liu, A.: How a consumer can measure elasticity for cloud platforms. In: International Conference on Performance Engineering. ACM/SPEC (2012)
3. Li, A., Yang, X., Kandula, S., Zhang, M.: Cloudcmp: comparing public cloud providers. In: SIGCOMM Conference on Internet Measurement. ACM (2010)
4. Mao, M., Humphrey, M.: A performance study on the vm startup time in the cloud. In: International Conference on Cloud Computing. IEEE (2012)
5. Yigitbasi, N., Iosup, A., Epema, D., Ostermann, S.: C-meter: A framework for performance analysis of computing clouds. In: International Symposium on Cluster Computing and the Grid. IEEE (2009)
6. Adar, E., Weld, D.S., Bershad, B.N., Gribble, S.S.: Why we search: visualizing and predicting user behavior. In: International Conference on World Wide Web. ACM (2007)
7. Sladescu, M., Fekete, A., Lee, K., Liu, A.: Event aware workload prediction: A study using auction events. In: International Conference on Web Information System Engineering (2012)
8. Sladescu, M., Fekete, A., Lee, K., Liu, A.: A polymorphic model for event associated workload bursts. In: International Workshop on Resource Management of Cloud Computing. IEEE (2013)
9. Islam, S., Keung, J., Lee, K., Liu, A.: Empirical prediction models for adaptive resource provisioning in the cloud. In: Future Generation Computer Systems (2011)
10. Bermolen, P., Rossi, D.: Support vector regression for link load prediction. Computer Networks 53(2), 191–201 (2009)
11. Urgaonkar, B., Shenoy, P., Chandra, A., Goyal, P., Wood, T.: Agile dynamic provisioning of multi-tier internet applications. ACM Transactions on Autonomous and Adaptive Systems 3(1), 1:1–1:39 (2008)
12. Lassnig, M., Fahringer, T., Garonne, V., Molfetas, A., Branco, M.: Identification, modelling and prediction of non-periodic bursts in workloads. In: International Conference on Cluster, Cloud and Grid Computing. IEEE/ACM (2010)
13. You, Y., Huang, G., Cao, J., Chen, E., He, J., Zhang, Y., Hu, L.: Geam: A general and event-related aspects model for twitter event detection. In: Lin, X., Manolopoulos, Y., Srivastava, D., Huang, G. (eds.) WISE 2013, Part II. LNCS, vol. 8181, pp. 319–332. Springer, Heidelberg (2013)
14. Sakaki, T., Okazaki, M., Matsuo, Y.: Earthquake shakes twitter users: real-time event detection by social sensors. In: International conference on World Wide Web. ACM (2010)
15. Arlitt, M., Jin, T.: A workload characterization study of the 1998 world cup web site. IEEE Network 14(3), 30–37 (2000)
16. Bodik, P., Fox, A., Franklin, M.J., Jordan, M.I., Patterson, D.A.: Characterizing, modeling, and generating workload spikes for stateful services. In: Symposium on Cloud Computing. ACM (2010)

17. Bhatia, S., Mohay, G., Schmidt, D., Tickle, A.: Modelling web-server flash events. In: International Symposium on Network Computing and Applications. IEEE (2012)
18. Quinlan, J.R.: C4.5: programs for machine learning, vol. 1. Morgan kaufmann (1993)
19. MacDonell, S., Gray, A.: A comparison of modeling techniques for software development effort prediction. In: International Conference on Neural Information Processing. Springer (1998)
20. Hyndman, R.J., Koehler, A.B.: Another look at measures of forecast accuracy. International Journal of Forecasting 22(4), 679–688 (2006)
21. Mituzas, D.: Wikipedia Visitor Statistics (2013), https://dumps.wikimedia.org/other/pagecounts-raw/ (online: accessed February 19, 2013)
22. Arlitt, M., Jin, T.: 1998 World Cup Web Site Access Logs (August 1998), http://www.sigcomm.org/ITA (online: accessed February 19, 2013)
23. Bagdonavièus, V., Kruopis, J., Nikulin, M.: Nonparametric Tests for Complete Data. John Wiley & Sons (2013)
24. Arlot, S., Celisse, A.: A survey of cross-validation procedures for model selection. Statistics Surveys 4, 40–79 (2010)
25. Menascé, D., Akula, V.: Towards workload characterization of auction sites. In: International Workshop on Workload Characterization. IEEE (2003)

High-Payload Image-Hiding Scheme Based on Best-Block Matching and Multi-layered Syndrome-Trellis Codes

Tao Han[1,2], Jinlong Fei[1,2], Shengli Liu[1,2], Xi Chen[1,2], and Zhu Yuefei[1,2]

[1] Zhengzhou Information Science and Technology Institute, Zhengzhou, China
[2] State Key Laboratory of Mathematical Engineering and Advanced Computing,
Zhengzhou, China
lhstslhsts@163.com, feijinlong@126.com, 475737@qq.com,
xycuckoo@tsinghua.org.cn, zyf0136@sina.com

Abstract. Image steganography has been widely used in the domain of privacy protection, such as the storage and transmission of the secret images. This paper presents a novel image-hiding scheme to embed the secret image into an innocent cover image. We use a block matching procedure to search for the best-matching block for each secret image block. Then the k-means clustering method is used to select k representative blocks for the not-well-matched secret image blocks. Moreover, the bases and indexes of the well-matched blocks together with the representative blocks are compressed by the Huffman coding. Finally, we use multi-layered syndrome-trellis codes (STC) to embed all the relevant data into multiple least significant bit (LSB) planes of the cover image. The results of experiments show that the proposed method outperforms some previous image-hiding methods.

Keywords: Privacy Protection, Image Steganography, Image Hiding, Best-Block Matching, Sydrome-Trellis Codes.

1 Introduction and Related Works

In recent years, the rapid development of computer technology and the Internet makes the storage and transmission of digital images become more and more convenient and common. From privacy protection's point of view, if the secret images (also called hidden images or hidden images), which belong to the individuals or enterprises, are stored and transmitted in the form of plaintext without any protective measures, some serious security problems may happen. Therefore, the secret images must be protected from being accessed illegally by the unauthorized users. However, traditional encryption techniques are not suitable for the storage and transmission of secret images, since the cipher texts are always meaningless binary streams, which obviously and easily arouse the suspicion of the malicious attackers. In order to protect the confidentiality of the contents of secret images or the concealment of the transmission of secret images, image steganography can be available for hiding the existence of the secret images. This technique, where both the cover objects and secret messages are digital images, is also called image hiding. The main process of image hiding is to embed the secret image into a cover image to

B. Benatallah et al. (Eds.): WISE 2014, Part II, LNCS 8787, pp. 336–350, 2014.

generate a stego image, which is not easily perceived by the attackers. In general, the technique of image hiding mainly pursues three goals: (1) the stego image must appear as similar as possible with the cover image, i.e., the modifications of the cover image caused by image hiding must be as few as possible; (2) the embedding capacity must be as large as possible, and large secret images may be embedded into a relatively small cover image; (3) the visual quality of the secret image extracted from the stego image must be as good as possible, i.e., the extracted secret image must maintain the content characters of the original secret image as many as possible.

So far, many image-hiding schemes have been proposed, where one kind is based on LSB replacement [1, 2, 3] and another kind is based on modulus operation [4, 5, 6]. However, some researchers have proven that the schemes using LSB replacement are not secure [7]. Furthermore, the embedding capacities of these methods are not large enough to carry the secret images larger than the cover images. Due to a certain robustness of the contents of digital images, it's acceptable that the extracted secret images contain a small amount of distortion, which is different from the condition that text messages are the embedding objects and the extracted messages must be the same as the original. Consequently, some image-hiding schemes [8, 9, 10, 11, 12, 13, 14, 15] have been proposed to enlarge the embedding capacity by using the technique of image compression, where the schemes based on vector quantization (VQ) attract more and more attention. This is because the VQ technique is a block-based quantization method with the advantages of simple structure, efficient decoding, and easy implementation [14]. Based on VQ and modulus operation, [10] presented an image-hiding scheme to compress the secret images by using VQ of index compression and then to embed the compressed data into the cover image. [11] proposed a VQ-like image-hiding method, which uses a block matching procedure to compress the image blocks with small distance, selects representative blocks from the blocks with large distance, encodes the indices and representative blocks, and then embeds all the encoded data into multiple LSB planes of the cover images. [12] presented an image-hiding scheme that utilized a two-way block matching procedure to find the highest similarity block for each block of the secret image and embed the bases and indexes obtained as well as the not-well-matched blocks into the LSB planes of the cover image. In addition, [13] improved the performance of [12] by means of a k-way block matching procedure. Based on the concept of vector quantization, [14] proposed a novel image-hiding scheme, which used a two-codebook combination with three-phase block matching to enhance the visual quality of the extracted secret image. [15] utilized JPEG2000 to compress the secret image and then embedded the compressed data by means of the tri-way pixel-value differencing [16], where the residual value coding is used to reduce the distortion of the extracted secret image.

Although the previous methods perform well, there is still room for the improvement of the performance of the image-hiding schemes. This is because the embedding efficiency of the methods is still not high. In this paper, a novel image-hiding method based on best-block matching and multi-layered STC is proposed. Firstly, a block matching process is used to find the best-matching block for each secret image block and record the index. Secondly, the k-means clustering method is used to deal with the not-well-matched blocks. Thirdly, all the relevant data are compressed by the Huffman coding. At last, the multi-layered STC is used to embed the compressed data. The experimental results show that the performance of the proposed method is superior to four previous image-hiding schemes.

The rest of this paper is organized as follows. Section 2 introduces the k-means clustering method and the sydrome-trellis codes with their multi-layered construction to be used in Section 3. Section 3 presents the proposed image-hiding scheme. Section 4 describes the experimental results and analysis. Finally, the conclusions are summarized in Section 5.

2   Preliminaries

2.1   The k-Means Clustering Method

It is well-known that the k-means clustering method is a kind of unsupervised learning method, which can solve the cluster problems simply and quickly. The basic idea of this method is that k data objects are selected as the original cluster centroids, and then other data objects are divided into different clusters by means of iteration according to such a principle that the similarity among the data objects inside the cluster is strong while the similarity among the data objects belonging to different clusters is weak. Obviously, the selection of k original cluster centroids may influence the clustering results. One strategy to improve the clustering results is that the data objects with weak similarity are chosen as k original cluster centroids and the other strategy is the iteration method.

Given a data set containing n data objects, assume that we want to classify the set into k clusters. We utilize the mean square error to measure the similarity between a data object and a clustering centroid, i.e.,

$$F = \frac{1}{k \times n} \sum_{j=1}^{k} \sum_{i=1}^{n} \left\| x_i^{(j)} - c_j \right\|^2 , \tag{1}$$

where $\left\| x_i^{(j)} - c_j \right\|^2$ denotes the distance between the i-th data point of the data object $x_i^{(j)}$ and the j-th cluster centroid c_j.

The steps of the k-means cluster method can be summarized as follow:
1. k data objects are selected as the original cluster centroids from n data objects.
2. Each data object is classified into the cluster that has the closest centroid.
3. Recalculate the k cluster centroids after all the objects are classified.
4. Repeat 2 and 3 until the cluster centroids no longer change or the maximum number of iterations is reached.

2.2   Sydrome-Trellis Codes and their Multi-Layered Construction

[17] proposed an efficient and fast-implemented method of steganographic coding, which is called sydrome-trellis codes (STC). For arbitrary given additive distortion model, the STC pursues to minimize the embedding distortion and can achieve asymptotically the theoretical bound of embedding efficiency, the performance of which is the best among the binary steganographic coding methods as far as we know. Based on the STC, a very famous and efficient spatial image steganography called HUGO (Highly Undetectable steGO) is proposed in [18].

As a matter of fact, the STC is a kind of specific matrix encoding, whose parity check matrix $\mathbf{H}$ of size $k \times n$ is comprised of a submatrix $\hat{\mathbf{H}}$ of size $h \times w$ in a certain concatenation form. The goal of the STC is that the messages $\mathbf{m} = (m_1, m_2, ..., m_k)$ are carried by modifying the binary cover $\mathbf{x} = (x_1, x_2, ..., x_n)$ into the stego $\mathbf{y} = (y_1, y_2, ..., y_n)$ and the total embedding distortion is minimized; therefore, the receiver can easily extract $\mathbf{m}$ through computing $\mathbf{H}\mathbf{y}^T$. The STC expresses each solution of $\mathbf{H}\mathbf{y}^T = \mathbf{m}^T$ as a path through the trellis and the length of each step of the path is defined as the embedding distortion $\{\rho_i(x_i, y_i)\}_{1 \leq i \leq n}$ caused by modifying $\{x_i\}_{1 \leq i \leq n}$ into $\{y_i\}_{1 \leq i \leq n}$. Therefore, the problem of minimizing the embedding distortion becomes a problem of finding the shortest path. Moreover, the shortest path can be quickly obtained by the Viterbi decoding algorithm. Note that the embedding distortion $\{\rho_i(x_i, y_i)\}_{1 \leq i \leq n}$ can be defined according to arbitrary principles.

Inspired by the double-layered embedding construction [19, 20], [17] also presented the multi-layered construction of STC, which is called multi-layered STC for short. Based on the multi-layered STC, some secure and efficient steganographic algorithms have been proposed in [21, 22, 23], which include the spatial domain, JPEG domain, and JPEG domain with side-information. Below, we take the double-layered STC as an example to describe the process of the multi-layered STC. First of all, the embedding rate is decompressed according to the chain rule of entropy and the embedding rate carried in the second LSB plane is estimated. Then, under this circumstance, compute the optimal changing probabilities of the second LSBs. Moreover, by means of the Flipping Lemma, the STC is used to embed a part of messages in the second LSBs. After the second LSBs are modified, the conditional changing probabilities of the LSBs can be calculated, where the elements with the second LSBs being modified are defined as wet elements (i.e., the element that are not allowed to change again). Finally, by means of the Flipping Lemma again, the STC is used to embed the remaining messages in the LSBs.

For the double-layered STC, the maximum modification amplitude of the cover elements is 1, i.e., the set of modification modes is $\{-1, 0, +1\}$, and the maximum theoretic embedding rate is $\log_2 3$. For the three-layered STC, the maximum modification amplitude of the cover elements is 2, i.e., the set of modification modes is $\{-2, -1, 0, +1, +2\}$, and the maximum theoretic embedding rate is $\log_2 5$. Similarly, for the p-layered STC, the maximum modification amplitude of the cover elements is p, i.e., the set of modification modes is $\{-p+1, \cdots, -1, 0, +1, \cdots, p-1\}$, and the maximum theoretic embedding rate is $\log_2(2p-1)$.

3   The Proposed Method

Assume that both the cover image $\mathbf{CI}$ of size $h_{\text{CI}} \times w_{\text{CI}}$ and the secret image $\mathbf{SI}$ of size $h_{\text{SI}} \times w_{\text{SI}}$ are q-bit images, for instance, $q = 8$ for the gray spatial images.

Besides, the stego image **ST** is generated after **SI** is embedded into **CI**. Let 1-LSB be the LSB plane, 2-LSB be the second LSB plane, and similarly, q-LSB be the most significant bit (MSB) plane. In addition, the parameter p denotes the number of LSB planes of the cover image used to carry the payload. In this paper, in order to preserve a good visual quality of the stego image **ST**, the maximum value of p is 4.

3.1   The Embedding Phase

Assume that the size of a block is $b = m \times n$. And, without loss of generosity, suppose that h_{CI}/m, w_{CI}/n, h_{SI}/m and w_{SI}/n are integers. Therefore, the numbers of the cover image blocks and the secret image blocks are $bnCI = (h_{CI} \times w_{CI})/(m \times n)$ and $bnSI = (h_{SI} \times w_{SI})/(m \times n)$ respectively. And then the block forms of the cover image and secret image are set to $\mathbf{CI} = \{\mathbf{CI}_j\}_{1 \le j \le bnCI}$ and $\mathbf{SI} = \{\mathbf{SI}_j\}_{1 \le j \le bnSI}$, respectively. In this paper, the distance between two $m \times n$ arbitrary blocks **A** and **B** is measured by their mean square error (MSE):

$$\text{MSE}(\mathbf{A}, \mathbf{B}) = \frac{1}{m \times n} \sum_{i=1}^{m} \sum_{j=1}^{n} \left(\mathbf{A}_{i,j} - \mathbf{B}_{i,j} \right)^2. \tag{2}$$

First of all, the secret image is divided into several non-overlapping blocks of size $b = m \times n$. For each block $\mathbf{SI}_j$ $(1 \le j \le bnSI)$, compute the mean value of $\mathbf{SI}_j$, i.e.,

$$\overline{\mathbf{SI}}_j = \frac{1}{m \times n} \sum_{k=1}^{m} \sum_{l=1}^{n} \mathbf{SI}_j(k,l), \tag{3}$$

where $\mathbf{SI}_j(k,l)$ denotes the pixel value in the position (k,l) of $\mathbf{SI}_j$. Let B_j be the nearest integer to $\overline{\mathbf{SI}}_j$ and then each $\mathbf{SI}_j$ can be expressed as a base-difference form $\mathbf{SI}_j = B_j + \mathbf{D}_j$, where the difference block is $\mathbf{D}_j = \{\mathbf{SI}_j(k,l) - B_j\}_{1 \le k \le m, 1 \le l \le n}$ and $\mathbf{D}_j$ will be used for the matching in the block matching process.

Then we construct a set of candidate blocks and must guarantee the reconstruction of the set in the extracting phase. Since p LSB planes of the cover image are utilized to carry the payload, $(q - p)$ MSB planes are used to generate the candidate set. Figure 1 shows the flowchart of the proposed method. We give the construction process of the candidate set in detail below.

Firstly, the $(p+1)$-LSB to q-LSB of the pixel $\mathbf{CI}_j(k,l)$ are concatenated with the $(p+1)$-LSB to $2p$-LSB of the pixel $\mathbf{CI}_{j+1}(k,l)$ to obtain the pixel value of a temporary block $\mathbf{T}_j$, i.e.,

$$\mathbf{T}_j(k,l) = \sum_{b=p+1}^{q} 2^{b-1} \times \mathbf{CI}_j^b(k,l) + \sum_{b=p+1}^{2p} 2^{b-1-p} \times \mathbf{CI}_{j+1}^b(k,l), \tag{4}$$

where $1 \le j \le bnCI -1$, $1 \le k \le m$, $1 \le l \le n$ and $\mathbf{CI}_j^b(k,l)$ denotes the value of the b-LSB of the pixel $\mathbf{CI}_j(k,l)$. According to Equation (5), we calculate the pixels of the $bnCI$-th temporary block $\mathbf{T}_{bnCI}$,

$$\mathbf{T}_{bnCI}(k,l) = \sum_{b=p+1}^{q} 2^{b-1} \times \mathbf{CI}_{bnCI}^b(k,l) + \sum_{b=p+1}^{2p} 2^{b-1-p} \times \mathbf{CI}_1^b(k,l). \tag{5}$$

The above process will construct $bnCI$ temporary blocks. In order to enlarge the block number of the candidate set, i.e, increase the possibility of finding the best-matching block for the secret image blocks, we use a strategy that the $(p+1)$-LSBs to q-LSBs of $\mathbf{CI}_j(k,l)$ ($1 \le j \le bnCI$) are circularly shifted to right by 1, 2, and 3 bits, and then another $3 \times bnCI$ temporary blocks $\mathbf{T}_{bnCI+1}, \mathbf{T}_{bnCI+2}, \cdots, \mathbf{T}_{4 \times bnCI}$ are similarly obtained on the basis of Equation (4) and (5). Figure 2 shows an example of circular shifting to right, where the 8-bit pixel is 10010111, $q=8$ and $p=4$.

After obtaining $4 \times bnCI$ temporary blocks, $\mathbf{T}_1, \mathbf{T}_2, \cdots, \mathbf{T}_{4 \times bnCI}$, we similarly calculate the average values of all the temporary blocks according to Equation (3), which are denoted as $\overline{\mathbf{T}}_1, \overline{\mathbf{T}}_2, \cdots, \overline{\mathbf{T}}_{4 \times bnCI}$. The integer nearest to $\overline{\mathbf{T}}_j$ ($1 \le j \le 4 \times bnCI$) is denoted as TB_j and then the temporary blocks can be denoted as $\mathbf{T}_j = TB_j + \mathbf{TD}_j$, where the temporary difference is $\mathbf{TD}_j = \{\mathbf{T}_j(k,l) - TB_j\}_{1 \le k \le m, 1 \le l \le n}$. If the distances between the current $\mathbf{TD}_j$ and $TH1$ previous candidate blocks are larger than a given threshold $TH2$, the current $\mathbf{TD}_j$ is assigned as a new candidate block. The above process repeats until all the temporary differences $\mathbf{TD}_j$ ($1 \le j \le 4 \times bnCI$) are traversed. Therefore, we can obtain the set containing t candidate blocks, which is denoted as $\mathbf{C} = \{\mathbf{C}_i\}_{1 \le i \le t}$, where $1 \le t \le 4 \times bnCI$.

After the candidate set is obtained, with regard to each difference block $\mathbf{D}_j$ ($1 \le j \le bnSI$) of the secret image, we search for the best-matching block in the candidate set $\mathbf{C} = \{\mathbf{C}_i\}_{1 \le i \le t}$, and the index corresponding to the best-matching block is obtained by the following equation: $index_j = \arg\min_i \{MSE(\mathbf{D}_j, \mathbf{C}_i), 1 \le i \le t\}$.

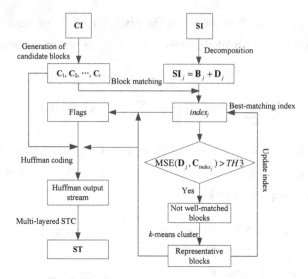

Fig. 1. The flowchart of the proposed method

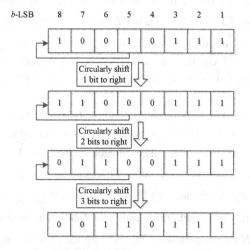

Fig. 2. An example of circularly shifting $(q - p)$ MSB planes to right

The above block matching procedure can find the best-matching block for each $\mathbf{D}_j$, but the distance between $\mathbf{D}_j$ and its best-matching block may be still large enough to affect the visual quality of the extracted secret image. If the distance between them is larger than a threshold $TH3$, the secret image block corresponding to $\mathbf{D}_j$ is referred to as a not-well-matched block. In order to reduce the effect of the not-well-matched blocks, we use the k-means clustering method to select k representative blocks and record them in the LSBs of the cover image. The value of the parameter k will be discussed later. In the block matching process, we use k representative blocks to substitute the first k unused candidate blocks. Moreover, for each candidate block, a flag is used to indicate whether this block is the best-matching block of some secret image blocks. If so, $\text{flag} = 1$; otherwise, $\text{flag} = 0$. For the not-well-matched blocks, the indexes of their best-matching blocks are set to their representative blocks obtained by the k-means clustering method.

Now we discuss the value of the parameter k. Since the storage space of the p-LSB planes of the cover image by means of the p-layered STC is $\log_2(2p-1) \times h_{CI} \times w_{CI}$, the base and the index of a well-matched block take $(s+q)$ bits, the number of secret image blocks is $(h_{SI} \times w_{SI})/b$, the flags take t bits, and each not-well-matched block takes $(s+b \times q)$ bits, we have

$$\log_2(2p-1) \times h_{CI} \times w_{CI} \geq (s+q)\left(\frac{h_{SI} \times w_{SI}}{b} - k\right) + (s+b \times q) \times k + t . \tag{6}$$

Therefore, the maximum number of the representative blocks is

$$k = \left\lfloor \frac{\log_2(2p-1) \times h_{CI} \times w_{CI} - \dfrac{(s+q) \times h_{SI} \times w_{SI}}{b} - t}{(b-1) \times q} \right\rfloor . \tag{7}$$

Although we can find the optimal threshold $TH3$ that corresponding to the best quality of the extracted secret image by means of the exhaustive method, the computation time is too high. To find a trade-off between the visual quality and the computation time, we will use an iteration method to find an approximate optimal value of $TH3$. In this paper, an initial value φ is set to $TH3$, a small $\Delta\varphi$ is added to φ in each iteration and then we implement the embedding process to estimate the MSE introduced to the secret image with the current threshold. Repeat the process until the MSE of the current embedding attempt is larger than the last two previous embedding attempts and the threshold corresponding to the smallest MSE among all of these embedding attempts is selected as $TH3$ to embed the secret image.

For the sake of reducing the amount of the data embedded into the cover image **CI**, we use the optimal notational system transform algorithm in [24] to convert the t-ary indexes into the binary form, and then the bases, indexes, flags, and k representative blocks are divided into a sequence of 8-bit data sections. Finally, the Huffman coding is used to compress the sequence and the p-layered STC is used to embed the parameters t, b, k, $TH1$, $TH2$, $TH3$, h_{SI}, w_{SI}, the Huffman table, and the Huffman output stream with its length into the p LSB planes of **CI**, where the distortions caused by modifying the cover pixels   in the p-layered STC are defined as follow.

Assume that the cover pixels are $\mathbf{x} = (x_1, x_2, ..., x_n)$, $q = 8$, i.e., $0 \leq x_i \leq 255$ $(i = 1, \cdots, n)$, and the stego pixels are $\mathbf{y} = (y_1, y_2, ..., y_n)$. Suppose that $p = 3$, and $\mathbf{r} = (r_1, r_2, ..., r_n)$ denotes 3 LSB planes of the cover pixels, i.e., $r_i = x_i (\text{mod} 8)$ $(i = 1, \cdots, n)$. Since 5 MSB planes of the cover pixels can't be modified in the embedding process, we must define the distortion of modifying x_i to y_i, which is recorded as $\rho_i(x_i, y_i)$ $(i = 1, \cdots, n)$, according to r_i. For the 3-layered STC, the maximum modification amplitude of the pixel x_i is 2, and the set of modification modes is $\{-2, -1, 0, +1, +2\}$, i.e., the value range of y_i is $I_i = \{x_i - 2, x_i - 1, x_i, x_i + 1, x_i + 2\}$.

In order to avoid the circumstances that some modification modes may cause large distortion of the cover image, the definition of the distortion $\rho_i(x_i, y_i)$ needs some special treatments. Now we give the details of the definition of $\rho_i(x_i, y_i)$ below.

1. In the condition of $y_i = x_i$, $\rho_i(x_i, y_i) = 0$.
2. In the condition of $y_i = x_i - 1$, if $r_i = 0$, $\rho_i(x_i, y_i) = \infty$; otherwise, $\rho_i(x_i, y_i) = 1$.
3. In the condition of $y_i = x_i + 1$, if $r_i = 7$, $\rho_i(x_i, y_i) = \infty$; otherwise, $\rho_i(x_i, y_i) = 1$.
4. In the condition of $y_i = x_i - 2$, if $r_i = 0$ or 1, $\rho_i(x_i, y_i) = \infty$; otherwise, $\rho_i(x_i, y_i) = 2^2$.
5. In the condition of $y_i = x_i + 2$, if $r_i = 6$ or 7, $\rho_i(x_i, y_i) = \infty$; otherwise, $\rho_i(x_i, y_i) = 2^2$.

Note that, since the STC is an adaptive steganographic coding, in the circumstances of $\rho_i(x_i, y_i) = 1$ and 2^2, $\rho_i(x_i, y_i)$ can be adaptively redefined to enhance the security of the proposed method according to the side information of the cover image such as image texture.

Next we describe the steps of the proposed method.

1. Choose the appropriate parameters, p, t, b, where p is the number of LSB planes used to embed the payload, t is the number of the candidate blocks and b is the block size.
2. According to the order from left to right and from top to bottom, divide the cover image and the secret image into several nonoverlapping blocks of size $b = m \times n$, respectively.
3. According to the block matching process, construct the set of t candidate blocks.
4. The following steps are used to obtain the approximate optimal value of the threshold $TH3$.
   (1) Set an initial value φ to $TH3$ and add $\Delta\varphi$ to φ in each embedding attempt.
   (2) For each secret image block, find the best-matching block from the candidate set.
   (3) If the MSE between a secret image block and its best-matching block is not larger than φ, the secret image block belongs to the set of the well-matched blocks; otherwise, it belongs to the set of the not-well-matched blocks.
   (4) For the not-well-matched blocks, the k-means clustering method is used to select k representative blocks.

(5) Calculate the total MSE caused by the current embedding attempt, including the MSE between the well-matched blocks and their best-matching blocks, and the MSE between the not-well-matched blocks and their representative blocks.

(6) Repeat the steps (2)-(5) until the total MSE of the current embedding attempt is larger than two previous consecutive embedding attempts. The threshold corresponding to the smallest MSE is set as $TH3$.

5. According to $TH3$, use the following steps to embed the secret image.

(1) For each secret image block, find the index of its best-matching block, denoted as *index*, and set the flag of the *index*-th candidate block to 1.

(2) If the MSE between this secret image block and its best-matching block is larger than $TH3$, this block is classified into the set of the not-well-matched blocks.

(3) For the not-well-matched blocks, use the k-means clustering method to select k representative blocks.

(4) Use k representative blocks to replace the first k unused candidate blocks, and their flags are set to 0. For each not-well-matched block, set its best-matching block to its representative block.

6. Use the optimal notational system transform algorithm in [24] to convert the t-ary indexes to the binary form, and use the Huffman coding to compress the bases, indexes, flags and representative blocks.

7. Compute the distortions introduced by modifying the cover pixels, which are used in the p-layered STC.

8. Use the embedding process of the p-layered STC to embed the parameters t, b, k, $TH1$, $TH2$, $TH3$, h_{SI}, w_{SI}, the Huffman table, and the Huffman output stream with its length into the p LSB planes of **CI**, and obtain the stego image **ST**.

3.2   The Extracting Phase

The following steps are used to extract the secret image from the stego image **ST**.

1. By means of the extracting process of the p-layered STC, the following values are extracted from p LSB planes of the cover image: the parameters t, b, k, $TH1$, $TH2$, $TH3$, h_{SI}, w_{SI}, Huffman table and Huffman output stream with its length.

2. According to the Huffman table, decode the Huffman output stream to get the bases, binary indexes, flags and representative blocks. Utilize the optimal notational system transform algorithm in [24] to convert the binary indexes to the t-ary form.

3. For each flag, if flag $= 1$, construct the candidate block as discussed in the embedding phase. Otherwise, select a representative block in sequence as a candidate block. Repeat this process until t candidate blocks are generated.

4. Repeat the following steps until all the blocks of the secret image are extracted:

(1) Sequentially get the next index that is not processed.

(2) If flag $= 1$, get the candidate block corresponding to this index, add the base to it, and the result is served as a secret image block in sequence; otherwise, the representative block is served as a secret image block.

4   Experimental Results and Analysis

In our experiments, the test images are 8-bit standard gray images[1] of size 512×512, shown in Figure 3. In addition, the peak signal to noise ratio (PSNR) is used to measure the similarity of two gray images, which is defined as follows:

$$PSNR = 10 \times \log_{10}\left(\frac{255 \times 255}{MSE}\right) \qquad (8)$$

where the definition of MSE is the same as Equation (2).

In this paper, the PSNR between the stego image and the cover image is defined as PSNR(ST) for short. The stego image corresponding to larger PSNR(ST) is more similar to the cover image, i.e., the existence of the embedded secret image is not easier to attract the attention of the attackers. Meanwhile, the PSNR between the extracted image and the secret image is defined as PSNR(EI). The extracted image corresponding to larger PSNR(EI) is more similar to the secret image, i.e., the extracted image preserves more content characters of the secret image.

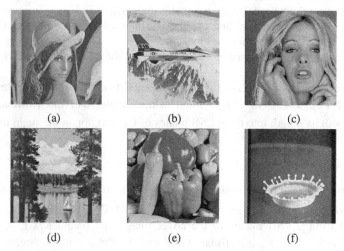

(a)   (b)   (c)

(d)   (e)   (f)

Fig. 3. Six test images: (a)Lena, (b)Jet, (c)Tiffany, (d)Sailboat, (e)Peppers, (f)Milk

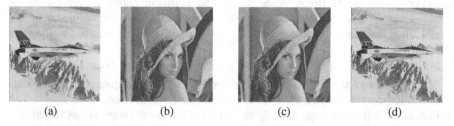

(a)   (b)   (c)   (d)

Fig. 4. Jet is embedded into Lena: (a) is the secret image, (b) is the cover image, (c) is the stego image, (d) is the extracted secret image

---

[1] http://sipi.usc.edu/database/

Table 1. The comparisons of PSNR(ST) and PSNR(EI) of five image-hiding schemes

[(PSNR(ST)/PSNR(EI))[11], (PSNR(ST)/PSNR(EI))[13], (PSNR(ST)/PSNR(EI))[14], (PSNR(ST)/PSNR(EI))(Proposed)]	Secret images			
Cover images	Jet	Lena	Tiffany	Sailboat
Jet		[(44.15/36.52), (45.16/40.01), (47.15/39.40), (**47.39/40.28**)]	[(43.81/35.85), (45.14/40.47), (47.46/38.21), (**47.68/40.53**)]	[(44.11/32.15), (45.16/36.53), (47.01/33.06), (**47.25/36.68**)]
Lena	[(44.14/38.13), (45.12/40.15), (47.05/39.62), (**47.32/41.00**)]		[(43.76/35.45), (45.10/40.63), (47.27/38.52), (**47.44/40.80**)]	[(44.10/32.33), (45.11/36.38), (47.09/32.98), (**47.38/36.63**)]
Tiffany	[(44.16/36.21), (45.15/39.92), (47.13/40.19), (**47.47/41.02**)]	[(44.08/35.02), (45.13/40.09), (47.33/39.66), (**47.69/40.28**)]		[(44.04/31.78), (45.16/36.16), (47.12/33.52), (**47.41/36.52**)]
Sailboat	[(44.13/36.70), (45.10/39.84), (46.95/39.15), (**47.22/40.56**)]	[(44.04/34.72), (45.11/39.58), (47.03/38.83), (**47.23/40.18**)]	[43.97/37.39], (45.07/40.12), (47.22/38.50), (**47.60/40.23**)]	

In experiment I, we attempt to embed a secret image into a cover image of the same size. 3 LSB planes of the cover image are used to carry the payload, i.e., $p = 3$, the block size is $b = 4 \times 4$, $TH1 = 3$, and $TH2 = 32$. Moreover, the threshold $TH3$ and the number of clustering centroids k will be different along with different cover images and secret images. Figure 4 shows an example of embedding Jet into Lena, where (a), (b), (c) and (d) are the secret image, the cover image, the stego image and the extracted secret image respectively. In this example, $\varphi = 32$, $\Delta\varphi = 1$, $k = 1683$, and $PSNR(ST) = 47.32$, from which we can see that the visual quality of the stego image is high and the attackers can't observe the existence of the hidden secret image. Simultaneously, $PSNR(EI) = 41.00$, from which we can see that the visual quality of the extracted secret image is also high and the extracted secret image preserves the content characters of the original secret image well. Table 1 shows the comparisons of PSNR(ST) and PSNR(EI) of five image-hiding schemes, from which we can see that, when the size of the secret image is the same as the cover image, both PSNR(ST) and PSNR(EI) in the proposed method are higher than those in [11, 13, 14] and then the visual qualities of both the stego images and the extracted secret images obtained by the proposed method are higher than [11, 13, 14].

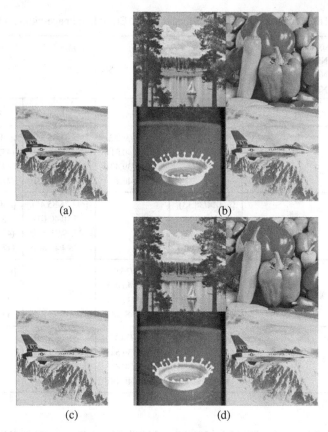

(a)   (b)

(c)   (d)

Fig. 5. The secret image of size 1024×1024 is embedded into Jet of size 512×512: (a) is the cover image, (b) is the secret image, (c) is the stego image, (d) is the extracted secret image

In experiment II, we attempt to embed a secret image that is four times larger than the cover image. 4 LSB planes of the cover image are used to carry the payload, i.e., $p = 4$, and the block size is $b = 8 \times 8$, $TH1 = 3$, and $TH2 = 32$. Figure 5 shows the results of experiment II. Figure 5(b) is the secret image of size 1024×1024 comprised of Sailboat, Peppers, Milk, and Jet. In this example, $\varphi = 200$, $\Delta\varphi = 1$, and $k = 718$. The stego image and the extracted secret image are shown in Figure 5(c) and 5(d) respectively. Table 2 shows the comparisons of PSNR(ST) and PSNR(EI) of five image-hiding schemes, from which we can see that, when the size of the secret image is four times larger than the cover image, both PSNR(ST) and PSNR(EI) in the proposed method are higher than those in [11, 13, 14] and then the visual qualities of both the stego image and the extracted secret image obtained by the proposed method are higher than [11, 13, 14].

Table 2. The comparisons of PSNR(ST) and PSNR(EI) of five image-hiding schemes when Figure 5(b) is embedded into Figure 5(a)

Image-hiding schemes	PSNR(ST)/PSNR(EI)
Scheme in [11]	37.79/29.63
Scheme in [13]	39.15/32.35
Scheme in [14]	36.85/30.86
Proposed scheme	**39.54/32.75**

5   Conclusion

The rapid development of the Internet makes the transmission of digital images over the Internet become a common communication mode. However, since we need to protect the contents of the secret images and ensure the concealment of the transmission, the conventional encryption techniques are not suitable. This paper proposes a new image-hiding method, which mainly contains the following steps: the generation of candidate blocks, the best-matching block process, the special handling of the not-well-matched blocks, the compression of the relevant data, and the embedding of the compressed data by means of the multi-layered STC. The experimental results show that the visual qualities of the stego images and the extracted secret images obtained by the proposed method outperform four previous methods. How to construct more suitable candidate set will be studied in the future.

Acknowledgement. This work is supported partly by the National Science-Technology Support Plan Project of China (No. 2012BAH47B01), partly by the Natural Science Foundation of China (No. 61170234, 61309007), and partly by the Municipal Science and Technology Innovation Team Project of Zhengzhou (No. 10CXTD150).

References

1. Chan, C., Cheng, L.: Hiding data in images by simple LSB substitution. Pattern Recognition 37(3), 474–496 (2004)
2. Chang, C., Hsiao, J., Chan, C.: Finding optimal least-significant-bits substitution in image hiding by dynamic programming strategy. Pattern Recognition 36(7), 1583–1595 (2003)
3. Wang, R., Lin, C., Lin, J.: Image hiding by optimal LSB substitution and genetic algorithm. Pattern Recognition 34(3), 671–683 (2001)
4. Chang, C., Chan, C., Fan, Y.: Image hiding scheme with modulus function and dynamic programming. Pattern Recognition 39(6), 1155–1167 (2006)
5. Thien, C., Lin, J.: A simple and high-hiding capacity method for hiding digit-by-digit data in images based on modulus function. Pattern Recognition 36(12), 2875–2881 (2003)
6. Wang, S.: Steganography of capacity required using modulo operator for embedding secret image. Applied Mathematics and Computation 164(1), 99–116 (2005)

7. Fridrich, J., Goljan, M.: Practical steganalysis of digital images: state of the art. In: Proceedings of the SPIE 4675, Security and Watermarking of Multimedia Contents IV, pp. 1–13. SPIE (2002)

8. Chung, K., Shen, C., Chang, L.: A novel SVD- and VQ-based image hiding scheme. Pattern Recognition Letters 22(9), 1051–1058 (2001)

9. Maniccam, S., Bourbakis, N.: Lossless compression and information hiding in images. Pattern Recognition 37(3), 475–486 (2004)

10. Hu, Y.: High-capacity image hiding scheme based on vector quantization. Pattern Recognition 39(9), 1715–1724 (2006)

11. Wang, R., Tsai, Y.: An image-hiding method with high hiding capacity based on best-block matching and k-means clustering. Pattern Recognition 40(2), 398–409 (2007)

12. Wang, R., Chen, Y.: High-payload image steganography using two-way block matching. IEEE Signal Processing Letters 13(3), 161–164 (2006)

13. Chen, S., Wang, R.: High-payload image hiding scheme using k-way block matching. In: Proceedings of the 6th International Conference on Intelligent Information Hiding and Multimedia Signal Processing, pp. 70–73. IEEE (2010)

14. Hsieh, Y., Chang, C., Liu, L.: A two-codebook combination and three-phase block matching based image-hiding scheme with high embedding capacity. Pattern Recognition 41(10), 3104–3113 (2008)

15. Lee, Y., Lee, J., Chen, W., Chang, K., Su, I., Chang, C.: High-payload image hiding with quality recovery using tri-way pixel-value differencing. Information Sciences 191, 214–225 (2012)

16. Chang, K., Chang, C., Huang, P., Tu, T.: A novel image steganographic method using tri-way pixel-value differencing. Journal of Multimedia 3(2), 37–44 (2008)

17. Tomas, F., Jan, J., Jessica, F.: Minimizing additive distortion in steganography using syndrome-trellis codes. IEEE Transactions on Information Forensics and Security 6(3), 920–935 (2011)

18. Pevny, T., Filler, T., Bas, P.: Using high-dimensional image models to perform highly unde-tectable steganography. In: Proceedings of the 12th International Conference on Information Hiding, pp. 161–177. Springer, Berlin (2010)

19. Zhang, X., Zhang, W., Wang, S.: Efficient double-layered steganographic embedding. Electronics Letters 43(8), 482–483 (2007)

20. Zhang, W., Zhang, X., Wang, S.: A double layered "plus-minus one" data embedding scheme. IEEE Signal Processing Letters 14(11), 848–851 (2007)

21. Vojtěch, H., Jessica, F., Tomáš, D.: Universal distortion function for steganography in an arbitrary domain. EURASIP Journal on Information Security 2014(1), 1–13 (2014)

22. Guo, L., Ni, J., Shi, Y.: Uniform Embedding for Efficient JPEG Steganography. IEEE Transactions on Information Forensics and Security 9(5), 814–825 (2014)

23. Vojtěch, H., Jessica, F.: Designing steganographic distortion using directional filters. In: Proceedings of the 4th International Workshop on Information Forensics and Security, pp. 234–239. IEEE (2012)

24. Chen, J., Zhang, W., Hu, J., Zhu, Y., Guo, D.: An efficient (k, p) notational system transform algorithm. Journal of Applied Sciences 31(6), 569–578 (2013)

Educational Forums at a Glance:
Topic Extraction and Selection

Bernardo Pereira Nunes[1], Ricardo Kawase[2], Besnik Fetahu[2],
Marco A. Casanova[1], and Gilda Helena B. de Campos[3]

[1] Department of Informatics – PUC-Rio – Rio de Janeiro, Brazil
{bnunes,casanova}@inf.puc-rio.br
[2] L3S Research Center – Leibniz University Hannover – Hannover, Germany
{kawase,fetahu}@L3S.de
[3] Department of Education – PUC-Rio – Rio de Janeiro, Brazil
gilda@ccead.puc-rio.br

Abstract. Web forums play a key role in the process of knowledge creation, providing means for users to exchange ideas and to collaborate. However, educational forums, along several others online educational environments, often suffer from topic disruption. Since the contents are mainly produced by participants (in our case learners), one or few individuals might change the course of the discussions. Thus, realigning the discussed topics of a forum thread is a task often conducted by a tutor or moderator. In order to support learners and tutors to harmonically align forum discussions that are pertinent to a given lecture or course, in this paper, we present a method that combines semantic technologies and a statistical method to find and expose relevant topics to be discussed in online discussion forums. We surveyed the outcomes of our topic extraction and selection method with students, professors and university staff members. Results suggest the potential usability of the method and the potential applicability in real learning scenarios.

1 Introduction

Over the past decade, the World Wide Web became an important source of information and knowledge. The diversity and engagement of independent users and communities contributed to the creation and proliferation of a rich set of content available in different communication channels (such as social media, real-time channels, blogs, forums, etc.) as well as in formats (such as text, audio and video).

In particular, online discussion forums have played a key role in the process of knowledge creation [13], providing means for its users to exchange ideas, form opinions, position themselves and collaborate. As an outcome of the importance of online discussion forums is Wikipedia[1], where for each Wikipedia article there is a forum-based page[2] that relies on the collaboration, discussion, consensus and collective effort of its users to keep Wikipedia constantly updated and curated.

---

[1] http://www.wikipedia.org
[2] http://en.wikipedia.org/wiki/Help:Using_talk_pages

B. Benatallah et al. (Eds.): WISE 2014, Part II, LNCS 8787, pp. 351–364, 2014.
© Springer International Publishing Switzerland 2014

Due to the benefits generated by users' participation in forums, most online courses combine educational materials and message boards. However, even though forums clearly leverage the creation of collective intelligence [19], the assessment of users' participation is still rather difficult [12,18]. Depending on the number of students and posts, manual assessment becomes impractical. Previous work addressed the problem of assessing the quality of students' participation [15,16]. However, they do not take into account whether a particular set of topics were addressed in a thread of a specific discipline.

Furthermore, different backgrounds in online discussion forums may lead a discussion to unforeseen directions, needing external support to realign the discussed topics of a thread. This task is often conducted by a tutor or moderator. But, as we will show in this paper, on average, 50% of forums discussing a specific subject with different audience or tutor/moderator cover distinct topics. This means even though online discussions are often different, a set of specific topics must be addressed to achieve the course goals. Therefore, if a given forum does not cover a set of expected topics, the assessment of the students might be hampered, since the acquired knowledge depends on the topics discussed in the forum.

In this paper, we combine semantic technologies and a statistical method to find, expose and recommend relevant topics as guidance to conduct debate forums. Briefly, with the help of semantic tools, the proposed method first performs Named-Entity Recognition (NER) and topic extraction, followed by a statistical approach that selects and ranks the most relevant topics of a forum thread. Finally, the method outputs the topmost representative topics discussed in a specific forum as well as a set of suggested topics to be discussed. We used 97 online forums from a Brazilian university to validate and assess our method.

Our main contribution in this work is the development of a well perceived semantic-based topic enrichment model for educational forums, in combination with its evaluation. Subsequently, this contribution accounts for positive effects in high-level assessment of tutor/moderator progress, topic recommendation and parity of knowledge acquisition by students in online forums.

The rest of this paper is organised as follows. Section 2 reviews related literature. Section 3 describes the use of forums in our context. Section 4 introduces the topic extraction and selection method. Moreover, we also extended Vygotsky's zone of proximal development to serve as a recommendation method. Section 5 presents the evaluation setup. Section 6 discusses the results obtained in the evaluation along with a brief analysis of the topics extracted from the forum threads. Finally, Section 7 discusses our outcomes and future work.

2 Related Work

Li and Wu [11] combine approaches involving sentiment analysis and text mining to detect hotspot forums within a certain time span. Their method assists users to make decisions and predictions over polarised groups of messages in online forums. Despite not performing topic extraction in the hotspot forums, the emotional polarity information for each topic extracted would help users on understanding how a given topic is addressed in a discussion.

Cong et al. [3] present an approach for finding question-answer pairs in online forums based on Labeled Sequential Patterns (LSPs) and graph-based propagation model. While the creation of patterns for interrogative sentences is made using part-of-speech tags, the answers are detected and ranked using KL-divergence language model. Again, our approach is complementary to their approach, since our approach would serve as a filter for finding question and answers based on topics. Conversely, our method would benefit of this approach by identifying key posts in an online discussion.

Online forums play a key role in the student skills development as shown by Scaffidi et al. [17]. Their study focuses on the types of posts that facilitate discussions and collaboration amongst novice developers. The study of user behavior in online forums help to promote active interaction amongst users and therefore the construction of collective knowledge. We believe that the introduction of new topics to be discussed by such community of users could trigger new discussions and hence new knowledge.

Desanctis et al. [6] provide an interesting discussion about e-venues for learning such as video-conferenced classrooms, online communities and group discussion spaces. Although each venue influences the learning process of a particular group, they all have in common the need to bring new discussions that promote the development of knowledge of the participants. For instance, online communities usually last more than private group discussion spaces, since new participants with fresh questions can drop in at anytime. Thus, in order to maintain the group discussion, the recommendation of new topics for discussion would foster longer interactions amongst participants and knowledge refreshment.

Evidently, online discussions can also be fueled by tutors responsible for bringing new topics and questionings for the discussion. Previous studies [4] have shown that tutored venues can improve both retention and performance of the participants. In this paper, we use the tool for assisting tutors on addressing new topics relevant to the discussion.

A highly relevant direction of work goes onto the topic extraction from forums' text. Hulpus et al. [10] extract a set of topics from a given textual resource. The topics correspond to a DBpedia sub-graph category. Furthermore, the relationship between the topics and textual resources are quantified using graph importance measures. In our case, we simply aim at providing students with discussions from the forums specific to a topic. Hence, simple tf-idf techniques offer us the efficiency on distinguishing the topic specific discussions. This simple, yet efficient technique offers the scalability over large corpora, and at the same we avoid exhaustive computations that rely on graph centrality measures like the ones in [10].

Finally, research on topic modelling and extraction has been addressed on various ways and for different purposes. A well known approach LDA [2] has seen a wide applicability on modelling and extracting topics from textual documents. Other approaches like [7] extract and rank topics with respect to their relevance to specific datasets, which are extracted through a named entity disambiguation process. In contrast to the previous approaches, in our case we aim at suggesting forum pages for discussion specific to online learning scenarios, hence, the problem becomes simpler with respect to filtering specific forum pages rather than exhaustive rankings of forum pages and their corresponding topics.

3 Motivation

To illustrate the motivation of our research, we describe two scenarios where participants of online discussion forums would benefit from our method. Both scenarios result from the need of the staff from a Brazilian university to assess the participations in forums and topics discussed.

Online discussion forums are fundamental in the learning process and most of the online courses take advantage of their use to meet specific goals. Assessing student participation in forums is not a simple task, and due to the high number of posts, it can become impracticable. Hence, in order to maintain the quality of teaching and student experience, the university staff members required a tool to track the discussion progress.

The first scenario described by the university staff members is that tutors constantly overlook the discussion of relevant topics in favor of a better flow. Although the discussion flow is of utmost importance, tutors must conduct the forum in such a way that specific topics must be addressed and, at the same time, preserve the discussion flow. Hence, the university staff members are interested in the analysis of forums to check if particular topics were covered in a thread. By doing this, they can ensure that all participants had similar experience and learning situations that can contribute to the next activities. In the case that a set of topics are not covered, they would like to intervene and extend the forum closure or create a new forum thread to discuss the missing topics.

The second scenario aims at fostering the discussion with suggestions that may assist students in the discussion. For many reasons, some forums lack interaction and students must be encouraged to participate. In this manner, university staff members believe that a recommendation tool would promote the discussion and help to reach the forum's goal.

The current work assists university staff members and students to have a better overview of what is happening in the forum to take the right action and create situations/activities that can improve the learning experience of the students.

4 Topic Extraction and Selection

In this section we present the main steps for a coherent process chain that semantically and statistically selects the most relevant discussed topics in a given online discussion forum. The process chain, depicted in Figure 1 is composed of three steps described as follows: (i) Entity Extraction and Enrichment; (ii) Topic Extraction; and (iii) Topic Selection. We also present in this section a simple topic recommendation method used to assist learners and tutors in the teaching-learning process.

4.1 Entity Extraction and Enrichment

When dealing with online discussion forums, we are essentially working with unstructured data, which in turn hinders data manipulation and the identification of atomic elements in texts. To alleviate this problem, information extraction (IE) methods, such as Named-Entity Recognition (NER) and name resolution, are employed. These tools

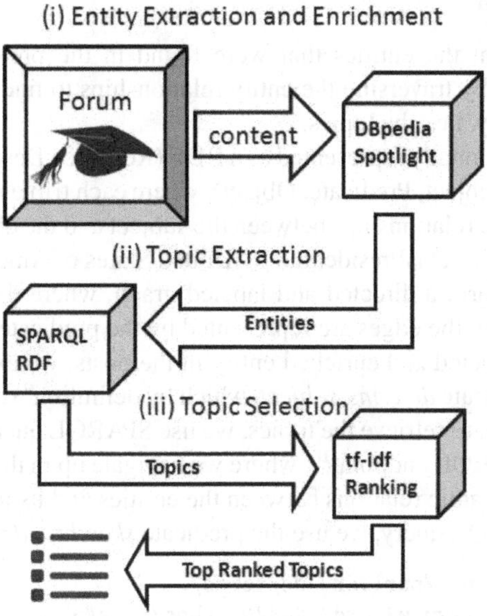

Fig. 1. Topic extraction process workflow

automatically extract structured information from unstructured data and link to external knowledge bases in the Linked Open Data cloud (LOD), such as DBpedia[3].

For instance, after processing the following sentence using an IE tool: "I agree with Barack Obama that the whole episode should be investigated.", the entity "Barack Obama" is annotated and classified as *person* and linked to the DBpedia resource <http://dbpedia.org/resource/Barack_Obama>, where structured information about him is available.

We use the DBpedia Spotlight tool[4] to extract and enrich entities found in the posts within a forum thread. DBpedia Spotlight adds markups with semantic information surrounding atomic elements (entities) in the forum posts (as in [14]). These entities are the ones found in DBpedia dataset, and each one contains structured information extracted from Wikipedia[1].

Note that our method is language independent as long as we have a solid repository of entities (such as DBpedia or Freebase[5]) and a proper annotation tool (such as Spotlight, Alchemy[6] or WikipediaMiner[7]). However, the set of entities that can be identified by the annotation process is limited to the number known entities in the dataset, in our case, the Portuguese DBpedia dataset. This dataset currently contains 736,443 entities[8].

---

[3] http://www.dbpedia.org
[4] http://dbpedia-spotlight.github.io/demo/
[5] http://www.freebase.com
[6] http://www.alchemyapi.com
[7] http://wikipedia-miner.cms.waikato.ac.nz
[8] http://wiki.dbpedia.org/Datasets39/DatasetStatistics

4.2 Topic Extraction

Given as starting point the entities that were found in the previous step, the topic extraction step begins by traversing the entity relationships to find a more general representation of the entity, i.e., the topics.

An entity is conventionally represented as a RDF (Resource Description Framework) triple in the form of (Subject, Predicate, Object), where each triple represents a fact, and the predicate names the relationships between the subject and the object. For example, a triple is ("Barack Obama", "isPresidentOf", "United States of America"). Furthermore, a set of RDF triples form a directed and labeled graph, where the nodes are a set of subjects and objects and the edges are represented by the predicate.

Thus, for each extracted and enriched entity in the posts, we explore their relationships through the predicate *dcterms:subject*, which by definition[9] represents the topic of the entity. In that sense, to retrieve the topics, we use SPARQL query language for RDF over the DBpedia SPARQL endpoint[10], where we navigate up in the DBpedia hierarchy to retrieve broader semantic relations between the entities and its topics. As it is shown in the following SPARQL query, we use the predicate *skos:broader*.

```
PREFIX dcterms: <http://purl.org/dc/terms/>
PREFIX skos: <http://www.w3.org/2004/02/skos/core#>

    SELECT DISTINCT ?l1 ?l2 ?l3 ?l4
    WHERE {
        <entity_uri> dcterms:subject ?l1 .
        ?l1 skos:broader ?l2 .
        ?l2 skos:broader ?l3 .
        ?l3 skos:broader ?l4
    } LIMIT 1000;
```

The variable *entity_uri* represents the entity in which we are interested in retrieving the topics extracted from the posts in a forum thread, while the variables *l1* to *l4* represent the topics that will be retrieved from the entity. Thus, given an entity, the topics of an entity are retrieved through the predicate *dcterms:subject* and *skos:broader*. The latter predicate is used to obtain a more general representation of the topic. This strategy will help us find the topics that best cover a forum thread.

Note that an entity/concept can be found in different levels of the hierarchical categories of DBpedia, and hence this approach would lead us to retrieve topics in different category levels. However, as in previous works [9,8], we take advantage of the co-occurrence of the topics in the different levels to find the most representative ones (see Section 4.3).

4.3 Topic Selection

Finally, in this last step, we select the most representative topics extracted from the posts that belong to a forum thread. For this, we rely on *tf-idf* (term frequency - inverse

---

[9] http://dublincore.org/documents/2012/06/14/dcmi-terms/
?v=terms#elements-subject

[10] http://pt.dbpedia.org/sparql - DBpedia SPARQL endpoint in portuguese.

document frequency) score to statistically measure the importance of a topic in a forum thread.

Typically, *tf-idf* is used on information retrieval and text mining to measure the importance of a word to a document in a collection. However, in this paper, we adapted this metric to take into account entities and topics extracted from the posts instead of words.

To select the most representative topics, we compute *tf-idf* score twice, one for the entities extracted from the forum thread (i.e. the most representative entities in the collection) and another for the topics extracted from the entities (see Section 4.2).

Basically, to compute the term frequency (*tf*), we count the number of occurrences of an entity e in a post $p \in P$. As for the inverse document frequency (*idf*), we compute the (*idf*) score by dividing the total number of posts $|P|$ by the number of posts containing the entity $|P_e|$, see Eq. 1.

$$tfidf(e, p, P) = tf(e, p) \times idf(e, P) \tag{1}$$

where *tf* is the raw frequency of a term in a post, and *idf* is the measure of commonness/rareness of an entity in a collection P. *tf* and *idf* can be computed by the equations 2 and 3, respectively.

$$tf(e, p) = frequency(e, p) \tag{2}$$

$$idf(e, P) = log(\frac{|P|}{|P_e|}) \tag{3}$$

After computing the *tf-idf* score for each entity, the topmost representative entities are selected. From the selected entities, the topics are extracted according to the process described in Section 4.2.

With the topics in hands, we then compute the *tf-idf* score over the topics extracted from the entities and decreasingly rank them. Again, the topmost representative topics for a given forum thread are selected. Note that the number of topics that represent a forum is chosen by the user (in our case, the top 10 relevant topics). Finally, the top ranked topics are selected to represent the forum thread topics.

4.4 Topic Recommendation

Another contribution of this paper lies in the recommendation of topics based on the Zone of Proximal Development (ZPD) introduced by Vygotsky [20]. Briefly, this concept of Vygotsky describes the distance between the independent performance of an individual to perform a certain task and the performance of the individual when assisted by more capable peers. Thus, as the ZPD concept suggests, the assistance of an external peer may improve the learners' skills.

In this paper we extended the ZPD concept to perform as a topic recommendation tool to learners participating in educational forums. As in our context forum threads occur simultaneously, we consider a sibling-forum[11] as the more capable peer. Hence, the topic recommendation is based on the topics discussed in the sibling-forums.

Following the technique presented in the previous sections, the topmost representative topics discussed in a sibling-forum that are missing in the actual forum thread are recommended as topic seeds to foster the discussion. Although simple, the recommendation assists learners and tutors on addressing topics overlooked in the current thread and to broaden the discussion to topics that they would not address without the indirect assistance of their peers. We would like to emphasize that tutors and learners can always opt to accept or not the recommendation.

5 Evaluation Setup

In this section, we present the evaluations performed to validate the applicability of our method in real scenarios. The first evaluation consists of a questionnaire to university staff and participants of the forums. The second evaluation consists of expert manual assessment of the generated topics performed by two educators.

5.1 Technology Acceptance Model Evaluation

Over the course of our study, real data from online discussion forums were used to perform a comprehensive evaluation of our method. It was evaluated using 97 online discussion forums containing in total 10,785 anonymised posts provided by the distance education department of a Brazilian university. All selected threads occurred at least twice concurrently. Furthermore, each professor assessed the suggested topics from forums conducted by themselves.

Our main objectives included a thorough assessment of the recommendation of topics based on previous online discussion forums as well as the assessment of the selected topics that cover a forum discussion. For this, we submitted 3 questionnaires to 11 students, 4 professors and 3 coordinators of the distance education department to gather different perspectives and views of the proposed method.

The questionnaires were divided into three different categories of questions, namely *perceived usefulness*, *perceived ease-of-use* and additional suggestions. Basically, the questions followed the Technology Acceptance Model (TAM) proposed by Davis [5], arguably the most influential "Technology Acceptance Theory".

Briefly, this theory states that there are two key aspects to measure users' intention to adopt a new technology, the *perceived usefulness* and *perceived ease-of-use*. Perceived usefulness (PU) refers to "the degree to which a person believes that using a particular system would enhance his or her job performance", while perceived ease of use (PEOU) refers to "the degree to which a person believes that using a particular system would be free of effort" [5].

Each questionnaire was divided in 6 PU questions, 6 PEOU questions and additional 3 opinion mining questions where we asked participants for further suggestions. Feedback for assertions such as *'The evaluation performed by teachers can be facilitated.'*,

---

[11] Two or more forums are considered as sibling-forums if they address the same subject, occur simultaneously and have different tutors and learners.

'The tool can broaden the discussion.', and *'Suggested topics are relevant to the topics discussed.'* were collected in a 5-point Likert scale fashion.

Note that, in the case of university staff members, the topics were assessed over two randomly chosen forum threads, since they did not participate on the forum. Thus, a list of topics discussed in the forum and a list of suggested topics for each forum thread was available for their evaluation. As they are staff members of the university, they also have access to the forum discussions in case they would need additional information.

5.2 Expert Assessment

In parallel to the evaluation presented in Section 5.1, we recruited two experts in the distance education department. These experts are part of the senior university staff and are directly involved in research and in the management of online courses. The main objective of this evaluation is to have a first look on the performance of the proposed method in terms of precision. In practice, two distinct aspects were evaluated by the experts. First, (i) they evaluated the correctness of the assigned topics for a given forum thread. Second, (ii) they evaluated the topics that were recommended to a given forum thread based on previous forum threads.

For this evaluation, we randomly selected 22 forum threads from our corpus to be inspected by the two experts. In total, both experts read all 2070 posts corresponding to 19,1% of the total number of posts of our corpus.

(i) First expert assessment: topic assignment. After reading through all the posts of the randomly selected forum threads, the experts were presented with the top 10 topics that were automatically identified by our method. Next, the experts were asked to mark which topics were correct, and which were incorrect assigned to the forum (precision).

(ii) Second expert assessment: topic recommendation. Again, after reading through each randomly selected forum thread, the experts were presented with topic recommendations. We recall that the topic recommendation was based on the Zone of Proximal Development introduced by Vygotsky (see Section 4.4). We used the topics discussed in sibling-forums threads to recommend topics to the current thread. We considered the sibling-forum thread as the more capable peer and recommended the missing topics to the current discussion in order to assist the teaching-learning process. Similar to the previous assessment, the experts were asked to mark which recommended topics might be relevant for the discussion. Additionally, they were asked how important (in a 5-point Likert scale) the recommended topics were.

6 Results

Following the evaluation strategy presented in the previous section, we first present the results obtained by the TAM model and subsequently the results obtained by the evaluation performed with distance learning experts.

6.1   Technology Acceptance Model Results

The results of the questionnaires are summarised in Figure 2. The error bar charts show that all participants reported a high positive perception for the proposed topics, the implications and the applicability of the results. In particular, professors had a slightly better acceptance, when compared to the other tiers of participants. The coefficient of internal consistency Cronbach's α of 0.65 for PU and 0.72 for PEOU indicated a good reliability of the results. These questionnaire results suggest the potential usability of our proposed topic extraction and selection method.

Regarding the suggestions included in the questionnaires, we observe that the most controversial question referred to whether or not the recommended topics should be available for professors, students or both. All professors suggested that the topics should be available only to them. All staff members suggested that topics should be available to both. Interestingly, students did not come to a common agreement. While the majority (64%) agreed that suggested topics should be available to professors and students, 36% opined that topic suggestions should be available only to professors.

We believe that the controversy is raised by the different backgrounds each group of participants had and the understanding they had of the topics. Staff members, who are not effectively involved in the online forums, assumed that the discussed topics should come out from an agreement between professors and students. On the other hand, the opinion of professors that a tool should present topics recommendation directly to them in fact reflects their need to control those around them. Finally, the split students' opinions lie in the fact that some students are still skeptical that online educational forums can smoothly evolve without proper moderation.

Unlike the questionnaires given to students and staff members, professors' questionnaire had an additional question regarding whether other professors can benefit from the suggested topics. The results reported that 75% of the professors strongly agree that *other* professors would take advantage of the suggested topics.

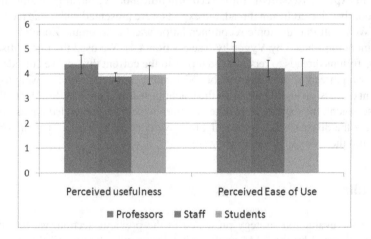

Fig. 2. Error Bars for survey questions regarding perceived usefulness and perceived ease-of-use

Finally, all staff members and professors (strongly) agree that the assessment of students would be facilitated if disparate forums addressed the same topics. Likewise, all staff members (strongly) agree that the proposed method would help in the assessment of the professor regarding the coverage of topics addressed in the forums. Nevertheless, 88% of all participants agree that the use of such method should be optional, and therefore, preserve the independence of tutors and learners.

6.2 Assessment Results by Distance Learning Experts

Table 1 depicts the results of the experts' assessment. The table expose individual (for each expert) and combined results. The combined results required the positive matching of both experts' opinion. Thus, for given a topic to be classified as *correct*, both experts must agree. If one of the experts marked as *incorrect*, the topic is automatically classified as incorrect.

In the (i) topic assignment assessment, each discussion forum was assigned with the 10 best ranked topics, thus the results are presented as precision (P@1, P@5 and P@10). In the (ii) topic recommendation assessment, since the recommendations were originated from sibling-forums, not all of them received the same number of recommendations. In average, each forum received 4.95 topic recommendations ($\sigma = 2.68$). This means that the number of suggested topics is equivalent to, in average, 50% of the topics discussed in sibling-forums. This result demonstrates that sibling-forums being conducted by different tutors and having different learners can take different directions. Thus, the topic recommendation method has shown to be extremely important to assist tutors in the guidance of the forum thread and to align the topics being coverage in sibling-forums.

The results given by the experts' assessment show a high precision achieved by our proposed method. It reaches close to 100% precision for the top 1 topic assignment and impressive results above 82% for top 5 and top 10 topics. For the harder task of topic recommendation, we also observe quality results with average precision above 73%. These results reinforce the findings from the Technology Acceptance Model evaluation, reaffirming the benefits of our topic extraction and recommendation method.

Table 1. Expert assessment results

	Topic Assignment			Topic Recommendation
	(P@1)	(P@5)	(P@10)	(Avg. Precision)
Expert 1	100	94.63	90.97	83.90
Expert 2	97.56	90.24	86.34	79.51
Combined	97.56	87.80	82.19	73.17

6.3 Discussion Evolution

It is noteworthy that by applying our proposed method, we are able to post-identify the top topics of a given forum discussion. However, for effective topic guidance support, it is important that the top topics are identified on the fly, during the progress of the discussion. To understand the convergence of the automatic identified topics, we incrementally generated topics for 88 forum discussions. The selected forums had at least 50 posts each.

We considered for this experiment a 10-post step granularity, i.e. after every 10 new posts in a discussion forum, we re-generated the list of top 10 topics and compare with the previous list. We used the overlap of topics in the lists (precision) as a metric for comparison. We defined a convergence of topics if the overlap between the lists is equal or greater than 90%.

Out of the 88 forum discussions that were used in this evaluation, in only 10 cases (11.4%) we observed that the identified topics diverge after converging above the threshold.

In average, the topics converge after 37.9% ($\sigma = 26.7$) of the posts in a discussion forum. In practice, 52.3% of the discussion forums have the assigned topics converging after 20 posts, and 79.5% after 30 posts.

From this analysis we infer that, with 30 posts as input, the method can provide descriptive topics with descent performance. This result is important for the setup of the method and deployment in real scenarios. The topic recommendation also fosters new discussions and open new directions in the discussion. Once again, we recall to ZPD concept to show that with external assistance the discussion can become richer.

7 Discussion and Outlook

We presented a method for automatically generating topics that represent a forum thread in distance learning environments. We combined semantic and statistical techniques in a coherent process chain to extract, select and rank the most relevant topics of a forum. Moreover, we also introduced a simple topic recommendation method based on Vygotsky's educational theory.

Our experiments showed that most professors, university staff and students are willing to use our proposed approach in future forums. Moreover, 75% of the professors reported that other professors would benefit from the suggested topics.

Reviewing a sample of 97 forum threads, we verified that, on the average, 50% of the topics discussed in disparate forums addressing the same subject are different. This situation resulted in a concern with regard to the topics addressed in the forums and the post assessment of the students. A priori, students in disparate forums covering the same subject should have a similar experience and learn the same topics.

Thus, providing a method to overview the topics discussed in different forums will help university staff members, such as course coordinators, to rapidly intervene in forums that topics are being overlooked. The topic recommendation method has proven useful for the alignment and diversification of the discussions between sibling-forums. As reported in Section 6.3 52.3% of the topics discussed in forum threads converge after 20 posts, and 79.5% after 30 posts. Thus, if the suggested topics are taken into consideration by tutors and learners during the discussion, it may last longer, active and cover the expected topics for the discussion.

In theory, the use of the proposed method would bring more control of what is being taught in a forum and, therefore, ensure quality. In practice, this can be different and some considerations arose out of the purpose of using the proposed method by a few interviewed respondent.

The first consideration lies in the freedom of the professors in guiding the forums. As every professor has its own teaching style and may also have a different point-of-view when they approach a subject, the concern of having to address specific topics in a forum might decrease the creativity and engagement of some professors. On the other hand, (assistant) professors may also take advantage of the suggested topics to guide the forum.

Another consideration with respect to the suggested topics is its availability to students. In the same time a topic suggestion may trigger an insight or make some students more confident, other students may stick only to the suggested topics. In the latter case, professors may take advantage of the students' participation and use it as a starting point to new discussions.

In general, the proposed method aims at assisting university staff members, professors and students to have a better overview of what is being discussed in the forum and, therefore, enable professors to take more informed actions to preserve discussion flow, improve students' experience and ensure topic coverage.

Our method also provides to the university staff members the possibility of assessing forum coverage, tracking what students are learning in different forums and, in some cases, detecting deviations in the forums. Adopting the method, depends on the instructional design of the course. The set-up of the course is crucial to determine which methods must be used and who will use it (professors, students or both).

As for future work, we plan to expand the method to accept external topic suggestions. For instance, professors involved in the course can also add topics to the discussion. Furthermore, we also plan to create a Moodle plugin.

References

1. Auer, S., Bizer, C., Kobilarov, G., Lehmann, J., Cyganiak, R., Ives, Z.: Dbpedia: A nucleus for a web of open data. In: Aberer, K., Choi, K.-S., Noy, N., Allemang, D., Lee, K.-I., Nixon, L.J.B., Golbeck, J., Mika, P., Maynard, D., Mizoguchi, R., Schreiber, G., Cudré-Mauroux, P. (eds.) ASWC 2007 and ISWC 2007. LNCS, vol. 4825, pp. 722–735. Springer, Heidelberg (2007)
2. Blei, D.M., Ng, A.Y., Jordan, M.I.: Latent dirichlet allocation. In: Dietterich, T.G., Becker, S., Ghahramani, Z. (eds.) NIPS, pp. 601–608. MIT Press (2001)
3. Cong, G., Wang, L., Lin, C.-Y., Song, Y.-I., Sun, Y.: Finding question-answer pairs from online forums. In: Proceedings of the 31st Annual International ACM SIGIR Conference on Research and Development in Information Retrieval, SIGIR 2008, pp. 467–474. ACM, New York (2008)
4. Cottam, J.A., Menzel, S., Greenblatt, J.: Tutoring for retention. In: Proceedings of the 42nd ACM Technical Symposium on Computer Science Education, SIGCSE 2011, pp. 213–218. ACM, New York (2011)
5. Davis, F.D.: Perceived usefulness, perceived ease of use, and user acceptance of information technology. MIS Quarterly, 319–340 (1989)
6. DeSanctis, G., Fayard, A.-L., Roach, M., Jiang, L.: Learning in online forums. European Management Journal 21(5), 565–577 (2003)

7. Fetahu, B., Dietze, S., Pereira Nunes, B., Antonio Casanova, M., Taibi, D., Nejdl, W.: A scalable approach for efficiently generating structured dataset topic profiles. In: Presutti, V., d'Amato, C., Gandon, F., d'Aquin, M., Staab, S., Tordai, A. (eds.) ESWC 2014. LNCS, vol. 8465, pp. 519–534. Springer, Heidelberg (2014)

8. Fetahu, B., Dietze, S., Pereira Nunes, B., Antonio Casanova, M., Taibi, D., Nejdl, W.: A scalable approach for efficiently generating structured dataset topic profiles. In: Presutti, V., d'Amato, C., Gandon, F., d'Aquin, M., Staab, S., Tordai, A. (eds.) ESWC 2014. LNCS, vol. 8465, pp. 519–534. Springer, Heidelberg (2014)

9. Fetahu, B., Dietze, S., Nunes, B.P., Taibi, D., Casanova, M.A.: Generating structured profiles of linked data graphs. In: Blomqvist, E., Groza, T. (eds.) International Semantic Web Conference. CEUR Workshop Proceedings, vol. 1035, pp. 113–116. CEURWS. org (2013)

10. Hulpus, I., Hayes, C., Karnstedt, M., Greene, D.: Unsupervised graph-based topic labelling using dbpedia. In: Leonardi, S., Panconesi, A., Ferragina, P., Gionis, A. (eds.) WSDM, pp. 465–474. ACM (2013)

11. Li, N., Wu, D.D.: Using text mining and sentiment analysis for online forums hotspot detection and forecast. Decision Support Systems 48(2), 354–368 (2010)

12. Mazzolini, M., Maddison, S.: Sage, guide or ghost? the e ect of instructor intervention on student participation in online discussion forums. Computers Education 40(3), 237–253 (2003)

13. Nonaka, I.: A dynamic theory of organizational knowledge creation. Organization Science 5(1), 14–37 (1994)

14. Pereira Nunes, B., Mera, A., Casanova, M.A., Kawase, R.: Boosting retrieval of digital spoken content. In: Graña, M., Toro, C., Howlett, R.J., Jain, L.C. (eds.) KES 2012. LNCS, vol. 7828, pp. 153–162. Springer, Heidelberg (2013)

15. Pendergast, M.: An analysis tool for the assessment of student participation and implementation dynamics in online discussion forums. SIGITE Newsl. 3(2), 10–17 (2006)

16. Romero, C., López, M.-I., Luna, J.-M., Ventura, S.: Predicting students' final performance from participation in on-line discussion forums. Comput. Educ. 68, 458–472 (2013)

17. Scaffidi, C., Dahotre, A., Zhang, Y.: How well do online forums facilitate discussion and collaboration among novice animation programmers? In: King, L.A.S., Musicant, D.R., Camp, T., Tymann, P.T. (eds.) SIGCSE, pp. 191–196. ACM (2012)

18. Shaul, M.: Assessing online discussion forum participation. IJICTE 3(3), 39–46 (2007)

19. Veerman, A.L., Andriessen, J.E.B., Kanselaar, G.: Collaborative learning through computer-mediated argumentation. In: Proceedings of the 1999 Conference on Computer Support for Collaborative Learning, CSCL 1999. International Society of the Learning Sciences (1999)

20. Vygotsky, L.: Mind in society. The development of higher psychological processes. Harvard University Press, Cambridge; edited by cole, michael et al. edition, 0 1978 / 1930

PDist-RIA Crawler: A Peer-to-Peer Distributed Crawler for Rich Internet Applications

Seyed M. Mirtaheri[1], Gregor V. Bochmann[1], Guy-Vincent Jourdan[1], and Iosif Viorel Onut[2]

[1] School of Electrical Engineering and Computer Science, University of Ottawa,
Ottawa, Ontario, Canada
staheri@uottawa.ca, {gvj,bochmann}@eecs.uottawa.ca
[2] Security AppScan® Enterprise, IBM
770 Palladium Dr, Ottawa, Ontario, Canada
vioonut@ca.ibm.com

Abstract. Crawling Rich Internet Applications (RIAs) is important to ensure their security, accessibility and to index them for searching. To crawl a RIA, the crawler has to reach every application state and execute every application event. On a large RIA, this operation takes a long time. Previously published *GDist-RIA Crawler* proposes a distributed architecture to parallelize the task of crawling RIAs, and run the crawl over multiple computers to reduce time. In GDist-RIA Crawler, a centralized unit calculates the next task to execute, and tasks are dispatched to worker nodes for execution. This architecture is not scalable due to the centralized unit which is bound to become a bottleneck as the number of nodes increases. This paper extends GDist-RIA Crawler and proposes a fully peer-to-peer and scalable architecture to crawl RIAs, called *PDist-RIA Crawler*. PDist-RIA doesn't have the same limitations in terms scalability while matching the performance of GDist-RIA. We describe a prototype showing the scalability and performance of the proposed solution.

Keywords: Web Crawling, Rich Internet Application, Peer-to-Peer Algorithm, Crawling Strategies.

1 Introduction

Crawling a web application refers to the process of discovering and retrieving client-side application states. Traditionally, a web crawler finds the initial state of the application through its URL, referred to as *seed URL*. The crawler parses this page, finds new URLs that belong to the application, and retrieves them. This process continues recursively until all states of the application are discovered.

Unlike traditional web applications, in modern web applications (referred to as *Rich Internet Applications* or RIAs), different states of the application are not always reachable through URLs. When the user interacts with the application locally, the client-side of the application may or may not interact with the server, and different states of the application are constructed on the client. In this realm,

B. Benatallah et al. (Eds.): WISE 2014, Part II, LNCS 8787, pp. 365–380, 2014.

it is no longer sufficient to discover every application URL. To crawl a RIA, the crawler has to execute all events in all states of the application.

In effect, to crawl a RIA, the crawler emulates a user session. It loads the RIA in a virtual web browser, interacts with the application by triggering user interface events such as clicking on buttons or submitting forms [2, 7, 11, 20]. Occasionally, the crawler cannot reach a target state from its current state merely by interacting with the website. When this is the case, the crawler has to reload the seed URL. Through interaction with the application, the crawler discovers all application states.

For a large RIA, executing all events is a very time-consuming task. One way to reduce the time it takes to crawl RIAs is to parallelize the crawl and run it on multiple computers (henceforth referred to as *nodes*). Parallel crawling of RIAs was first explored by *Dist-RIA Crawler* [24]. It was proposd to run the crawl in parallel on multiple nodes with a centralized unit called *coordinator*. Dist-RIA Crawler partitions the task of crawling a RIA by assigning different events to different nodes. In this algorithm, all nodes visit all application states, however, each node is only responsible for the execution of a subset of the events in each state. Together the nodes execute all events on all states. When a node discovers a new state, it informs the coordinator, and the coordinator informs other nodes about the new state. The coordinator is also the one detecting termination.

In the context of crawling RIAs, the *crawling strategy* refers to the strategy that the crawler uses to choose the next event to execute [7]. Efficiency of crawling is effected by the crawling strategy. Two of the most efficient crawling strategies are the greedy [26] and the probabilistic [10] strategies. The greedy strategy finds the un-executed event which is the closest to the current state of the crawler. The probabilistic strategy, uses the history of event executions and chooses an event that maximized the likelihood of finding a new state. Running simple strategies such as breath-first and depth-first search does not require knowledge of all application graph transitions. To run efficient strategies, however, it is crucial to have access to all known application graph transitions.

In Dist-RIA Crawler only the application states are sent to the other nodes. This limits the ability of the nodes to run efficient crawling strategies. To address this shortcoming, *GDist-RIA Crawler* [22] offers a coordinator-based approach to run the crawling strategy. It calculates the tasks to be done, and then dispatches the tasks to the nodes. Individual nodes execute the task and update the coordinator about their findings. Although this approach can apply any crawling strategy, it is not scalable, since the coordinator will eventually become a bottleneck as number of nodes increases.

In this paper, we propose *PDist-RIA Crawler*, a peer-to-peer and scalable architecture to crawl RIAs. Unlike Dist-RIA and GDist-RIA crawlers, all nodes are peer and homogeneous in the PDist-RIA Crawler. Nodes broadcast transitions information to the other nodes, therefore efficient crawling strategies can be implemented in this architecture. Termination is handled through a peer-to-peer ring-based protocol.

This paper contributes to the literature of web crawling by enhancing previously published works: Dist-RIA and GDist-RIA Crawlers. Additionally, performance of different operations during the crawling of RIAs are measured. These measurements are used to justify some of the decisions made in designing PDist-RIA Crawler. Finally, an implementation was used to get performance measurements.

We assume that nodes are independent from each other and there is no shared memory. Nodes can communicate between each other through message passing. Both the nodes and the communication medium are reliable. We assume that the target RIA is deterministic, that is: execution of an event from a state always leads to the same target state. We assume that the number of events and the number of states in the web application are finite. Finally, we assume that on average, there are more events per state than there are nodes in the system, i.e. the application graph is dense.

The rest of this paper is organized as follow: In Section 2 we describe some of the related works. In Section 3 we describe PDist-RIA Crawler. In Section 4 we describe some experimental results on the time it takes to perform different operations. In Section 5 we evaluate the performance of PDist-RIA Crawler against GDist-RIA Crawler. Finally, this paper is concluded in Section 6.

2 Related Works

Web crawling has a long and interesting history [23, 25]. Parallel crawling of the web is the topic of extensive research in the literature [3, 4, 12, 16, 18]. Parallel crawling of RIAs, on the other hand, is a new field. Dist-RIA crawler [24] studies running breath-first search in parallel. GDis-RIA crawler [22] uses a centralized architecture to run the greedy strategy and to dispatch jobs to other nodes. Hafaiedh et al. [15] propose a fault tolerant ring-based architecture to crawl RIAs concurrently using the greedy strategy with multiple coordinators. To the best of our knowledge, the PDist-RIA Crawler is the first work that does not have coordinators and where nodes are homogeneous.

Using efficient crawling strategies is another way to reduce the time it takes to crawl RIAs. Duda et al. studied the breath-first search strategy [11, 13, 19], and Mesbah et al. studied the depth-first search strategy [20, 21]. *Model Based Crawling* (MBC) increases the efficiency of the crawler by using a model of the application to choose the next task to execute. Examples of MBC are *Hypercube model* [2, 9], *Menu model* and *Probabilistic model* [6, 8, 10].

Identification of client-side states is an important aspect of crawling RIAs. Strict equivalence of DOMs is determined through hashing the serialized DOM [11, 13, 19]. Less strict approaches use an edit distance [20, 21] and elements on the page [1]. In this paper we use strict equivalence of DOMs and use hash of serialized DOM to identify the state.

Ranking states and pages is an approach to better utilize limited resources available in crawling a website [25]. In the context of traditional web applications, the *PageRank* [5] algorithm emulates a user session, calculates the probability of a hypothetical user reaching a page, and uses this probability to rank a page. The *AjaxRank* [13] algorithm applies the same technique to RIA crawling. In this paper we assume that all states have the same rank.

3   Overview of the PDist-RIA Crawler

Nodes of the PDist-RIA Crawler partition the search space through events in the page (the same way as in Dist-RIA Crawler). Nodes divide the events in each state deterministically and autonomously, such that every event belongs to one node. Nodes are responsible to go to all states and execute their events. New states and transitions, discovered through executing events, are shared with all other nodes through broadcasting.

3.1   Algorithm

As depicted in Figure 1, nodes in the PDist-RIA Crawler can be in one of the following states: *Initial, Awake, Working, Idle,* and *Terminated.* Nodes start in the *Initial* state. In this state, nodes start up a headless browser process. The crawling starts when the node with node identifier 0 broadcasts a message that moves all nodes to the *Working* state. In this state nodes start running the crawling algorithm. Nodes find the next event to execute locally and deterministically.

If a node has nothing to do it goes to the *Idle* state. In this state the node waits for the termination state token or a new state. If a new state becomes available, it goes back to the *Working* state. If the termination token arrives, the node runs the termination algorithm to determine the termination status. The termination algorithm is described in detail in the next section.

3.2   Termination

A peer-to-peer protocol is run to determine termination. The protocol runs along with the crawling algorithm throughout the crawling phase. This protocol works by passing a token called *termination state* token around a ring overlay network that goes through all nodes. The termination token contains the following objects:

- List of state IDs for discovered application states: This list has an element per discovered state. As the token goes around the ring, more state IDs are added to this list.
- The number of known application states for each node: When the token visits a node, the node counts the number of application states it knows and stores this number in this list.

Using the information stored in the token, the termination algorithm described below decides whether to pass the token to the next node, or initiate termination.

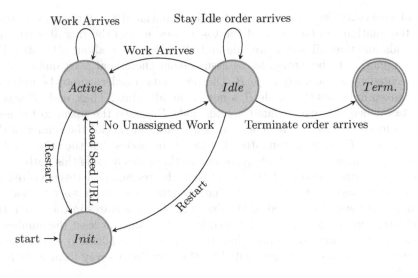

Fig. 1. The Node Status state diagram

Termination Algorithm. When the token arrives at a node, the node is either in Working state or in Idle state. If it is in Working state it will continue executing tasks and hold on to the token until it goes into the Idle state. In the Idle state the node performs the following steps:

1. The node updates the token with the new application states it knows about that do not yet exist in the token.
2. The node updates the number of the states it knows about.
3. If the status of the node is not indicated to be Idle in the token, the node updates its status to Idle in the token. This situation happens if this is the very first time the node takes the token. Initially, the status of all nodes is set to Active in the token. As the token goes around the ring, it can only pass a node if the node is in the idle state. Thus after one round of going around the ring, all node statuses will be Idle.
4. The node loops through the list of node states in the token and if it finds at least one node that is in Working state, the node passes the token to the next node in the ring.
5. If all nodes are in the Idle state in the token, the node loops through the list of number of known states for all nodes in the token and compares the number of known states for all nodes against the number of application state IDs in the token. If there is at least one node that does not know about all states discovered, the node passes the token to the next node.
6. If the last two steps do not pass the token to the next node, the node concludes that crawling is over and it initiates a termination by broadcasting a termination order to all the nodes.

Proof of Correctness: Let us assume that the algorithm is not correct and the termination is initiated while there are still events to be executed. Without loss of generality, let us assume that node A initiated the termination order. The termination can only start if the token goes around the ring at least once and finds out that all nodes are idle and all nodes know about all states. For the termination to be wrong, let us assume that there is at least one event to be executed by a node, say node B. The termination order cannot be initiated if the token indicates that node B is not in the idle state. Thus, node B was in idle state when the token visited it after node B passed the token to the next node, a message was sent to it with a new state. Let us call the sender of the message node C. Node C can either be one of the nodes that the token visited on its way from node B to node A, or one of the nodes outside this path.

Node C cannot be one of the nodes that the termination token visited on its way from node B to node A. If that was the case, on its visit to node C the new state would be added to the list of application states in the token, the termination order would not be initiated by node A since at least the number of states known by node B is lower than the number of application states known in the token. So node C is not visited by the token on its way from node B to node A.

For the same reason that was stated for node B, node C was idle at the time when the token visited it and another node sent it a message with a new state. The sender, henceforth referred to as node D cannot be on the way from node C to node A, for the same reason that node C cannot be on the way from node B to node A.

This reasoning does not stop at node D and it continues indefinitely. Since the number of nodes that are not on the way of the token from the sender node to node A is finite eventually we run out of nodes to be potential senders, and thus the initial message telling node B about a new state could have never been initiated. Thus the termination algorithm is proven to be correct by contradiction.

4   Performance Measurements

In this section we measure the performance of certain operations used by the crawling algorithm. These measurements are used to justify some of the design decisions made in Section 3.2.

4.1   Time to Transmit Messages

In PDist-RIA Crawler, communication happens through message passing. In this section we measure the efficiency of message passing. Figure 2 shows stack-bars of the time it takes to send a message from one node to another as a function of message size, in logarithmic scale. Each message was sent 100 times and the distribution of the time is demonstrated by the corresponding stack-bar.

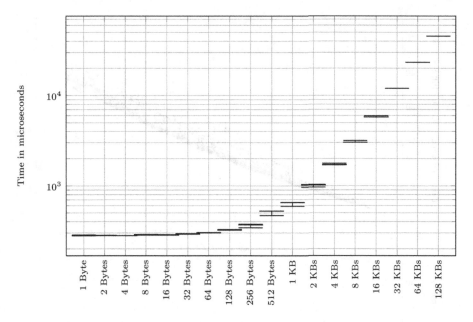

Fig. 2. Cost of sending messages between nodes

Given the measured network delays, we can calculate the overhead of sending different messages during the crawling process:

- **Overhead of sending state information:** A message to inform another node about a new state discovery contains state identifiers, number of events, a parent state identifier, and an event identifier in the parent state that leads to the discovered state. These items, along with the message header take between 128 to 256 bytes. Thus on average, the network delay to inform another node about a state is from 324 to 369 microseconds.
- **Overhead of sending transition information:** A message to inform another node about a transition contains source identifiers, target state identifiers, and event identifier. Similar to the *State* message, a transition message takes between 128 to 256 bytes. Thus on average network delay to inform another node about a transition is expected to be from 324 to 369 microseconds.
- **Termination Token:** The size of termination token varies depending on the number of crawler nodes, the number of discovered states, and the number of visited states by crawler nodes. In the worst case, the token contains state identifiers for all application states, and all nodes have visited all states. As explained in Section 3.2, among the test applications used in this paper, *Dyna-Table* with 448 states has the largest number of application states. Assuming we are crawling this application with 20 nodes, the termination token can get as large as 8 kilobytes. Thus in the worst case scenario where the token is at its largest size, the average time it takes to send the token from a node to another is less than 3 milliseconds.

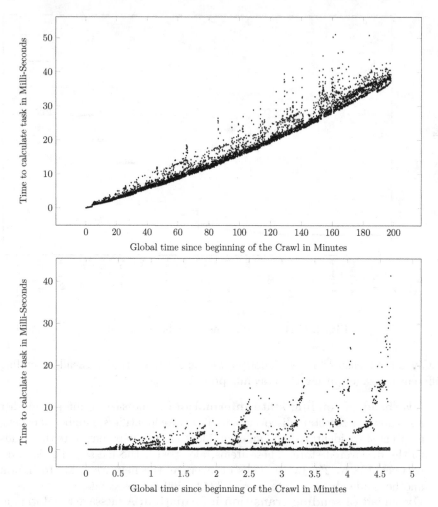

Fig. 3. Time to calculate next to execute as crawling proceeds using Breath-First (top) and Greedy (bottom) strategies for Dyna-Table application

4.2   Time to Calculate the Task to Execute

When the crawling strategy specifies a target event to execute, the crawlers often have to execute additional events to reach to the application state where the target event resides. We call this path of events and the target event a *task*. After execution of an event, the crawler has to calculate the next task to execute. Different crawling strategies run different algorithms to calculate the next task to execute. In this section we measure the time it takes to calculate the next task to execute for different crawling strategies.

Figure 3 shows the time it takes to calculate the next task using breath-first search and greedy strategies. In this figure, the y-axis shows the time to calculate the next event, and the x-axis represents the clock since the beginning of the crawl. As the figure shows:

- The time to calculate the next task to execute, tends to rise steadily in the case of the breath-first search strategy. In this algorithm, after finishing each task, the crawler looks for the next task to execute by looking into the seed URL and then the most immediate children for a task to perform. As the crawl proceeds the algorithm should go deeper in the graph before it can find a new task. Thus the cost of running the algorithm tends to increase as the crawl proceeds.
- The time it takes to calculate the next task to execute is generally lower in the greedy algorithm. In this algorithm, the crawler does not have to start from the seed URL and check all immediate children before going to further children for a task. Thus the greedy algorithm has a higher chance of finding tasks earlier.

As Figure 3 shows, in the majority of the cases it takes a few milliseconds to calculate the next task to execute.

4.3 Number of Events in Tasks

Executing the task which often involves executing several client-side events, possibly interacting with the server, and possibly performing a reset, often takes much longer than calculating the next task to execute. In this section we measure the number of events in the tasks calculated by different crawling strategy.

Different crawling strategies create tasks with different number of events. Execution of an event can start an interaction with the application server. This interaction (often in form of an asynchronous request to the server) is often the most time consuming aspect of executing the task. Therefore, the number of events in each task is a good indicator of the time it takes to execute the task. By measuring the number of events in the tasks, in effect we forecast the time it takes to execute the tasks.

Figure 4 shows the number of events in tasks created using breath-first search and greedy strategies. In these figures, the y-axis shows the number of events, and the x-axis represents the clock since the beginning of the crawl. As the figures show:

- As the crawling proceeds the length of events in tasks increases using the breath-first search strategy. In Dyna-Table application this number can be as high as 14 events.
- The greedy strategy represent a very efficient strategy, where often very few events exist in each task. The rare worst case scenario happens with 7 events in a task.

The time it takes to execute individual events depends highly on the target application and the server hosting it. Execution of JavaScript events, that do not trigger an asynchronous call to the server, is substantially faster than the events that interact with the server. For example, in the Dyna-Table web application,

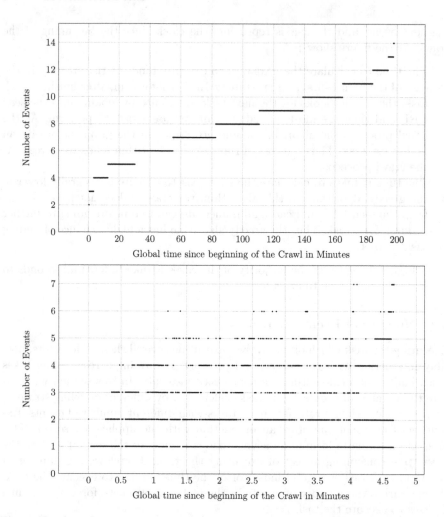

Fig. 4. Number of events to execute before executing a new event as crawling proceeds using Breath-First (top) and Greedy (bottom) strategies, for Dyna-Table application

execution of events that do not interact with the server always take less than 20 milliseconds. In the same application, events that do interact with the server often take more than 85 milliseconds to execute, and can take up to 1.3 seconds.

As the experimental results in the section show:

- The time required for communicating a state or transition between two nodes is less than a millisecond.
- The time required for calculating the next event to execute is often less than 100 milliseconds.
- The time required for executing a JavaScript event is often between 10 to 1000 milliseconds. A task is often composed of several JavaScript events, as the result executing a task can take up to several seconds.

The time it takes to broadcast a message is orders of magnitude smaller than the time it takes to calculate a task or execute it. It is thus reasonable to expect a good performance from a peer-to-peer architecture where nodes broadcast states and, when necessary, transitions. As the number of nodes increases, the number of messages per second increases too, and the network is bound to become a bottleneck. For example, based on the experimental measurements presented, we expect that when both states and transitions are broadcasted, by crawling the Dyna-Table application with 434 nodes or more, the network becomes a bottleneck. Before reaching this point, however, a good performance speedup is expected with this architecture.

Based on this observation, we devise a peer-to-peer architecture. This architecture takes advantage of low network delay and relies on broadcasting the information needed. Assuming that the number of events per state is larger than the number of nodes, through elimination of any centralized unit, this architecture does achieve a better scalability.

5   Evaluation

5.1   Test-Bed

For the experimental results discussed in this chapter, the nodes and the co-ordinator are implemented as follow: The JavaScript engine in the nodes is implemented using PhantomJS 1.9.2. Strategies are implemented in the C programming language and GCC version 4.4.7 is used to compile them. All crawlers use the Message Passing Interface (MPI) [27] as the communication mechanism. MPI is an open standard communication middleware developed by a group of researchers with background both in Academia and industry. MPI aims at creating a communication system that is interoperable and portable across a wide variety of hardware and software platforms. Efficient, scalable, and open source implementations of MPI are available such as OpenMPI [14] and MPICH [17]. MPICH version 3.0.4 is used to implement the communication channel in our experiments. All nodes, as well as the coordinator in case of GDist-RIA crawlers, run on Linux kernel 2.6.

The nodes are hosted on Intel® Core(TM)2 Duo CPU E8400 @ 3.00GHz and 3GB of RAM. The coordinator is hosted on Intel® Xeon® CPU X5675 @ 3.07GHz and 24GB of RAM. The communication happens over a 10 Gbps local area network. We ran each experiment three times and the presented numbers are the average of those runs.

To compare the relative performance of the crawlers, we implemented GDist-RIA and PDist-RIA crawlers using the same programming language and used MPI as communication channel. In all cases the C programming language is used to calculate the next task to do. In order to compare the performance of crawlers we crawled two different target web applications using the two architectures. In both cases we used the breath-first search and the greedy strategies.

5.2 Target Applications

Two real world target applications (Figure 5) are chosen to measure the performance of the crawlers. The target web applications are chosen based on their size, complexity and client side features they use.

Dyna-Table is a real world example of a JavaScript widget, with asynchronous call ability, that is incorporated into larger RIAs. This widget helps developers to handle large interactive tables. It allows to show a fixed number of rows per page, to navigate through different pages, to filter content of a table based on given criteria, and to sort the rows based on different fields. This application was developed using the *Google Web Toolkit* and has 448 states and a total of 5, 380 events.

Periodic-Table is an educational open source application that simulates an interactive periodic table. This application allows the user to click on each element and show the user information about the element in a pop-up window. The application can display the periodic table in two modes: the small mode, and the large mode. The two modes are identical in terms of functionality, however, they offer two very different interfaces. Once an element is clicked and the pop-up window shows up for the element, other elements can be clicked or the pop-up window can be closed. This application is developed in PHP and JavaScript, and it has 240 states and 29, 040 events.

Table 1 shows some information about the graph of the target web applications. As the table shows: Dyna-Table is a large size RIA with a small number of events per page, and Periodic-Table is a large size RIA with a large number of events per page. In Periodic-Table, all states have more events per page than the number of web crawlers and the application graph is dense.

Table 1. Target Applications graph summary

Application Name	Number of States	Number of Transitions	Average Events per Page
Dyna-Table	448	5,380	12.01
Periodic-Table	240	29,040	121.00

5.3 Results

Figure 6 shows the time it takes to crawl the Dyna-Table and Periodic-Table applications with different number of nodes using different crawlers. As the figure shows the overhead of the peer-to-peer architecture is not tangible. In fact, even with 20 nodes, the GDist-RIA Crawler does not scale as well as PDist-RIA Crawler. This is particularly noticeable for the greedy strategy. In this strategy, tasks take less time to execute, and therefore nodes ask the coordinator for new tasks to execute more frequently, making the coordinator a bottleneck. Therefore, the new method is not less efficient than the best known method to date, when that best known method to date is not overloaded. In additionally, the new method beats the old method squarely when the other becomes overloaded.

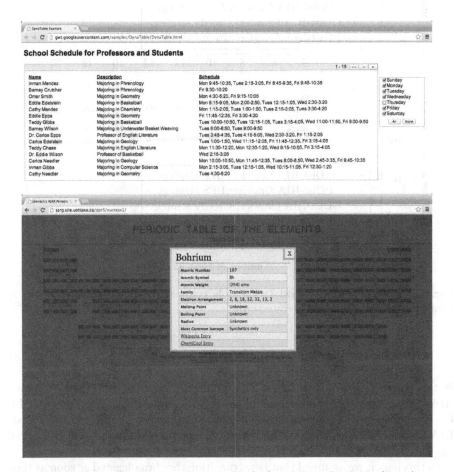

Fig. 5. Target Application: Dyna-Table (up) and Periodic-Table (down)

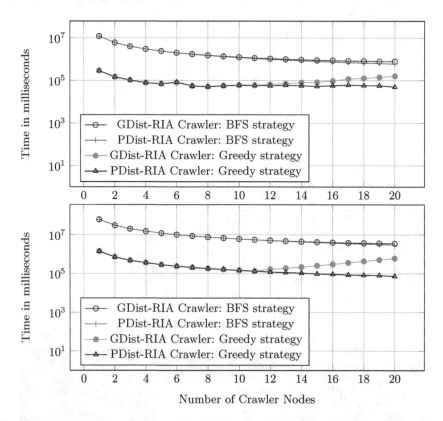

Fig. 6. The total time to Crawl Dyna-Table (top) and Periodic-Table (bottom) in parallel using using different architectures

6   Conclusion and Future Improvements

In this paper we introduced PDist-RIA Crawler, a peer-to-peer crawler to crawl RIAs. To design this crawler, we measured the time it takes to execute different operations, and based on the measured values we engineered the PDist-RIA Crawler. This crawler was implemented and its performance was compared against the GDist-RIA crawler.

In this paper, we assumed that new transitions are broadcasted as soon as they become available. A study of the impact of this assumption is missing from this paper. More formally, the impact of the following two assumptions is missing: Firstly, to utilize the network better, it may be more efficient not to broadcast the new transitions as they become available, but to broadcast them in batches, or broadcast them at given time intervals. Secondly, not all transitions have a major impact on reducing the time it takes to crawl the RIA. Sharing only a sub-set of transitions, instead of all transitions, may not increase the time it takes to crawl the RIA, while it reduces network traffic.

Acknowledgements. This work is largely supported by the IBM® Center for Advanced Studies, the IBM Ottawa Lab and the Natural Sciences and Engineering Research Council of Canada (NSERC). A special thank to Sara Baghbanzadeh.

References

1. Amalfitano, D., Fasolino, A.R., Tramontana, P.: Experimenting a reverse engineering technique for modelling the behaviour of rich internet applications. In: IEEE International Conference on Software Maintenance, ICSM 2009, pp. 571–574 (September 2009)
2. Benjamin, K., von Bochmann, G., Dincturk, M.E., Jourdan, G.-V., Onut, I.V.: A strategy for efficient crawling of rich internet applications. In: Auer, S., Díaz, O., Papadopoulos, G.A. (eds.) ICWE 2011. LNCS, vol. 6757, pp. 74–89. Springer, Heidelberg (2011)
3. Boldi, P., Codenotti, B., Santini, M., Vigna, S.: Ubicrawler: A scalable fully distributed web crawler. In: Proc. Australian World Wide Web Conference, vol. 34(8), pp. 711–726 (2002)
4. Boldi, P., Marino, A., Santini, M., Vigna, S.: Bubing: Massive crawling for the masses
5. Brin, S., Page, L.: The anatomy of a large-scale hypertextual web search engine. In: Proceedings of the Seventh International Conference on World Wide Web 7, WWW7, pp. 107–117. Elsevier Science Publishers B. V, Amsterdam (1998)
6. Choudhary, S., Dincturk, E., Mirtaheri, S., Bochmann, G.V., Jourdan, G.-V., Onut, V.: Model-based rich internet applications crawling: Menu and probability models
7. Choudhary, S., Dincturk, M.E., Mirtaheri, S.M., Jourdan, G.-V., Bochmann, G.v., Onut, I.V.: Building rich internet applications models: Example of a better strategy. In: Daniel, F., Dolog, P., Li, Q. (eds.) ICWE 2013. LNCS, vol. 7977, pp. 291–305. Springer, Heidelberg (2013)
8. Choudhary, S., Dincturk, M.E., Mirtaheri, S.M., Jourdan, G.-V., Bochmann, G.v., Onut, I.V.: Building rich internet applications models: Example of a better strategy. In: Daniel, F., Dolog, P., Li, Q. (eds.) ICWE 2013. LNCS, vol. 7977, pp. 291–305. Springer, Heidelberg (2013)
9. Dincturk, E., Jourdan, G.-V., Bochmann, G.V., Onut, V.: A model-based approach for crawling rich internet applications. ACM Transactions on the Web (2014)
10. Dincturk, M.E., Choudhary, S., von Bochmann, G., Jourdan, G.-V., Onut, I.V.: A statistical approach for efficient crawling of rich internet applications. In: Brambilla, M., Tokuda, T., Tolksdorf, R. (eds.) ICWE 2012. LNCS, vol. 7387, pp. 362–369. Springer, Heidelberg (2012)
11. Duda, C., Frey, G., Kossmann, D., Matter, R., Zhou, C.: Ajax crawl: Making ajax applications searchable. In: Proceedings of the 2009 IEEE International Conference on Data Engineering, ICDE 2009, pp. 78–89. IEEE Computer Society, Washington, DC (2009)
12. Edwards, J., McCurley, K., Tomlin, J.: An adaptive model for optimizing performance of an incremental web crawler (2001)
13. Frey, G.: Indexing ajax web applications. Master's thesis, ETH Zurich (2007), http://e-collection.library.ethz.ch/eserv/eth:30111/eth-30111-01.pdf
14. Gabriel, E., et al.: Open MPI: Goals, concept, and design of a next generation MPI implementation. In: Kranzlmüller, D., Kacsuk, P., Dongarra, J. (eds.) EuroPVM/MPI 2004. LNCS, vol. 3241, pp. 97–104. Springer, Heidelberg (2004)

15. Hafaiedh, K., Bochmann, G., Jourdan, G.-V., Onut, I.: A scalable p2p ria crawling system with partial knowledge (2014)
16. Heydon, A., Najork, M.: Mercator: A scalable, extensible web crawler. World Wide Web 2, 219–229 (1999)
17. Karonis, N.T., Toonen, B., Foster, I.: Mpich-g2: A grid-enabled implementation of the message passing interface. Journal of Parallel and Distributed Computing 63(5), 551–563 (2003)
18. Li, J., Loo, B., Hellerstein, J., Kaashoek, M., Karger, D., Morris, R.: On the feasibility of peer-to-peer web indexing and search. In: Kaashoek, M.F., Stoica, I. (eds.) IPTPS 2003. LNCS, vol. 2735, pp. 207–215. Springer, Heidelberg (2003)
19. Matter, R.: Ajax crawl: Making ajax applications searchable. Master's thesis, ETH Zurich (2008),
    http://e-collection.library.ethz.ch/eserv/eth:30709/eth-30709-01.pdf
20. Mesbah, A., Bozdag, E., Deursen, A.V.: Crawling ajax by inferring user interface state changes. In: Proceedings of the 2008 Eighth International Conference on Web Engineering, ICWE 2008, pp. 122–134. IEEE Computer Society Press, Washington, DC (2008)
21. Mesbah, A., van Deursen, A., Lenselink, S.: Crawling ajax-based web applications through dynamic analysis of user interface state changes. TWEB 6(1), 3 (2012)
22. Mirtaheri, S.M., Bochmann, G.V., Jourdan, G.-V., Onut, I.V.: Gdist-ria crawler: A greedy distributed crawler for rich internet applications
23. Mirtaheri, S.M., Dinçtürk, M.E., Hooshmand, S., Bochmann, G.V., Jourdan, G.-V., Onut, I.V.: A brief history of web crawlers. In: Proceedings of the 2013 Conference of the Center for Advanced Studies on Collaborative Research, pp. 40–54. IBM Corp. (2013)
24. Mirtaheri, S.M., Zou, D., Bochmann, G.V., Jourdan, G.-V.,, I.V.: Dist-ria crawler: A distributed crawler for rich internet applications. In: 2013 Eighth International Conference on P2P, Parallel, Grid, Cloud and Internet Computing (3PGCIC), pp. 105–112. IEEE (2013)
25. Olston, C., Najork, M.: Web crawling. Foundations and Trends in Information Retrieval 4(3), 175–246 (2010)
26. Peng, Z., He, N., Jiang, C., Li, Z., Xu, L., Li, Y., Ren, Y.: Graph-based ajax crawl: Mining data from rich internet applications. In: 2012 International Conference on Computer Science and Electronics Engineering (ICCSEE), vol. 3, pp. 590–594 (March 2012)
27. Snir, M., Otto, S.W., Walker, D.W., Dongarra, J., Huss-Lederman, S.: MPI: the complete reference. MIT Press (1995)

Understand the City Better: Multimodal Aspect-Opinion Summarization for Travel

Ting Wang and Changqing Bai

Wireless Communication Research Lab,
Beijing University of Posts and Telecommunications,
Beijing 100876, P.R.China
{tina437213,bcq}@bupt.edu.cn

Abstract. Every city has a unique taste, and attracts tourists from all over the world to experience personally. People like to share their opinions on scenic spots of a city via the Internet after a wonderful journey, which has become a kind of important information source for people who are going to make their travel planning. Confronted with the ever-increasing multimedia content, it is desirable to provide visualized summarization to quickly grasp the essential aspects of the scenic spots. To better understand the city, we propose a novel framework termed multimodal aspect-opinion summarization (**MAOS**) to discover the aspect-opinion about the popular scenic spots. We devolop a three-step solution to generate the multimodal summary in this paper. We first select important informative sentences from reviews and then identify the aspects from the selected sentences. Finally relevant and representative images from the travelogues are picked out to visualize the aspect opinions. We have done extensive experiments on a real-world travel and review dataset to demonstrate the effectiveness of our proposed method against the state-of-the-art approaches.

Keywords: aspect-opinion, summarization, incremental learning.

1 Introduction

Recent years have witnessed the increasing prosperity of travel, which makes travel-related information attract much attention. People who want to visit one city need information about the scenic spots and their characteristics as much as possible for formulating their travel plans. Meanwhile, with the popularity of Web 2.0, more and more people are willing to record and share their travel experiences and some specific opinions on the Web. Textual reviews and travelogues with large scale of photos are two kinds of popular user-generated content (UGC) on the web. Although the information in a single textual review and travelogue may be noisy or biased, the content contributed by numerous travelers as a whole could reflect people's overall preference and understanding of the travel city, thus it can serve as a reliable knowledge source for scenic spot

B. Benatallah et al. (Eds.): WISE 2014, Part II, LNCS 8787, pp. 381–394, 2014.

summarization. Faced with sustained increasing massive travel-related information on Internet, efficient multimodal travel summarization is highly desired to
facilitate people to obtain accurate information and prepare for their journeys.

To some extent, the multimodal scenic spot summarization for travel can be
formulated as a multi-document summarization (MDS) task [1], which has been
extensively studied in information retrieval. Over recent decades, a variety of
methods have been proposed. Notable approaches include the frequency-based
methods, e.g., Sum-Basic [13] and MEAD [15], and semantic-based methods
such as PLSA [5] or LDA [7]. Moreover, other methods such as the graph-based
methods(e.g., LexPageRank [2]) and machine learning-based methods [17] have
also been proposed. However, previous MDS techniques are designed for well
organized texts, such as the news articles and they just extract or abstract the
whole summary from large number of documents. In travel-related UGC, a distinguished natural scene, entertainment, food or culture etc can be considered as
one **aspect** of a scenic spot (e.g., Kunming Lake for the Summer Palace, Temple
for the Forbidden City). And people's opinions or descriptions (beautiful, mag-
nificent, interesting, etc.) towards a certain aspect of the scenic spot are jointly
called **aspect-opinion**, which are crucial to summarize the characteristic information about the scenic spot. Different from the traditional general summary, we
want to generate a summary composed of many aspect-opinions, which can help
people to explicitly understand the scenic spot from many interesting, hot or useful angles. Some other work in the travel domain only focuses on the exploiting
textual metadata for semantic knowledge mining, e.g., place tag extraction [16],
landmark recognition [22] and travel route recommendation [8]. However, little
work has substantially combined the aspect mining with the visual information.
In this work, we study the problem of multimodal aspect-opinion summarization (MAOS) aiming to exploiting the intuitive aspect-opinion and combining it
with visual information effectively. Specifically, we consider two types of UGC,
namely reviews and travelogues. By reviews, we refer to the text contributed by
travelers to share their opinions or record travel experiences on blogs, forums, or
Web 2.0 communities like Dazhongdianping[1]; By travelogues we refer to articles
with images and surrounding textual logs, namely contexts, which are released
on online albums such as BaiduLvyou[2].

However, either review mining for aspect-opinions or image selection for aspect visualization is a non-trivial task because user-generated reviews and travelogues, like other UGC, are noisy and unstructured data. The challenges derived
from three-fold: 1) Noisy aspects: most reviews are inevitably unstructured and
contain much noise. 2) Multiple viewpoints and expression habit: tourists from
all around the world may have opposite opinions (good or bad, like or hate) toward the same thing in the scenic spot and different expression style may affect
the comprehension. 3) Noisy image context: the context content adjacent to each
image written by users is possibly biased and inaccurate.

---

[1] http://www.dianping.com
[2] http://lvyou.baidu.com

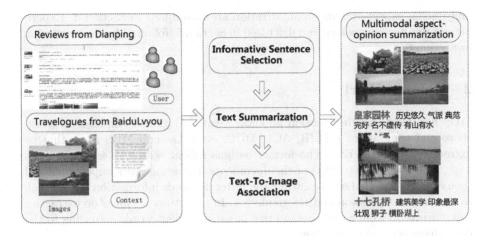

Fig. 1. The proposed framework for multimodal aspect-opinion summarization

We propose the MAOS framework to address these issues (see Fig. 1). The input is a set of reviews and travelogues with images about each specific scenic spot. The multimodal summarization for each scenic spot is automatically generated. To begin with, for each scenic spot, we build a vocabulary V barely composed of nouns and adjectives, and calculate the entropy of each sentence in reviews about the scenic spot. Then the significant sentences with high information entropy are chosen as the input dataset for the following process, which can reduce the noisy sentences from source. Moreover, we exploit textual content based on the informative sentences to discover multiple aspects by an incremental learning method. In this process, we extract the aspect salient words from aspect sentences and update when a new sentence is added to the current aspect, which can restrain the noise words and protrude the salient words that can express the aspect properly. In the end, we conduct the text to image association by utilizing the similarity between the review aspect sentences and the contexts in the travelogues and vote to select the representative images for the textual aspect. After all the three key steps, we obtain multimodal aspect-opinion summary composed of aspect salient words and representative images.

The contributions of this work can be summarized as follows:

- We address the problem of multimodal summary with many aspect-opinions, which offers a novel vision for multimedia data mining in the travel domain and enables a variety of applications.
- We propose an incremental learning scheme to discover the aspect-opinions, which can automatically mine the underlying aspects by exploiting the salient words and combining textual review sentences in a principled way.
- We build the text-to-image association by exploiting the similarity between the review aspect sentences and the context around the images and then vote to decide which aspect the image should be assigned to.

The rest of the paper is organized as follows. Section 2 surveys some related work. Section 3 formally defines the problem. The details of generating multimodal aspect-opinion summarization are introduced in Section 4. Experimental results are reported and discussed in Section 5, followed by the conclusion in Section 6.

2   Related Work

MDS has drawn much attention all these years and gained emphasis in workshops and conferences (SIGIR, ACL, DUC, etc.). General MDS can either be extractive or abstractive. The former assigns salient scores to semantic units (e.g., sentences, paragraphs) of the documents indicating their importance and then extracts top ranked ones, while the latter demands information fusion, such as sentence compression and reformulation. In this study we focus on extractive summarization from aspect perspective based on salient words according to the data character and application.

Centroid-based method is one of the most popular extractive summarization methods. MEAD and NeATS [10] are such implementations, using position, term frequency and theme, etc. MMR [4] algorithm is used to remove redundancy. Most recently, the graph-based ranking methods have been proposed to rank sentences or passages based on the "vote" or "recommendation" between each other. LexPageRank uses algorithms similar to PageRank and HITS to compute sentence importance. Wan et al. [20] improved the graph-ranking algorithm by differentiating intra-document and inter-document links between sentences. Cluster information has been incorporated in the graph model to better evaluate sentences [19]. Li et al. [9] used a structural SVM to learn for sentence selection in MDS. However, all these approaches are for traditional MDS and they just give some general descriptions as a whole not from the multiple aspects.

Probabilistic topic models, e.g., probabilistic latent semantic analysis (PLSA) and latent Dirichlet allocation (LDA), which have been successfully applied to a variety of text mining tasks in recent years, owing to their powerful capability of discovering topics from texts and representing documents in a low-dimensional space spanned by the topics. However, to the best of our knowledge, the existing models are not applicable for our objective because they do not consider or address the limitations of short reviews and travelogue data.

Besides, the problem of aspect-opinion mining has recently attracted increasing attention in the service and product domain. Most of the early works studying on this problem are feature-based approaches (e.g., [11,12]). These approaches normally apply some constraints on high-frequency noun phrases to identify product aspects. As a result they usually produce too many non-aspects and miss low-frequency aspects [6]. In addition, feature-based approaches require the manual tuning of various parameters which makes them hard to port to another dataset. Our work pays more attention to mining the salient words and representative images so as to produce multimodal aspect-opinion summary for the scenic spot automatically.

3   Problem Formulation

The input to our multimodal aspect-opinion summarizer is a collection of reviews and travelogues $P = \{R, T\}$ related to the same scenic spot. And R means pieces of reviews, and $T = \{C, I\}$ means the travelogues uploaded by travelers, while I means the images and C means the contexts surrounded the images in the travelogues. Besides, the context and images are organized in travelogues as user's writing order. The objective of our framework is to get textual aspect-opinions and obtain representative images. The output of our approach is some informative aspect-opinions consisting of salient aspect words and representative images corresponding to the aspect for each scenic spot.

4   Multimodal Aspect-Opinion Summarization

Our approach contains the following three components: 1) Informative sentence selection. We select sentences with high entropy which can obtain the important and significant sentences. 2) Text summarization. We discover multiple textual aspects for each scenic spot from the informative review sentences. 3) Text to image association. Through the contexts nearest the images and similarity between the aspect sentences and context, we vote to select the representative images associated to the corresponding aspect.

4.1   Informative Sentence Selection

As text pre-processing procedure, we firstly build individual vocabulary dictionary V for each scenic spot by segmenting Chinese words using FudanNLP[3] package from all reviews R. We filter stop words, meaningless words, time and number related words, camera related words, some general frequent words and select significant words. We give a premise that the information about the scenic spot is uniformly distributed over the review dataset. Besides, it is well-known that the information entropy can reflect that the average amount of information it brings when you know the results of a stochastic event. So we calculate the entropy for each word w.

$$H(w) = -p(w)logp(w) \tag{1}$$

where $p(w) = \frac{tf(w,R)}{\sum_{w'} tf(w',R)}$ denotes the probability of word w in review sentence set R and tf denotes the term frequency for word w. Then entropy of each sentence s in R can be calculated as summation of entropy for each word w_k in s.

$$H(s) = \sum_k H(w_k) \tag{2}$$

Therefore, we select the significant sentences with high entropy as follows:

$$S = \{s|H(s) > \varepsilon\} \tag{3}$$

where ε is the threshold value for sentence selection and S is the selected informative sentence set from reviews R.

---

[3] https://code.google.com/p/fudannlp/

4.2   Text Summarization

Aspect-opinion mining suffers from large variance of noisy sentences and various expressions. To deal with these difficulties, we propose a novel incremental learning method whose procedure is in Algorithm 1 to mine the potential aspect-opinions from S for each scenic spot. The learning scheme consists of two parts: salient words extraction and sentence aspect assignment.

Algorithm 1. Incremental Learning for Multiple Aspects mining in a Scenic Spot

---

Input: A data set of review sentences S after informative sentence selection and sentence feature vector X over the vocabulary V in a scenic spot

Output: multiple aspects in a scenic spot $A = \{a_k\}_{k=1}^{K}$

repeat

   Score aspect for updated aspect a

   Determine the aspect of the input sentence s using the existing aspects with the corresponding distribution

   Augment the dataset with the newly input sentence

until sentences S exhausted;

Salient Words Extraction: Given all sentences S in a scenic spot P and a couple of sentences $S_a = \{s_i\}_{i=1}^{N} \in S$ of one aspect a, we aim to extract salient words S_W for this aspect from the whole associated words T_a with sentences S_a. These salient words including nouns and adjectives should be able to well describe the aspect. We propose our technique to extract salient words by making two assumptions [16] as follows:

- If a word w well describes a semantic concept c, then the probability of observing the semantic concept c among sentences S_a of aspect a is larger than the probability of observing it among all review sentences S about P.
- A word w is semantically representative if its parasitic sentences are textually similar to each other, containing a common semantic concept c such as an object, a scene or some opinions towards them.

Based on these two assumptions, we present a formulation that simultaneously integrates the above two assumptions in a single framework. Considering the first assumption, we measure the observation probability between $w_i \in W$ and sentences S_a of aspect a with word co-occurrence.

$$p(w_i|S_a) = \frac{N(w_i \cap S_a)}{N(S_a)} \tag{4}$$

where $N(S_a)$ denotes the total number of sentences in aspect a, while $N(w_i \cap S_a)$ is the number of sentences associated with word w_i in aspect a. Analogously, we calculate the observation probability for word w_i in all sentences S in a scenic spot. $p(w_i|S) = \frac{N(w_i \cap S)}{N(S)}$, where $N(S)$ is the total number of all sentences and $N(w_i \cap S)$ is the number of sentences containing word w_i in all sentences. Furthermore, we have the constraint between $p(w_i|S_a)$ and $p(w_i|S)$ as

$$F(w_i, S_a) = f(x), \quad x = p(w_i|S_a) - p(w_i|S) > 0 \tag{5}$$

where we introduce the increasing logistic function $f(x) = \frac{1}{1+e^{-x}}$. Therefore, the separation of a salient word set W in sentences S_a is given by $\sum_{w_i \in W} F(w_i, S_a)$. Considering the second assumption, we adopt pairwise similarity constraints to measure the cohesion for sentences of aspect a. We denote $x_u \in X_{w_i}$ as the textual feature vector of sentence s_u associated with word w_i in sentences S_a. s_w is the set of sentences containing word w_i in sentences S_a and X_{w_i} are corresponding feature vectors of S_{w_i}. We adopt the classical law of cosines [21] to measure the semantic similarity between sentence pairs in sentences S_{w_i}

$$sim(S_{w_i}) = \frac{1}{|X_{w_i}|^2} \sum_{u,v=1}^{|X_{w_i}|} \cos(x_u, x_v) \tag{6}$$

where $|X_{w_i}|$ is the cardinality of X_{w_i} and cos is the cosines law. Accordingly, we define the cohesion of word w_i with S_a as

$$G(w_i, S_a) = f(\tau), \quad \tau = sim(S_{w_i}) \tag{7}$$

where $f(\tau)$ is the same with the above $f(x)$. The cohesion of a salient word set W in sentences S_a is computed as $\sum_{w_i \in W} G(w_i, S_a)$. Based on the definitions of terms regarding separation and cohesion for salient word set W, we present formulation for extracting salient words W as follows:

$$W^* = \arg\max\{\frac{1}{|W|} \sum_{w_i \in W} \varphi(w_i)\}, \varphi(w_i) = \lambda F(w_i, S_a) + (1 - \lambda)G(w_i, S_a) \tag{8}$$

where $|W|$ is the number of the selected words. $\lambda \in [0, 1]$ is a trade-off parameter to module two contributions; $\varphi(w_i)$ is the saliency score of word w_i in aspect a. However, it is computationally intractable to solve Eq.(8) directly since it is a non-linear integer programming (NIP) problem. Alternatively, we resort to a greedy strategy which is simple but effective to solve the problem. In reality, for each aspect in a scenic spot, only a small set of words are salient and valuable. Therefore, we perform a pre-filtering step to obtain the words W_F with large values of $F(w_i, S_a)$. This can reduce the computational cost to favor the salient words extraction. For salient words extraction, we first select the word $w \in W_F$ with largest value of $F(w, S_a)$ and then choose the next word w_i from $W_F \backslash W^*$ by solving $\arg\max_{w_i} \varphi(w_i)$. The salient words set is updated by $W^* = W^* \cup \{w_i\}$ and obtained until $W_F = \emptyset$. The whole procedure for salient words extraction is summarized in Algorithm 2. In the incremental learning procedure, when an aspect is augmented by adding a new sentence it is not necessary to process all the words of the aspect. Instead, we only need to compute the salient scores of the words associated with the added sentence and update the salient words for the aspect based on the previous salient words, which can largely reduce the computational complexity for aspect mining.

Sentence aspect assignment: Given salient words and sentences for each existing aspect in a scenic spot, we aim to assign a aspect label for a new input sentence. We develop a two-step approach to address the problem. We first use a salient word correlation method to find the most potential correlated aspects from the existing aspect sets. Then, we comprehensively analyse the relevance between the sentence and aspect on purpose of automatically predict the most probable aspect for the new input sentence. We elaborate this process as follows.

Algorithm 2. Extracting Salient Words

Input: aspect a with S_a , parameter λ, W_a in a scenic spot;
Output: W^*;
  Select words W_F with large values of $F(w, S_a)$;
  Select the word $w \in W_F$ with largest value of $F(w, S_a)$
  $W^* = W^* \cup \{w\}$
  repeat
    Choose the next word from $W_F \backslash W^*$ by solving
    $\arg\max_{w_i}\{\lambda F(w_i, S_a) + (1 - \lambda)G(w_i, S_a)\}$
    $W^* = W^* \cup \{w_i\}$
  until $W_F = \emptyset$ or $|W^*| = n$;
  return W^*

- **Candidate aspects.** We try to discover the candidate aspects based on measuring the word relatedness between the associated words W_s of the input sentence s and the salient words W_a of aspect a. it can be represented as:

$$C(W_s, W_a) = \sum_{w_i \in W_a} H(w_i|W_s) \cdot p(w_i|W_a) \tag{9}$$

where $H(w_i|W_s)$ is 1 if there exists a salient word $w_i \in W_a$ in W_s otherwise 0. $P(w_i|W_a) = \frac{\varphi(w_i)}{\sum_j \varphi(w_j)}$ is the salient weight of word in aspect a computed by the salient score in salient words extraction stage. Then we get the candidate aspects

$$A_{candidate} = \{a|\quad C(W_s, W_a) > e\} \tag{10}$$

where e is the threshold value set for all aspects. If an input sentence is not correlated with all existing aspects, we assign a new aspect label a_{new} for it.

- **Sentence label.** After we get some candidate aspects to reduce the scope, we now consider the sentence label problem. Users prefer to see the targeted aspects about the scenic spot. The more related and concentrated the sentences are, the better the aspect is. Therefore we denote Φ_{S_a} as the word distribution of current aspect a and it can be estimated as:

$$p(w|\Phi_{S_a}) = \frac{tf(w, \Phi_{S_a})}{\sum_{w \in W} tf(w, \Phi_{S_a})} \tag{11}$$

where $tf(w, \Phi_{S_a})$ denotes the term frequency of word w in S_a . We employ the Kullback-Leibler (KL) divergence to measure the distance of two distributions D1 and D2 as

$$D_{KL}(D_1||D_2) = \sum_w p(w|D_1) \log \frac{p(w|D_1)}{P(w|D_2)} \tag{12}$$

Therefore, the relevance between the input sentence and the current aspect a could be measured as:

$$U_R(s, a) = D_{KL}(\Phi_{S_a} \cup s||\Phi_{S_a}) \tag{13}$$

since we favor a small distance for $U_R(s, a)$, so we choose the aspect a from $A_{candidate}$ with min $U_R(s, a)$ as the finally excellent aspect for s. In summary, the aspect of a new input sentence s is determined a_s by as:

$$a_s = \begin{cases} a^* & if \quad A_{candidate} \neq \emptyset \\ a_{new} & otherwise \end{cases} \qquad (14)$$

When s is assigned to one aspect, we need to update the salient words in the aspect meanwhile.

4.3   Text to Image Association

To visualize the aspect-opinion, now we consider the problem of text to image association. According to our observation, people usually give some simple textual description C location adjacent to the uploaded image I in the travelogue. Therefore, we can build the text to image association through the extracted aspect-opinion sentences and the textual description about the image in travelogue. We use a majority voting scheme to select the correlated images to the aspect a. The detail process is as follows:

- **Image clustering.** We first extract image I which contains some description context c_I in the travelogue to formulate the image set I_p and context C_p about the scenic spot p. Then we apply spectral clustering to group I_p into visually diverse clusters $L_P = \{l_1, l_2, ..., l_{|l|}\}$ based on the visual content feature vector. Hence we labeled the corresponding l_i to each context c_I whose adjacent image I is clustered to l_i .
- **Aspect image cluster voting.** For each sentence s in S_a, we look for the most similar context sentences c_I from C_p according to the cosines theory. Then we vote to decide the potentially associated image cluster L_a based on the similar context label . The simple algorithm is shown in Algorithm 3.

Algorithm 3. Vote for Representative Aspect Image Cluster

Input: aspect a with S_a, C_p, L_p in a scenic spot;
Output: L_a;
   $N(lc_I) = 0$;   # initial count
   for each s in S_a **do**
      for each c_I in C_p **do**
         $score(s, c_I) = \cos(s, c_I)$
      end for
      choose the top c_I, $N(lc_I) = N(lc_I) + 1, (lc_I \in L_p)$
   end for
   rank $N(lc_I)$ and choose the top label as L_a
   return L_a

Until now, we have got the associated aspect-opinion image cluster label L_a for each aspect. Then we select the representative images for each aspect-opinion via the affinity propagation method [3] from image cluster L_a.

5   Experiments

We have conducted extensive experiments to evaluate the effectiveness and usefulness of the proposed framework.

5.1   Dataset and Experimental Settings

We construct the dataset by choosing 7 most well-known scenic spots in the capital city — Beijing. Our dataset was collected by crawling reviews from Dazhong-dianping and travelogues from Baidulvyou. We use each scenic spot's name as the search word and all queried reviews and travelogues are collected together with their associated information. The character of reviews from Dazhongdian-ping is that each piece of review is short and may include only 2 or 3 sentences which depict some aspect-opinion about the scenic spot and lack visual images. Comparatively, the travelogues from Baidulvyou are less informative and noisy in opinions but more abundant in high-quality images and corresponding context. For each scenic spot, we initially get average 4000 reviews and 300 travelogues. A pre-filtering process is performed on the dataset to remove duplicate reviews, images and travelogues which are not related to the scenic spot, etc. We also restrict to images that they should have completely associated textual contexts. After the pre-processing, the dataset for each scenic spot contains 3000+ reviews and 3000+ images with the surrounding contexts.

To represent the image content, we extract five types of visual features to form an 809-dimension vector for each image, including 81-dimension color moment, 37-dimension edge histogram, 120-dimension wavelet texture feature, 59-dimension LBP feature [14] and 512-dimension GIST feature [18]. We concentrate on a vocabulary of 5000 words extracted from reviews for usage in the evaluation of aspect-opinion mining. The tradeoff parameter λ in Eq.(8) is empirically set to 0.6. The threshold parameter e in Eq.(10) is chosen as 0.10 respectively, through qualitative cross validation measuring and is fixed in subsequent experiments. We extract the top 15 salient words to represent each aspect.

5.2   Summarization Performance

In the following we will evaluate the performance of multimodal aspect-opinion summary.

Quantitative Evaluation: We compare the following methods in this process:

- k-means, the well-known clustering method, which considers the similarity values from instances to k centers. We apply k-means on each scenic spot review sentences represented by the corresponding feature vector.
- SC(Spectral Clustering), which exploits pairwise similarities of data instances. Spectral clustering requires computing pairwise similarity among n data instances to construct the similarity matrix. We use textual sentence feature of each scenic spot to compute relevant similarity matrix.
- IL(Incremental Learning). Our proposed method for scenic spot aspect-opinion.

We apply these three methods on the reviews of the top-7 scenic spots for comparison. K-means and spectral clustering need to specify the number of clusters in advance. We set the number of clusters for each scenic spot according to its corresponding scale of reviews. Consequently, we set the number of clusters to be 30 for the Summer Palace and 20 for the other scenic spots. We manually

Table 1. Evaluation of aspect-opinion mining for the popular top-7 scenic spots

Scenic spot	k-means	SC	IL
The Forbidden City	0.656	0.683	0.765
The Summer Palace	0.666	0.733	0.813
798 Art Zone	0.566	0.675	0.783
South Luogu Lane	0.476	0.647	0.714
Badaling Great Wall	0.685	0.676	0.819
Beijing National Stadium	0.566	0.611	0.755
Beijing Zoo	0.419	0.533	0.790
Overall	0.576	0.651	0.777

build the ground truth for each scenic spot. Sentences that describe about one specific point such as internal scene spot, activity , event are divided into several groups for representing each aspect in a scenic spot. For each scenic spot, let $A = \{a_1, a_2, ..., a_J\}$ denote the discovered aspects using certain algorithms, and $\hat{A} = \{\hat{a}_1, \hat{a}_2, ..., \hat{a}_k\}$ denote its manually labeled aspect groups. We then borrow the purity metric [21] from traditional clustering problem to evaluate the quality of aspect-opinion mining, which is defined as

$$Purity(A, \hat{A}) = \frac{1}{N} \sum_j \max_K |a_j \cap \hat{a}_k|$$

where N is the total number of sentences in the ground truth dataset. A higher purity score means a better mining for scenic spot aspects. Table 1 present the comparison results. We have following observations. 1) Spectral clustering outperforms k-means in terms of mean purity. It coincides with previous conclusion that spectral clustering is more effective in finding clusters than k-means. 2) Our incremental learning method significantly and consistently outperforms spectral clustering and k-means. The average purity score obtained by our method is 0.777, while the average purity score obtained by spectral clustering and k-means is 0.651 and 0.576 respectively. This is because that the reviews contain a large number of noisy words and sentences. Our incremental learning method can differentiate and discard the noisy words and sentences to some extent, which can improve the quality of aspect mining. In addition, the number of cluster is not preset which is automatically adaptive in accordance with the incremental learning process.

User Study: Fig. 2 shows the three outstanding aspects with salient words and exemplary images of the Top-7 scenic spots in Beijing detected by our approach. Clearly, these aspects are related to the nature scene, landmark, animals, street scene, food, culture, and so on. It can be seen that the extracted images and words are consistent and the discovered aspects are adequate to represent the scenic spots. We conduct a small-scale user study to evaluate the effectiveness of the proposed method and user experience of the novel visualization form. Three criteria are considered: (1) consistency, the level of consistency between the visual content and salient words (0: Not consistent, 10:Very consistent); (2) relatedness, the extent that the mined aspects relate to the scenic spot (0: Not related, 10: Very related); and (3) satisfaction, how satisfactory are the

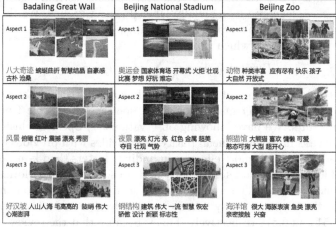

Fig. 2. Multimodal aspect-opinion summary

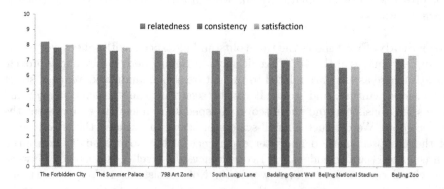

Fig. 3. User study evaluation results

aggregated multimodal aspect-opinion summary (0: Not satisfied, 10: Very satisfied) We invited 20 participants for the user study experiment. The aspects depicted in Figure 2 are selected for evaluation. The results are averaged over all participants for each aspect-opinion and shown in Fig. 3. It is obvious that the participants have given positive feedback to the novel multimodal aspect-opinion scheme, which further validates the potential of MAOS in advanced travel exploration and related applications.

6 Conclusion

In this paper, we have presented a framework of multimodal aspect-opinion summarization for travel by leveraging the rich textual and visual information in large number of user-generated reviews and travelogues. A novel incremental learning method is introduced to scenic spot aspect mining and voting method for image association. The mined salient words and corresponding images formulate the aspect-opinions to generate representative and comprehensive summary, which help understand the city better from various pesperctives. Based on a large collection of reviews and travelogues, experimental results show the effectiveness of the proposed MAOS framework.

For the future work, we plan to take full advantage of the content of images to improve the performance of image selection and user experience. Besides, it is an interesting direction to take various granularity levels of locations into consideration, so as to better meet application needs.

References

1. Bian, J., Yang, Y., Chua, T.-S.: Multimedia summarization for trending topics in microblogs. In: Proceedings of the 22nd ACM International Conference on Conference on Information and Knowledge Management, pp. 1807–1812. ACM (2013)
2. Erkan, G., Radev, D.R.: Lexpagerank: Prestige in multi-document text summarization. In: EMNLP, vol. 4, pp. 365–371 (2004)
3. Frey, B.J., Dueck, D.: Clustering by passing messages between data points. Science 315(5814), 972–976 (2007)
4. Goldstein, J., Kantrowitz, M., Mittal, V., Carbonell, J.: Summarizing text documents: sentence selection and evaluation metrics. In: Proceedings of the 22nd Annual International ACM SIGIR Conference on Research and Development in Information Retrieval, pp. 121–128. ACM (1999)
5. Gong, Y., Liu, X.: Generic text summarization using relevance measure and latent semantic analysis. In: Proceedings of the 24th Annual International ACM SIGIR Conference on Research and Development in Information Retrieval, pp. 19–25. ACM (2001)
6. Guo, H., Zhu, H., Guo, Z., Zhang, X., Su, Z.: Product feature categorization with multilevel latent semantic association. In: Proceedings of the 18th ACM Conference on Information and Knowledge Management, pp. 1087–1096. ACM (2009)
7. Haghighi, A., Vanderwende, L.: Exploring content models for multi-document summarization. In: Proceedings of Human Language Technologies: The 2009 Annual Conference of the North American Chapter of the Association for Computational Linguistics, pp. 362–370. Association for Computational Linguistics (2009)

8. Hao, Q., Cai, R., Wang, C., Xiao, R., Yang, J.-M., Pang, Y., Zhang, L.: Equip tourists with knowledge mined from travelogues. In: Proceedings of the 19th International Conference on World Wide Web, pp. 401–410. ACM (2010)

9. Li, L., Zhou, K., Xue, G.-R., Zha, H., Yu, Y.: Enhancing diversity, coverage and balance for summarization through structure learning. In: Proceedings of the 18th International Conference on World Wide Web, pp. 71–80. ACM (2009)

10. Lin, C.-Y., Hovy, E.: From single to multi-document summarization: A prototype system and its evaluation. In: Proceedings of the 40th Annual Meeting on Association for Computational Linguistics, pp. 457–464. Association for Computational Linguistics (2002)

11. Liu, B., Hu, M., Cheng, J.: Opinion observer: analyzing and comparing opinions on the web. In: Proceedings of the 14th International Conference on World Wide Web, pp. 342–351. ACM (2005)

12. Moghaddam, S., Ester, M.: Opinion digger: an unsupervised opinion miner from unstructured product reviews. In: Proceedings of the 19th ACM International Conference on Information and Knowledge Management, pp. 1825–1828. ACM (2010)

13. Nenkova, A., Vanderwende, L.: The impact of frequency on summarization. Microsoft Research, Redmond, Washington, Tech. Rep. MSR-TR-2005-101 (2005)

14. Ojala, T., Pietikäinen, M., Harwood, D.: A comparative study of texture measures with classification based on featured distributions. Pattern Recognition 29(1), 51–59 (1996)

15. Radev, D.R., Jing, H., Styś, M., Tam, D.: Centroid-based summarization of multiple documents. Information Processing and Management 40(6), 919–938 (2004)

16. Rattenbury, T., Naaman, M.: Methods for extracting place semantics from flickr tags. ACM Transactions on the Web (TWEB) 3(1), 1 (2009)

17. Shen, D., Sun, J.-T., Li, H., Yang, Q., Chen, Z.: Document summarization using conditional random fields. IJCAI 7, 2862–2867 (2007)

18. Torralba, A., Murphy, K.P., Freeman, W.T., Rubin, M.A.: Context-based vision system for place and object recognition. In: Proceedings of the Ninth IEEE International Conference on Computer Vision, pp. 273–280. IEEE (2003)

19. Wan, X., Yang, J.: Multi-document summarization using cluster-based link analysis. In: Proceedings of the 31st annual international ACM SIGIR Conference on Research and Development in Information Retrieval, pp. 299–306. ACM (2008)

20. Wan, X., Yang, J., Xiao, J.: Single document summarization with document expansion. In: Proceedings of the National Conference on Artificial Intelligence, vol. 22, p. 931. AAAI Press, MIT Press, Menlo Park, CA (2007)

21. Zhao, Y., Karypis, G.: Criterion functions for document clustering: Experiments and analysis. Technical report, Technical report (2001)

22. Zheng, Y.-T., Zhao, M., Song, Y., Adam, H., Buddemeier, U., Bissacco, A., Brucher, F., Chua, T.-S., Neven, H.: Tour the world: building a web-scale landmark recognition engine. In: IEEE Conference on Computer Vision and Pattern Recognition, CVPR 2009, pp. 1085–1092. IEEE (2009)

Event Processing over a Distributed JSON Store: Design and Performance

Miki Enoki[1], Jérôme Siméon[2], Hiroshi Horii[1], and Martin Hirzel[2]

[1] IBM Research, Tokyo, Japan
{enomiki,horii}@jp.ibm.com
[2] IBM Research, New York, USA
{simeon,hirzel}@us.ibm.com

Abstract. Web applications are increasingly built to target both desktop and mobile users. As a result, modern Web development infrastructure must be able to process large numbers of events (e.g., for location-based features) and support analytics over those events, with applications ranging from banking (e.g., fraud detection) to retail (e.g., just-in-time personalized promotions). We describe a system specifically designed for those applications, allowing high-throughput event processing along with analytics. Our main contribution is the design and implementation of an in-memory JSON store that can handle both events and analytics workloads. The store relies on the JSON model in order to serve data through a common Web API. Thanks to the flexibility of the JSON model, the store can integrate data from systems of record (e.g., customer profiles) with data transmitted between the server and a large number of clients (e.g., location-based events or transactions). The proposed store is built over a distributed, transactional, in-memory object cache for performance. Our experiments show that our implementation handles high throughput and low latency without sacrificing scalability.

Keywords: Events Processing, Analytics, Rules, JSON, In-Memory Database.

1 Introduction

In-memory computing reverses traditional data processing by embedding the compute where data is stored, instead of moving the data to where the compute happens. Analysts predict the market for in-memory computing to grow 35% between now and 2015 [18]. This is in part because in-memory computing can radically change the time-frame for completing some data analysis tasks. In addition, it allows for systems to combine on-the-fly processing of events with analytics, which are increasingly important for business applications. We describe a system designed around a distributed in-memory data store that supports such applications. Our work is part of a broader project, *Insight to Action* (I2A), whose goal is to support high-throughput transaction processing along with decision capabilities based on large-scale distributed analytics.

Event streaming systems have been a subject of intense research and development in the last 10 years. Existing distributed event streaming systems include both research prototypes [1,5] and commercial products [12,20]. Those systems usually operate with a limited amount of state, and allow users to combine operators that process one or

B. Benatallah et al. (Eds.): WISE 2014, Part II, LNCS 8787, pp. 395–404, 2014.

more input events before passing a new event to the next operator. The I2A architecture is organized around an agent model in which each agent can process a fraction of the incoming events. One key difference compared to prior event-based systems is the presence of a distributed in-memory data store that can be accessed by those agents. The store is used both to allow agents to hold on to some state (e.g., information about a specific customer or product useful when processing a specific event), as well as to enable analytics that can inform decisions on those events (e.g., by computing the average order amount for customers in the same area).

This paper focuses on I2A's distributed store, which is one of the main novel parts of the overall system. The store handles both event storage as well as data obtained from systems of record (e.g., relational backends). Another novelty is the store's ability to handle flexible data through the JSON format. This is particularly important in this context as information is usually obtained from existing and usually heterogeneous data sources. Furthermore, it allows to easily blend events with stored data.

This paper makes the following technical contributions:

- It describes a system that supports combined workloads for events and analytics over an in-memory distributed JSON store.
- It illustrates the benefits of the JSON data model to unify representations on the client, on the wire, and in the store, and to support better flexibility in the system.
- It describes the architecture for the distributed JSON store, which leverages the MongoDB API [14] on top of an in-memory distributed object cache (WXS [22]).
- It reports experimental results on throughput, latency, and scalability.

2   Insight to Action

This section provides background on the Insight to Action (I2A) system on top of which the contributions of this paper rest.

Processing events with analytics. The starting point for this paper is the I2A system, which is currently under development at IBM. I2A combines event processing with analytics over a distributed store. The idea behind the name is that analytics discover insights and event processing performs actions. Both analytics and event processing use the same distributed in-memory store. Combining both events and analytics in a single system fosters ease of use (no need to configure multiple systems) and performance (no need to move data back and forth).

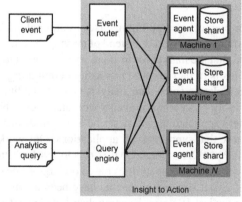

Fig. 1. Architecture for processing events and analytics over a distributed in-memory store

```
 1  rule NormalTakeOff {
 2    when {
 3      toe: TakeOffEvent(sched+20 >= now);
 4      flight: toe.flight();
 5    } then {
 6      update flight.departed = true;
 7      update flight.on_time = true;
 8    }
 9  }
10  rule LateTakeOff {
11    when {
12      toe: TakeOffEvent(sched+20 < now);
13      flight: toe.flight();
14    } then {
15      update flight.departed = true;
16      update flight.on_time = false;
17      emit new FlightDelayEvent(
18          flight.fl_no());
19    }
20  }
```

```
21  rule RAvg {
22    when {
23      late_count: aggregate {
24        f: Flight(on_time = false);
25        fn: f.fl_no();
26      }
27      groupby { fn }
28      do { count { f }; }
29    } then {
30      insert new FlightStat(
31          fn, late_count);
32    }
33  }
```

Fig. 2. Sample rules for I2A event processing   **Fig. 3.** Sample rule for I2A analytics

The top part of Fig. 1 illustrates event processing in I2A. Each arriving client event contains a key associating it with an entity. Each machine runs a store shard storing entities, as well as an event agent. The system uses the key in the event to route it to the machine where the corresponding entity is stored. The event agent acts upon the event by reading and writing entities in the local store shard, and by emitting zero or more output events (not shown in the figure).

The bottom part of Fig. 1 illustrates analytics in I2A. Analytics can be either user-initiated, or scheduled to repeat periodically. In the latter case, the time period between analytics recomputation is on the order of minutes, in contrast to the expected latency for event processing, which is in the sub-second range. That is because an analytics query typically scans all entities on all machines, unlike an event, which only accesses a few entities on a single machine. A query engine coordinates the distributed analytics, and combines the results (i.e., insight). The results are then either reported back to the user, or saved in the store for use by future events (i.e., action). While this section deliberately does not specify the data formats used, Section 3 will argue that JSON is a good choice.

Scenario. To illustrate I2A, we consider a simple example inspired by a real scenario from the airline industry. For each flight, the system receives events, such as when passengers book or cancel, when the crew is ready, when the flight takes off, etc. For each of those events, an agent updates the flight entity in its local store shard. Agents are written using JRules [21]. Due to space limitations, we only show simple rules that illustrate the main features: how to process events, and how to run analytics.

The rules in Fig. 2 implement the agent handling events for a flight taking off. JRules uses a condition-action model similar to that of production systems [10]. The condition is inside the **when** clause of the rule, and the action in the **then** clause. The first example applies to take-off events that were scheduled at most 20 minutes before their actual time ($sched+20 >= now$), while the second applies when the take-off was at least 20 minutes late. In both cases, the action part of the rule updates the flight information with the corresponding status, storing it in the shard so that other rules may access it. The second rule also emits a new `FlightDelayEvent` that can be used by other rules, for instance, to notify passengers or handle connecting flights.

Periodic analytics jobs can be run over all the data in the store, e.g., to compute summary information about all flights. For instance, the rule in Fig. 3 computes the number of late departures for every flight number. The rule can access all flight information in the store and compute aggregation using the **aggregate** expression, which here groups information by flight number, for flights that have been marked as delayed. This summary information is then replicated back out to each of the shards, making it available for future event processing. For instance, if a particular flight is frequently delayed, the can trigger a review, or it can be taken into account when re-booking passengers. As a more complex example, assume that the system receives an event about a passenger missing their flight. The agent handling this event can consult available summary status information to look for alternative flights. The system reports those alternatives back to the passenger, who can then re-book, leading to another event.

While this particular example focuses on the airlines industry, I2A is obviously not restricted to this domain. There are many use cases in different industries where insights from distributed analytics can drive actions in low-latency event processing.

3   Leveraging JSON

One important aspect of our work is the built-in support for JSON as a data model throughout the system, and how it allows to more easily integrate new information as it becomes available.

JSON End-to-End. The JavaScript Object Notation (JSON) has gained broad acceptance as a format for data exchange, often replacing XML due to its relative simplicity and compactness. Originally used for its easy integration with the client through JavaScript, it is now increasingly also used in both the back-end, e.g., through JSON databases [14], and the middleware, e.g., through Node.js [15].

Fig. 4 shows where the MongoDB API is integrated inside the I2A architecture. Both the event agent and the query engine can access data in the store through that JSON-centric API. From an application's perspective, the system can be seen as a JSON-centric development platform, which can load JSON data into the store, process incoming JSON events (e.g., through a REST API), execute rules on those events, and respond with JSON events. Fig. 4 also

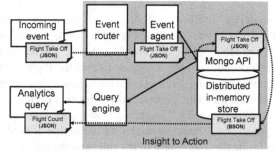

Fig. 4. JSON support and usage inside I2A

shows the life-cycle of the JSON data from an incoming event (flight take-off), which is then routed to the appropriate event agent, stored, and possibly queried back.

Because both external data and internal representation use JSON, we can pass most of the data through the system without expensive conversions common in similar architectures (to Java POJOs, to records in a relational store, etc.). Such conversions are particularly costly in distributed systems, which often need to (de)serialize objects for communication. Our architecture avoids most of that cost.

Flexibility. Another benefit of JSON is its flexibility, building on previous approaches for semi-structured data and Web data exchange formats such as XML. Because it does not require the user to declare the content of the documents or messages ahead of time, JSON supports data with some level of heterogeneity (e.g., missing or extra fields), and to extend the data with new information as it becomes available without having to recompile or deploy the application, or without having to reload the data.

In our example, the following JSON documents may correspond respectively to the delayed departure event and to the entities for the airports of departure and arrival. Some of those events and entities may include distinct information, for instance, the US airports include the state. If an event has extra fields not needed by a rule, the rule fires as normal. If an event lacks fields required by a rule, the rule simply does not fire.

    event { fl_no: "II32", from: "JFK", to: "CDG", departed: "21:43 GMT", on_time: false }
    event { fl_no: "II32", from: "JFK", to: "CDG", landed: "04:56 GMT", on_time: false }
    airport { code: "JFK", country: "USA", state: "NY", city: "New York" }
    airport { code: "CDG", country: "France", city: "Paris" }

Besides entities required for the applications, such as airport data in our example, the store records events as they are being processed, providing historical records that can be useful for analytical purposes. As the application evolves, there is often a need to integrate more information. In our example application, we can imagine that we might want to include weather information, or twitter data that may be used for notification. External services often provide data in an XML or JSON format, which can be easily integrated into the I2A architecture, and stored directly in the I2A distributed repository. For instance, the following JSON document may be provided by a location-based service providing weather alerts.

    alert { weather: "Snow", location: "New York", until: "19:00 GMT" }

That information can then be used to augment events being returned by the system. For instance, the delayed departure event could be enriched with information indicating that the reason for the delay is weather-related, with the corresponding alert information transmitted back to the user.

4 Integrated Distributed JSON Store

This section introduces the architecture of the distributed JSON store for I2A.

4.1 MongoDB

MongoDB is an open-source document-oriented database for JSON [14]. Internally it uses BSON (Binary JSON), a binary-encoded serialization of JSON documents. Developers interact with MongoDB by using a language driver (supported languages include Java, C/C++, Ruby, and NodeJS). Clients communicate with the database server through a standard TCP/IP socket. MongoDB supports only atomic transactions on individual rows, and it scales horizontally based on auto-sharding across multiple machines.

4.2 Integration with Insight to Action

JSON Store for I2A. Though MongoDB supports a distributed JSON store, it does not have the full transactional semantics that I2A requires for consistent event processing. Our challenge is how to reuse MongoDB's API without losing the transactionality and scalability advantages. Therefore, instead of using MongoDB directly, we used WebSphere eXtreme Scale (WXS [22]) as a distributed in-memory store with the MongoDB API, as shown in Fig. 5 (a), which depicts more details of each of the machines shown in Fig. 1. WXS is an elastic, scalable, in-memory key/value data store. It dynamically caches, shards, replicates, and manages the application data and business processing across multiple servers.

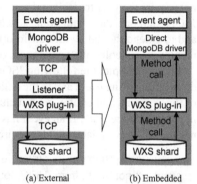

(a) External   (b) Embedded

Fig. 5. From external to embedded interactions using the MongoDB interface over the WXS store

Fig. 6 shows a sequence diagram for a read query by an event agent. A query is described using a JSON format such as {"name": "Miki Enoki", "department": "S77" }. The query selects the JSON documents that contain "Miki Enoki" as the name and "S77" as the department. The event agent communicates with the listener using the MongoDB driver for TCP and sends the query as serialized BSON data to the listener. The listener is a server application on WXS that intercepts the MongoWire Protocol messages. The query is deserialized as BSON data by the listener and then converted into a query for the WXS Object Query API to read the response data from the WXS shards by using the WXS plug-in. In the WXS shards, each query agent processes a query and then returns the requested data via TCP. The results are collected by the WXS plug-in, and serialized for transmission to the event agent.

WXS supports sharding for a distributed store, which allows running multiple WXS shards within multiple machines in an I2A instance for scaling out with sharded data. With a distributed store, each query of the event agent has to be routed to the corresponding WXS shard to access the appropriate data.

Embedded JSON Store. In I2A, since high-performance data processing is important for both event processing and analytics, we developed the embedded store shown in Fig. 5 (b) by eliminating the TCP communications from Fig. 5 (a). Even without eliminating the TCP communications, performance is improved when the event agent

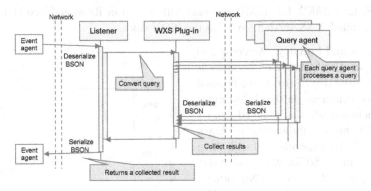

Fig. 6. Sequence Diagram of Query Processing

accesses only the WXS shards on its own machine. This assumption is true thanks to the fundamental design of I2A: since the I2A architecture includes an event router, we do not need to fall back on WXS for routing to the proper shard. In addition to insuring the communications are local, additional performance improvement comes from eliminating the TCP connections between the clients and the listener, and between the listener and the WXS store. Without the TCP connections, we can avoid the data serialization, which was needed to convert the BSON objects into byte array objects and vice versa. We wrote a direct MongoDB driver (class `DirectDB`) as a replacement for class `com.mongodb.DB`. The original class `com.mongodb.DB` sends query request to the Listener via TCP. Our replacement class `DirectDB`, in contrast, directly hands query requests to the WXS plug-in via a method call. WXS is also written in Java, so we can run all of the components in one JVM by using method calls in each machine in the cluster, as shown in Fig. 5 (b). Similarly, the event agent can handle low-latency read and write events by using method calls.

5 Experimental Evaluation

This section explores the throughput, latency, and scale-out of our JSON store for I2A.

Evaluation Methodology. We evaluated the effectiveness of our embedded JSON store with YCSB (Yahoo! Cloud Serving Benchmark) [8]. YCSB is a framework and common set of workloads for evaluating the performance of various key-value stores. I2A routes events to WXS shards storing the corresponding entity based on an entity key in the event. Therefore, we use YCSB to emulate an event processing workload. Some parameters are configurable by the user, such as the number of records, number of operations, or the read/update ratio. We fixed the number of records and operations to 100,000 and 5,000,000, respectively, and experimented with varying read/update ratios. As described in Section 4, events are processed through the MongoDB API in the JSON store, so we used the MongoDB driver included in YCSB for the benchmark client. The key and value are document ID and its JSON document respectively. For the embedded JSON store, we used our Direct MongoDB driver to directly access WXS with an internal method call. The environment was as follows: a 2-CPU Xeon X5670 (2.93GHz,

L1=32KB, L2=256KB, L3=12MB, 6 cores) with 32 GB of RAM and Red Hat Linux 5.5. We installed WXS version 8.6.0.2 as the JSON store.

Throughput and Latency. We compared the throughput and latency of the embedded JSON store (Fig. 5 (b)) to the original (Fig. 5 (a)) with a single shard. The number of client threads was changed from 1 to 200. The highest throughput is shown in the Fig. 7. We experimented with three read/update ratios 50/50, 95/5, and 100/0, which correspond to the YCSB official workloads a, b, and c, respectively. I2A handles the events for event processing and analytics. The event agent updates the entities in the JSON store corresponding to the incoming events, and then reads the related entities or the analytics results. For the an-

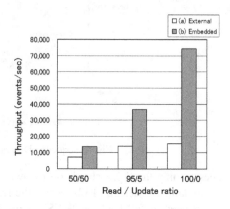

Fig. 7. Throughput

alytics, the events in the store are read periodically. Therefore, the data access pattern of I2A is read intensive. As seen in Fig. 7, the throughput of the embedded store was better for all read/update ratios. In particular, the result for the read-only scenario was about 4.8 times higher than the original.

Fig. 8 shows latency results for read accesses. The embedded store reduced the average latency by 87%. The 95-percentile embedded latency was also much smaller than the external latency. This indicates that eliminating the TCP/IP communication is highly effective for I2A.

	External	Embedded
Average	640 μs	48 μs
Minimum	302 μs	32 μs
Maximum	66,639 μs	40,731 μs
95-percentile	623 μs	66 μs

Fig. 8. Read latency

Scalability. Next, we evaluated the scalability of our embedded store in a distributed environment. We measured the maximum throughput while changing the number of machines (nodes) for each workload. As described in Section 4, each event agent accesses only a local WXS shard in the same node, so each benchmark client runs in its own WXS node. Fig. 9 shows the results. In all of the workloads, the throughputs scaled well as the number

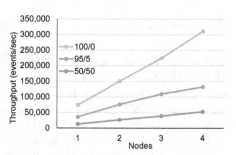

Fig. 9. Scale-out results for distributed store

of nodes increased. The throughputs with four nodes with read/update ratios of 50/50, 95/5, and 100/0 are respectively 3.9, 3.6, and 4.1 times higher than that with one node. This demonstrates the good scalability property of our proposed distributed JSON store for I2A. Since we do not need to fall back on WXS for routing to a data shard in our proposed store, it made a contribution for scalability.

6   Related Work

Stream processing systems, such as StreamBase [20], or InfoSphere Streams [12], support high-speed aggregation, enrichment, filtering, and transformation of streams of events. They analyze data in motion, as we do, but in addition, we can also analyze data at rest in a distributed store. Complex event processing (CEP) uses patterns over sequences of simple events to detect complex events. Distributed CEP engines include Cayuga [5], and CEP engines can also be distributed by embedding them in general streaming systems [11].

Several projects from the database area are closely related to our work. Ceri and Widom use production rules in a distributed database [6]. JAM uses Java agents to run batch analytics over a distributed database [19]. Kantere et al. discuss using triggers in databases that are not just distributed, but even (unlike our work) federated [13]. None of these works consider JSON support, or directly address the issue of performing analytics of data in motion alongside batch analytics of data at rest.

I2A is based on production rule languages, similar to [10,21], but allows integration with a distributed store. A related area is business process management systems (BPMSs), which use an event-driven architecture to coordinate human activities along with automated tasks. In contrast to distributed BPMSs [4], we also address the integration with a store and with batch analytics. From a database perspective, Datalog is the most commonly used rules language. It is storage centric and can be distributed [2]. To the best of our knowledge no work around Datalog addresses the integration of events with batch analytics, or includes support for JSON. A notable exception is JSON rules [16]; however, that work embeds rules in a browser rather than a distributed store.

The Percolator [17] runs distributed observers (similar to database triggers) over BigTable [7] (a distributed store). However, it does not directly address low-latency event processing; furthermore, it does not use JSON. Distributed stores that support JSON include MongoDB [14] and CouchDB [3]. Our work goes one step further by also offering rule-based event processing. Finally, our work is orthogonal to the question of query languages for JSON (e.g., JSONiq[9]) which could be easily integrated in our approach.

7   Conclusion

The *Insight to Action* (I2A) system embeds event processing into a distributed in-memory store. This enables event processing on large amounts of data without paying the penalty of going to disk or multiplexing all computation on a single machine. This paper is about the store component of I2A, specifically, about using JSON for this store. The advantages of JSON for the I2A store are that it helps simplify the system by using the same data model in all layers, and that it increases flexibility when schemas change. The challenge was how to reuse familiar APIs without losing the scalability advantages. We present our architecture for solving these challenges, along with performance results. Overall, I2A with a JSON store enables simple, flexible, and scalable stateful event processing. We are still actively developing I2A, and investigating several improvements, notably efficient execution strategies for aggregations, and how to improve freshness for the analytics without interfering with transaction performance.

References

1. Abadi, D.J., Ahmad, Y., et al.: The design of the Borealis stream processing engine. In: Conference on Innovative Data Systems Research (CIDR), pp. 277–289 (2005)
2. Alvaro, P., Condie, T., Conway, N., Elmeleegy, K., Hellerstein, J.M., Sears, R.: BOOM analytics: Exploring data-centric, declarative programming for the cloud. In: European Conference on Computer Systems (EuroSys), pp. 223–236 (2010)
3. Anderson, J.C., Lehnardt, J., Slater, N.: CouchDB: The definitive guide. O'Reilly (2010)
4. Bonner, A.J.: Workflow, transactions and Datalog. In: Symposium on Principles of Database Systems (PODS), pp. 294–305 (1999)
5. Brenna, L., Gehrke, J., Johansen, D., Hong, M.: Distributed event stream processing with non-deterministic finite automata. In: Conference on Distributed Event-Based Systems (DEBS) (2009)
6. Ceri, S., Widom, J.: Production rules in parallel and distributed database environments. In: Conference on Very Large Data Bases (VLDB), pp. 339–351 (1992)
7. Chang, F., Dean, J., Ghemawat, S., Hsieh, W.C., Wallach, D.A., Burrows, M., Chandra, T., Fikes, A., Gruber, R.E.: Bigtable: a distributed storage system for structured data. In: Operating Systems Design and Implementation (OSDI), pp. 205–218 (2006)
8. Cooper, B.F., Silberstein, A., et al.: Benchmarking cloud serving systems with YCSB. In: Symposium on Cloud Computing (SoCC), pp. 143–154 (2010)
9. Florescu, D., Fourny, G.: JSONiq: The history of a query language. IEEE Internet Computing 17(5), 86–90 (2013)
10. Forgy, C.L.: OPS5 user's manual. Technical Report 2397, Carnegie Mellon University (CMU) (1981)
11. Hirzel, M.: Partition and compose: Parallel complex event processing. In: Conference on Distributed Event-Based Systems (DEBS), pp. 191–200 (2012)
12. Hirzel, M., Andrade, H., et al.: IBM Streams Processing Language: Analyzing big data in motion. IBM Journal of Research and Development (IBMRD) 57(3/4), 7:1–7:11 (2013)
13. Kantere, V., Kiringa, I., Zhou, Q., Mylopoulos, J., McArthur, G.: Distributed triggers for peer data management. In: Meersman, R., Tari, Z. (eds.) OTM 2006. LNCS, vol. 4275, pp. 17–35. Springer, Heidelberg (2006)
14. MongoDB NoSQL database, http://www.mongodb.org/ (retrieved December 2013)
15. Node.js v0.10.24 manual & documentation (2013), http://nodejs.org/
16. Pascalau, E., Giurca, A.: JSON rules: The JavaScript rule engine. In: Knowledge Engineering and Software Engineering (KESE) (2008)
17. Peng, D., Dabek, F.: Large-scale incremental processing using distributed transactions and notifications. In: Operating Systems Design and Implementation (OSDI), pp. 251–264 (2010)
18. Rivera, J., van der Meulen, R.: Gartner says in-memory computing is racing towards mainstream adoption. Press Release (April 2013),
http://www.gartner.com/newsroom/id/2405315
19. Stolfo, S.J., Prodromidis, A.L., Tselepis, S., Lee, W., Fan, D.W., Chan, P.K.: JAM: Java agents for meta-learning over distributed databases. In: Conference on Knowledge Discovery and Data Mining (KDD), pp. 74–81 (1997)
20. Streambase, http://www.streambase.com/ (retrieved December 2013)
21. WODM: IBM Operational Decision Manager, http://www-03.ibm.com/software/products/en/odm/ (retrieved December 2013)
22. WXS: IBM WebSphere eXtreme Scale (2013), http://www.ibm.com/software/products/en/websphere-extreme-scale/ (retrieved November)

Cleaning Environmental Sensing Data Streams Based on Individual Sensor Reliability

Yihong Zhang, Claudia Szabo, and Quan Z. Sheng

School of Computer Science
The University of Adelaide, SA 5005, Australia
{yihong.zhang,claudia.szabo,michael.sheng}@adelaide.edu.au

Abstract. Environmental sensing is becoming a significant way for understanding and transforming the environment, given recent technology advances in the Internet of Things (IoT). Current environmental sensing projects typically deploy commodity sensors, which are known to be unreliable and prone to produce noisy and erroneous data. Unfortunately, the accuracy of current cleaning techniques based on mean or median prediction is unsatisfactory. In this paper, we propose a cleaning method based on incrementally adjusted individual sensor reliabilities, called *influence mean cleaning* (IMC). By incrementally adjusting sensor reliabilities, our approach can properly discover latent sensor reliability values in a data stream, and improve reliability-weighted prediction even in a sensor network with changing conditions. The experimental results based on both synthetic and real datasets show that our approach achieves higher accuracy than the mean and median-based approaches after some initial adjustment iterations.

Keywords: Internet of Things, data stream cleaning, sensor reliability.

1 Introduction

In environmental sensing, sensors are deployed in physical environments to monitor environmental attributes such as temperature, humidity, water pressure, and pollution gas concentration. With the emergence of the Internet of Things (IoT), which connects billions of small devices such as sensors and RFID tags to the Internet, environmental sensing is becoming a significant means towards understanding and transforming the environment [12]. Many IoT-inspired environmental sensing projects have emerged recently, including the Air Quality Egg[1] and the Cicada Tracker[2].

In most environmental sensing projects, commodity sensors are deployed to minimize the cost. Commodity sensors, however, are widely known to be unreliable and prone to producing noisy and erroneous data [2, 8]. Data cleaning is therefore an important issue in environmental sensing, especially when critical realtime decisions need to be made based on the collected data. Recent works

---

[1] http://airqualityegg.com/
[2] http://project.wnyc.org/cicadas/

B. Benatallah et al. (Eds.): WISE 2014, Part II, LNCS 8787, pp. 405–414, 2014.

have proposed solutions to extract the truthful information from noisy sensor data [8, 14, 16]. A common approach to automatically predict truthful readings is by aggregating spatially correlated readings, using either mean [8, 16] or median [14]. However, it is documented that such approaches have not achieved satisfying accuracy [4].

Intuitively, knowing individual sensor reliability can improve prediction accuracy by, for example, giving unreliable sensors less weight when aggregating the readings. In this paper, we propose a sensor data cleaning technique based on incrementally adjusted individual sensor reliabilities. We adopt a *data-centric* approach to sensor reliabilities, and identify potential sensor malfunctioning through *faulty data*. There are two types of sensor malfunctioning exist, namely, *systematic* and *random* [3]. Faulty data caused by *systematic malfunctioning* typically can be fixed by a single change in the calibration parameter, as proposed by several works [3, 7]. In this paper, we focus on the *random malfunctioning*, which can be caused by unpredictable issues such as sensor damage or battery exhaustion.

Our proposed reliability-based sensor data cleaning method, called *influence mean cleaning* (IMC), weights the mean prediction based on individual sensor reliabilities, and incrementally updates sensor reliabilities based on the readings in each data collecting iteration. We validate our approach extensively by using both synthetic and real datasets. The experimental results show that IMC can significantly improve prediction accuracy over the traditional mean and median methods. When there are sensor condition changes in the network, our method also accurately captures different types of changes, and allows the predictions to adjust to new sensor conditions quickly.

The remainder of this paper is organized as follows. In Section 2, we overview the related work. In Section 3, we present the proposed IMC, which consists of a weighted mean prediction and an incremental reliability update model. We report the experimental results with the simulated and real datasets in Section 4. In Section 5, we provide some concluding remarks.

2   Related Work

A data-centric approach to detect sensor faults has been studied in several research projects. Ni et al. [10] investigated different types of sensor malfunctioning (e.g., battery exhaustion and hardware malfunction) and associated faulty data patterns with them. Sharma et al. [11] identified three faulty data patterns in a number of real datasets, and proposed techniques for their detection. In the evaluation of their techniques, they injected faulty data patterns into known clean data, and used the original clean data as the ground truth. We adopted this data synthesis method when designing our experiments.

A large number of works exist on data cleaning in wireless sensor networks [8, 14–16]. Most of the proposed techniques are based on the assumption that sensor readings are aggregated when transmitted, and individual sensor readings are not available or difficult to obtain. The IoT inspired environmental sensors,

however, are assumed to be connected to the Internet directly, like those used by
Devarakonda et al. in [4]. Such direct Internet connection of individual sensors
allows individual sensor readings to be accessed and preserved, which creates an
opportunity for studying individual sensor behaviors.

Data source reliability has appeared in truth prediction in information retrieval. In the Web environment, it is not unusual to have multiple data sources
that may have different views on a same fact. Most of the existing works are
based on probabilistic inference [6, 13]. The probability-based solutions for truth
finding, however, are ineffective for environmental sensing data, where sensor
reliabilities can be influenced by unpredictable external factors over time. We
argue that our incremental update approach is more effective for reflecting unpredictable changes of sensor reliabilities in continuous sensing data streams.

3   Reliability-Weighted Prediction

In this section, we first discuss generic faulty data patterns in real sensor datasets
and introduce our proposed data cleaning procedure, which we will explain in
two parts: the reliability-based prediction called *influence mean cleaning* (IMC),
and the incremental reliability update model.

3.1   The Faulty Data Patterns and the Cleaning Procedure

Sensors can produce noisy and erroneous data when operating in less than ideal
working conditions. Fig. 1 shows three patterns of faulty data found in real sensor
data that may be caused by sensor malfunctioning. According to the research
by Ni et al. [10], *high volatility*, characterized by a sudden rise of variance in
the data, can be caused by hardware failure or a weakening in battery supply.
Single spikes, occasional unusually high or low readings occurred in a series of
otherwise normal reading, can be caused by battery failure. *Intense* single spikes
that occur with high frequency, may indicate hardware malfunction.

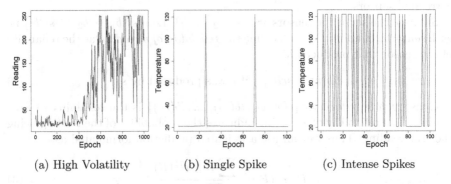

(a) High Volatility   (b) Single Spike   (c) Intense Spikes

Fig. 1. Faulty sensor data patterns

Our intention is not to detect the type and cause of sensor faults, but to calculate a representative reliability value for each individual sensor that can be used to improve prediction accuracy. In engineering, reliability is defined as *"the probability that a device will perform its intended function during a specified period of time under stated conditions"* [5]. The intended function of an environmental sensor is to generate readings according to the environmental feature that it is monitoring. Consequently, when a sensor produces a reading, the sensor reliability indicates the probability that this reading is the same as the presumed true value.

Our proposed *influence mean cleaning* (IMC) predicts true readings based on incrementally updated individual sensor reliabilities. The general procedure of applying our approach to a sensing data stream is depicted in Fig. 2. Following the data-centric approach, we do not assume any prior hardware information that can be used to infer the reliability of individual sensors, and our approach allows initial reliabilities to be set arbitrarily. The continuous operation of the cleaning method consists of iterations. In each iteration, new sensing data are collected, predictions of true readings are made, and the reliabilities are updated by comparing individual readings to the prediction. The procedure repeats as the data being continuously collected.

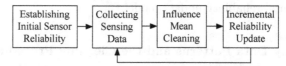

Fig. 2. Continuous cleaning procedure on a sensing data stream

3.2   Influence Mean Cleaning

In an environmental sensing application, the true reading value can be predicted as the mean of the readings made by a group of spatially correlated sensors:

$$P_{MEAN}(R) = \frac{1}{k}\sum R \tag{1}$$

where $R = \{r_1, r_2, ..., r_k\}$ is the set of k readings produced by the spatially correlated sensors.

Suppose the set of the sensors are $\{s_1, s_2, ..., s_l\}$. Let $\{srlb_1, srlb_2, ..., srlb_l\}$ be each sensor's reliability. We can define the *reliability of a reading* as the reliability of the sensor that produced it:

$$rlb(r) = srlb_i, \quad \text{if } r \text{ was produced by } s_i \tag{2}$$

Consequently, $rlb(R) = \{rlb(r_1), rlb(r_2), ..., rlb(r_k)\}$ is the reliability of each reading. The reliability of a reading indicates the probability of the reading being the true reading. Thus we can use a weighted prediction formulated as:

$$P_{IMC}(R) = \frac{\sum R \times rlb(R)}{\sum rlb(R)} \tag{3}$$

We call the prediction defined by Equation (3) *influence mean*, in the sense that it does not aggregates specific readings, but the *influences* of the sensors on the prediction, which are determined by their reliabilities.

3.3 Incremental Reliability Update

After the prediction is made, the reliability update compares individual readings with the prediction. Since the reading value is typically a real number, it is rare to get two readings exactly the same. Therefore, to compare a reading value with the prediction, we use a tolerance threshold *tol*. If the difference between the reading value and the prediction is within the threshold, the reading is considered as *consistent* with the prediction. We define the *consistency* of a reading $r \in R$ as the following:

$$cons(r) = \begin{cases} 1, & |r - P_{IMC}(R)| \le tol \\ 0, & otherwise \end{cases}$$ (4)

We calculate the reliability of a sensor as the percentage of the readings made by the sensor that are consistent with the prediction, from the total number of readings it has made during an observation period:

$$srlb = \frac{1}{n} \sum_{i=1}^{n} cons(r_i)$$ (5)

where $\{r_1, r_2, ..., r_n\}$ are the readings made by the sensor. When applying the method to continuous streams, the observation period is usually a moving time window with a fixed length. In practice, the choice of observation period length usually depends on the type of temporary interference that can occur in the deployment.

Now we can derive an incremental reliability update formula. Suppose that after making n readings, the reliability calculated for a sensor using Equation (5) is $srlb$. If the sensor has made another reading since then, the new reliability $srlb'$ can be calculated as:

$$srlb' = \frac{1}{n+1} \sum_{i=1}^{n+1} cons(r_i)$$

Substituting Equation (5) into above formula will give:

$$srlb' = srlb \times \frac{n}{n+1} + \frac{1}{n+1} cons(r_{n+1})$$ (6)

Equation (6) can be used as an incremental formula for calculating the new reliability given the current reliability and a new reading. Substituting Equation (4) into the formula gives a *reward or penalty function*, which lets the sensor gain or lose some reliability based on its new reading:

$$srlb' = \begin{cases} srlb + \dfrac{1 - srlb}{n+1}, & \text{if } cons(r_{n+1}) = 1 \\ srlb - \dfrac{srlb}{n+1}, & \text{if } cons(r_{n+1}) = 0 \end{cases}$$ (7)

4   Experimental Analysis on Synthetic and Real Dataset

In this section, we describe our experiments for testing our approach on synthetic and real datasets. In both cases, we first obtained a set of clean data, then injected faulty data to simulate sensor malfunctions.

4.1   Influence Meaning Cleaning in a Synthetic Dataset

Our first experiment simulated a scenario of attaching sensors to motor vehicles to monitor air pollution in urban areas. Such a scenario has been run in several projects such as OpenSense, which put air quality sensors on trams in Zurich [9], and Common Sense, which put air quality sensors on street sweepers in San Francisco [1]. The dataset in such projects usually contains sensor readings and time and location of the sensor readings. In addition, each reading is associated with a sensor, which changes its location frequently.

We first simulated a pollution map. The pollution map consists of 100×100 location points, and the corresponding pollution information at each point, as shown in Fig. 3a. The size of a dot on the map indicates the pollution level: the larger the dot, the higher the pollution level at the corresponding location. The maximum pollution level is 1, and the minimum pollution level is 0. We then simulated 20 mobile sensors. In each data collection iteration, a sensor made 50 readings at random locations on the map, and a total of 1,000 readings were made, similar to the readings shown in Fig. 3b. These are clean readings, as they are exactly the same as the ground truth pollution level at their report locations. We used the faulty data injection method introduced in [11] to simulate sensor malfunctioning. We injected *high volatility* faults, similar to those shown in Fig. 1a, which are commonly found in air quality sensors.

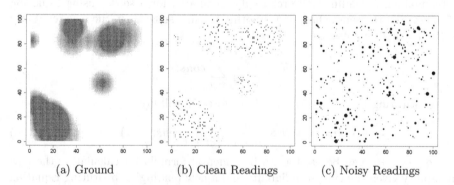

| (a) Ground | (b) Clean Readings | (c) Noisy Readings |

Fig. 3. The ground truth pollution map and generated readings

We ran the IMC procedure shown in Fig. 2. First we set an initial reliability of 1 for all sensors. In each iteration, we generated a set of noisy readings similar to the one shown in Fig. 3c. To divide the readings into spatially correlated groups, we divided the map into 100 10×10 blocks, and the readings whose coordinates fall within the same block were grouped. In each iteration, one prediction was made for each block. The ground truth pollution level of each block is calculated as the mean of pollution levels of all location points in the block. We also recorded the predictions of mean and median methods in each iteration. The mean prediction is defined in Equation (1). The median is defined as $P_{MED}(R) = median(R)$, where R is the set of readings in one block.

We measured the precision and the mean square error for the predictions made by three methods in each iteration, as shown in Fig. 4. We notice that the performance of the mean and median methods remain stable over the iteration. The performance of IMC, however, quickly improves in the first 20 iterations, before it becomes stable. The reason is that the reliability update process is picking up appropriate individual reliabilities, thus allowing the reliability-weighted method to become more accurate. After 20 iterations, IMC steadily outperforms the other two methods. For instance, in the last ten iterations of the 50 iteration run, the average precisions for mean, median and IMC are 0.6, 0.78 and 0.85, respectively, while the average mean square error are 0.017, 0.011, and 0.006, respectively. In the long run, the IMC has the potential to have nearly a 10 percent higher precision than the mean and median methods.

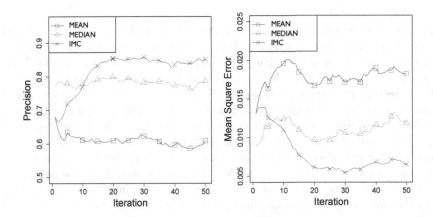

Fig. 4. The precision and mean square error of three prediction methods

4.2 Influence Mean Cleaning in a Real Dataset

We tested our approach on a real dataset provided by the Intel Berkeley Research Lab, called *Intel Lab data*[3]. The data contains environmental readings, such as humidity and temperature, reported by 52 Mica2Dot sensors. The sensors were installed in an indoor area, and had the same reporting frequency of once per 31 seconds. In our experiment, we chose a portion of temperature data in the Intel Lab data made by nine motes with ID 1, 2, 3, 4, 6, 7, 8, 9, 10. These nine

---

[3] http://db.csail.mit.edu/labdata/labdata.html

sensors were installed next to each other in a continuous open area, and can be considered as spatially correlated. We chose a study period of roughly 75 hours, between March 2 and March 5, 2004. There were 9,000 report epochs in this period. We visually confirmed that the readings produced by the nine sensors for these epochs are clean, as shown in Fig. 5a.

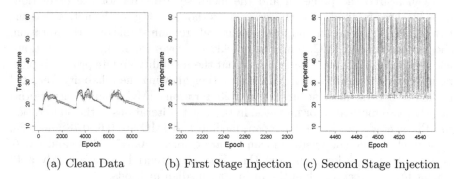

(a) Clean Data   (b) First Stage Injection   (c) Second Stage Injection

Fig. 5. The data used for testing. (b) and (c) show the injected faulty data

To generate the noises, we injected faulty data into the dataset. We injected *intense spikes*, which can be found in other parts of the Intel Lab data. To imitate sensor condition changes over time, we injected the faults in three stages. First we grouped the sensors into three groups: the first group contained sensors 1, 4, 8, the second group contained sensors 2, 6, 9, and the third group contained sensors 3, 7, 10. We made the second group of sensors fail in stage one and two, and the third group of sensors fail in stage two and three. So in the first fault stage, which lasted from epoch 2250 to 4500, readings from the sensors in the second group were injected with faulty data, as shown in Fig. 5b. In the second fault stage, which lasted from epoch 4500 to 6750, readings from the sensors in the second and third groups were injected with faulty data, as shown in Fig. 5c. In the third stage with remaining epochs, only readings from the sensors in the third group were injected with faulty data. When being injected with faulty data, each reading had a probability of 0.5 to be replaced by a spike sensor value (60 in our case).

Similar to our experiments with the synthetic dataset, we ran the IMC procedure with the generated noisy data, as well as the mean and median prediction. Since these nine sensors were considered as a single spatially correlated group, only one prediction was generated in each iteration. We recorded the predictions made by three methods in each iteration. When updating reliabilities, we used a modified version of reward or penalty function, by adding a constant value of 0.01 to the penalty defined in Equation (7). The higher penalty is chosen to mitigate the effect of extremeness of faulty values. How to dynamically change the reward or penalty amount in the case of unpredictable extreme faulty values is a topic of future work.

We measured the performance of the three prediction methods as the square error of the prediction, given the ground truth as the mean of the clean data. Fig. 6a shows predictions of the three methods and the ground truth over 9,000 epochs. Fig. 6b shows the square error of the three methods over 9,000 epochs. To avoid showing high volatility in the graph, the data was smoothed by 100 epochs before plotting.

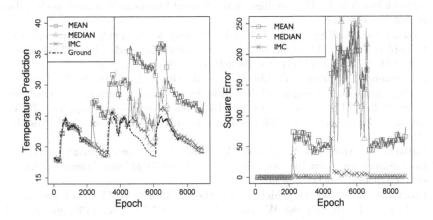

Fig. 6. The prediction and square error of three prediction methods

As shown in the figures, IMC is affected the least by the intense faults in the second stage, and produces only small errors comparing to other methods. At the beginning of the first and second fault stages where the portion of faulty sensors increases, the performance of IMC experiences sudden declines, but can always recover in a short time. This was because our method adjusted the sensor reliability to the new sensor conditions. In the third fault stage, the performance of IMC improves from the second stage, and becomes similar to what it is in the first stage. This adjustment shows that our reliability update process not only detects sensor faults, but also captures sensors' recovery from the faults.

5   Conclusion and Future Work

In this paper, we propose a sensor reliability-based method for sensor data cleaning, called influence mean cleaning (IMC). Our experiment results show that for noisy datasets with different types of faulty data patterns, IMC can achieve higher accuracy than mean and median methods over time. By updating the reliability incrementally, our method can properly discover the latent sensor reliability values. The experimental results with the real dataset from Intel Lab show that our method can capture both sensor malfunctioning and recovery. While individual sensor reliability is largely overlooked in current sensor network research, we show that individual sensor reliabilities can be leveraged to create positive impacts. In the future, we plan to investigate the performance of our approach in datasets with mixed or changing faulty data patterns.

References

1. Aoki, P.M., Honicky, R.J., Mainwaring, A., Myers, C., Paulos, E., Subramanian, S., Woodruff, A.: A vehicle for research: Using street sweepers to explore the landscape of environmental community action. In: Proceedings of the SIGCHI Conference on Human Factors in Computing Systems (2009)
2. Buonadonna, P., Gay, D., Hellerstein, J.M., Hong, W., Madden, S.: Task: Sensor network in a box. In: Proceeedings of the Second European Workshop on Wireless Sensor Networks (2005)
3. Bychkovskiy, V., Megerian, S., Estrin, D., Potkonjak, M.: A collaborative approach to in-place sensor calibration. In: Zhao, F., Guibas, L.J. (eds.) IPSN 2003. LNCS, vol. 2634, pp. 301–316. Springer, Heidelberg (2003)
4. Devarakonda, S., Sevusu, P., Liu, H., Liu, R., Iftode, L., Nath, B.: Real-time air quality monitoring through mobile sensing in metropolitan areas. In: Proceedings of the 2nd ACM SIGKDD International Workshop on Urban Computing (2013)
5. Enrick, N.L.: Quality, reliability, and process improvement. Industrial Press Inc. (1985)
6. Galland, A., Abiteboul, S., Marian, A., Senellart, P.: Corroborating information from disagreeing views. In: Proceedings of the Third ACM International Conference on Web Search and Data Mining (2010)
7. Hasenfratz, D., Saukh, O., Thiele, L.: On-the-fly calibration of low-cost gas sensors. In: Picco, G.P., Heinzelman, W. (eds.) EWSN 2012. LNCS, vol. 7158, pp. 228–244. Springer, Heidelberg (2012)
8. Jeffery, S.R., Alonso, G., Franklin, M.J., Hong, W., Widom, J.: Declarative support for sensor data cleaning. In: Fishkin, K.P., Schiele, B., Nixon, P., Quigley, A. (eds.) PERVASIVE 2006. LNCS, vol. 3968, pp. 83–100. Springer, Heidelberg (2006)
9. Li, J.J., Faltings, B., Saukh, O., Hasenfratz, D., Beutel, J.: Sensing the air we breathe-the opensense zurich dataset. In: Proceedings of the Twenty-Sixth AAAI Conference on Artificial Intelligence (2012)
10. Ni, K., Ramanathan, N., Chehade, M.N.H., Balzano, L., Nair, S., Zahedi, S., Kohler, E., Pottie, G., Hansen, M., Srivastava, M.: Sensor network data fault types. ACM Transactions on Sensor Networks 5(3), 25:1–25:29 (2009)
11. Sharma, A.B., Golubchik, L., Govindan, R.: Sensor faults: Detection methods and prevalence in real-world datasets. ACM Transactions on Sensor Networks 6(3) 23, 23:1–23:39 (2010)
12. Sheng, Q.Z., Li, X., Zeadally, S.: Enabling next-generation RFID applications: Solutions and challenges. IEEE Computer 41(9), 21–28 (2008)
13. Wang, D., Kaplan, L., Le, H., Abdelzaher, T.: On truth discovery in social sensing: A maximum likelihood estimation approach. In: Proceedings of the 11th International Conference on Information Processing in Sensor Networks (2012)
14. Wen, Y.J., Agogino, A.M., Goebel, K.: Fuzzy validation and fusion for wireless sensor networks. In: Proceedings of the ASME International Mechanical Engineering Congress (2004)
15. Zhang, Y., Meratnia, N., Havinga, P.: Outlier detection techniques for wireless sensor networks: A survey. IEEE Communications Surveys Tutorial 12(2), 159–170 (2010)
16. Zhuang, Y., Chen, L., Wang, X., Lian, J.: A weighted moving average-based approach for cleaning sensor data. In: Proceedings of the 27th International Conference on Distributed Computing Systems (2007)

Managing Incentives
in Social Computing Systems with PRINGL

Ognjen Scekic, Hong-Linh Truong, and Schahram Dustdar

Distributed Systems Group, Vienna University of Technology, Austria
{oscekic,truong,dustdar}@dsg.tuwien.ac.at

Abstract. Novel web-based socio-technical systems require incentives for efficient management and motivation of human workers taking part in complex collaborations. Incentive management techniques used in existing crowdsourcing platforms are not suitable for intellectually-challenging tasks; platform-specific solutions prevent both workers from comparing working conditions across different platforms as well as platform owners from attracting skilled workers. In this paper we present PRINGL, a domain-specific language for programming complex incentive strategies. It promotes re-use of proven incentive logic and allows composing of complex incentives suitable for novel types of socio-technical systems. We illustrate its applicability and expressiveness and discuss its properties and limitations.

Keywords: rewards, incentives, social computing, crowdsourcing.

1 Introduction

Human participation in web-based socio-technical systems has overgrown conventional crowdsourcing where humans solve simple, independent tasks. Emerging systems ([1, 2, 3]) are attempting to leverage humans for more intellectually challenging tasks, involving longer lasting worker engagement and complex collaboration workflows. This poses the problems of finding, motivating, retaining and assessing workers, as well as making the virtual labor market more competitive and attractive to workers. Paper [4] highlights a number of important research areas that need to be investigated in order to build such systems. Incentive management has been identified as one of the important parts of this initiative. However, contemporary incentive management in social computing systems usually imply a hard-coded and completely system-specific solution [5]. Such approach is not portable, and prevents reuse of common incentive logic. That hinders cross-platform application of incentives and reputation transfer.

Motivation. Our ultimate goal is to develop a general framework for automated incentive management for the emerging social computing systems. Such an *incentive management framework* could be coupled with different workflow or crowdsourcing systems, and, based on monitoring data they provide, would perform incentivizing measures and team adaptations. In this way, incentive management could be offered as a service in the cloud.

B. Benatallah et al. (Eds.): WISE 2014, Part II, LNCS 8787, pp. 415–424, 2014.

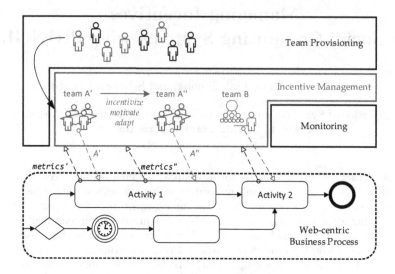

Fig. 1. Operational context of incentive management systems

Figure 1 visualizes the context in which an incentive management framework is supposed to operate: A complex business process execution employs crowd-sourced team(s) of human experts to perform various workflow activities. The teams are provisioned by a dedicated service (e.g., SCU [6]) that assembles teams based on required elasticity parameters, such as: worker skills, price, speed or reputation. However, choosing appropriate workers alone does not guarantee the quality of subsequent team's performance. In order to monitor and influence the behavior of workers during and across activity executions an incentive scheme needs to be enacted. This is the task of the incentive management framework. It enacts the incentive scheme by applying rewards or penalties in a timely manner to induce a wanted worker behavior, thus effectively performing runtime team adaptations (e.g., Fig. 1: $A' \to A''$).

Contribution. In [7] we presented a framework for low-level incentive management – PRINC. Although PRINC allowed monitoring of metrics and application of basic incentive mechanisms for social computing systems, it lacked a comprehensive, human-readable way of encoding incentive strategies, motivating us to design PRINGL[1] – a novel domain-specific language (DSL) for modeling incentives for socio-technical systems. In this paper we illustrate how real-world incentive mechanisms for social computing systems can be modeled in PRINGL.

Paper Organization. Section 2 gives an overview of PRINGL's design and intended usage. Section 3 introduces some of PRINGL's basic language constructs, describes the implemented language metamodel and discusses the advantages and limitations of the proposed approach. Section 4 presents the related work. Section 5 concludes the paper.

---

[1] **PR**ogrammable **IN**centive **G**raphical **L**anguage

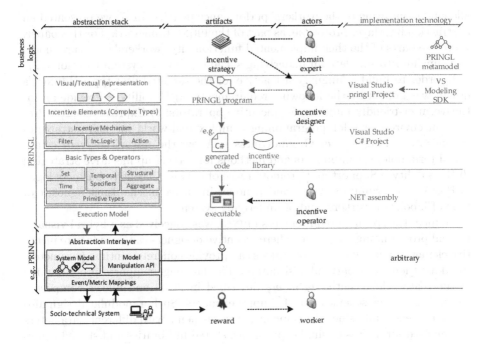

Fig. 2. Overview of PRINGL's architecture and usage

2 PRINGL Overview

Designing an incentive scheme is itself a challenging task usually performed by domain experts for a particular work type or company. However, as shown in [5, 8] most real-world incentive strategies used in social computing environments can be composed of modelable and reusable bits of incentive logic. PRINGL is a domain-specific language intended to be used by two types of *users* (Figure 2): a) *incentive designers* – domain experts that design and implement incentive strategies for different organizations (in particular crowdsourcing and socio-technical platforms); and b) *incentive operators* – members of the organizations responsible for managing the every-day running and adaptation of the scheme. While incentive designers may need to concern themselves with implementation details of the underlying system in order to adapt general incentive mechanisms for it, incentive operators want to manage the incentive scheme by using a simple and intuitive user interface without knowing implementation internals.

Figure 2 shows an overview of PRINGL's architecture and usage. An incentive designer models an incentive scheme using PRINGL's visual system-independent syntax. The PRINGL-encoded scheme gets translated into a system-specific executable able to exchange monitoring and incentive events with a social computing system through an *abstraction interlayer*. We use the term abstraction interlayer to denote any middleware sitting on top of a socio-technical system, exposing to external users a simplified model of its employed workforce and allowing monitoring of the workers' performance metrics. In [7] we presented an abstraction interlayer prototype, as part of the PRINC framework. For this paper, we re-use parts of the then implemented functionality (workers' structure model and timeline) to simulate an underlying social computing system (Section 3.2).

In order to build a language attractive for the targeted user types, PRINGL's design was guided by the following requirements: *a*) Usability – Provide an intuitive, user-friendly interface for incentive operators; *b*) Expressiveness – Provide an environment for programming complex real-world incentive strategies for incentive designers; *c*) Groundedness – Allow the use of *de facto* established terminology, components and methods for setting up incentive strategies; *d*) Reusability – Support and promote reuse of existing incentive business logic; *e*) Portability – Support system-independent incentive mechanisms, agnostic of type of labour or workers, and of underlying systems.

To meet the specified requirements PRINGL was conceived as a hybrid visual/-textual programming language, where incentive designers can encode core incentive elements, while incentive operators can provide concrete runtime parameters to adapt them to a particular situation. The language supports programming of the real-world incentive elements described in [5, 8] and allows composing complex incentive schemes out of simpler elements. Such a modular design also promotes reusability since the same incentive elements with different parameters can be used for a class of similar problems, stored in libraries and shared across platforms. PRINGL allows incentive designers to model natural-language, realistic incentive strategies (i.e., business logic) into a *platform-independent specification* through a number of incentive elements represented by a visual syntax (graphical elements with code snippets). The designer programs new incentive elements or reuses existing ones from an *incentive library* to compose new, more complex ones. Once the entire incentive scheme is specified, PRINGL translates it into a *platform-specific* code in a common programming language that can be further compiled into executable or library assemblies. The assemblies can then be used by incentive operators to execute and manage incentive enactment (Figure 2).

3   Modeling Incentives with PRINGL

The **incentive elements** are the basic functional units of a PRINGL program. Due to space constraints a detailed, conventional description of PRINGL's visual syntax and programming model cannot be presented here. Instead, in this section we briefly describe the functionality of the principal language constructs. The interested reader is encouraged to visit the PRINGL homepage[2] containing the full PRINGL specification, as well as other useful links and documents.

3.1   PRINGL Language Constructs

Incentive Logic. These constructs encapsulate different aspects of business logic related to incentives in reusable bits. They can be thought of as library-storable functions with predefined signatures allowing only certain input and

---

[2] http://dsg.tuwien.ac.at/research/viecom/PRINGL

output parameters. They are invoked from other PRINGL constructs, including other IncentiveLogic elements. Implementation is dependent on the abstraction interlayer, but not necessarily on the underlying socio-technical platform, meaning that many libraries can be shared across different platforms, promoting reusability of proven incentives, uniformity and reputation transfer. The Designer is encouraged to implement incentive logic elements as small code snippets with intuitive and reusable functionality. Depending on the intended usage, incentive logic elements have different subtypes: **A**ction, **S**tructural, **T**emporal, **P**redicate, **F**ilter. The subtype prescribes different input parameters and allows PRINGL to populate some of them automatically (marked with auto). Similarly, different subtypes dictate different return value types. These features encourage high modularization and uniformity of incentive logic elements. Incentive logic elements are denoted by a diamond shape surrounding the letter indicating the subtype, e.g., ◁T▷ and ◁P▷ in Fig. 3, bottom.

Worker Filter. Its function is to identify, evaluate and return matching workers for subsequent processing based on user-specified criteria. The criteria are most commonly related (but not limited) to worker's past performance and team structure. The workers are matched from the input collection of Workers that is provided by the PRINGL environment at runtime. By default, all the workers in the system are considered. The output is a collection of matching workers. We use a right-pointed shape ▷ to denote filters (Fig. 3, top left). A Composite-WorkerFilter definition consists of graphical elements representing instances of previously defined WorkerFilters (Fig. 3, top right).

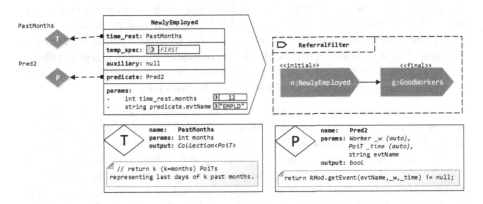

Fig. 3. A CompositeWorkerFilter for referral bonuses

Example: A company wants to introduce employee referral process[3] in which an existing employee can recommend new employee candidates and get rewarded if the newly employed candidates spend a year in the company having exhibited satisfactory performance. In order to pay the referral bonuses the company needs

---

[3] http://en.wikipedia.org/wiki/Employee_referral

to: a) identify the newly employed workers; and b) asses the worker performance of those workers. Let us assume that the company already has the business logic for assessing the workers implemented, and that this logic is available as the library filter GoodWorkers. In this case, we need to define one additional simple filter NewlyEmployed, and combine it with the existing GoodWorkers filter. In Figure 3 we show how the new composite ReferralFilter is constructed. The F⟩ instance n:NewlyEmployed makes use of: a) ⟨T⟩ PastMonths returning time points representing end-of-month for the given number of months (12 in this particular case); and b) predicate ⟨P⟩ Pred2 checking if the employee got hired 12 months ago. Pred2's general functionality is to check whether the abstraction interlayer (RMod) registered an event of the given name at the specified time.

Rewarding Action. Its function is to notify the abstraction interlayer that a concrete action should be taken against specific workers at a given time, or that certain specific actions should be forbidden to some workers during a certain time interval. In order to perform the action, the runtime environment needs to know to which workers the action applies, so a worker filter needs to be applied. In some cases, the workers that are rewarded/punished may be the same as initially evaluated ones. In that case we can reuse the original filter used for evaluation. In other cases, workers may be rewarded based on the outcome of evaluation of other workers (e.g., team managers for the performance of team members). The runtime also needs to determine the timing for action application. We use temporal specifiers (see PRINGL specification[4]) to determine the exact time moment(s). The output of a RewardingAction is a Collection<Worker> containing affected workers, i.e., those to which the action was successfully applied. To execute the action PRINGL needs to invoke the appropriate action in the abstraction interlayer which will then send out a system-specific message to the underlying system. We use a trapezoid shape /A\ to denote RewardingAction elements (Fig. 4, bottom right). Similarly to composite filters, a CompositeRewardingAction definition consists of graphical elements representing instances of previously defined RewardingActions (Fig. 4, bottom left).

Example: Consider a company that wants to reward workers either with free days or with a monetary reward. The choice is left to the worker. Free days are offered first. Only workers that refuse the free days will be given monetary rewards. We define a new composite rewarding action BonusOrDays (Figure 4) that, for the sake of demonstration, assumes the existence of a RewardAtEndProject action to award monetary bonuses, as well as a newly-defined action FreeDays to award free working days to the workers. The output of a:FreeDays is the set of workers who accepted the 3 free days offered. However, due to a complement edge (↛) connecting a and b, the output set of a is subtracted from the original input set. Therefore, the input of b:RewardAtEndProject are only those workers who declined to accept working days as award, and want to be evaluated at the end of project and paid a bonus according to their performance.

Incentive Mechanism. This is the main structural and functional incentive element used to express complex incentive schemes. It combines the previously

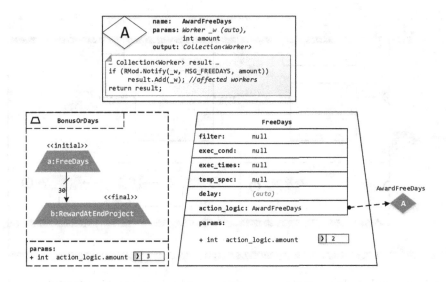

Fig. 4. A CompositeRewardingAction letting the workers choose one of the rewards

defined constructs (incentive elements) to select, evaluate and reward workers. As a self-sufficient and independent unit, it does not have any inputs or outputs. It can be stored and reused through instantiations with different runtime parameters. It also has dedicated GUI elements for definition and instantiation, as well as a shorthand notation – IM. Due to spatial restrictions, a full example of design and usage of incentive mechanisms is provided in supplement materials[4].

3.2 Implementation

Figure 2 (Section 2) shows the overview of implemented components. PRINGL's language metamodel prototype was implemented[4] in Microsoft's Modeling SDK for Visual Studio 2013 (MSDK). MSDK allows defining visual DSLs and translating them to an arbitrary textual representation. Using MSDK we generated a Visual Studio plug-in providing a complete IDE for developing PRINGL projects. In it, an incentive designer can create a dedicated Visual Studio PRINGL project and implement/model real-world strategies using the visuo-textual elements introduced in this paper (see Figure 5). The graphical elements provided in the implemented Visual Studio PRINGL environment, although not as visually appealing as those presented in this paper, functionally and structurally match them fully. PRINGL models are stored in .pringl files that get automatically transformed to the corresponding C# (.cs) equivalents. The generated code can then be used in the rest of the project as regular C# code or compiled in .NET assemblies (e.g., libraries or executables).

---

[4] Source code, screenshots and additional info available at:
http://dsg.tuwien.ac.at/research/viecom/PRINGL/

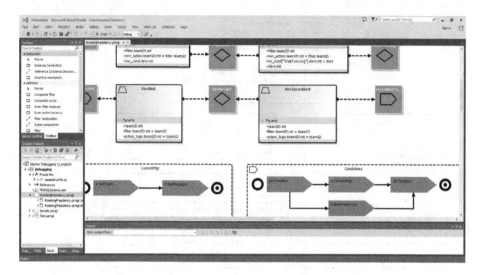

Fig. 5. Screenshot. Implementing a realistic incentive scheme using PRINGL Visual Studio environment.

Figure 5 shows a screenshot of the implementation of the rotating presidency example[4] using the VS PRINGL IDE. The entire scheme was modeled using the generated PRINGL tools, demonstrating the feasibility of the proposed architectural design. The C# code obtained from the implemented model can be used to produce arbitrary incentive management applications, using PRINC as the acting interlayer (see Section 2).

3.3   Discussion

Advantages. In case of a conventional social computing platform (cf. [8]), the business logic necessary for enacting a platform-specific incentive scheme would likely need to be implemented and tested anew; subsequent changes would require changing the source code. With PRINGL, however, the incentive designer is likely to implement and test the basic incentive elements only once. PRINGL encouraging a modular design of incentive schemes composed of many small, easily testable components. With the basic elements available, composing complex ones can be done in a matter of minutes by visual modeling, copy-pasting and simple editing of fields and parameters. Once defined, the incentive elements can be stored in libraries and shared across different social-computing platforms promoting reusability and portability. PRINGL's composite actions enable the incentive designer to create tailored rewarding actions for different personality types or worker roles by combining a number of available rewarding mechanisms. As an additional benefit, by using standardized PRINGL incentive elements, comparing incentives across different social computing platforms becomes much easier. This is one of the fundamental requirements necessary to establish fair working conditions and sustainable virtual careers of crowdsourcing workers [4]. PRINGL's programming model was designed to support modeling of real-world incentive strategies from [5], thus addressing the desired Groundedness and Expressiveness requirements from the Section 2.

Limitations. So far PRINGL has been tested only in simulation environment with simple provisioning engines. However, it is important to point out that our goal is *not* to invent novel incentive mechanisms, nor to compare or improve existing ones. Rather, the focus is on functionally validating PRINGL's design and expressiveness in modeling documented, existing incentive mechanisms, thus not requiring evaluation with human subjects. As there are no known similar languages, a comparative qualitative evaluation was not possible. Also, at this moment PRINGL is limited to supporting worker-centric incentives only. Incentive adaptations currently require human intervention.

4 Related Work

Previous research on incentives for socio-technical systems is dispersed and problem-specific. It can be roughly categorized in two groups. One group seeks to find optimal incentives in formally defined environments through precise mathematical models [9]. Although successfully used in microeconomic models, these incentive models do not fully capture the diversity and unpredictability of human behavior that becomes accentuated in socio-technical systems. The other group examines the effects of incentives by running experiments on existing crowdsourcing platforms and rewarding real human subjects with actual monetary rewards. For example, in [10] the authors examine the effects of incentives by running experiments on existing crowdsourcing platforms and rewarding real human subjects with actual monetary rewards. In [11] the authors compare the effects of lottery incentive and competitive rankings in a collaborative mapping environment. In [2] the focus is on pricing policies that should elicit timely and correct answers from crowd workers. The major limitation of this research approach ([12]) is that the findings are applicable only for a very limited range of simple activities, such as image tagging and text translation. Two surveys of commonly used incentive techniques today can be found in [5, 8]. To the best of our knowledge, there have been no previous attempts of formalizing a general approach to incentive management for socio-technical systems.

5 Conclusions and Future Work

In this paper we introduced a domain-specific language named PRINGL for programming incentives for socio-technical systems. PRINGL allows the incentives to stay decoupled of the underlying systems. It fosters a modular approach in composing incentive strategies that promotes code reusability and uniformity of incentives, while leaving the freedom to incentive operators to adjust the strategies to their particular needs helping cut down development and adjustment time and creating a basis for development of standardized but tweakable incentives. This in turn leads to more transparency for workers and creates a basis for an incentive uniformity across companies; a necessary precondition for worker reputation transfer. In future, we plan to include support for artifact-centric incentives, and integrate PRINC into a general programming model for Hybrid Collective Adaptive Systems (HDA-CAS).

Acknowledgements. This work is supported by the EU FP7 SmartSociety project under grant №600854.

References

1. Ahmad, S., Battle, A., Malkani, Z., Kamvar, S.: The jabberwocky programming environment for structured social computing. In: Proceedings of the 24th Annual ACM Symposium on User Interface Software And Technology, UIST 2011, vol. 53 (2011)
2. Barowy, D.W., Curtsinger, C., Berger, E.D., McGregor, A.: Automan: A platform for integrating human-based and digital computation. SIGPLAN Not. 47(10), 639–654 (2012)
3. Minder, P., Bernstein, A.: *CrowdLang*: A Programming Language for the Systematic Exploration of Human Computation Systems. In: Aberer, K., Flache, A., Jager, W., Liu, L., Tang, J., Guéret, C. (eds.) SocInfo 2012. LNCS, vol. 7710, pp. 124–137. Springer, Heidelberg (2012)
4. Kittur, A., Nickerson, J.V., Bernstein, M., Gerber, E., Shaw, A., Zimmerman, J., Lease, M., Horton, J.: The future of crowd work. In: Proceedings of the 2013 Conference on Computer Supported Cooperative Work, CSCW 2013, p. 1301 (2013)
5. Scekic, O., Truong, H.L., Dustdar, S.: Incentives and rewarding in social computing. Communications of the ACM 56(6), 72 (2013)
6. Dustdar, S., Bhattacharya, K.: The social compute unit. IEEE Internet Computing 15(3), 64–69 (2011)
7. Scekic, O., Truong, H.-L., Dustdar, S.: Programming incentives in information systems. In: Salinesi, C., Norrie, M.C., Pastor, Ó. (eds.) CAiSE 2013. LNCS, vol. 7908, pp. 688–703. Springer, Heidelberg (2013)
8. Tokarchuk, O., Cuel, R., Zamarian, M.: Analyzing crowd labor and designing incentives for humans in the loop. IEEE Internet Computing 16(5), 45–51 (2012)
9. Laffont, J.J., Martimort, D.: The Theory of Incentives. Princeton University Press, New Jersey (2002)
10. Mason, W., Watts, D.J.: Financial incentives and the "performance of crowds". In: Proceedings of the ACM SIGKDD Workshop on Human Computation, HCOMP 2009, pp. 77–85. ACM, New York (2009)
11. Ramchurn, S.D., Huynh, T.D., Venanzi, M., Shi, B.: Collabmap: crowdsourcing maps for emergency planning. In: Proceedings of 5th ACM Web Science Conference, Paris, France, pp. 326–335 (May 2013)
12. Adar, E.: Why i hate mechanical turk research (and workshops). In: Proc. of CHI 2011 Workshop on Crowdsourcing and Human Comp. ACM, Vancouver (2011)

Consumer Monitoring of Infrastructure Performance in a Public Cloud

Rabia Chaudry[1,3], Adnene Guabtni[2,3], Alan Fekete[1,3],
Len Bass[2,3], and Anna Liu[2,3]

[1] School of Information Technologies, University of Sydney, Sydney, Australia
[2] School of Computer Science and Engineering, University of New South Wales,
Sydney, Australia
[3] NICTA, Sydney, Australia
{firstname.lastname}@nicta.com.au

Abstract. Many web information systems and applications are now run
as cloud-hosted systems. The organization that owns the information
system or application is thus a consumer of cloud services, and often
relies on the cloud provider to monitor the virtual infrastructure and
alert them of any disruption of the offered services. For example, Amazon
Web Services' cloud disruptions are announced by the cloud provider on
a dedicated RSS feed so that the consumers can watch and act quickly.

In this paper, we report on a long-running experiment for the monitoring and continuous benchmarking of a number of cloud resources on
Amazon Cloud from a consumer's perspective, aiming to check whether
the service disruptions announced by the cloud provider are consistent
with what we observe. We evaluate the performance of cloud resources
over several months. We find that the performance of the cloud can vary
significantly over time which leads to unpredictable application performance. Our analysis shows also that continuous benchmarking data can
help detect failures before any announcement is made by the provider,
as well as significant degradation of performance that is not always connected with Amazon service disruption announcements.

Keywords: Cloud Monitoring, Failure Detection, Benchmarking.

1 Introduction

Applications such as web information systems are increasingly being deployed
on cloud computing infrastructure spread across data centres around the globe
to avail themselves of virtualized high performance, highly available computing,
large scale storage and high bandwidth communication. But since the underlying physical resources are being shared among many applications with changing
demands, the effective capacity each application sees can vary significantly despite the efforts of cloud providers to guarantee expected quality (stated through
service level agreements or SLAs) [15,9,21,24,13].

The cloud's unpredictable performance behavior has prompted efforts towards
monitoring the resources of interest by both the providers and the tenants of the

B. Benatallah et al. (Eds.): WISE 2014, Part II, LNCS 8787, pp. 425–434, 2014.

cloud platform. The providers tend to perform monitoring of the infrastructure and the resources they provide to meet the promised SLAs to the tenants. If a problem is detected, the providers make announcements regarding the event to inform the public. The tenants have limited visibility into the underlying infrastructure, and instead they are usually focused towards monitoring the applications installed on their instances. For coping with major events affecting the infrastructure, they rely on the announcements made by the providers. However, the providers' announcements are not always as rapid as possible – the providers have incentive to avoid false alarms and so may be rather conservative in announcing when a potential problem first becomes suspected.

In this paper, we report on a long-running experiment for the monitoring of the performance of cloud resources available to a tenant, particularly those usually involved in major cloud failures, such as networking and storage[1]. Unlike typical observations of the tenant's actual running applications, we use intensive micro-benchmarks that quantify the performance of the virtualized cloud resources including cpu and memory, the throughput and latency of the network, and the storage throughput. A key novelty of our approach lies in a distributed "lease" driven scheduler ensuring isolation of benchmarking activities and avoiding unreliable measurements. The benchmarking results of the networking and storage services are calibrated to further improve the reliability of the results by filtering out the effects of multi-tenancy of the physical servers. Another contribution is a way to analyze the time series measurements to see how well they correlate with the provider's announcements of service disruptions.

We have applied our approach to benchmark the performance of cloud resources in AWS (Amazon Web Services) over seven months. The results show clearly that sometimes performance degradation has been evident before any announcement and there have been other long-term changes in performance, especially of the storage tier (EBS), which were not reflected in announcements. If consumers do monitoring as we propose, they will have a valuable source of information that could enable early reactions including migrating their applications to other availability zones or providers and/or adding instances to compensate for degraded performance. In the remainder of the paper, section 2 provides a literature review, section 3 presents the key challenges of this work, section 4 introduces the data and empirical strategy adopted, section 5 outlines the findings and implications for cloud consumers and finally, section 6 concludes the paper and discusses the limitations and future research directions.

2   Related Work

Many studies have been reported in the literature on performance benchmarking and/or monitoring of cloud infrastructure. We review in this section the most

---

[1] Summaries of significant AWS disruptions in the US East Region:
   http://aws.amazon.com/message/65648/,
   http://aws.amazon.com/message/67457/,
   https://aws.amazon.com/message/680342/

relevant studies based on their intended goal. These include studies relating to comparison of various cloud offerings and architectures, study of performance patterns and variations over time and across resource types, evaluation of resource provisioning strategies and performance interference in cloud environments.

Table 1. Structured Literature Review

Reference	Data Source[1]	Environment[2]	Detection Technique[3]	Prediction Technique[4]	Purpose[5]
[6]	HM	Lab (PM)	AC,AD	T, W	FD
[11]	HM	Lab (PM)	AC,AD	T, W	FD
[18]	HM	Lab (PM)	AC,AD	T	FD
[20]	HM	Lab (PM)	AC,AD	T, W	FD
[19]	HM	Lab (PM)	AC,AD	T, W	FD
[17]	RB[6] (C,M,S,N,H)	Cloud (Pvt)	AD	OL	FD
[26]	HM	Lab (PM)	ST	T, W	FD
[2]	HM	Lab (Sim)	PB	T, W	FD
[22]	HM	Lab (Sim)	AD	T, W	FD
[12]	RB (C,M,S)	Cloud (Pub)	N/A	N/A	PE
[13]	RB (C,M,S)	Cloud (Pub)	N/A	N/A	PE
[3]	AB[7] (QoS)	Cloud (Pub)	N/A	N/A	PE
[14]	RB (C)	Cloud (Pub)	N/A	N/A	PE
[24]	RB (C)	Cloud (Pub)	N/A	N/A	PE
[10]	AB (DB)	Cloud (Pub)	N/A	N/A	PE
[29]	AB (DB)	Cloud (Pub)	N/A	N/A	PE
[23]	AB (QoS)	Lab (VM)	N/A	N/A	PE
[16]	RB (M,S,N)	Lab (VM)	N/A	N/A	PE
[9]	RB (M,S)	Cloud (Pub)	N/A	N/A	PE
[21]	RB (C,M,S,N)	Cloud (Pub)	N/A	N/A	PE
[5]	RB (C,D,S)	Cloud (Pub)	N/A	N/A	PE

The above table summarizes the related work on performance evaluation of the cloud platform and failure detectors. Some studies propose failure detectors for monitoring and detecting failures of processes with low overhead and fast detection times [6,11,18,20,19,17,26,2,22]. These failure detectors often employ the lightweight heartbeat monitoring strategy to monitor the health of the processes rather than benchmarking the physical resources. The limitation of such approach is that it does not provide any early insights into the health of the monitored process. In fact, it identifies a process as failed when it has already crashed. On the other hand, many studies target performance evaluation of cloud services through benchmarking the cloud infrastructure [12,13,3,14,24,10,29,23,16,9,21,5].

In more recent studies, the performance interference in cloud environments has been investigated and performance isolation has gained a lot of interest from the research community [1,25,4,27,16,28]. These studies propose that sustained

---

[1] HM=Heartbeat Monitoring, RB=Resource Benchmarking, AB=Application Benchmarking
[2] PM=Physical Machine, VM=Virtual Machine, Sim=Simulation, Pub=Public, Pvt=Private
[3] AC=Accrual, AD=Adaptive, ST=Self-tuning, PB=Probablistic
[4] T=Threshold, W=Window, OL=Outlier
[5] FD=Failure Detector, PE=Performance Evaluation
[6] C=CPU, M=Memory, S=Storage, N=Network, H=Hypervisor
[7] QoS=Quality of Service, DB=Database

background load, co-scheduling and co-location of CPU-bound and latency-sensitive tasks all contribute towards poor response times and degraded system performance.

3   Key Challenges

Cloud computing is based on the principle of multi-tenancy using virtualization technologies. As a consequence, cloud performance is highly variable [13,21,15] and unpredictable [9,21,24]. In particular, instance level resources (e.g. CPU, Memory) are hard to benchmark in a cloud environment due to performance disruptions caused by the multiple tenants (cloud instances) sharing the same underlying physical server, introducing a bias in benchmarking results. Unlike instance level resources, cloud storage services, load balancers and networking services are used simultaneously by a large number of instances within an entire cloud data center. As a consequence, any performance disruptions of networked resources are likely to affect a large number of instances, or even the entire data center. In fact, failures of networked resources represented the vast majority of announced AWS failures over the past few years, primarily around networking, API and storage services such as Elastic Block Store (EBS) or Simple Storage Service (S3).

Although networked resources, such as EBS, are thought to be independent of the instance's compute performances, Wang *et al.* [25] demonstrated empirically that compute performance (at the instance level) could affect network performance, and therefore EBS or S3 could also be affected as they are networked resources. Furthermore, the benchmarking activity itself would have to be conducted from within EC2 instances and its results would also be affected by the multi-tenancy of physical servers underlying EC2. It is therefore the focus of this paper to benchmark both instance-level and non-instance level resources, in order to detect critical failures at the region or availability zone level and factor the multi-tenancy bias in the analysis.

4   Data and Empirical Strategy

We propose a framework to deal with benchmarking cloud infrastructure as discussed in the previous section. Figure 1 illustrates the data and empirical strategy to schedule benchmarking activities, collect data, check the quality of data by estimating the potential bias due to the multi-tier cloud environment, caliber the data to filter out such bias, detect significant cloud infrastructure disruptions and finally match the detected disruptions with AWS failure announcements.

The benchmarking activities test, at regular interval, the compute, network and storage capabilities of Amazon EC2 instances and their underlying network and storage resources. Firstly, the storage benchmarking activity tests EBS performance using Flexible IO Tester[8] for measuring read and write performance of EBS volumes. The protocol used for this benchmarking activity consists of

---

[8] Flexible IO Tester, http://freecode.com/projects/fio

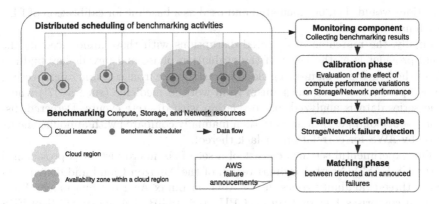

Fig. 1. Data and empirical strategy

executing two consecutive 10 seconds tests for read/write, both involving data blocks of 1MB. Secondly, the compute benchmarking activities include benchmarking CPU and Memory performance using Sysbench[9]. Memory is tested for read/write iops in a 10 seconds test, while CPU is benchmarked for prime number generation in a 10 seconds test too. Thirdly, the network benchmarking activity tests the performance of the network using IPerf[10]. Unlike CPU or Memory benchmarking, network tests require two instances connected to each other, and lasts 10 seconds, measuring the bandwidth.

A benchmarking agent in charge of executing/scheduling the benchmarking activities on the EC2 instances involved, ensures reliable results for all four benchmarking activities. To do so, they have to be executed in isolation, meaning that they cannot execute concurrently on the same instance. For that reason, we use "leases" [8]. Before an instance performs a benchmark, it obtains a lease from its benchmarking agent and the potential participating paired instances (if it is a network benchmark). A "map" of all the leases is constructed and maintained in real time in a centralized transactional service.

Collecting the results produced by the benchmarking agents is done by a key component of the conceptual framework: the monitoring system. The monitoring system receives the results of every execution of benchmarking activities. As a result, time series datasets are produced for every resource being benchmarked.

The results of CPU, Memory and Network benchmarking activities are used for calibrating the EBS benchmarking results to filter out the effect of the shared physical server. This is achieved by analyzing the benchmarking results using regression analysis to reveal whether or not compute performance significantly impacts the network performance, and if that is the case, for how much, i.e what is the maximum disruption (deviation from mean) that could be caused by compute performance variation. A maximum observable side effect of compute

---

[9] SysBench: a system performance benchmark, http://sysbench.sourceforge.net/
[10] Iperf, http://iperf.sourceforge.net/

variations would then be calculated and used as a baseline for calibration of EBS benchmarking results.

Finally, the matching of detected disruptions with those announced by the cloud provider is at the heart of the evaluation phase and unveils key findings about the benefits of monitoring cloud infrastructure from consumer perspective. We adapted a matching mechanism from work in medicine (such as [7]) where time series data is analyzed for correlation with discrete events. Detected disruptions of cloud performance are then manually compared with announcements made by AWS for correlation, or lack thereof.

The experimental settings consist of a set of six micro instances provisioned in the US-EAST (North Virginia) region of the Amazon Cloud and distributed across three availability zones, using Amazon Linux AMI (64-bit) image. Each of them comprises 1 virtual core (vCPU), 0.613 GiB of memory, 8GiB of EBS storage for the OS and an additional 1GiB EBS volume for disk benchmarking. The choice of micro instances was ideal to get the worse effect of multi-tenancy.

5   Results

The empirical data was obtained from the various benchmarking activities executed every 5 minutes over a 7 months period from 1st April to 11th November 2013. Figure 2 illustrates the time series decomposition of one of the obtained datasets (EBS read latency in availability zone 1a) with the x-axis showing time in days since the start of the benchmarking. It is interesting to note that the seasonal component reflects the periodic variations corresponding to a normal daily variation. It is also clear that after the first 150 days or so, a significant deterioration of the trend began, and by about day 170, the underlying EBS latency trend had shifted to around 200,000, where it remained for the next couple of months. This type of performance shift can be caused by major changes in the cloud infrastructure. In this particular example, we found that, around the same time as the performance shift, AWS introduced a new type of EBS volume offering guaranteed IOPS which might have had a negative impact on existing EBS volumes (non guaranteed IOPS). This is an example of a long-lasting shift in cloud infrastructure behavior, despite a consistent configuration from the consumer; if the consumer is aware of this, they would adapt their deployment to shift work, or acquire additional resources, in order to maintain the application-level service quality they require. In addition, the residual component of the timeseries data shows relatively sudden spikes, sometimes corresponding to several standard deviations away from the mean. Note that the data is fine-grained (one data point every 5 minutes) and therefore, a zoomed time window of 24 hours is provided as a close-up example of typical spikes in EBS performance.

The results of a multi-regression analysis of the effect of CPU and Memory performance on the Network performance, summarized in Table 2, reveal that the effect of CPU and memory on network performance is statistically significant in 5 benchmarks out of 6. This confirms the findings of [25] on the impact of virtualization on the networking performance of Amazon Elastic Cloud Computing (EC2). Table 2 also shows the calibration threshold which corresponds to

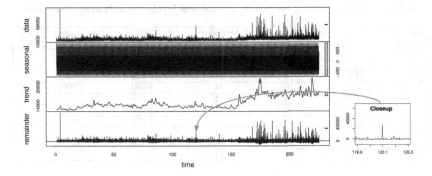

Fig. 2. Time series decomposition of one dataset measuring EBS read latency

the maximum network performance variation that could be caused by CPU and Memory performance variation. It is calculated based on the regression coefficients. A key finding is that such variation does not exceed 3 standard deviations (SD) of the networking performance. This is a crucial finding for the calibration and failure detection phase.

Table 2. Regression analysis results and estimation of calibration thresholds

	CPU		Memory		Likelihood ratio test		Calibration
	B	p	B	p	Chi square	p	threshold
Instance 1 + 2 (AZ 1a - 1a)	872.8	<0.001	23.95	<0.001	537.75	<0.001	0.74 SD
Instance 3 + 4 (AZ 1c - 1c)	4845	<0.001	-31.46	<0.001	1023.6	<0.001	1.38 SD
Instance 5 + 6 (AZ 1d - 1d)	2372	<0.001	0.201	0.94	1011.5	<0.001	1.33 SD
Instance 1 + 3 (AZ 1a - AZ 1c)	-24.56	0.696	13.53	<0.001	62.161	<0.001	0.22 SD
Instance 1 + 5 (AZ 1a - AZ 1d)	1823	<0.001	27.61	<0.001	1501.8	<0.001	1.33 SD
Instance 4 + 6 (AZ 1c - AZ 1d)	3351	<0.001	-3.786	0.123	4163.6	<0.001	2.91 SD

The calibration consists of using the calibration threshold as a lower cap to the EBS latency residual component so that only significant increases in latency are considered as failures and transformed into discrete events. All 6 EBS datasets are then calibrated according to their respective thresholds (which depend on their respective standard deviations). All resulting calibrated positive residual components are combined into one single dataset that represents all significant variations potentially related to cloud infrastructure failures. Finally, the failure detection and matching with AWS announcements are performed using a tolerance window of 6 hours around every announced failure. If failures are detected within the 6 hours window, it is a match. Note that when several disruptions of EBS performance are detected within the 6 hours window, they are considered as one single detection.

Table 3 shows the results of the failure detection and matching with AWS announcements for various values of the calibration threshold, starting at 3 Standard Deviations to as high as 12 SDs. The number of detected failures is reported

Table 3. Failure detection results for various levels of calibration threshold

Calibration Threshold	# Detected failures	# Matched failures	Early detections	Earliest (minutes)	Latest (minutes)
3 SD	218	11	8	-175	160
4 SD	177	8	4	-100	160
5 SD	157	8	4	-100	160
6 SD	143	6	4	-100	160
7 SD	129	6	4	-100	160
8 SD	115	6	4	-100	160
9 SD	106	6	4	-100	160
10 SD	101	4	3	-100	160
11 SD	90	4	3	-100	160
12 SD	79	3	3	-20	-

as well as the number of matches with AWS announcements. The results show that, even with very high thresholds of up to 12 Standard Deviations, it is still possible to detect failures. Furthermore, failures detected at that threshold level are all detected much earlier than the announcement made by AWS with the earliest detected up to 100 minutes prior to the first announcement made by Amazon. We also note that the number of detected failures, even for very high thresholds, is significantly higher than announced failures.

6 Conclusion and Future Work

In this paper, we have introduced novel ways to schedule and analyze micro-benchmarking results, that a consumer can carry out to reveal how well the cloud provider's infrastructure is performing. Our results show that there are sometimes substantial changes in the performance of the cloud, that these are often connected to failures that the provider announces. We have also shown that, with sensitive rules for detecting failures, a consumer may be alerted to problems before the provider makes any announcement; this can allow early response. We have also found substantial long-lasting performance degradation that was not announced.

If consumers do continuous benchmarking as we propose, they can benefit from a valuable source of information about the infrastructure hosting their applications enabling potential early reactions. However, our experiments required dedicated instances continuously running benchmarking tests. Such experimental setting cannot be provisioned by regular cloud consumers due to the induced cost. In future work, we will focus on the cost-effectiveness of the benchmarking, by allowing regular workload to be executed concurrently to the benchmarking activities. The idea is to use additional calibration techniques for adapting to existing workload and suppressing its effect on the benchmarking results. That would allow cloud consumers to use their existing production instances to conduct continuous benchmarking of cloud infrastructure without additional costs.

Acknowledgment. NICTA is funded by the Australian Government through the Department of Communications and the Australian Research Council through the ICT Centre of Excellence Program.

References

1. Barker, S.K., Shenoy, P.: Empirical evaluation of latency-sensitive application performance in the cloud. In: Proceedings of the First Annual ACM SIGMM Conference on Multimedia Systems, MMSys 2010, pp. 35–46. ACM, New York (2010)
2. Chen, W., Toueg, S., Aguilera, M.K.: On the quality of service of failure detectors. IEEE Trans. Comput. 51(5), 561–580 (2002)
3. Chhetri, M.B., Chichin, S., Vo, Q.B., Kowalczyk, R.: Smart cloudbench – automated performance benchmarking of the cloud. In: 2013 IEEE Sixth International Conference on Cloud Computing (CLOUD), pp. 414–421 (June 2013)
4. Chiang, R.C., Howie Huang, H.: Tracon: Interference-aware scheduling for data-intensive applications in virtualized environments. In: Proceedings of 2011 International Conference for High Performance Computing, Networking, Storage and Analysis, SC 2011, pp. 47:1–47:12. ACM, New York (2011)
5. Dejun, J., Pierre, G., Chi, C.-H.: Ec2 performance analysis for resource provisioning of service-oriented applications. In: Dan, A., Gittler, F., Toumani, F. (eds.) ICSOC/ServiceWave 2009. LNCS, vol. 6275, pp. 197–207. Springer, Heidelberg (2010)
6. Dfago, X., Urbn, P., Hayashibara, N., Katayama, T.: The accrual failure detector. In: RR IS-RR-2004-010. Japan Advanced Institute of Science and Technology, 66–78 (2004)
7. Giannini, D., Paggiaro, P.L., Moscato, G., Gherson, G., Bacci, E., Bancalari, L., Dente, F.L., Di Franco, A., Vagaggini, B., Giuntini, C.: Comparison between peak expiratory flow and forced expiratory volume in one second (fev1) during bronchoconstriction induced by different stimuli. Journal of Asthma 34(2), 105–111 (1997)
8. Gray, C., Cheriton, D.: Leases: An efficient fault-tolerant mechanism for distributed file cache consistency. SIGOPS Oper. Syst. Rev. 23(5), 202–210 (1989)
9. Iosup, A., Yigitbasi, N., Epema, D.: On the performance variability of production cloud services. In: Proceedings of the 2011 11th IEEE/ACM International Symposium on Cluster, Cloud and Grid Computing, CCGRID 2011, pp. 104–113. IEEE Computer Society, Washington, DC (2011)
10. Kossmann, D., Kraska, T., Loesing, S.: An evaluation of alternative architectures for transaction processing in the cloud. In: Proceedings of the 2010 ACM SIGMOD International Conference on Management of Data, SIGMOD 2010, pp. 579–590. ACM, New York (2010)
11. Lakshman, A., Malik, P.: Cassandra: A decentralized structured storage system. SIGOPS Oper. Syst. Rev. 44(2), 35–40 (2010)
12. Li, A., Yang, X., Kandula, S., Zhang, M.: Cloudcmp: Shopping for a cloud made easy. In: Proceedings of the 2Nd USENIX Conference on Hot Topics in Cloud Computing, HotCloud 2010, p. 5. USENIX Association, Berkeley (2010)
13. Li, A., Yang, X.: Srikanth Kandula, and Ming Zhang. Comparing public-cloud providers. IEEE Internet Computing 15(2), 50–53 (2011)
14. Makhija, V., Herndon, B., Smith, P., Roderick, L., Zamost, E., Anderson, J.: Vmmark: A scalable benchmark for virtualized systems. VMware Inc, CA, Tech. Rep. VMware-TR-2006-002, September (September 2006)
15. Dave Mangot. Measuring ec2 system performance (2009), http://bit.ly/48Wui (May 2009)

16. Novaković, D., Vasić, N., Novaković, S., Kostić, D., Bianchini, R.: Deepdive: Transparently identifying and managing performance interference in virtualized environments. In: Presented as part of the 2013 USENIX Annual Technical Conference (USENIX ATC 2013), San Jose, CA, pp. 219–230. USENIX (2013)

17. Pannu, H.S., Liu, J., Guan, Q., Fu, S.: Afd: Adaptive failure detection system for cloud computing infrastructures. In: IPCCC 2012, pp. 71–80 (2012)

18. Ren, X., Dong, J., Liu, H., Li, Y., Yang, X.: Low-overhead accrual failure detector (2012)

19. Satzger, B., Pietzowski, A., Trumler, W., Ungerer, T.: A new adaptive accrual failure detector for dependable distributed systems. In: Proceedings of the 2007 ACM Symposium on Applied Computing, SAC 2007, pp. 551–555. ACM, New York (2007)

20. Satzger, B., Pietzowski, A., Trumler, W., Ungerer, T.: Variations and evaluations of an adaptive accrual failure detector to enable self-healing properties in distributed systems. In: Lukowicz, P., Thiele, L., Tröster, G. (eds.) ARCS 2007. LNCS, vol. 4415, pp. 171–184. Springer, Heidelberg (2007)

21. Schad, J., Dittrich, J., Quiané-Ruiz, J.-A.: Runtime measurements in the cloud: Observing, analyzing, and reducing variance. Proc. VLDB Endow. 3(1-2), 460–471 (2010)

22. Kelvin, C., So, W., Sirer, E.G.: Latency and bandwidth-minimizing failure detectors. SIGOPS Oper. Syst. Rev. 41(3), 89–99 (2007)

23. Turner, A., Fox, A., Payne, J., Kim, H.S.: C-mart: Benchmarking the cloud. IEEE Trans. Parallel Distrib. Syst. 24(6), 1256–1266 (2013)

24. Walker, E.: Benchmarking amazon ec2 for high-performance scientific computing. LOGIN 33(5), 18–23 (2008)

25. Wang, G., Eugene Ng, T.S.: The impact of virtualization on network performance of amazon ec2 data center. In: Proceedings of the 29th Conference on Information Communications, INFOCOM 2010, Piscataway, NJ, USA, pp. 1163–1171. IEEE Press (2010)

26. Xiong, N., Vasilakos, A.V., Wu, J., Richard Yang, Y., Rindos, A., Zhou, Y., Song, W.-Z., Pan, Y.: A self-tuning failure detection scheme for cloud computing service. In: 2012 IEEE 26th International Parallel & Distributed Processing Symposium (IPDPS), pp. 668–679. IEEE (2012)

27. Xu, Y., Musgrave, Z., Noble, B., Bailey, M.: Bobtail: Avoiding long tails in the cloud. In: Proceedings of the 10th USENIX Conference on Networked Systems Design and Implementation, nsdi 2013, pp. 329–342. USENIX Association, Berkeley (2013)

28. Zhang, X., Tune, E., Hagmann, R., Jnagal, R., Gokhale, V., Wilkes, J.: Cpi2: Cpu performance isolation for shared compute clusters. In: SIGOPS European Conference on Computer Systems (EuroSys), Prague, Czech Republic, pp. 379–391 (2013)

29. Zhao, L., Liu, A., Keung, J.: Evaluating cloud platform architecture with the care framework. In: 2010 17th Asia Pacific Software Engineering Conference (APSEC), pp. 60–69 (November 2010)

Business Export Orientation Detection through Web Content Analysis

Desamparados Blazquez, Josep Domenech, Jose A. Gil, and Ana Pont

Universitat Politecnica de Valencia
Cami de Vera, s/n. 46022 Valencia, Spain

Abstract. Economic indicators are essential for economic studies, forecasts and economic policy designs. To meet their objective, they should be available in a frequent and timely fashion. However, official data are usually released with a long delay. Web-based economic indicators can be made available on real-time basis, thus contributing to alleviate this lag. Across all the economic indicators, those related to the export orientation of the companies are particularly interesting because of the growing importance of international trade to most developed countries.

This paper proposes a prediction system that analyzes corporate websites to produce web-based economic indicators for the export orientation of the companies. To validate our approach, we compared the prediction accuracy of our model to a baseline model made by manually retrieving the web indicators from 350 corporate websites. Our results showed that the proposed prediction system captures most of the prediction accuracy of the model with manual web indicators, but achieved at a minimum processing cost.

1 Introduction

The massive spread of the WWW has nowadays empowered millions of small businesses to sell their goods and services to worldwide customers in a 24 hours a day, seven days a week basis. Web and on-line platforms have made it possible for companies, independently of their size, to enter new markets and increase their export sales due to the removal of some geographical constraints and the instant communication all over the world [1].

The relation between the development of the Internet and the trade growth has attracted a number of recent studies, revealing that the Internet stimulates trade [2]. Focusing on the WWW, the work by Freund and Weinhold [3] revealed that the growth in the number of websites in a country predicted its export growth in the following year. In short, whether it is dealing with individual clients, or with corporate buyers or suppliers, the WWW is useful to increase firm's visibility and potential customers. Furthermore, it can be used to broaden market reach and to improve operational efficiency. This is due to its capacity to make communications and transactions easier and less expensive, leading to important efficiency gains [4,5].

---

[1] This work has been partially supported by the Spanish Ministry of Economy and Competitiveness under Grant TIN2013-43913-R.

B. Benatallah et al. (Eds.): WISE 2014, Part II, LNCS 8787, pp. 435–444, 2014.

Export-oriented businesses allow the economies to increase their competitiveness, since firms become more proactive and adaptable. For these reasons, exports figure prominently in the minds of policy-makers [6] with programs that promote not only this business activity but also the presence of SMEs in the Internet. In the present situation where the economic and technological context changes continuously, the prompt availability of any economic indicator is essential for economic studies, forecasts and economic policy designs. However, most current indicators are expensive to collect because of a required manual processing, and generally all of them suffer important lags and born outdated. Web-based economic indicators can be made available on real-time basis, which contributes to alleviate this lag [7].

This paper proposes to use the content hosted in the WWW as a source of information to produce timely economic indicators. Especifically, in this work we focus on obtaining indicators for firm export orientation. To this end, we have developed a prediction model based in a selection of features from corporate websites. The variables used by this model were automatically captured by a predictive system designed for retrieving economic information from corporate websites. Our prediction model has been tested and compared against a baseline model based on manually retrieved variables.

The remainder of the paper is organized as follows. Section 2 reviews some related research on linking web activities to the economy and on the website features that are expected to be related to the export orientation. Section 3 describes the experimental environment and the data used to carry out the performance analysis. Section 4 analyzes the prediction performance of the proposed system. Finally, Section 5 draws some concluding remarks and provides directions for future work.

2   Background

First web-based economic indicators relied on the reports generated by Google Trends (GT), which provides the evolution of the popularity of certain keywords in Google Search. Models based in these data were successfully applied to predict unemployment-related magnitudes [8], economic uncertainty [9], and sales in a variety of markets [10,11]. Although GT can supply useful hints on the economic activity at an aggregate level, its ability for characterizing individual firms is limited because it only provides data about what the user demands.

Some other research works have used corporate websites to infer economic characteristics of the companies. In this line, Overbeeke and Snizek [12] reviewed company websites to find indicators of corporate culture, while Meroño-Cerdan and Soto-Acosta [13] related web content to firm performance. Departing from a selection of website features, Blazquez and Domenech [14] described some web-based indicators for firm export orientation. Unfortunately, these proposals rely on a manual retrieval of the proposed web indicators, making them inappropriate for continuously monitoring the economy.

When analyzing the WWW variables that are related to business export behavior, it has to be taken into account that experimented firms are usually more likely to export, as they have had time to increase knowledge and accumulate useful resources for internationalization [15]. Firms with more experience on the Internet could follow this same pattern towards exporting. The date in which a domain name is registered suggests the approximate date in which a company started to go online [16], despite the temporal gap between a domain name is registered and a website is implemented [17]. Hence, the domain name's age is related to the firm experience in the Internet. As older firms usually own older domains, having an older domain could be indicative of a higher propensity to export.

The domain name is the main identifier of a company in the Internet. The top-level domain (TLD), as part of the firm Internet name, is either an ISO country code (e.g. .es for Spain) or a generic code (e.g. .com). According to Murphy and Scharl [18], using a country code or a generic one reflects local or global interests, respectively. Thus, its election could be related to firm strategic orientation. Current exporters or companies which have the intention to start exporting in the near future would prefer to choose any generic domain code, which have a more international profile, to establish its presence on the Internet. Therefore, a generic top-level domain could be positively related to firm export orientation.

Web contents can also provide some useful information to infer firm activities. For instance, the presence of different language versions or some specific keywords in the corporate website can be very useful to predict the firm export orientation.

Offering websites in more than one language is usually related to greater marketing effectiveness. Moreover, deploying multi-language websites helps firms to succeed in reaching their target markets and to better deal with clients and suppliers, as the cultural barrier of language disappears and users feel more confident. In fact, offering a multi-lingual website helps firms to gain a competitive advantage in the global market, enabling them to reach a higher number of potential customers [19,20]. Therefore, the availability of a corporate website in more than one language could be related to the foreign target market of companies. Across all languages, English seems the most natural option for exporting firms from non-English speaking countries, as it is the most used language in international businesses.

Business strategies could emerge on the web and this can be monitored by the presence of key terms, as demonstrated by recent research [21]. For this reason, if a firm is selling abroad or intends to reach new markets, it is likely that information about these matters is given in its corporate website. Consequently, the presence of trade-related keywords in a corporate website could be positively related to firm export orientation.

3 Experimental Environment

3.1 Framework

In order to automatically extract and analyze the content information of the selected corporate websites to produce the export indicators, we use the prediction system presented in [7]. Figure 1 shows the architecture of this system, which consists of three main modules: the capture module, the analysis module and the production module.

The **capture module** acts basically as a crawler that parses and downloads all the website contents from the corporate sites provided as input. It has been implemented as a modified version of HTTrack [22], which is a robot that parses recursively the links found in the initial URI. This crawler provides certain support for discovering and

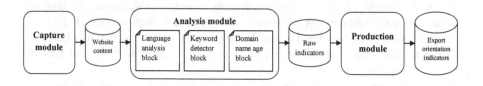

Fig. 1. Architecture for a prediction system for retrieving firm export indicators

interpreting *javascript* hyperlinks, as well as for parsing *Flash* content. Both technologies are relatively common in corporate websites.

The **analysis module** examines the content downloaded by the capture module to produce some raw indicators potentially related to the firm's economic variable under study, i.e., the export orientation in this case. This module is composed of several independent blocks, each one computing related indicators.

The *language analysis block* detects the language in which every HTML file in the site is written. Its output is the number of resources in each considered language. Here, the list of HTML resources is obtained from the *Content-type* HTTP response header and the language is recognized by means of the *aspell* spell checker.

The *keyword detector block* departs from a list of keywords and counts the number of occurrences of each keyword in the text of the website. It provides counting not only for strict matching (i.e., exact coincidence), but also for wide matching, which is performed by applying a word stemmer to the website contents.

The *domain name age block* makes a request to a *whois* server to find the date in which the provided domain name was registered. Although this block does not use the downloaded contents, it has been included in this module for convenience.

Finally, the **production module** takes as input all the raw indicators generated by the analysis module to compute the web-based indicators for predicting, in this case of study, the export orientation of the firm. For this purpose, statistical methods to estimate the probability of exporting given the raw indicators are used. More details on these methods are provided in Section 4.2.

3.2   Data

The sample for this study, which was randomly retrieved from the SABI[1] database, included 350 manufacturing companies (NACE Rev. 2 codes 10-33) with corporate website established at the Region of Valencia, in Spain. Then, from each website, the following web-based variables were manually retrieved and coded during the last months of year 2012:

– Domain name's age (DOM_AGE_i): Continuous variable measured as the number of years since the corporate website domain was registered.
– English version (EN_i): Binary variable that takes value 1 if the corporate website had a functional English version available.

---

[1] SABI: Sistema de Análisis de Balances Ibéricos. It is published by Bureau van Dijck.

- Export-related keywords ($KEYWORDS_i$): Binary variable with value 1 if the website contained any term associated to exportation and 0 otherwise. A word list[2] containing key terms potentially connected to export orientation was prepared and later searched for by querying Google with each term at each website using the advanced search tool.

This set of web-based variables was supplemented with some economic characteristics of the firms. This information was collected from the companies' financial statements available in SABI, the records of exporters of the Spanish Institute for Foreign Trade (ICEX) and the High Council of Chambers of Spain. Data from year 2012 were used because it was the most recent year for which this information was available (with a delay of more than one year). The following variables were included:

- Firm's industry ($INDUSTRY_i$): Vector of binary variables for two-digit NACE Rev. 2 codes used to control for specific industry effects. It included 14 variables, from which 13 corresponded to different industry categories with at least 10 companies in the sample. The remaining one gathered all those firms in sectors with less than 10 companies in the sample. Ensuring that each variable controls for 10 or more firms allows us to avoid overfitting.
- Export orientation ($EXPORT_i$): Binary variable that takes value 1 if the firm is enrolled in exporting activities (48% of companies in the sample were exporters) and 0 otherwise. It is the dependent variable in the prediction models.

4 Results

4.1 Baseline Model

This section describes the prediction model based on manually retrieved variables. It will be later used to check the performance of the prediction model with the variables produced by the proposed system. To do so, a logistic regression with the manually retrieved variables that varied across exporters and non-exporters was built and afterwards reproduced with the automatically retrieved ones. The estimations of both models are compared in Section 4.3 in order to determine the validity of our proposal.

About the statistical methods, logistic regression was applied because it is the appropriate when the dependent variable is binary, as in this study. The variables selected were those that varied with an admissible level of significance ($p < 0.05$) between both groups of firms and that were not highly correlated [23]. According to these criteria, the web-based model was defined as follows:

$$EXPORT_i = \beta_0 + \beta_1 DOM\_AGE_i + \beta_2 EN_i + \beta_3 KEYWORDS_i + \gamma INDUSTRY_i$$

That is, the probability of being enrolled in exporting activities ($EXPORT_i$) depends on the age of the domain name (DOM_AGE_i), the availability of an English

---

[2] The terms included in the word list (mostly Spanish) were: Continental; continente; continentes; export; exporta; exportación; exportaciones; exportamos; exportando; exporter; extranjero; globalización; internacional; internacionales; internacionalización; mundial; países

Table 1. Prediction of export orientation with manually retrieved WWW variables

Variables	β	SE	p-value
DOM_AGE_i	0.068	0.039	0.083
EN_i	2.186	0.333	0.000
$KEYWORDS_i$	1.203	0.311	0.000
$INDUSTRY_i$	–	–	0.004
$(Constant)$	-1.717	0.489	0.000
Hosmer-Lemeshow	0.112		
Prediction accuracy	81.4%		

version (EN_i), the occurrence of export-related keywords ($KEYWORDS_i$) and the specific sector in which the firm operates ($INDUSTRY_i$). The coefficients β_1, β_2, β_3 and γ indicate the relative influence of each feature on the prediction of the category of the dependent variable, while their signs show the direction of their relationship (positive or negative) with it. β_0 is a constant.

The parameters of the model were estimated by means of the maximum-likelihood method, whose results are reported in Table 1. For each variable, the table shows the estimated regression coefficients (β) in the logistic function, the standard error for these estimations, and the p-value associated to each estimation (small values indicate significantly non-zero coefficients).

As Table 1 shows, the domain name's age effect is positive and statistically significant, thus increasing the probability of exporting. The availability of an English version of the website also contributes to inferring the export orientation. Similarly, the presence of export-related keywords is also connected to export orientation, as it significantly increases the probability of exporting. In addition, the industry effect is statistically significant, indicating the expected differences across sectors. The model performs relatively well, as it is pointed out by the prediction accuracy (81.4%) and the Hosmer-Lemeshow test, which in this case indicates that the model is adequate to explain the data.

4.2 Construction and Validation of Automatic Web-Based Indicators

The baseline model described above relied on two website features (*EN* and *KEYWORDS*) that were manually retrieved. This subsection describes the supervised learning methods applied to estimate the manually retrieved variables from the raw indicators generated by the analysis module of the prediction system.

English Version. The detection of the foreign language version of the website from the related raw indicators (number of HTML documents in each language) relied on the ratio of documents in the foreign language (English) to the number of documents in the local language (Spanish). The rationale behind this is that the English version could be functional although not all sections are translated.

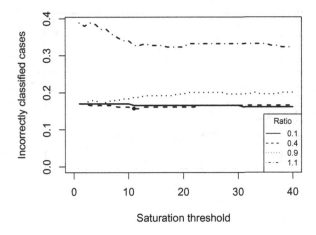

Fig. 2. Cross-validation test error for a range of parameter values of the automatic English version indicator

Table 2. Prediction performance of the automatic English version indicator

English version	$AUTO = 0$	$AUTO = 1$
$MANUAL = 0$	50.7%	8.3%
$MANUAL = 1$	7.4%	33.6%

As our crawler's ability for detecting duplicate content is limited, a saturation parameter was included to avoid the uncontrolled growing of the number of apparently-different documents when the site is dynamic or using cookies. This saturation threshold was defined as the maximum number of files to be considered in each language, so that the number of documents saturates at this level.

Both language ratio and saturation threshold parameters were tuned by a 10-fold cross-validation method, whose results are shown in Figure 2. The saturation threshold was varied from 1 to 40 documents, while the language ratio ranged from 0.1 to 2.0. However, for the sake of clarity, only a few of these values are shown in the figure. Results indicated that the optimal value for the saturation threshold is 11 HTML files, while the optimum for the language ratio is 0.4. These parameter values were used to compute the estimated *EN* variable.

When compared to the manually retrieved indicator, the estimated one provides an overall prediction accuracy of 84.3%, as Table 2 shows. A detailed analysis on the classification errors reveals that false positives occur when some error messages are found in the HTML text (generated by the web server or a server-side scripting language). False negatives are found to be usually caused by crawling errors (i.e., not all pages are downloaded).

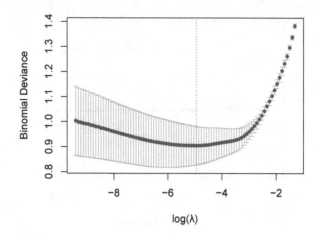

Fig. 3. Cross-validation test error (with 5% confidence intervals) for a range of λ parameter values of the LASSO Method for computing the automatic *KEYWORDS* indicator.

Table 3. Prediction performance of the automatic *KEYWORDS* indicator

KEYWORDS	$AUTO = 0$	$AUTO = 1$
$MANUAL = 0$	52.8%	1.7%
$MANUAL = 1$	13.1%	32.3%

Presence of Export-Related Keywords. The automatic *KEYWORDS* indicator was built from the raw indicators that included the number of occurrences applying strict and wide matching algorithms to every word in the list of terms related to business exports. As the number of features is large, a least square regression with least absolute shrinkage and selection operator (LASSO) was employed to find a more parsimonious model. The shrinkage parameter (λ) required by this method was tuned by means of a 10-fold cross validation procedure, whose results are presented in Figure 3. This procedure resulted in the selection of a model to estimate the presence of the export-related keywords with 15 features.

The prediction performance of this model is summarized in Table 3. As one can observe, the proposed system works relatively well since the prediction accuracy is 85.1%. Most misclassifications come from false negatives, which are found in 13.1% of the corporate websites. A more detailed analysis shows that most of these false negatives are due to the incomplete crawling of the site.

4.3   Predicting Export Orientation from Automatic Web Indicators

The automatic export-related indicators (*EN* and *KEYWORDS*) computed in the previous subsection are now employed to finally estimate the business export orientation. To do so, two manually retrieved variables used in the baseline model are replaced by the automatically retrieved ones.

The estimation results of this new model bring a prediction accuracy of 78.2%, which is slightly below the 81.4% from the baseline model. This means that most (96%) of the prediction power of the model based in manually retrieved indicators has been successfully reproduced, but achieved at a minimum processing cost (as no human intervention is required). Notice that the manual model acts as an upper bound for the prediction performance of the automatic model because no new variables are considered.

5   Conclusions

The prompt availability of up-to-date economic indicators is of crucial importance for forecasting and steering economic policies. Unfortunately, the official data are typically collected at low frequency and published with long delays (more than one year for the case of the firm export orientation). Delayed economic indicators slow down the process of public policy decision making, which ultimately impacts on the evolution of the economy.

As the Internet presence of large, medium and small companies is nowadays a reality, the use of corporate websites as an updated source of timely information is increasingly applicable to any kind of studies. More precisely, many economic studies and reports require from the business export orientation of a particular industry, as long as it is an activity of growing importance for many countries.

In this line, this work has focused on how the firm export orientation can be inferred by just obtaining and analyzing some specific features from corporate websites. Moreover, as these features are automatically gathered and processed by our system, it is possible to automatically produce useful information about the firm export orientation from a wide set of firms in real time. In this way, the long delay of the publication of the official records can be alleviated.

To validate our approach, we compared the prediction accuracy of our model to a baseline model made by manually retrieving web indicators from 350 corporate websites. Our results showed that the proposed prediction system captures most of the prediction accuracy of the model with manual web indicators, but achieved at a minimum processing cost. As for future work, we plan to use some website features other than those manually retrieved to increase the prediction accuracy beyond the one obtained by the baseline model.

References

1. Sinkovics, N., Sinkovics, R.R., Jean, R.J.B.: The internet as an alternative path to internationalization? International Marketing Review 30, 130–155 (2013)
2. Choi, C.: The effect of the internet on service trade. Economics Letters 109, 102–104 (2010)
3. Freund, C.L., Weinhold, D.: The effect of the internet on international trade. Journal of International Economics 62, 171–189 (2004)
4. Dholakia, R.R., Kshetri, N.: Factors impacting the adoption of the internet among smes. Small Business Economics 23, 311–322 (2004)

5. Vivekanandan, K., Rajendran, R.: Export marketing and the world wide web: perceptions of export barriers among tirupur knitwear apparel exporters - an empirical analysis. Journal of Electronic Commerce Research 7, 27–40 (2006)
6. Girma, S., Greenaway, D., Kneller, R.: Does exporting increase productivity? a microeconometric analysis of matched firms. Review of International Economics 12, 855–866 (2004)
7. Domenech, J., de la Ossa, B., Pont, A., Gil, J.A., Martinez, M., Rubio, A.: An intelligent system for retrieving economic information from corporate websites. In: IEEE/WIC/ACM International Joint Conferences on Web Intelligence (WI) and Intelligent Agent Technologies (IAT), pp. 573–578. IEEE, Macau (2012)
8. Choi, H., Varian, H.: Predicting initial claims for unemployment benefits (2009)
9. Dzielinski, M.: Measuring economic uncertainty and its impact on the stock market. Finance Research Letters 9, 167–175 (2011)
10. McLaren, N., Shanbhogue, R.: Using internet search data as economic indicators. Bank of England Quarterly Bulletin (Q2 2011)
11. Hand, C., Judge, G.: Searching for the picture: forecasting uk cinema admissions using google trends data. Applied Economics Letters 19, 1051–1055 (2012)
12. Overbeeke, M., Snizek, W.E.: Web sites and corporate culture: A research note. Business & Society 44, 346–356 (2005)
13. Meroño-Cerdan, A.L., Soto-Acosta, P.: External web content and its influence on organizational performance. European Journal of Information Systems 16, 66–80 (2007)
14. Blazquez, D., Domenech, J.: Inferring export orientation from corporate websites. Applied Economics Letters 21, 509–512 (2014)
15. Fernández, Z., Nieto, M.J.: Impact of ownership on the international involvement of smes. Journal of International Business Studies 37, 340–351 (2006)
16. Scaglione, M., Schegg, R., Murphy, J.: Website adoption and sales performance in valais' hospitality industry. Technovation 29, 625–631 (2009)
17. Murphy, J., Hashim, N.H., O'Connor, P.: Take me back: Validating the wayback machine. Journal of Computer-Mediated Communication 13, 60–75 (2007)
18. Murphy, J., Scharl, A.: An investigation of global versus local online branding. International Marketing Review 24, 297–312 (2007)
19. Samiee, S.: Global marketing effectiveness via alliances and electronic commerce in business-to-business markets. Industrial Marketing Management 37, 3–8 (2008)
20. Escobar-Rodríguez, T., Carvajal-Trujillo, E.: An evaluation of spanish hotel websites: Informational vs. relational strategies. International Journal of Hospitality Management 33, 228–239 (2013)
21. Arora, S.K., Youtie, J., Shapira, P., Gao, L., Ma, T.: Entry strategies in an emerging technology: a pilot web-based study of graphene firms. Scientometrics 95, 1189–1207 (2013)
22. Roche, X.: Httrack (2014), http://www.httrack.com
23. Nassimbeni, G.: Technology, innovation capacity, and the export attitude of small manufacturing firms: a logit/tobit model. Research Policy 30, 245–262 (2001)

Towards Real Time Contextual Advertising

Abhimanyu Panwar[1], Iosif-Viorel Onut[2], and James Miller[1]

[1] Electrical and Computer Engineering ,University of Alberta, Edmonton, Canada
{panwar1,jimm}@ualberta.ca
[2] IBM Canada Ltd., Ottawa, Canada
vioonut@ca.ibm.com

Abstract. The practice of placement of advertisements on a target webpage which are relevant to the page's subject matter is called contextual advertising. Placement of such ads can lead to an improved user experience and increased revenue to the webpage owner, advertisement network and advertiser. The selection of these advertisements is done online by the advertisement network. Empirically, we have found that such advertisements are rendered later than the other content of the webpage which lowers the quality of the user experience and lessens the impact of the ads. We propose an offline method of contextual advertising where a website is classified into a particular category according to a given taxonomy. Upon a request from any web page under its domain, an advertisement is served from the pool of advertisements which are also classified according to the taxonomy. Experiments suggest that this approach is a viable alternative to the current form of contextual advertising.

Keywords: Classification, Algorithms, Performance.

1 Introduction

Worldwide online advertisement revenues have grown to 117 billion US Dollars [9]. They have been growing at a steady rate of almost 20% each year. Contextual advertising (CA) contributes to these revenues. Usually advertisements (ads) are shown in the form of textual and banner ads. These ads are delivered by an ad-network to the publisher. Placement of such ads which are contextually related to the webpage are believed to increase user experience [11], [12], [14], [15]. Moreover, this practice brings revenue to the publisher website, the ad-network and advertiser.

Ad-networks fulfill the act of mediator between the publisher and the advertiser. An ad-network hosts a repository of ads and bears the responsibility of selecting suitable ads from this repository. Traditionally, the selection of suitable ads is done on the fly when a request is made from the target web page to the ad-network. The ad-network analyses the webpage content and selects the "best matching" ad from the repository. This process takes place while the browser is busy rendering the webpage. Ideally, the ads should be displayed along with the surrounding content of the webpage without any time delay. However, we have empirically found that the ads are displayed later than that of the surrounding content of the web page. This delay

B. Benatallah et al. (Eds.): WISE 2014, Part II, LNCS 8787, pp. 445–459, 2014.

can be attributed to the fact that the ad-network has to process the request made by the webpage online. This latency not only reduces the quality of the user experience but defeats the purpose of advertising which is to capture the attention of the user.

In order to avoid these latency issues, we propose an offline method to serve the purpose of contextual advertisement for a website. Obviously this model can also accommodate subsets of websites, but the number of sub-sites must be finite (#sub-sites << #pages). We build a taxonomy; and classify the website, or sub-site, into one of the nodes of the taxonomy. The ads present in the ad-network repository are also classified into the same taxonomy; in fact we argue that the taxonomy should be built from these ads not from web pages. For delivery to the target web page, only ads with the same classification are considered. The website is crawled and classified offline by the ad-network. Whenever a request is made from any of the constituting webpages of a website to the ad-network, the ad-network having already classified the website – delivers a suitable ad based on its category. In this way, the processing time of a request is greatly reduced as compared to the traditional approach, since only a look-up operation is required. As a result, simultaneous display of ads with other content on the webpage leads to enhanced user experience. In this way, an ad captures the user attention on equal terms with the other content of the webpage which will eventually lead to more clicks on the ad, and therefore, increasing revenue.

The contributions of this paper are:

- We propose an offline approach with minimal latency for the selection of ads to serve the purpose of contextual advertisement. This approach increases the user experience in contrast to the traditional approach.
- We design novel schemes for the purpose of the classification of a website. We show that a website can be represented by the information extracted from its home page and the pages situated at one level crawling depth, to form feature vectors for the purpose of classification.
- We show that an intersection of Web Accessibility guidelines and Search Engine Optimization (WASEO) clues can be used to represent a web page to form feature vectors for classification.
- Finally, we demonstrate that the approach is viable, by executing an initial empirical trial which shows that we are able to classify real-world web pages into abstract classification classes.

The rest of the paper is organized as follows. In Section 2, we discuss the current model of contextual advertising and the latency problems associated with it. We present our approach in detail in Section 3. Experimental design, settings and analysis are given in Section 4. Section 5 details the outcomes from the experiments. A discussion of the outcomes is presented in Section 6. And, we provide conclusions about the approach in Section 7.

2   Current Model of Contextual Advertising

When a user visits a webpage, embedded JavaScript communicates to an ad-network and requests that an ad be rendered at a specific location and of a specific size.

The ad-network having received the request analyzes the contents of the webpage, estimates the "central theme", and chooses a suitable ad from its repository [16], [17].

Table 1. Rendering times of the primary ad. T1 = Time in seconds when neighboring content is rendered. T2 = Time in seconds when advertisement is rendered. C1 = Percentage of webpage loaded at T1. C2 = Percentage of webpage loaded at T2.

Website	T* in seconds			C* in percentage			
	T1	T2	T2-T1	C1	C2	C2-C1	Alexa Rank
amazon.com	1.7	3.2	1.5	73	82	9	12
indiatimes.com	3.2	5	1.8	29	47	18	103
yahoo.com	2.2	4.3	2.1	87	96	9	4
wsj.com	5.5	8	2.5	84	97	13	211
usatoday.com	3.3	5.3	2	78	95	17	268
ebay.com	5	7.3	2.3	64	76	12	24
washingtonpost.com	3.8	4.6	0.8	56	81	25	295
msn.com	1.9	2.7	0.8	44	71	27	33
sourceforge.net	2.1	3.2	1.1	89	97	8	166
bbc.co.uk	2.7	3.1	0.4	54	59	5	59
Average			1.53			14.3	

We conducted an experiment to find issues with the delivery of ads from this process. We selected 10 random publisher websites from Alexa 500 and conducted the test on their home page. We measured the time delay between the rendering of the ad and that of the surrounding content of the webpage (Table 1). The ad considered in the experiment is the one which appears above the fold. This ad is of primary importance, since it catches the user's attention as soon as the page starts loading. We refer to this ad as the "Primary Ad". We found that the primary ad is displayed later than its surrounding content by an average delay of 1.53 seconds. This amount of delay is enough for the user to divert his attention to somewhere else on the page. It may also happen that the user may not even notice the ad or gets irritated by the fact that there is a blank rectangle in the browser window. This kind of scenario defeats the original purpose of CA. Processing time of a request by the ad-network is crucial since the operation has to be done on the fly. The ad-network has to reply back within a fraction of a second. Current state of the art research methods for the ad-network to analyze a webpage include steps such as page summarization [21], [22], [12], [23], conversion of the summarized page into feature vector, classification into a node of given taxonomy [10, 11, 12, 13], ad-matching via keywords [11] or link analysis [13]etc. Page summarization includes steps such as parsing a webpage and getting values of specific tags like Title, headings etc. Many papers, e.g. [10], propose to get information from the parent pages of the target webpage as well. This step proves to be the most expensive of all. In essence these additional tactics would all significantly increase the time delay presented in Table 1.

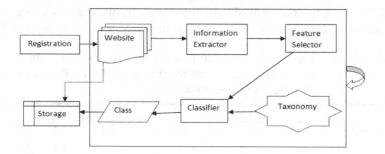

Fig. 1. Architecture of offline processing of a publisher's website by the ad-network

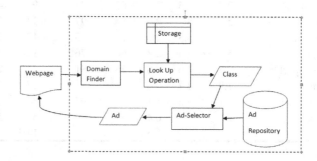

Fig. 2. Architecture of real time processing of the ad request performed by the ad-network

3   A New Approach

Considering the disadvantages of the existing methods, we introduce a new solution to mitigate against these problems. We propose a novel approach where most of the analysis work is done offline. Moreover, we introduce a new way to improve the relevance of the ads retrieved from the ad repository. We tackle this problem in two phases. Figure 1 and 2 depicts the overall architecture of the proposed approach. Figure 1 outlines the process performed offline; and figure 2 gives out details of the work done online.

3.1   Phase 1: Offline Processing

In the first phase, we extract information from and classify the website into one of the nodes of a given taxonomy. This process is done offline. The website URL coupled with its category information is stored in a Hash Map. As soon as a publisher registers for the services of an ad-network, the ad-network performs a semantic analysis of the contents of the website. This phase consists of 4 modules:

1. The Information Extractor: This module extracts information from a website to represent it as feature vectors for classification. For this purpose, a website is simply considered as a tree of documents (web pages).

Web Page Representation. A webpage has both visible and hidden components. HTML inside <title> and <body> is rendered visually in the browser, whereas meta-information consisting of meta-keywords and meta-description remains hidden to the normal user. Users make the assessment about the topic and function of the web page by analyzing the visible information. Hidden information is exploited by search engines for indexing and presenting ranked results of search queries. We present a novel approach to extract both types of information present in a webpage.

Search Engine Optimization (SEO) as a helper to find features. A number of properties such as meta fields inside the head tag, title, headings in the text inside the <body> are extensively used for SEO [2][20]. Moreno *et al.* [19] state that there is a significant overlap between SEO techniques and Web Content Accessibility Guidelines (WCAG) [18]. WCAG are guidelines issued by theW3C which website content developers follow to allow a web site's content to be accessible to individuals with disabilities. A positive correlation has been found between search engine rankings and web accessibility by Elgharabawy *et al.* [3]. The common points of interest between these two are the use of Image alt text, presence of meaningful meta-description attribute, title of the webpage, text of internal and external anchor tags and text of heading tags. For example, headings, title, anchor words, images and animations visually stand out or these are read out loud by screen readers for users with visual disability. We exploit the information from these common points of interest to represent the website in terms of feature vectors for the classification.

Topic modeling for web pages. Topic modeling discovers topics, a collection of words from a text corpus. Latent Dirichlet Allocation (LDA), Blei *et al.* [8], extracts groups of co-occurring words and reports them as topics. The intuition behind this is that if a document is about a particular subject, then in order to generate the document, the writer has had to select topic mixtures consisting of words related to the subject. LDA back-tracks this generative process to discover topics of interest. We propose that the process of writing the text inside the web page will follow this generative process. This LDA approach is language-neutral; this allows the approach to be extended to cover any language which can be found in Unicode; and hence any web-site. Figure 3 shows a partial view of a TD bank web page. This web page is about options to pay back a Graduate studies loan which is the subject of the web page. Table 2 shows the topics discovered on this web page. Each topic is a collection of seven co-occurring words. The first topic is about *bank facilities* and consists of words like trust, plans, branch and investment etc. The second topic is *graduate school programs* which contains words such as school, medical and dental etc. It can be observed that words inside a topic convey the information about it and are co-occurring. These topic words are exploited to represent a webpage in terms of feature vectors.

From web page to web site representation: We consider two approaches to achieve this task (1) Home page only, (2) Home page plus the direct ("one hop only") children of the Home page. Deeper crawls lead to more disadvantages than advantages.

Fig. 3. Partial view of https://www.tdcanadatrust.com/products-services/banking/student-life/start-studies/graduate.jsp.

Table 2. Topics retrieved by LDA when applied on TD Canada trust Web page as shown in Figure 1

Topics	Topic Words
Topic 1	Waterhouse, trust, bank, investment, plans, branch, paying
Topic 2	Td, professional, school, medical, flexible, dental, affordable
Topic 3	line , students, payments, options, dominion, programs, life
Topic 4	Student, credit, Canada, interest, services, solutions, grad
Topic 5	Studies, customized, graduation , window, career , fund personal

The amount of information collected grows exponentially with crawling depth. Moreover as the "distance" of a web page increases from the root page (homepage), the web page tends not to provide relevant information about the subject of the web site.

2. Feature Selector: The input for this module is the text corpora prepared by the "Information Extractor". We represent each website as a feature vector; therefore, the text corpora corresponding to each website is treated according to a "Bag of Words" model. We compute the TF-IDF [4] metric on the text corpora to turn textual information into numeric information.

3. Classifier: To predict the category or class of a website, it has to be classified into one of the nodes of a relevant taxonomy. This module is essentially a state of the art classifier.

4. Storage: The ad-network stores the website information such as its home page URL, domain name and its class information in the database. This will be used in the online phase by the ad-network.

In today's era, the content of the websites change very frequently. For example, news websites, blogs and shopping websites keep changing their content asynchronously. As a result, the class information of the website will change in the future. Moreover, refinement in the taxonomy will lead the previous class information of the website useless. Therefore, to keep up with the changes in the content of the website or in the taxonomy, phase 1 is repeated periodically to store the latest and correct information about the class of the website in the "Storage".

3.2   Phase 2: Online Processing

This part deals with the real time processing of a request and delivery of the suitable ad to the web pages. This phase serves as the front end of the ad-network, while Phase 1 serves as the back end. It consists of 3 modules.

1. Domain-name Extractor: This module receives requests from the webpage which contains information such as the URL and the type of ad being requested.

2. Lookup Module: This module crosschecks whether the derived domain name is authorized to use the services of the ad-network. After authentication, it fetches the required information from "Storage".

3. Ad-Selector: All the ads supplied by the advertisers have been pre-classified into the given taxonomy. Based on the ad-type requested (dimensions of the ad, banner type etc.), this module selects an ad randomly from the ad repository corresponding to the class of the domain name of the requesting web page.

In contrast to approaches in [10, 11, 12], we serve ads based on the content of the entire website, or sub site. Since a website is inherently designed to serve a purpose and user visits several pages at the same time, ads served based on the subject of the entire website are bound to be more relevant than that of the single webpage. It is worthwhile noting that, by this approach, real time computation performed in the process of ad delivery by the ad network has been minimized.

4   Experiments

We have conducted a number of experiments to assess the feasibility of the proposed approach. We prepared a dataset of 500 home pages from the Alexa top 500 [1] websites (most popular websites). Alexa tracks 30 million websites worldwide and exploits traffic data from millions of users. It ranks the most successful sites under several categories; we selected the top ranked global websites for this study. To prepare the dataset for this study, we selected only those websites whose content is written in English because of the requirement to manually classify the sites reliably.

Taxonomy: We recruited 5 graduate students to form a taxonomy which encompassed the dataset. The team was also assigned the task of labelling the domain names with appropriate class names. These tasks were achieved in 3 steps:

(1) Discussions were held among team members to agree upon a relevant taxonomy which covers the prepared dataset. The team was instructed that the resulting taxonomy must explain the dataset according to the subject of the websites. Once the taxonomy was finalised, the team was asked to write definitions for each category inside the taxonomy.

(2) A document was produced containing the taxonomy and description of its nodes.

(3) All the 5 members were asked to label the dataset individually according to this document in a multiclass, hard classification manner. In other words, a web site is assigned only one specific class label. When opinions varied between the students, simple majority voting was used to determine the class of a web site.

The resulting taxonomy consists of five nodes 1) Business, 2) Information, 3) Shopping, 4) Entertainment and 5) Networking; Figure 4 shows the distribution of the websites among the five nodes.

- Business: Banks, corporations buying and selling goods or services in bulk as an organisation.
- Information: News, Wikis, Search Engines, Tutorials etc.
- Shopping: Buying of goods and services by an individual for personal use without any business motives.
- Entertainment: Actions or events for the purpose of delight and pleasure in the form of audio, video and games etc.
- Networking: Social networking communities for both professional as well as non-professional purposes.

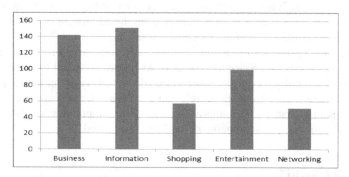

Fig. 4. The distribution of webpages amongst the five nodes

4.1   Design of Experiments

We study the comparative performance of the website representation techniques as well. To represent a website in terms of feature vectors for classification, we propose five schemes. The purpose for creating these schemes is to find out the best combinations of features which can represent a web page and a website and provide accurate classification results.

Crawl One Level Deep: We propose that a website can be effectively represented by its home page plus the pages directly accessible from home page located at one level crawling depth. During this process, we discard media (images etc.) and external links. There are three schemes under this category: 1). *CrawlDeepContent*, 2). *CrawlDeepContentMeta* and 3). *CrawlDeepWASEO*.

Only Home Page: We propose that the home page of a website is the forefront of it. Home page must correctly summarize the purpose, subject and function of the web site. There are two schemes under this category: 1). *HomeContentMeta* and 2). *HomeWASEO*.

In total, we have created 5 schemes to represent a web site:
1. CrawlDeepContent: Crawl one level deep and perform LDA on every page visited to collect topics. Consider meta-keywords and meta-descriptions of

only the Home page. We deliberately do this because in many web sites, on pages other than home pages, meta-keywords and meta-descriptions are automatically generated and not manually inserted.

2. CrawlDeepContentMeta: Crawl one level deep and perform LDA on every webpage's body text. Retrieve the meta-keywords, meta-description and title of every web page visited.

3. CrawlDeepWASEO: Crawl one level deep and consider title, meta-keywords, meta-description, anchor text, image alt text and headings text of all the web pages visited.

4. HomeContentMeta: Access only the home page and extract topics provided by LDA by running it on the text inside the body tag and extract the title, meta-keywords and the meta-description.

5. HomeWASEO: Visit only the home page and extract the title, meta-keywords, meta-description, anchor text, image alt text and headings text.

Evaluation: In order to test the performance of each website representation scheme, we trained a classifier on the dataset using the defined taxonomy. We built an ad-repository where each ad has been classified according to the given taxonomy. To calculate the performance of the proposed approach, we use a simple binary (correctly classified / incorrectly classified) evaluation methodology.

5   Results

We trained three different classifiers (Naïve Bayes, SVM and Naïve Bayes Multinomial [5, 6, 7]) for each of the five schemes creating 15 CA systems for comparison. Prior to that, we performed feature selection using Information Gain filter on the feature vectors prepared by the website representation schemes [5, 6]. We apply 10 fold cross validation on the dataset. Table 3 shows the accuracy, precision, recall and F-score of the CA systems clustered. Results show that CA systems using the *CrawlDeepWASEO* scheme perform the best among all the systems with a maximium accuracy of 84.03% (SVM classifier). The other three systems which provide accuracy of more than 80% and F-score of more than or equal to 0.8 are *CrawlDeepContentMeta*, *HomeContentMeta* and *HomeWASEO*. Furthermore, the schemes which gathered information from Web Accessibility and SEO clues, *CrawlDeepWASEO* and *HomeWASEO*, perform better than the schemes which were based on the content of the pages of a web site. It can be noted that CA systems which implemented SVM classifier performed, in general, better than the other classifiers. Among the content based schemes, *CrawlDeepContent*, *CrawlDeepContentMeta* and *HomeContentMeta*, the best performance is achieved by CA system under the scheme *HomeContentMeta* with an accuracy of 80.61%. *CrawlDeepContent* performs worst among this category of schemes.

A Confusion matrix (Table 4) allows us to have a closer look at the performance of a classifier. Table 4 shows the confusion matrix for the SVM classifier under the *CrawlDeepWASEO* scheme (the best performing {scheme, classifier}). Instances of the actual class are represented in a row while a column shows the instances predicted

Table 3. Performance results for CA systems on Schemes proposed using different classification algorithms

Scheme	Algorithm	Accuracy	Precision	Recall	F-Score
CrawlDeepContent	Naïve Bayes	71.24	0.72	0.69	0.71
	SVM	77.51	0.80	0.71	0.75
	NB Multinomial	70.37	0.69	0.68	0.67
CrawlDeepContentMeta	Naïve Bayes	65.36	0.67	0.65	0.65
	SVM	82.34	0.82	0.82	0.82
	NB Multinomial	72.47	0.74	0.72	0.72
CrawlDeepWASEO	Naïve Bayes	70.21	0.71	0.69	0.70
	SVM	84.03	0.81	0.80	0.80
	NB Multinomial	72.77	0.75	0.72	0.72
HomeContentMeta	Naïve Bayes	70.51	0.76	0.70	0.72
	SVM	73.85	0.76	0.72	0.73
	NB Multinomial	80.61	0.82	0.80	0.80
HomeWASEO	Naïve Bayes	71.35	0.76	0.71	0.70
	SVM	81.47	0.82	0.81	0.81
	NB Multinomial	78.29	0.79	0.78	0.78

Table 4. Confusion matrix for SVM classifier under the scheme = CrawlDeepWASEO. Here a, b, c, d and e denote the class labels. a = entertainment, b = shopping, c = information, d = business and e = networking

Confusion Matrix					
a	b	c	d	e	Classified as
86	0	11	1	1	a
0	51	0	5	1	b
6	2	119	15	9	c
7	15	5	110	5	d
2	0	5	3	41	e

in that class. On observing row 1 and column 1, we find that 86 websites have been correctly classified as *entertainment* class, while 13 websites belonging to *entertainment* category have been misclassified. Out of these 13, 11 have been wrongly classified as *information* and 1 each have been misclassified as *business* and *networking*. Interestingly, not a single website which belonged to *entertainment* class has been misclassified as *shopping* and vice-versa. This shows that the classifier distinctly classifies *entertainment* and *shopping* classes. Similarly, no website with the actual class as *shopping* was predicted to be *information* and no website belonging to *networking* category is classified as *shopping*. Several websites in the *information* class are classified as *business* and vice versa i.e. the classifier gets confused between these two classes. This makes sense because some websites in the *business* category have content pertaining to the *information* class as well. For example, oracle.com is labelled as a *business* website. Oracle.com conducts business but it also provides tutorials, lectures, support of various types, thereby making it a website belonging to the *Information* category as well. Therefore, it can be established that in the adopted taxonomy, several classes have ambiguous and overlapping definitions with one another which lead to misclassifications and thereby poor performance of the classifier. These instances were caused disagreement between the "human classifiers".

We decided to modify the existing taxonomy to observe the impact of the labelling process. We hypothesize that the current labelling process may have caused many of the misclassifications. Hence we introduced two "mutually exclusive" labels. By mutually exclusive, we mean that the class definitions have "discrete boundaries" and do not overlap with each other. In order to achieve this task, we called upon our team and assigned them the task of modifying the existing taxonomy so that the resulting taxonomy has classes with lesser ambiguity. It was achieved in two steps. Firstly, the two most ambiguous classes were located in the existing taxonomy. After discussions and examining the labelled dataset, the team reported *Information* and *Entertainment* classes to be the most ambiguous. This observation is also supported by the confusion matrix as shown in Table 3. In the second step, these classes were replaced by two classes which were believed to be non-overlapping. By examining the Alexa list of popular websites, the team agreed upon including *Government* and *Banks* as the two new classes. Finally, a similar procedure was followed in forming a new taxonomy as discussed in the previous section. The new taxonomy (from now onwards we refer it as Tx2.0 and older taxonomy as Tx1.0) now contains five classes; 1) Business, 2) Government, 3) Shopping, 4) Banks and 5) Networking. Figure 5 shows the distribution of the websites among the five nodes.

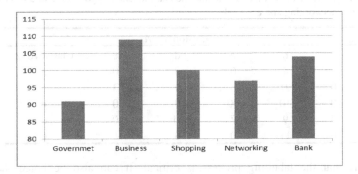

Fig. 5. The distribution of webpages amongst the five nodes in Tx2.0

We reconstructed the CA system on the Tx2.0 and re-performed the experiments. Table 5 shows the performance of CA systems with Tx2.0. It can be observed that performance of the CA system under all the schemes (consider the best performing classifier) has increased in comparison to the results from Tx1.0. This proves our

Table 5. Performance results for CA systems with Tx2.0 on Schemes proposed using different classification algorithms

Scheme	Algorithm	Accuracy	Precision	Recall	F-Score
CrawlDeepContent	Naïve Bayes	71.83	0.72	0.71	0.71
	SVM	81.69	0.85	0.81	0.82
	NB Multinomial	74.37	0.79	0.74	0.74
CrawlDeepContentMeta	Naïve Bayes	71.93	0.73	0.71	0.71
	SVM	83.92	0.84	0.84	0.83
	NB Multinomial	74.21	0.74	0.74	0.74
CrawlDeepWASEO	Naïve Bayes	74.62	0.74	0.74	0.74
	SVM	89.30	0.89	0.89	0.89
	NB Multinomial	83.65	0.85	0.83	0.84
HomeContentMeta	Naïve Bayes	72.65	0.83	0.65	0.73
	SVM	82.36	0.81	0.82	0.81
	NB Multinomial	81.23	0.82	0.80	0.80
HomeWASEO	Naïve Bayes	73.52	0.72	0.70	0.70
	SVM	84.75	0.85	0.84	0.84
	NB Multinomial	81.59	0.82	0.81	0.81

Table 6. Confusion matrix for SVM classifier under the scheme = CrawlDeepWASEO with taxonomy Tx2.0. Here a, b, c, d and e denote the class labels. a = government, b = business, c = shopping, d = networking and e = bank.

Confusion Matrix					
a	b	c	d	e	Classified as
87	4	0	0	0	a
0	91	6	9	3	b
0	10	90	0	0	c
0	9	7	81	0	d
0	11	0	0	93	e

hypothesis that the labelling process is critical. It can also be noted that SVM classifier under *CrawlDeepWASEO* scheme again gives the best classification results with the classification accuracy of 89.30%. This observation again confirms our hypothesis that feature vectors prepared based on the WASEO clues perform better than that of based on generic content based approaches.

Table 6 shows the confusion matrix for the experiment based on Tx2.0 and SVM classifier under *CrawlDeepWASEO* scheme (the best performing {scheme, classifier}). It is interesting to note by observing column one that no non-*Governemnt* website was misclassified into the *Government* class. Moreover, only 4 websites out of 91 belonging to *Government* were misclassified into *Business*. A similar effect can be noted by observing websites for the *Bank* class. This shows that newly introduced classes, Government and Bank, are mutually exclusive with each other and the other classes in the taxonomy. Therefore, while implementing a CA system, care must be taken to build the taxonomy to make "distinct" categories for optimal classification.

6   Discussion

We have established through the experiments that the presence of categories which are well defined and have non-overlapping boundaries, in the taxonomy are crucial for the CA system to perform with the optimal accuracy. Moreover, each category label must have enough documents in order to provide adequate information to the learning algorithm. Therefore, extra care must be taken while deciding the category labels and their definitions. However, previous studies on CA have exploited the ODP (http://www.dmoz.org/) taxonomy [10][13]. ODP is a large-scale taxonomy with more than 200,000 categories. However the majority of categories have very few webpages labeled under them i.e. the distributions is highly skewed towards few categories [13]. Commercially available Yahoo directory (http://dir.yahoo.com/) also has a skewed distribution. Liu *et al.* [24] reveals that 76% of the categories have fewer than 5 webpages under them. Training a classifier based on such type of highly skewed taxonomies is bound to produce sub minimal classification performance, no matter what classifier algorithm is used, the reason being that there is not enough information for the classifier to learn for the majority of classes. Moreover, [24] has shown that large scale taxonomies having number of classes in the order of 100,000's tend to produce poor quality of classification results. A study done by [24] on Yahoo directory shows that the F1 score obtained by the SVM classifier is less than 0.5, even if the number of documents belonging to a category are more than 100. In addition, the time complexity to implement such a classifier system is huge. Lee *et al.* [13], to diagnose the problem of the skewed distribution of the ODP, pruned the taxonomy and reduced total number of nodes to ~5000 but ultimately found out that due to inherent nature of tagging of ODP, the skewness could not be removed from it. Therefore existing large scale taxonomies having skewed distribution of webpages are not suitable candidates to implement a commercial CA system. While this may seem obvious, because of the overheads involved, all current work just reuses these taxonomies developed for other purposes. It is believed that this is creating a situation which is unsolvable, limits the advancement of the field, and is an inaccurate statement of the problem.

We performed small scale experiments for the CA systems based upon a proposed approach. The class labels have non-overlapping boundaries and the distribution of the websites was uniform with ~100 websites per class label. This kind of set up enabled us to achieve ~90 % classification accuracy. However, real world systems are large with millions of publishers and with taxonomies potentially having thousands of nodes. For our system to scale up to the size of commercial systems, we propose to use a taxonomy based upon the advertisements, in contrast to the existing approaches which exploit taxonomies based on the websites such as ODP and Yahoo Directory.

The size of World Wide Web is very big and it is practically infeasible to build a taxonomy which covers each and every webpage and website. However, the number of ads in an ad-network repository is smaller and less diverse in subject matter. Therefore, a taxonomy with class labels based on the ads will not only have fewer class labels but they will be more relevant and well defined. Using such a taxonomy, it is believed that a commercial CA system can be constructed according to the approach discussed in this paper.

7   Conclusion

We have empirically established that the ads are displayed later than that of their surrounding content on the webpages of popular publishers. Such a situation leads to a degraded user experience and reduced revenues for the parties interested in the ad business. To solve these problems, we introduced a novel approach to implement a CA system. This approach aims to deliver ads in real time. We deliver ads based on the content of entire website or a finite number of sub-sites. The ad-network classifies a website into one of the nodes of a given taxonomy and selects an ad corresponding to this category to deliver to the requesting webpage. In the approach presented, the information extraction and classification processes are performed offline.

We also propose novel schemes for the information extraction from the websites to form feature vectors. We studied the comparative performance of these schemes by employing state of the art classifiers on the feature vectors prepared according to them. *CrawlDeepWASEO*, a scheme which exploits information from SEO and web accessibility clues performed best of all and produced classification accuracy of 84.03% in the experiment on Tx1.0 and of 89.30% on the Tx2.0. We report that schemes based on WASEO clues perform better than that of generic content based schemes. We studied the effect of the presence of well-defined classes in a taxonomy on the performance of a classification system. Finally, in order to implement a commercial CA system using the approach presented in this paper, we propose to use a taxonomy which defines its tagging methodology based on advertisements rather than the webpages and websites of the World Wide Web.

References

1. Alexa - Actionable Analytics for the web, http://www.alexa.com
2. http://static.googleusercontent.com/media/www.google.com/en//webmasters/docs/search-engine-optimization-starter-guide.pdf
3. Elgharabawy, M.A., Ayu, M.A.: Web content accessibility and its relation to Webometrics ranking and search engines optimization. In: 2011 International Conference on Research and Innovation in Information Systems (ICRIIS). IEEE (2011)
4. Choi, B., Yao, Z.: Web Page Classification*. In: Foundations and Advances in Data Mining, pp. 221–274. Springer, Heidelberg (2005)
5. Aggarwal, C.C., Zhai, C.: A survey of text classification algorithms. Mining text data, pp. 163–222. Springer US (2012)
6. Sebastiani, F.: Machine learning in automated text categorization. ACM Computing Surveys (CSUR) 34(1), 1–47 (2002)

7. Wu, X., et al.: Top 10 algorithms in data mining. Knowledge and Information Systems 14(1), 1–37 (2008)

8. Blei, D.M., Ng, A.Y., Jordan, M.I.: Latent dirichlet allocation. The Journal of Machine Learning Research 3, 993–1022 (2003)

9. Google to Rake in 33% of Online Ad Revenues This Year., http://www.statista.com/topics/1176/online-advertising/chart/1409/global-online-ad-revenue

10. Vargiu, E., Giuliani, A., Armano, G.: Improving contextual advertising by adopting collaborative filtering. ACM Transactions on the Web (TWEB) 7(3), 13 (2013)

11. Broder, A., et al.: "A semantic approach to contextual advertising.". In: Proceedings of the 30th Annual International ACM SIGIR Conference on Research and Development in Information Retrieval. ACM (2007)

12. Anagnostopoulos, A., et al.: Just-in-time contextual advertising. In: Proceedings of the Sixteenth ACM Conference on Conference on Information and Knowledge Management. ACM (2007)

13. Lee, J.-H., et al.: Semantic contextual advertising based on the open directory project. ACM Transactions on the Web (TWEB) 7(4), 24 (2013)

14. Chatterjee, P., Hoffman, D.L., Novak, T.P.: Modeling the clickstream: Implications for web-based advertising efforts. Marketing Science 22(4), 520–541 (2003)

15. Wang, C., et al.: Understanding consumers attitude toward advertising. In: Eighth Americas Conference on Information Systems (2002)

16. About contextual targeting – AdWords Help, https://support.google.com/adwords/answer/2404186?hl=en&ref_topic=3121944

17. Yahoo! Bing Network Contextual Ads powered by Media.net, http://contextualads.yahoo.net/features.php

18. Introduction to Understanding WCAG 2.0, Understanding WCAG 2.0, http://www.w3.org/TR/UNDERSTANDING-WCAG20/intro.html

19. Moreno, L., Martinez, P.: Overlapping factors in search engine optimization and web accessibility. Online Information Review 37(4), 564–580 (2013)

20. Pringle, G., Allison, L., Dowe, D.L.: What is a tall poppy among web pages? Computer Networks and ISDN Systems 30(1), 369–377 (1998)

21. Buyukkokten, O., et al.: "Efficient web browsing on handheld devices using page and form summarization." . ACM Transactions on Information Systems 20(1), 82–115 (2002)

22. Kolcz, A., Prabakarmurthi, V., Kalita, J.: Summarization as feature selection for text categorization. In: Proceedings of the Tenth International Conference on Information and Knowledge Management. ACM (2001)

23. Shen, D., et al.: Web-page classification through summarization. In: Proceedings of the 27th Annual International ACM SIGIR Conference on Research and Development in Information Retrieval. ACM (2004)

24. Liu, T.-Y., et al.: Support vector machines classification with a very large-scale taxonomy. ACM SIGKDD Explorations Newsletter 7(1), 36–43 (2005)

On String Prioritization
in Web-Based User Interface Localization

Luis A. Leiva and Vicent Alabau

PRHLT Research Center, Universitat Politècnica de València, Spain
{luileito,valabau}@prhlt.upv.es

Abstract. We have noticed that most of the current challenges affecting user interface localization could be easily approached if string prioritization would be made possible. In this paper, we tackle these challenges through Nimrod, a web-based internationalization tool that prioritizes user interface strings using a number of discriminative features. As a practical application, we investigate different prioritization strategies for different string categories from Wordpress, a popular open-source content management system with a large message catalog. Further, we contribute with WPLoc, a carefully annotated dataset so that others can reproduce our experiments and build upon this work. Strings in the WPLoc dataset are labeled as relevant and non-relevant, where relevant strings are in turn categorized as critical, informative, or navigational. Using state-of-the-art classifiers, we are able to retrieve strings in these categories with competitive accuracy. Nimrod and the WPLoc dataset are both publicly available for download.

Keywords: Localization, L10n, Internationalization, i18n, Translation.

1 Introduction

Today most applications are looking forward to being available in more than one language, mainly to reach a global audience, to gain competitive advantage, or just because of legal requirements. In general, there is an increasing and stringent need to adapt any type of software so that it can meet the cultural and linguistic needs of every customer. For instance, according to a recent survey [7], a few years ago web companies would have to translate content into 37 languages to reach 98% of Internet users, and now it takes 48 languages to reach the same amount of users.

Taking a different tack, the "translate frequently and fast" mantra is central to companies seeking to increase global market share. As discussed in the Drupal Translation project,[1] with high-quality translation enterprises can tailor web content to consumers around the world; but when it comes to the "frequently and fast" part of the equation, enterprises run into problems. In fact, one place where localization has always had big problems is within graphical user interfaces [10,15]. Later on we discuss the most important of these problems and show that they could be easily approached if string prioritization would be made possible. Consequently, we have developed a method to support this goal.

B. Benatallah et al. (Eds.): WISE 2014, Part II, LNCS 8787, pp. 460–473, 2014.
© Springer International Publishing Switzerland 2014

1.1   Research Goals and Contributions

This paper presents Nimrod, a particular instantiation of our method for PHP-based software. Nimrod is a standalone open-source internationalization tool, however we integrate it on Wordpress to illustrate its capabilities. By means of an intuitive administration panel, webmasters can gain control over different string sorting features. We first investigate the best combination of these features, according to a manually annotated dataset. Next, we approach string prioritization as an information retrieval task, where relevant messages have to be differentiated from the non-relevant ones on the basis of a featurized string representation. Further, we explore different state-of-the-art classifiers to retrieve different categories among the relevant messages. The results show that ours is a valuable new method to improve web-based software localization. Nimrod and our dataset are both publicly available for download.

This paper is organized as follows. Section 2 provides the research background and discusses related work. Section 3 describes our system implementation. Section 4 evaluates the system. Section 5 concludes this paper and provides opportunities for future work.

2   Research Background

When developing software for a global market, applications must follow a two-step process: first internationalization (i18n), then localization (L10n). Internationalization consists of decoupling translatable text out of the application source code, basically by wrapping each message or "resource string" with a translation-capable function. After internationalization, the user interface (UI) is ready to support the requirements of different locales, i.e., specific languages and countries of the target audience; in short, the linguistic preferences that the user wants to see in their UI.

Localization in turn comprises 2 sub-levels: first *language translation*, then *aesthetic adaptation*; being the former the core activity due to its importance and associated costs [6,8,13]. Indeed, most companies are well aware of both sub-levels but eventually focus on translation due to time and budget limitations [9,12]. While aesthetic adaptation can improve the user experience [16], localized applications *must* speak the language of its users. What is more, nowadays that the Internet is pervasive, people use web browsers more than any other class of desktop software. Therefore, multilingual websites and web-based applications, much like any other type of software, are crucial to every player in the industry [12,18,19]. Thus, as businesses continue to globalize, localizing web-based UIs becomes more compelling. Last but not least, localization is a unique opportunity of preserving a language [13,14].

2.1   Related Work

Previous work on how to prioritize UI localization strings is actually scarce. So far the closest attempts we have found are tools that focus on string extraction but little to none perform string prioritization. TranStrL [21] is an Eclipse plugin that takes the source code of a Java application and automatically produces a list of untranslated strings. This solves one of the current challenges UI localization (see Section 2). However, TranStrL is only available for the Java platform. Smartling[2] provides an Objective-C library that achieves the same effect by adding minimal modifications to the source code of the applications. Again, this is not applicable to web-based software. Finally, Globalyzer[3] and World Wide Navi[4] provide a suite of desktop tools to analyze, test, and fix internationalization issues in different programming languages. Unfortunately, none of these tools allows to prioritize localization strings. Neither do current web-based localization tools such as Launchpad,[5] Pootle,[6] Verbatim,[7] or Transifex;[8] not at least to cope with a proper prioritization method as discussed in the next section.

As a mechanism for string prioritization, the GNU gettext manual [10] encourages translators to "use and peruse [sic] the program like a user would do and then use the suite of msg* command line tools to translate most urgent messages first." This is a somewhat improved approach, but actually impractical for two reasons. First, by following this approach the roles of translators, end users, and programmers are tightly coupled, which is rather an exception than the norm. Second, string prioritization is done manually with command line tools, which is time consuming. Messages can be sorted according to translation status (untranslated messages first) and frequency (most frequent messages first), based on the analysis of calls to the gettext() function and the like, e.g., t() in Drupal or translate() in WordPress, two of the most popular web content management systems (CMS) today. However, we believe that a more informed method is necessary. Then, one could perform either manual or machine translation (or a combination of both) of the prioritized strings, completing thus the UI localization workflow.

2.2 User Interface Localization Challenges

From the previous discussion, it follows that string prioritization has been overlooked. In order to stress the importance of this topic, here we identify a series of key issues in software localization that are affected by the lack of string prioritization. We believe that understanding these will be useful to the research community and for others trying to build i18n/L10n tools.

Where do We Start? Some applications have quite large message catalogs, e.g., most web CMS have well over 3000 messages, but an important number of them are seldom used on the UI. For instance, we have observed that in Wordpress this amounts to roughly half of the total messages. Usually, a translator sometimes has only a limited amount of time per week to spend on a package. Thus, it seems reasonable to start working on the strings that are most frequent.

The Trouble with String Sorting Order. Of particular importance is the fact that strings in a message catalog are sorted according to the source code files instead of the UI views. Actually the order of the strings in the source code is quite apart from what it is shown on the UI. Moreover, the strings that appear on the main UI view are usually more visible than those that appear on less frequented UI views. Therefore, a method to prioritize strings on the basis of the UI view where they appear would be quite a feat for current localization technology. For instance, it would allow more software iterations and faster development cycles by allowing translators to focus on the context of a single UI view.

Lack of Contextualization. Message catalogs lack of a proper localization context, which is a two-fold problem. On the one hand, there is not enough context due to the aforementioned sorting order according to source code files. On the other hand, there is no *visual* context available, because UI strings are decoupled from the source code. At best, translators can trust the comments developers may have left to them, such as `"# Translators: This message is related to...".` Unfortunately, these comments are scarce overall. Thus, those strings with useful comments and/or pointers to the UI should be translated in the first place.

Application Updates. Often a new software version comes with a new functionality and new messages attached to it. Moreover, some of the previous strings can be updated. Even if only a few words have changed in the original string, the translator may not see them with current localization tools; therefore she has to proofread the entire message. Consequently, new strings and string updates should be made more prominent to the translator, so that she can localize those messages earlier.

String Obsolescence. Sometimes a software patch removes old widgets from the UI but their associated strings are still included in the message catalog. Then, when localizing the software into a new language there is no way to tell those strings apart and thus they would be unnecessarily translated. Therefore, being able to identify this "dead code" would allow translators to work only on what is really needed.

Supporting Agile Localization. On another line, some companies such as Adobe are releasing updated versions of their product multiple times per day [20], making it imperative for localization to catch up and keep improving its agility. In consequence, software localizers should be able to translate most important messages first, e.g., those that are most visible to the user or that require a special attention, like error messages.

3 System Overview

With the aforementioned localization challenges in mind, we have developed Nimrod, a PHP internationalization tool that prioritizes resource strings through *progressive filtering*. The tool is based on PHP's built-in gettext library. To begin with, it selects those strings that are exclusively required to build the UI. Next, it takes into account a number of features to assign a different importance to each string, depending on the context of the string in the UI according to e.g.

visibility, frequency, semantics, or the interaction received by the user. Meanwhile, the remainder strings are considered to be less imperative and thus are left untouched, as they appear in the original message catalog. In sum, our tool selectively picks the most relevant candidates among all strings available in the message catalog for early translation, by moving the relevant strings to the top of the catalog file.

At a lower level, our tool exposes two functions to the developer: _gt() and _gx(), where the only difference between the two is that the latter allows the developer to specify a particular gettext context. This is so because in gettext all strings are indexed according to source string (msgid) and context (msgctxt), if available. Both Nimrod functions augment the PHP gettext() function and its shorthand equivalent _() with a special code, so that whenever the function is invoked, the localized message is complemented with UI-based information where available, such as size or contrast of the UI element where each string belongs to. This UI information is compiled by the web browser using injected JavaScript code. Since Nimrod preserves the usual localization information (e.g., developer comments or name of source files), it is able to reproduce the original message catalog together with feature-rich information.

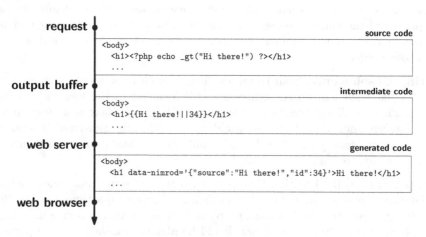

Fig. 1. Nimrod's processing pipeline

This complementary information is logged in a dedicated database, in JSON files, by means of an asynchronous Ajax call invoked on page load. To achieve this, the HTML code is parsed before being sent to the web browser, by means of PHP's output buffering capabilities; see Figure 1. Without output buffering this would not be possible, as the page would be sent into pieces as PHP processes each HTTP request and thus the document object model (DOM) would not be ready for manipulation. Then, the website administrator can access a control panel (Figure 2) and generate message catalogs on the fly, where strings can be sorted according to the following priority features (Section 3.1).

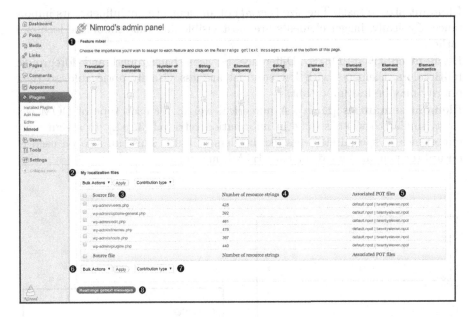

Fig. 2. Integration with Wordpress admin site. ❶ The importance degree of each feature is in $[-100, 100]$ %. Values are stored in a cookie. ❷ Localization files are shown in a dedicated table. ❸ Source files are available for each browsed URL. ❹ Number of resource strings, according to `gettext()` function calls. ❺ Each source file is associated at least with one gettext domain. ❻ Selected source files can be downloaded or inspected individually. ❼ Preferences can be considered coming from either the currently logged user or all admin users. ❽ This button creates one PO file per domain, taking into account all these preferences.

3.1   Prioritization Features

The following is a succinct description of the computed features.

Automatic comments [int]. Number of autogenerated comments, aimed at giving element examples where each message appears (Figure 3).

Developer comments [int]. Number of comments given by the programmer in the source code, directed at the translator.

Number of references [int]. Number of source files where each message appears. Theoretically, the higher the more their importance.

String frequency [int]. Number of times each message appears on the UI, as different elements may have the same string.

Element frequency [int]. Number of UI elements where each message appears. Strings appearing in multiple elements may require a special attention.

String visibility [bool]. Whether the string is shown on the UI or not. Page TITLE and most of the BODY elements are visible to the user.

Element size [float]. The size (*height*) of an element may influence its importance. Typically, bigger elements are most visible and thus should be localized earlier.

Element contrast [float]. Difference between background and foreground RGB colors. Elements with low contrast are less visually noticeable, so their priority may be less important.

Element semantics [float]. A numerical weight for each type of DOM element, similar to what is done in information retrieval [3, p.2], to assign more importance to special elements like headers or links. This can also be used to reward (or penalize) good (or bad) HTML markups.

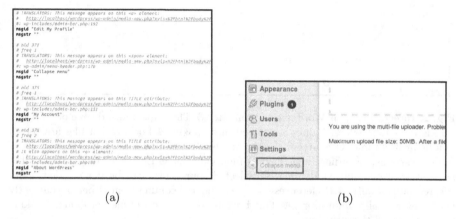

(a)   (b)

Fig. 3. Messages are augmented with automatic descriptions of the UI elements (3a), together with URLs to highlight such UI elements in the original web page (3b)

In addition, the message catalog can be extended with behavioral information that suggests which strings are more relevant to the user; e.g., mouse or touch events. Although Nimrod allows these events to be captured, they are not considered for analysis at the moment, since they are too dependent on website usage and webmaster activity.

4 Evaluation

We aim to assess to what extent Nimrod can be effectively used to prioritize the localization of a real website by end users. To do so, first we study the combination of feature presets (Figure 2) that best represent a number of different string categories (Section 4.4). Then, we convey an experiment to perform a fine-grained categorization of strings based on their relevance (Section 4.5).

4.1   Experimental Setup

We developed `WP_Nimrod`, a plugin for Wordpress on top of Nimrod. We chose Wordpress for evaluation because it is one of the most popular web CMS, is open-source software (GPL) written in PHP, and is already internationalized with gettext. After a clean Wordpress installation on our web server, we just had to activate the plugin. Basically, the `WP_Nimrod` plugin traces the calls to Wordpress' `translate()` function and stores in a JSON file the strings of each requested web page, together with the features mentioned in the previous section.

4.2   Procedure

We manually browsed all pages at the Wordpress admin site, by clicking on all links of the rightmost navigation menu, including links in submenus; see Figure 4. In total, 37 admin pages were browsed. All translatable strings from these admin pages were automatically registered, removing duplicates, resulting in 1679 messages. These account for 47.5% of the messages in the latest version of the official Wordpress PO file (v3.8.1, 3533 messages), which reveal that an important number of strings are seldom used on the admin site. In the following, we briefly describe the dataset we gathered from the remaining strings, named "Wordpress Localization dataset", or WPLoc for short. The dataset will be made publicly available after publication.

Fig. 4. Wordpress admin menu

4.3   Dataset

The WPLoc dataset comprises 1679 manually annotated messages from Wordpress admin site; see Table 1. These messages are distributed in a long tail: 1100 strings were logged only once, 390 strings appeared in more than 10 cases, and 49 strings appeared in more than 100 cases. Anecdotally, the most frequent message is "Add New", appearing 540 times. It is worth pointing out that 93% of the strings appear in one single element, and that 70% of the messages have no visibility on the UI. This is the case of strings in drop-down menus, hidden lists, or help paragraphs that are revealed after an explicit user click/touch. Indeed, the most frequent HTML tags among all UI elements are DIV, LABEL, OPTION, and table cells. Of utmost importance is the fact that just 4 strings have human-generated comments, which reveals that Wordpress developers do not leave enough localization comments for translators.

On the UI, different types of strings may appear, such as error messages, call-to-actions, labels, etc. Thus, each message in the WPLoc dataset was manually labeled according to the following categories:

1. **Relevant:** Messages that must be understood to make use of the UI.
2. **Non-relevant:** Less important strings, such as month names or cities.

Both categories are well balanced, with 841 relevant strings and 838 strings otherwise. In addition, a finer-grained categorization was achieved by dividing the relevant strings in 3 subclasses:

1.1. **Critical:** Most imperative messages, such as those related to website maintenance or error messages.
1.2. **Informative:** Those messages that facilitate the understanding of the basic functionality of the website, such as paragraphs.
1.3. **Navigational:** Messages that inform about browsing actions, such as menu items.

Table 1. Message categories used for evaluation

Group Label	Importance	No. Strings
Critical	relevant	92
Informative	relevant	437
Navigational	relevant	312
Others	non-relevant	838
Total		1679

4.4 Feature Analysis for User-Driven Prioritization

We investigated how the previous string categories are characterized by the prioritization features described in Section 3.1. This would give an intuition to webmasters on how they could take control over the panel shown in Figure 2 and sort the strings according to their wishes. A straightforward approach to finding such characterization consists in standardizing all sample features (to have zero-mean and unit-variance) and computing the group centroids. Without feature standardization, the features would be in different ranges and the importance of each feature would be biased toward features with high values. Table 2 summarizes these values for each message category, normalized to -100 and 100 as in Nimrod's control panel (Figure 2). This study provides a better idea on whether a given feature affects prioritization, either positively or negatively.

Table 2. Feature combinations, or "mixer preset" weights (see Figure 2), that prioritize different UI string categories. All weights are bounded to $[-100, 100]\%$, the sign indicating positive or negative influence.

Group	A.C	D.C	N.R	S.F	E.F	S.V	E.Si	E.C	E.Se
Informative	14.7	-14.3	4.4	-24.4	0.6	41.0	40.8	34.6	23.6
Navigational	-6.3	-14.3	8.8	43.2	15.5	12.2	18.6	57.2	22.7
Critical	-14.9	-14.3	-8.1	-13.8	-31.9	4.7	11.2	-8.2	-32.2
Other	-3.7	14.4	-4.7	-1.8	-2.6	-26.5	-29.4	-38.5	-17.2

Column names follow the same order given in Prioritization Features (Section 3.1); e.g., **A.C**: automatic comments, ..., **E.Se**: Element semantics.

Looking at Table 2, it can be observed that navigational messages show a high contrast, whereas critical messages are represented by a lower contrast. Although this may sound contradictory, it must be noted that in Wordpress most navigational elements have a white font over dark background, while critical messages often use a red font color over light red background, which would diminish its contrast. Thus, perhaps a more appropriate metric such as color saliency based on visual human perception would be a better option to measure element contrast.

On the other hand, it was found that string visibility and element size are more prominent in informative messages rather than in critical messages. This may be due to the fact that informative texts are often very descriptive and thus might have a higher size on the UI. However, critical messages are typically exceptions that are rarely shown, and thus they remain hidden most of the time. Something similar happens to string and element frequency: critical messages appear scarcely and often at the same places. Conversely, navigational messages appear in almost any page and often in different menu entries. In this regard, informative messages appear less frequently than its critical counterparts, probably because the former are more specific and descriptive. Additionally, developer comments appear only in the "other" category, hence their weights are uniform across categories. Finally, the number of references was found to have little importance to discriminate among the analyzed categories.

4.5   String Retrieval for Fine-Grained Prioritization

Given that we had defined a well-established ground truth, another experiment to analyze string prioritization possibilities consisted in performing a more detailed, automatic string retrieval task. As in the previous experiment, strings belong to the classes (and subclasses) of the WPLoc dataset, and the same set of features is used.

As a practical application, we decided to use several state-of-the-art classifiers. All classifiers were evaluated using the open-source Weka machine learning suite [11], which will allow others to easily reproduce our experiments. The following results present a subset of the algorithms that, to our knowledge, are representative of a broad range of techniques: Logistic Regression [4], Multilayer Perceptron [17], KStar [5], Random Forest [2], and Bagging [1] with REPTree.

The experimentation was formulated as a retrieval problem, where we search for the strings that match a given class. Thus, *precision* indicates, among all retrieved strings, the percentage of strings that belong to the given class, whereas *recall* indicates the percentage of strings of a given class that were effectively retrieved. Finally, *F-measure* is the harmonic mean of precision and recall, and it can be interpreted as a weighted accuracy. All experiments were run on the whole set of 1679 strings with a 10-fold cross-validation setup.

Table 3. Retrieval results of *relevant* messages, in percentage

Classifier	Precision	Recall	F-measure
Logistic Regression	73.8	83.9	78.6
Multilayer Perceptron	75.0	**94.4**	83.6
KStar	**80.6**	88.0	84.1
Random Forest	79.7	87.4	83.4
Bagging (REPTree)	78.5	90.7	**84.2**

First of all, Table 3 shows results when retrieving those strings that were classified as relevant. On the one hand, *KStar* achieves the best precision at the expense of a worse recall. On the other hand, *Multilayer Perceptron* obtains the best recall at the expense of a lower precision. In terms of F-measure, *Bagging* and *KStar* performed better than their peers. Actually, what is best depends on the particular user needs. For instance, high precision with low recall involves less strings to be translated. Concretely, *KStar* retrieves 918 strings, *Multilayer Perceptron* 1058 and *Bagging* 971. Thus, if the goal is to reduce localization costs, then *KStar* is a reasonable option. Nevertheless, *Multilayer Perceptron* retrieves most of the relevant strings, which is probably a more important technique for a more imperative scenario, i.e., to reduce localization time.

Table 4 disgregates the results for the different subclasses of relevant messages. Overall, critical messages present very low recall scores in all of the tested classifiers. We suspect that this is caused by the low number of strings that fall under this category (5.5% in the WPLoc dataset). Thus, the class prior probability is low and few samples are used to train each classifier. On the other hand, we observed that informative messages are easier to retrieve with *KStar*, with a 75.5% of recall. However, its precision suggests that half of the retrieved strings would not be informative. Finally, navigational messages achieve higher precision scores, probably also because of the class prior imbalance.

Intuitively, critical messages should be easier to retrieve, since typically they would contain words related to errors and warnings, while navigational messages should be short and semantically related to UI actions. Thus, we decided to expand the feature set with a weighted bag of words of each resource string, where weights are string frequencies. The results are presented in Table 4 and marked with an asterisk (*). Significant improvements can be observed in all cases, which remarks the importance of the text for relevance identification. Concretely, a qualitative analysis of Bagging's decision trees showed that the word 'error' is used to discriminate critical messages, while 'updated' is present in informative messages and 'link' in navigational messages.

It is worth mentioning that the bag-of-words analysis may not extrapolate to *any* website. Indeed, different websites and CMS do have different UI strings, and thus a classifier should be trained for each website or CMS. Nevertheless, it is possible to use any of the other classifiers tested so far, as they provide competitive accuracy. In any case, this study puts forward the fact that different classifiers are feasible to improve website localization.

4.6 General Discussion

The needs of the localization industry change often over time. Today's development cycles are typically shorter, with quick turnaround times. This trend encourages software localization for a prompt revision.

Table 4. Results for message retrieval of the different relevant subclasses, in percentages. Classifiers marked with (*) were expanded with a weighted bag of words of each resource string as additional features.

Classifier	Critical			Informative			Navigational		
	Prec	Rec	F_1	Prec	Rec	F_1	Prec	Rec	F_1
Logistic Regression	0.0	0.0	0.0	53.0	48.5	50.7	42.7	15.1	22.3
Multilayer Perceptron	69.2	9.8	17.1	51.3	70.9	59.6	49.4	39.7	44.0
KStar	66.7	17.4	27.6	51.3	**75.5**	61.1	*67.8*	37.8	48.6
Random Forest	47.4	*19.6*	27.7	*56.2*	67.7	61.4	63.1	*48.7*	*55.0*
Bagging (REPTree)	*76.2*	17.4	*28.3*	55.1	73.5	*62.9*	60.8	45.2	51.8
Random Forest*	58.2	**34.8**	**43.5**	62.2	65.4	63.8	64.2	**63.8**	**64.0**
Bagging (REPTree)*	**83.3**	27.7	41.0	**63.3**	73.0	**67.8**	**68.2**	57.1	62.1

Often it is necessary to localize just the essential parts of a UI, either because of competitive advantage or economical reasons. For instance, an advanced word processor may comprise an important number of menus and options, but actually only a few of these are used by regular users. Then, localizing just what is most important would allow to reach emerging markets or even introduce a new product sooner than the competence. Unfortunately, following the typical localization workflow, it is difficult to decide which elements should be localized earlier. Moreover, typically websites and web applications do change over time, and so it should be possible to perform localization in an incremental fashion. Therefore, a solution to prioritize UI elements for localization is necessary. Nimrod is our contribution to tackle this topic.

5   Conclusion and Future Work

Software localization is both costly and a slow process, partially affected by the lack of string prioritization. We have shown that it is possible to automate the prioritization of UI strings, so that web-based software can be quickly and frequently translated. Our method augments string information in a message catalog with UI features, such as widget size, visibility, or color, so as to selectively pick the most relevant strings in the first place.

For future work we will incorporate our evaluation findings in our string prioritization tool. Concretely, we will add preset configurations that would allow to retrieve critical, informative, and navigational messages earlier. We also plan to consider UI usage information for analysis, like mouse or touch events, in order to gain more knowledge on the best prioritization schema. We hope this work will be useful to researchers and companies interested in UI localization. Nimrod is open source and can be downloaded at http://personales.upv.es/luileito/nimrod/.

Notes

[1] http://www.drupaltranslate.com
[2] http://smartling.com
[3] http://lingoport.com/globalyzer
[4] http://kokusaika.jp/en/product/wwnavi.html
[5] http://launchpad.net
[6] http://pootle.translatehouse.org
[7] http://www.verbatimsolutions.com
[8] http://www.transifex.com

Acknowledgments. This work is supported by the 7th Framework Program of EU Commision under grants 287576 (CASMACAT) and 600707 (tranScriptorium). The motivation of choosing "Nimrod" to name our tool is left as an additional exercise for the reader.

References

1. Breiman, L.: Bagging predictors. Machine Learning 24(2) (1996)
2. Breiman, L.: Random forests. Machine Learning 45(1) (2001)
3. Cascia, M.L., Sethi, S., Sclaro, S.: Combining textual and visual cues for content-based image retrieval on the world wide web. In: IEEEWorkshop on Content-Based Access of Image and Video Libraries, CBAIVL (1998)
4. le Cessie, S., van Houwelingen, J.: Ridge estimators in logistic regression. Applied Statistics 41(1) (1992)
5. Cleary, J.G., Trigg, L.E.: K*: An instance-based learner using an entropic distance measure. In: 12th International Conference on Machine Learning (1995)
6. Collins, R.W.: Software localization for internet software: Issues and methods. IEEE Software 19(2) (2002)
7. DePalma, D.A., Hegde, V., Pielmeier, H., Stewart, R.G.: The language services market. An annual review of the translation, localization, and interpreting services industry (2013), http://commonsenseadvisory.com
8. Dunne, K.J. (ed.): Perspectives on Localization. John Benjamins Publishing Company (2006)
9. Esselink, B.: A Practical Guide to Localization. John Benjamins Publishing Company (2000)
10. Gettext: The GNU gettext manual. version 0.18.2. (1995), http://www.gnu.org/
11. Hall, M., Frank, E., Holmes, G., Pfahringer, B., Reutemann, P., Witten, I.H.: The WEKA data mining software: An update. SIGKDD Explorations 11(1) (2009)
12. Hogan, J.M., Ho-Stuart, C., Pham, B.: Key challenges in software internationalisation. In: Workshop on Australasian Information Security, Data Mining and Web Intelligence, and Software Internationalisation (ACSW Frontiers) (2004)
13. Keniston, K.: Software localization: Notes on technology and culture. Working Paper #26, Massachusetts Institute of Technology (1997)
14. Leiva, L.A., Alabau, V.: An automatically generated interlanguage tailored to speakers of minority but culturally in uenced languages. In: Proceedings of the SIGCHI Conference on Human Factors in Computing Systems (CHI) (2012)
15. Leiva, L.A., Alabau, V.: The impact of visual contextualization on UI localization. In: Proceedings of the SIGCHI Conference on Human Factors in Computing Systems (CHI) (2014)
16. Reinecke, K., Bernstein, A.: Improving performance, perceived usability, and aesthetics with culturally adaptive user interfaces. ACM Transactions on Computer-Human Interaction (TOCHI) 18(2), 8:1–8:29 (2011)
17. Rosenblatt, F.: The perceptron: a probabilistic model for information storage and organization in the brain. Psychological Review 65(6) (1958)
18. Sun, H.: Building a culturally-competent corporate web site: an exploratory study of cultural markers in multilingual web design. In: Proceedings of the 19th Annual International Conference on Computer Documentation (SIGDOC) (2001)
19. De Troyer, O., Casteleyn, S.: Designing localized web sites. In: Zhou, X., Su, S., Papazoglou, M.P., Orlowska, M.E., Jeffery, K. (eds.) WISE 2004. LNCS, vol. 3306, pp. 547–558. Springer, Heidelberg (2004)
20. VanReusel, J.F.: Five golden rules to achieve agile localization (2013), http://blogs.adobe.com/globalization/
21. Wang, X., Zhang, L., Xie, T., Mei, H., Sun, J.: TranStrL: An automatic need-to-translate string locator for software internationlization. In: Proceedings of IEEE 31st International Conference on Software Engineering (ICSE) (2009)

Affective, Linguistic and Topic Patterns
in Online Autism Communities

Thin Nguyen, Thi Duong, Dinh Phung, and Svetha Venkatesh

Deakin University, Australia
{thin.nguyen,thi.duong,dinh.phung,svetha.venkatesh}@deakin.edu.au

Abstract. Online communities offer a platform to support and discuss health issues. They provide a more accessible way to bring people of the same concerns or interests. This paper aims to study the characteristics of online autism communities (called Clinical) in comparison with other online communities (called Control) using data from 110 Live Journal weblog communities. Using machine learning techniques, we comprehensively analyze these online autism communities. We study three key aspects expressed in the blog posts made by members of the communities: sentiment, topics and language style. Sentiment analysis shows that the sentiment of the clinical group has lower valence, indicative of poorer moods than people in control. Topics and language styles are shown to be good predictors of autism posts. The result shows the potential of social media in medical studies for a broad range of purposes such as screening, monitoring and subsequently providing supports for online communities of individuals with special needs.

Keywords: weblog, web mining, affective computing, mental health.

1 Introduction

Online communities have reshaped health care in several ways, such as enabling physicians and patients to crowd-source. Physicians can exchange practices at Sermo[1], an online community with more than 200,000 physicians in 68 specialties as of January 2014; or patients can compare conditions, treatments and symptoms with hundred thousands of other patients at PatientsLikeMe[2]. Online networking has been integrated into medical intervention, especially in mental health and wellbeing promotion programs. For example, the utility of online community can be integrated into a walking program to help reduce the attrition rate [29]; or Facebook can be used to treat stress and depression [15].

Autism Spectrum Disorder (ASD) is a neurological disorder whose affected by have social communication difficulties, experience sensory sensitivities, have narrow interests and repetitive behaviour [1]. It is a trying condition, requiring continuous support from parents, relatives, friends and society. Internet offers

---

[1] http://www.sermo.com/
[2] www.patientslikeme.com

B. Benatallah et al. (Eds.): WISE 2014, Part II, LNCS 8787, pp. 474–488, 2014.

another avenue for support [18]. It is stated that "for many autistics the Internet is Braille" [4]. On this platform, autism communities refer to online communities involving users affected by or interested in ASD, related to or providing supports for people with ASD. The communities have been recently studied in [26], examining the topics people in these communities are interested in and the way they express their arguments in terms of language style. However, the affective aspect within these communities has not been studied. Indeed, several studies in psychiatry have found a link between autism with mood disorders [11,33].

This paper examines two questions: First, is the sentiment different between autism and other communities? If so, how? For this affective aspect, we use the moods tagged to blog posts and the sentiment conveyed in the content as the sentiment information. An emotion bearing lexicon, Affective Norms for English Words (ANEW) [7], is used to score the affective information. Second, what are the discriminatory features of autism posts? To do this we consider the topics discussed and the linguistic styles expressed in online diaries of autism communities as the feature space.

To extract a set of discriminatory features, a popular approach is to conduct a separate statistical test on the hypothesis of equality between autism and control groups for every feature in consideration, as in [26]. For a feature, if the hypothesis is rejected, there exists a significant difference between the two groups in the use of that feature. The greater the difference is, the more discriminatory the feature is. Since this approach selects features in isolation, it neglects interactions amongst features. Thus, it may select a few good features but they tend to be strongly correlated, and thus not highly desirable. We address this problem in this work, using logistic regression, selecting a set of features simultaneously [14]. We can then interpret the differences in topics and language style between autism and control groups based on the weight in the regression model.

We present an analysis of a large scale cohort of data from nearly 2,000 people in 10 autism communities in a clinical population set in contrast with a control group of other 100 standard communities, focusing on mood, topics and psycholinguistic processes expressed in the content of posts. The way to collect data unobtrusively from online sources, such as online autism communities in this paper, provides a valuable alternative approach to research in mental health, avoiding the issue of privacy as in traditional data collection for clinical studies.

The main contributions of this work are: (1) to introduce a relatively comprehensive view of online autism communities, including three aspects of the subscribers: sentiment information, topics of interest and language style. Of these, to our knowledge, the sentiment information of the communities is explored for the first time; and (2) to propose an efficient approach to select features and do regression simultaneously, providing a set of powerful predictors of autism posts. The result shows the potential of the new media in screenings and monitoring of online autism communities. In addition, it demonstrates the applications of machine learning in medical practice and research.

2   Related Work

Studies on mental health in online setting have been conducted, especially on depression. For example, the content and the diurnal trends of posting were used to predict depressed individuals in Twitter [10]. Also for tweets, it was found that the non-depressed join Twitter for consuming and sharing information, whereas the depressed join to gain social awareness [27]. For Facebook, subscribers' status updates were found to be powerful predictors of major depressive episodes [22]; it was claimed that Facebook can be a screening tool for major depressive disorders [34]. Likewise, suicidal thoughts and intentions were spotted in MySpace [9].

To characterize an online community, two popular ways have been used are through the topics its members are interested in and via the way they express the topics – the language style. For example, both topic modeling and language style, extracted by the Linguistic Inquiry and Word Count (LIWC) package [28], were used as features to predict the life satisfaction of US counties [32]. Also, in the use of LIWC features, depressed students were found to use more first person singular pronouns and more negative emotions words in their essays than non-depressed students [30]. For personal blogs, topics and linguistic features were found to have great predictive power in sentiment classification and in classifying bloggers by age groups or by the degree of social connectivity [24,25].

Few studies have been published on online autism communities. An example is [23] where language style, expressed in the blogs of 40 individuals with autism, was analyzed using the LIWC package. While they focus on the posts produced by ASD people in their individual blogs, we targets posts written in the context of community. Another work is [26] where topics and language style were used to discriminate autism communities from control ones. However, the affective aspect was not considered in these works.

Beyond online community setting, technology-based intervention for autism is a growing research topic. Focusing on visual support, many systems have demonstrated efficacy in early treatment of mental disorders including autism (e.g., [2,31]). Examples of proprietary software include *DTTrainer*[3] and the Picture Exchange Communication System (PECS) [5]. Video has also been used, in the form of social stories, which are concrete, idiosyncratic video narratives that teach social skills [20]. Computer-assisted intervention has proven effective in teaching language, reducing inappropriate verbalization, and improving functional communication and generalization [6,17]. *Teachtown*[4] provides graded online and paper-based lessons rooted in ABA theory, but suffers from a restricted set of stimuli and impoverished adaptation to response. Toby Playpad [37] is perhaps the most comprehensive application available to day to accelerate learning for children with autism which can adjust lessons depending on the children's progress. Even though social networking aspects in these applications have not yet implemented, they can readily be added to further widen interactions and reach out to community.

---

[3] http://www.dttrainer.com

[4] http://www.teachtown.com

Table 1. Clinical communities: Ten autism communities and their biographies collected from Live Journal used in this paper

Community	Descriptions from Live Journal
asd-families	This is a community for people who are related to somebody with Autistic Spectrum Disorder.
ask-an-aspie	Is there an Aspie in your life? Do you need help understanding the autistic point of view?
asperger	This is a support community for people with Asperger Syndrome.
aspie-trans	Transgender, transsexual and gender queer bring with it their own special challenges. Aspergers and Autism bring different challenges. This is a haven for those who are living with both.
aspient	A community for people involved with someone who has apserger's or high-functioning autism.
autism-spectrum	This Journal is for anyone affected by Autism or other Spectrum Disorders.
autism	This is a community for anyone who has been affected by autism.
autistic-abuse	This journal is a place to read and discuss public cases of abuse against autistic persons.
bsperger	Tired of People Blaming Their Aspergers.
spectrum-parent	Being a parent is often challenging, but parenting kids on the autism spectrum.

Table 2. Control communities: 100 communities in ten categories

Category	Community
Advice-Support	add_a_writer, addme25_and_up, baristas, boys_and_girls, i_am_thankful, iworkatborders, thenicestthings, todayirealized, walmart_employe, weddingplans
Creative-Expression	___quotexwhore, 20sknitters, adayinmylife, aesthetes, amateur_artists, behind_the_lens, charloft, color_theory, naturesbeauty, sew_hip
Entertainment-Music	beatlepics, bjorkish, broadway, just_good_music, news_jpop, patd, relaxmusic, rilokiley, theater_icons, thecure
Fandom	chuunin, house_cameron, miracle_____, ncisficfind, patdslashseek, rpattz_kstew, sgagenrefinders, sgastoryfinders, sheldon_penny, time_and_chips
Fashion-Style	beauty101, corsetmakers, curlyhair, dyedhair, egl, egl_comm_sales, madradstalkers, ourbedrooms, ru_glamour, vintagehair
Food-Travel	bentolunch, davis_square, eurotravel, filmsg, newyorkers, ofmornings, picturing_food, seattle, trashy_eats, world_tourist
Gaming-Technology	computer_help, computerhelp, gamers, htmlhelp, ipod, macintosh, thesims2, webdesign, worldofwarcraft, wow_ladies
Parenting-Pets	altparent, baby_names, breastfeeding, cat_lovers, clucky, dog_lovers, dogsintraining, naturalbirth, note_to_cat, parenting101
Politics-Culture	birls, blackfolk, classics, ftm, nonfluffypagans, ontd_political, poor_skills, prolife, queer_rage, transnews
Television	_we_are_lost, battlestar_blog, calmallamadown, doctorwho, glee_tv, gleeclub, gundam00, lword, theoffice_us, topmodel

3   Method

3.1   Datasets

We collected data from Live Journal website, querying for all communities interested in autism[5]. As of January 2014, about hundred communities were matched the query; however, a majority of these communities have few posts. Ten of them whose largest number of posts were chosen into this study and their biographies are listed in Table 1, which is referred to as Clinical communities. The earliest community was created in 2001, thus our dataset spans over 10 years with a total of 1,910 active users, who have made at least one post in the communities. 10,000 newest posts made by these ten autism communities were chosen into the content analysis. We chose latest posts to avoid introductory posts often made in early stages of community life.

We then construct a Control dataset. For diversity, for each of ten Live Journal categories shown in the Live Journal community directory[6], first ten communities were chosen. This process results in 100 communities in the Control dataset, shown in Table 2. For each community, 100 latest posts were crawled, making up 10,000 posts, the same amount as in the Clinical dataset.

(a) Distribution of moods (in cloud visualization) tagged by Clinical group

(b) Distribution of moods tagged by Control group

Fig. 1. Mood usage by Clinical and Control groups

3.2   Affective Information

To characterize a community or a blog post, three sets of features are derived from the content: affective words, topics and language style.

For the affective aspect, firstly, we use the mood tags as a sentiment bearing source. Live Journal provides a mechanism for users to tag their posts from a list of 132 pre-defined mood labels[7]. A cloud visualization of moods tagged to the blog posts made by Clinical and Control groups is illustrated in Figure 1. In addition to the mood tags, we use the Affective Norms for English Words (ANEW) [7], an emotion bearing lexicon, to extract the sentiment conveyed in the content. A cloud visualization of ANEW words used in the content of blog posts made by Clinical and Control groups is shown in Figure 2.

---

[5] http://www.livejournal.com/interests.bml?int=autism

[6] http://www.livejournal.com/browse/

[7] http://www.livejournal.com/moodlist.bml?moodtheme=140, retrieved April 2014.

(a) Distribution of ANEW words (in cloud visualization) used in the content of blog posts made by Clinical group

(b) Distribution of ANEW words used in the content of blog posts made by Control group

Fig. 2. ANEW usage by Clinical and Control groups

We then represent a community by its use in the affective aspect. For mood tags, let $\mathcal{M} = \{sad,\ happy,\ ...\}$ be the predefined set of moods where $|\mathcal{M}| = 132$ is the total number of moods provided by Live Journal. Each community is then represented as a 132-dimension mood usage vector $\boldsymbol{m}_j$ whose k^{th} element is the number of times the kth mood in $\mathcal{M}$ is tagged within this community. For ANEW, each jth community is represented with a 1,034-dimension ANEW feature vector $\boldsymbol{a}_j$, whose k^{th} element is the number of times the kth ANEW word is used in the content of the blog posts made by users belonging to the community.

To measure the affective score for each community in these representations, we use the rating in term of valence and arousal for words in ANEW lexicon [7]. The valence of affective words is on a scale of 1, *very unpleasant*, to 9, *very pleasant*. The arousal is measured in the same scale, 1 for *least active* and 9 for *most active*.

To estimate the distance among communities in the affective aspect, we project the affective-based representations for communities onto the 2D plane, using t-SNE [36], a locality-preserving dimensionality reduction method.

3.3 Topics and Language Style

For topics, we use latent Dirichlet allocation (LDA) [3], a Bayesian probabilistic topic modeling tool, to examine the topics discussed in the clinical and control communities. LDA learns the probabilities p (vocabulary | topic), that are used to describe a topic and assigns a topic to each word in every document. For the inference part, we implemented Gibbs inference detailed in [16]. LDA [3] requires that the number of topics be specified in advance. A commonly accepted rule of thumb is to choose the number of topics (k) in the scale of $c^*\log V$ where V is the vocabulary size and c is a constant (often $= 10$), making 50 a fitting value for k in this work. We ran the Gibbs for 5000 samples and used the last Gibbs sample to interpret the results.

For language style, we use the LIWC package [28] in which English words are grouped into psycholinguistic processes, such as linguistic, social, affective, cognitive, perceptual, biological, relativity, personal concerns and spoken categories.[8]

---

[8] http://www.liwc.net/descriptiontable1.php, retrieved May 2014.

3.4 Feature Selection and Classification

Denote by $\mathcal{B}$ the corpus of all N blog posts. A blog post $d \in \mathcal{B}$ is denoted as $\mathbf{x}^{(d)} = [\ldots, x_i^{(d)}, \ldots]$, a vector of features. When topics are the features, $x_j^{(d)}$ is the probability of topic j in document d. If LIWC processes are the features, $x_i^{(d)}$ represents the quantity of the process i in document d. Our experimental design examines the effect of these two features in classifying a blog post into one of the two types of communities, Clinical vs Control, referred to as *blog post classification*. Given a blog post $d \in \mathcal{B}$, we are interested in predicting if the post belongs to a Clinical or Control community based on the textual features $\mathbf{x}^{(d)}$.

We are interested not only in which sets of features perform well in the classification but also in which features in the sets are strong predictive of autism. For this purpose, the least absolute shrinkage and selection operator (Lasso) [14], a regularized regression model, is chosen. Lasso does logistic regression and selects features simultaneously, enabling an evaluation on both the classification performance and the importance of each feature in the classification. Particularly, in a prediction of autism posts, Lasso assigns positive and negative weight to features associated with autism and control posts, respectively. To those features irrelevant to the prediction, Lasso assigns zero weight. Thus, by examining the weight, we can learn the importance of each feature in the prediction.

The regularization (or penalty) parameter (λ) in the regression model is chosen such that it is the largest number and the accuracy is still within one standard error of the optimum (1se rule). This way prevents overfitting since not too many features are included in the model while the accuracy of classification is still assured.

10-fold cross validations are run on the feature sets, that is, for each of 10 runs, one held-out data fold is used for testing and other nine folds for training. Classification accuracy, the proportion of correctly classified cases, is used to evaluate the performance.

4 Result

4.1 Affective Analysis

Mood Tags. Figure 3 shows the difference in the affective aspects between Control and Clinical communities. On the mood tags, as shown in Figure 3a, a majority of moods preferred by autism communities are in low valence, including *confused* (valence: 3.21), *frustrated* (2.48), *depressed* (1.83) and *angry* (2.85). On the other hand, people in the Control group favor tagging high valence moods, such as *happy* (valence: 8.21). This matches well with the reported emotional state of autism people [19,21], suggesting that those affected by ASD have more negative moods – they are more anxious and annoyed than the average population. This leads to a lower value in the average of the valence of moods tagged in the Clinical group in comparison with Control communities, as shown in Figure 3b.

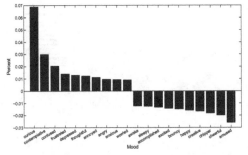
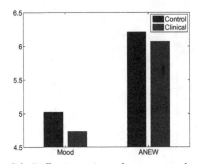

(a) Moods whose the difference above zero line are in favor of Clinical and those below zero line are in favor of Control people

(b) Differences in valence score between Clinical and Control.

Fig. 3. Difference in the use of mood tags and affective words between Control and Clinical

To visualize the distance among communities in the use of mood, we use t-SNE [36] to project the mood-based representation onto the 2D plane, plotted in Figure 4a. Only nine clinical communities are present in the figure since members of *bsperger* community do not tag moods to their posts. The figure shows a clear separation in the use of mood between Control and Clinical communities, barring two exceptions: *autism* and *asperger* communities.

ANEW Words. Similarly, in the use of sentiment bearing words in the content, the average valence of ANEW words in the blog posts made by Clinical is lower than that of Control communities, as shown in Figure 3b. In this ANEW-based representation, a cluster of Autism communities is well separated from their Control counterparts, as shown in Figure 4b, again except for *autism* and *asperger* communities.

4.2   Blogpost Classification and Feature Selection

Classification Result. We observe that the prediction of autism posts is best achieved when using topics as features, achieving an accuracy of 87.6%. The LIWC features also perform well, gaining an accuracy of 79.1% in a 2-class prediction (50% for a random guess since the input is balanced). The affective feature set, ANEW, gains an accuracy of 77.4%, slightly lower than LIWC performance. Without the need for a feature selection stage, the results for the predictions using psycholinguistic styles (through LIWC) and affective information (through ANEW) are comparable to topics, but at a lighter computational cost.

The process of feature selection for the model to predict autism posts is illustrated in Figure 5a. When λ decreases, the accuracy likely increases but at the same time, the number of features chosen into the model increases. For example, the optimal model of linguistic features to predict autism posts consists of

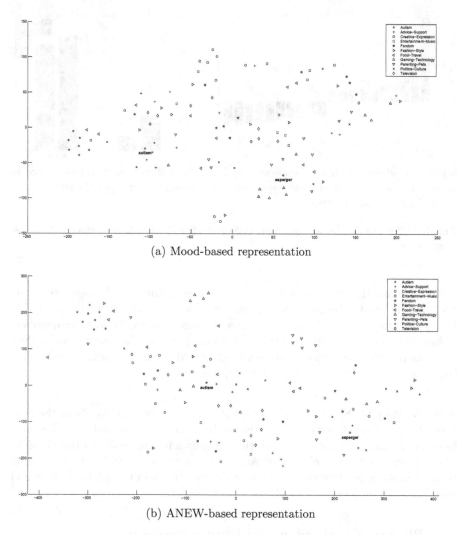

(a) Mood-based representation

(b) ANEW-based representation

Fig. 4. Visualization of the affective-based communities using t-SNE projection [36] (Best viewed in colors)

65 variables, whereas the *1se* model, which gains an accuracy within one standard error of the optimum, has only 50 variables. We choose the *1se* models for the classification and the models using topic, language style and affective information cues as features are shown in Figure 5.

Topics. For topic, 46 of 50 features were selected by Lasso into the *1se* model to predict autism posts, shown in Figure 5b. The topics whose high positive and negative weights in the model are shown in Table 3. As seen in the tables, the topics discussed in autism groups are found to be highly discriminative, focusing

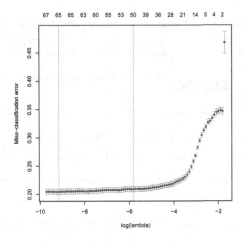

(a) Feature selection by Lasso to predict autism blog posts using language style (LIWC) as predictors

T37 T7 T4 T50 T6 T24 T26 T17 T36 T20 T34 T3 T2 ⋯ ⋯

⋯ ⋯ ⋯ ⋯ ⋯ ⋯   ⋯ ⋯ ⋯ ⋯ T37 T32 T1 T21 T16 T28

T48 T14 T11 T8 T45 T25

(b) A regression model of topical features

Humans Family Work Anxiety

Dictionary words Hear Health Insight

More−than−6−letter words   Nonfluencies   Prepositions   Negative emotion   Impersonal pronouns

⋯ ⋯ ⋯ ⋯ ⋯   ⋯ ⋯ 1st person 2nd person Body

Numbers Inhibition Sadness Positive emotion Leisure Home Ingestion

Achievement Perceptual processes Relativity

Money Religion Sexual See

(c) A regression model of linguistic (LIWC) features

social depressed teacher depression

stress concentrate child obsession education severe

upset person habit wonder mother noisy people violent kids hug

overwhelmed   interest   useful   scream   scared   theory   manner   brother   relaxed   boy   ship   beautiful

river   dress   out   adventure   iron   hurricane   pencil   queen   option   puppy   village   black   disaster

baby breast thankful wedding abortion

(d) A regression model of affective information (ANEW) features

Fig. 5. Regression models to predict autism blog posts. The size of variables reflects the absolute magnitude of corresponding coefficients ($|\beta|$); the color indicates the sign of the coefficients: red for positive and blue for negative.

Table 3. Powerful predictors in the model to predict autism blog posts based on topics

Var.	Word cloud	Coef.	Var.	Word cloud	Coef.
T37	**autism** autistic spectrum	38.57	T25	story read remember tony stories rodney fics advance ryan fic williams	-10.09
T7	**social** skills eye contact lack self conversation emotional difficulty	26.43	T45	color text background left image size font border style code important padding photos broken margin	-8.68
T4	**asperger** diagnosed syndrome diagnosis aspergers disorder	23.64	T8	doctor title rose chapter author rating summary characters	-8.2
T6	**autism** brain study research children university	12.45	T48	shipping condition price items white item black worn brand feedback waist lolita bought sale paypal size length free pink	-7.94
T20	**school** teacher college special students student teachers	6.65	T14	episode guild run post thread raid discussion members level running season open discuss join	-7.74

strongly on the topics of autism and health care, whilst the control group use more generic topics such as reading and shopping.

In particular, "*social skills*" (topic T7) is among the powerful positive predictors of autism posts. It likely mirrors one of the main concerns autism people face in real life: communication. As shown in the topic, they "*lack eye contact*" and often have "*self conversation*". In addition, *school* is a dominant topic in autism communities. It may reflect the challenges of individuals with autism in pursuing education due to their communication and social competencies, sensory issues and unique learning styles. The popularity of *school* topic shows needs for more information and discussion of these individuals and their carers in making informative education choices (e.g., autism specific schools, mainstream schools or home schools).

All these clear discrimination gains a high accuracy in classifying a post into either autism or control when topics are used as features.

Language Styles. For LIWC, 54 of 68 features were chosen into the 1se model to predict autism posts, shown in Figure 5c. Of the selected features, *biological* process obtains the maximum positive weight, despite three of its four sub-groups – *body*, *sexual* and *ingestion* – are assigned negative weight. It is because the rest sub-group – *health* (e.g., clinic, flu, pill) – has the intensity much higher in autism posts than in Control's.

Interestingly the next largest positive weight in the model to predict autism posts is assigned to two *social* process: *family* (e.g., daughter, husband, aunt) and *humans* (e.g., adult, baby, boy). According to [35], the *social* process somehow correlates with social concerns and social support. However, not all of the process are positive predictors of autism posts. Indeed, the rest social process, *friend* (e.g., buddy, friend, neighbor), is a negative signal of autism posts. This may suggest autism communities are focusing on their more pressing needs of the social relations of individuals with their families than with outsiders.

Table 4. A regression model of linguistic features to predict autism blog posts

Variable	Examples	Coef.	Variable	Examples	Coef.
(Intercept)		-6.96	Causation	Because, effect, hence	-0.24
Biological processes	Eat; blood; pain	13.11	Quantifiers	Few, many, much	-0.37
Humans	Adult; baby; boy	9.42	Articles	A, an, the	-0.45
Family	Daughter, husband, aunt	7.79	Time	End, until, season	-0.51
Dictionary words		7.76	Present tense	Is, does, hear	-0.53
Anxiety	Worried, fearful, nervous	7.36	Assent	Agree, OK, yes	-0.57
Work	Job, majors, xerox	7.15	Motion	Arrive, car, go	-0.59
Hear	Listen, hearing	5.8	1st pers plural	We, us, our	-0.61
Words >6 letters		5.13	Discrepancy	Should, would, could	-0.95
Insight	Think, know, consider	4.88	Death	Bury, coffin, kill	-1.33
Nonfluencies	Er, hm, umm	4.31	2nd person	You, your, thou	-1.6
Prepositions	To, with, above	3.6	Friends	Buddy, friend, neighbor	-2.04
Negative emotion	Hurt, ugly, nasty	2.87	Numbers	Second, thousand	-2.21
Impersonal pronouns	It, it's, those	1.19	Positive emotion	Love, nice, sweet	-4.37
Swear words	Damn, piss, fuck	1.11	Inhibition	Block, constrain, stop	-4.72
Word count		1.07	Leisure	Cook, chat, movie	-4.89
Fillers	Blah, Imean, youknow	0.87	Home	Apartment, kitchen, family	-4.92
Tentative	Maybe, perhaps, guess	0.78	Sadness	Crying, grief, sad	-5.05
3rd pers plural	They, their, they'd	0.56	Achievement	Earn, hero, win	-6.72
Adverbs	Very, really, quickly	0.44	Relativity	Area, bend, exit, stop	-7.65
Past tense	Went, ran, had	0.39	Perceptual processes	Observing, heard, feeling	-8.01
Certainty	Always, never	0.24	Money	Audit, cash, owe	-9
Conjunctions	And, but, whereas	0.23	Body	Cheek, hands, spit	-9.98
Words per sentence		0.02	See	View, saw, seen	-12.32
Exclusive	But, without, exclude	0.01	Religion	Altar, church, mosque	-12.97
Inclusive	And, with, include	-0.03	Sexual	Horny, love, incest	-15.82
3rd pers singular	She, her, him	-0.16	Ingestion	Dish, eat, pizza	-19.74
Future tense	Will, gonna	-0.21			

On the *affective* process, as expected, positive weight is assigned to *negative emotion* (e.g., hurt, ugly, nasty) and *anxiety* words (e.g., worried, fearful, nervous), meaning that the intensity for the features is higher in Clinical posts than in Control's. This partly supports the findings in [19,21], where anxiety is known to be an indicator of autism. It was also known that parents of children with ASD are prone to stress more than those of children with other disability [13]. However, it seems counter-intuitive that *sadness* (e.g., crying, grief, sad) is a negative signal of autism posts, meaning that bloggers in autism communities mention *sad* words in the content less frequently than do people in the control group. On the other hand, *positive emotion* (e.g., love, nice, sweet) is assigned negative weight, implying a lower intensity for the feature in Clinical posts than in Control's.

On the *personal concerns*, while *work* words (e.g., job, majors, xerox) is a positive signal of autism posts, *leisure, home, achievement, money* and *religion* is assigned negative weight in the prediction of autism posts. This partly contrasts with the signals of depression found in [10]: the depressed mention more about *religious* concepts.

Affective Information. For ANEW, the words selected into the 1se model to predict autism posts are shown in Figure 5c. Similar to topic case, the ANEW word *social* – probably in the topic of *"social skills"* – was chosen into the model. It has the largest positive weight on predicting the posts made by autism communities. Likewise, *school* related terms, such as *teacher* and *education*, have large positive weights in prediction of autism posts. In addition, *concentrate* was in. It may link to one of the symptoms of autism: having difficulties with concentration [38]. Also, *depression, depressed* and *stress* are among the powerful predictors of posts made by autism communities. These words can be in discussions among autistic people themselves or from their parents, confirming the finding of the presence of parenting stress and depression in those who have children with autism spectrum disorder [8,12].

5   Conclusion

We have analyzed online autism communities from Live Journal on three aspects: mood, psycholinguistic processes and topics in comparison with other standard communities. Our analysis has demonstrated that online autism communities are highly discriminative in these aspects when compared with a control dataset of standard communities, as shown in gaining approximately 90 percent classification accuracy. The result suggests that mood, psycholinguistic processes and topics can be considered the markers of mental health communities, illustrating potentials for employing computer-mediated communications into health care.

References

1. American Psychiatric Association, 5th edn. Diagnostic and Statistical Manual of Mental Disorders. American Psychiatric Publishing, Arlington (2013)
2. Bernard-Opitz, V., Sriram, N., Nakhoda-Sapuan, S.: Enhancing social problem solving in children with autism and normal children through computer-assisted instruction. Journal of Autism and Developmental Disorders 31(4), 377–384 (2001)
3. Blei, D.M., Ng, A.Y., Jordan, M.I.: Latent Dirichlet allocation. Journal of Machine Learning Research 3, 993–1022 (2003)
4. Blume, H.: "Autism & The Internet" or "It's The Wiring, Stupid" (1997), http://web.mit.edu/comm-forum/papers/blume.html (retrieved May 2014)
5. Bondy, A., Frost, L.: A picture's worth: PECS and other visual communication strategies in autism. Woodbine House (2001)
6. Bosseler, A., Massaro, D.W.: Development and evaluation of a computer-animated tutor for vocabulary and language learning in children with autism. Journal of Autism and Developmental Disorders 33(6), 653–672 (2003)

7. Bradley, M.M., Lang, P.J.: Affective norms for English words (ANEW): Instruction manual and affective ratings (1999)
8. Carter, A.S., de, F., Martínez-Pedraza, L., Gray, S.A.: Stability and individual change in depressive symptoms among mothers raising young children with asd: Maternal and child correlates. Journal of Clinical Psychology 65(12), 1270–1280 (2009)
9. Cash, S.J., Thelwall, M., Peck, S.N., Ferrell, J.Z., Bridge, J.A.: Adolescent suicide statements on MySpace. Cyberpsychology, Behavior, and Social Networking 16(3), 166–174 (2013)
10. De Choudhury, M., Gamon, M., Counts, S., Horvitz, E.: Predicting depression via social media. In: Proceedings of the International AAAI Conference on Weblogs and Social Media (2013)
11. Crane, L., Goddard, L., Pring, L.: Autobiographical memory in adults with autism spectrum disorder: The role of depressed mood, rumination, working memory and theory of mind. Autism 17(2), 205–219 (2013)
12. Davis, N.O., Carter, A.S.: Parenting stress in mothers and fathers of toddlers with autism spectrum disorders: Associations with child characteristics. Journal of Autism and Developmental Disorders 38(7), 1278–1291 (2008)
13. Dunn, M.E., Burbine, T., Bowers, C.A., Tantleff-Dunn, S.: Moderators of stress in parents of children with autism. Community Mental Health Journal 37(1), 39–52 (2001)
14. Friedman, J., Hastie, T., Tibshirani, R.: Regularization paths for generalized linear models via coordinate descent. Journal of Statistical Software 33(1), 1 (2010)
15. George, D.R., Dellasega, C., Whitehead, M.M., Bordon, A.: Facebook-based stress management resources for first-year medical students: A multi-method evaluation. Computers in Human Behavior 29(3), 559–562 (2013)
16. Griffiths, T.L., Steyvers, M.: Finding scientific topics. Proceedings of the National Academy of Sciences 101(90001), 5228–5235 (2004)
17. Hetzroni, O.E., Tannous, J.: Effects of a computer-based intervention program on the communicative functions of children with autism. Journal of Autism and Developmental Disorders 34(2), 95–113 (2004)
18. Jordan, C.J.: Evolution of autism support and understanding via the world wide web. Intellectual and Developmental Disabilities 48(3), 220–227 (2010)
19. Kim, J.A., Szatmari, P., Bryson, S.E., Streiner, D.L., Wilson, F.J.: The prevalence of anxiety and mood problems among children with autism and asperger syndrome. Autism 4(2), 117–132 (2000)
20. Lorimer, P.A., Simpson, R.L., Myles, B.S., Ganz, J.B.: The use of social stories as a preventative behavioral intervention in a home setting with a child with autism. Jnl. of Positive Behavior Interventions 4(1), 53 (2002)
21. Mazefsky, C.A., Conner, C.M., Oswald, D.P.: Association between depression and anxiety in high-functioning children with autism spectrum disorders and maternal mood symptoms. Autism Research 3(3), 120–127 (2010)
22. Moreno, M.A., Christakis, D.A., Egan, K.G., Jelenchick, L.A., Cox, E., Young, H., Villiard, H., Becker, T.: A pilot evaluation of associations between displayed depression references on Facebook and self-reported depression using a clinical scale. The Journal of Behavioral Health Services & Research 39(3), 295–304 (2012)
23. Newton, A.T., Kramer, A.D.I., McIntosh, D.N.: Autism online: a comparison of word usage in bloggers with and without autism spectrum disorders. In: Proceedings of the ACM Conference on Human Factors in Computing System (CHI), pp. 463–466 (2009)

24. Nguyen, T., Phung, D., Adams, B., Venkatesh, S.: Prediction of age, sentiment, and connectivity from social media text. In: Bouguettaya, A., Hauswirth, M., Liu, L. (eds.) WISE 2011. LNCS, vol. 6997, pp. 227–240. Springer, Heidelberg (2011)
25. Nguyen, T., Phung, D., Adams, B., Venkatesh, S.: Towards discovery of influence and personality traits through social link prediction. In: Proceedings of the International AAAI Conference on Weblogs and Social Media, pp. 566–569 (2011)
26. Nguyen, T., Phung, D., Venkatesh, S.: Analysis of psycholinguistic processes and topics in online autism communities. In: IEEE International Conference on Multimedia and Expo (2013)
27. Park, M., McDonald, D., Cha, M.: Perception differences between the depressed and non-depressed users in Twitter. In: Proceedings of the AAAI International Conference on Weblogs and Social Media (2013)
28. Pennebaker, J.W., Booth, R.J., Francis, M.E.: Linguistic Inquiry and Word Count (LIWC) [Computer software]. LIWC Inc. (2007)
29. Richardson, C.R., Buis, L.R., Janney, A.W., Goodrich, D.E., Sen, A., Hess, M.L., Mehari, K.S., Fortlage, L.A., Resnick, P.J., Zikmund-Fisher, B.J., et al.: An online community improves adherence in an Internet-mediated walking program. Part 1: results of a randomized controlled trial. Journal of Medical Internet Research 12(4), e71 (2010)
30. Rude, S., Gortnera, E.-M., Pennebaker, J.: Language use of depressed and depression-vulnerable college students. Cognition & Emotion 18(8), 1121–1133 (2004)
31. Schreibman, L., Whalen, C., Stahmer, A.C.: The use of video priming to reduce disruptive transition behavior in children with autism. Journal of Positive Behavior Interventions (2000)
32. Schwartz, H., Eichstaedt, J., Kern, M., Dziurzynski, L., Lucas, R., Agrawal, M., Park, G., Lakshmikanth, S., Jha, S., Seligman, M., Ungar, L.: Characterizing geographic variation in well-being using tweets. In: Proceedings of the International AAAI Conference on Weblogs and Social Media (2013)
33. Simonoff, E., Jones, C.R., Pickles, A., Happé, F., Baird, G., Charman, T.: Severe mood problems in adolescents with autism spectrum disorder. Journal of Child Psychology and Psychiatry 53(11), 1157–1166 (2012)
34. Jeong, Y.S., Nhi-Ha, T., Shyu, I., Chang, T., Fava, M., Kvedar, J., Yeung, A.: Using online social media, Facebook, in screening for major depressive disorder among college students. International Journal of Clinical Health & Psychology 13(1), 74–80 (2013)
35. Tausczik, Y.R., Pennebaker, J.W.: The psychological meaning of words: LIWC and computerized text analysis methods. Journal of Language and Social Psychology 29(1), 24–54 (2010)
36. Van der Maaten, L., Hinton, G.: Visualizing data using t-SNE. Journal of Machine Learning Research 9(2579-2605), 85 (2008)
37. Venkatesh, S., Phung, D., Greenhill, S., Duong, T., Adams, B.: TOBY: Early intervention in autism through technology. In: Proceedings of the ACM Conference on Human Factors in Computing System (CHI), Paris, France, pp. 3187–3196 (April 2013)
38. West, L., Waldrop, J., Brunssen, S.: Pharmacologic treatment for the core deficits and associated symptoms of autism in children. Journal of Pediatric Health Care 23(2), 75–89 (2009)

A Product-Customer Matching Framework for Web 2.0 Applications

Qiangqiang Kang, Zhao Zhang, Cheqing Jin*, and Aoying Zhou

Institute for Data Science and Engineering, Software Engineering Institute,
East China Normal University, Shanghai, China
qqkang@ecnu.cn,
{zhzhang,cqjin,ayzhou}@sei.ecnu.edu.cn

Abstract. Finding matching customers for a product is critical in many applications, especially in the e-commerce field. In this paper, we propose a novel product-customer matching framework to handle this issue, which consists of two components: data preprocessing and query processing. During the data preprocessing phase, a generation rule is proposed to learn the user's preference. With the spread of the web 2.0 applications, users like to rate some products they have experienced in the social applications, e.g. Dianping and Yelp. Hence, it is possible to construct users' preferences based on their rating information. In the query processing phase, we first propose Reverse Top-k-Ranks Query, which integrates reverse top-k query and reverse k-ranks query, to find some users to match the query product, and then devise an efficient method (BBPA) to handle this new query. Finally, we evaluate the efficiency and effectiveness of our matching framework upon real and synthetic datasets, showing that our framework works well in finding matching users for a query product.

Keywords: Matching Framework, User Preference, Reverse Query.

1 Introduction

Finding potential customers for a product is critical in many applications, especially the e-commerce. In general, there exist two phases to complete the matching from products to customers: one is the generation of user interest and the other is the product-driven queries. In the traditional way, user interest can be collected by the questionnaire, which costs a lot of human resources. However, with the spread of web 2.0, many user generated contents (UGC) are available, providing a data source to generate the users' preferences. In this study, we employ a simple rule to generate the user interests based on UGC and pay more attention to the product-driven query, which is actually reverse ranking query. This query is widely employed to find matching customers for a given product.

In the existing works [1–4], it is popular to describe user interest by using a weight vector w. The value at each entry is non-negative, and the sum of all entries is 1, i.e, $\sum_{i=1}^{d} w^i = 1$. Given a query product q, the inner product of q

---

* Corresponding author.

B. Benatallah et al. (Eds.): WISE 2014, Part II, LNCS 8787, pp. 489–504, 2014.

and w is used to describe the degree that a user likes the product. This score function is called a linear model [3]. However, researchers ignore some particular information of a product (e.g. whether a restaurant has free wifi), which are boolean attributes. It is worth noting that the users have the veto power for the boolean attributes. For example, a user only goes to a restaurant with free wifi.

In order to guarantee the users' veto power, we extended the existing linear model to compute the degree how a user likes a product. In the new model, the numeric attributes still employ the original linear model. The score value of this part is $\sum_{i=1}^{d} p^i \cdot w^i$ where p^i represents the ith attribute of p. For the boolean attributes, V_u represents a user's vote (the value is 0 or 1) and B_p is a product's boolean property (the value is also 0 or 1). The score value is $AND(B_p, V_u)$ which is the logical AND operation. If a user votes accept for the product's boolean attribute, then $AND(B_p, V_u)$ will be 1. Otherwise, the value is 0. In this study, we define the score function as $\frac{1}{AND(B_p, V_u)} \cdot \sum_{i=1}^{d} p^i \cdot w^i$. Without loss of generality, we assume that the smaller score values are preferable in this study.

Recently, reverse top-k query [5] is proposed to return a number of customers who regard a given product as one of the k most favorite products, based on the original linear model. Although a few hot products can be returned to some customers via this query, a large proportion of products cannot find any matching customers because such products are out of the top-k favorite list for any customers. Thereafter, reverse k-ranks query [6] is proposed to return a number of customers who like the product most, also based on the original linear model. This query can only return k customers for any product. However, "hot" products actually have more potential customers than k. Due to the limitation of the two queries, we integrate reverse top-k query and reverse k-ranks query to construct a new query, named Reverse Top-k-Ranks Query, increasing the coverage ratio of the returned customers. In our new query, the extended linear model is employed.

Example 1. *Figure 1 illustrates a simple example about Reverse Top-k-Ranks Query. All the shops do not offer free wifi (the value of the boolean attribute is 1) and all the users do not ask for free wifi except Jim. First, we compute the score value and rank value of all customer to p_4 based on the extended linear model, as shown in the above table. When k is set to 2, we can see that reverse top-k query does not return any user and reverse k-ranks returns Jim and Joe. Our query returns Joe and Lily. Then we set k = 4 and the results are still different between the existing refer query and our query, since our query have more coverage and employ an extended linear model in consideration of boolean properties.*

In this paper, we propose a product-customer matching framework to find appropriate customers for a given product. This framework mainly contains two components: data preprocessing and query processing. During the data preprocessing phase, we devise a novel rule to generate user preference. In the query processing part, we devise a Batch-Based Pruning Approach (BBPA) to handle Reverse Top-k-Ranks Query efficiently. Although similar to reverse top-k query and reverse k-ranks query in semantics, our query cannot be handled by existing methods directly. BBRA, the most efficient method to handle reverse-top-k

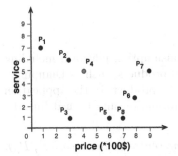

user	vote for wifi	w[price]	w[service]	Score	Rank
Jim	0	1	0	∞	-
Joe	1	0.7	0.3	4.3	3
Lily	1	0.4	0.6	4.6	3
Mary	1	0.5	0.5	4.5	4
Kate	1	0.3	0.7	4.7	4
Lucy	1	0.2	0.8	4.8	4
Alice	1	0	1	5	5
...	...	...	...	...	...

	score functin	k=2	k=4
Reverse Top-k	Linear Model	null	Jim, Joe, Lily, Mary, Kate, Lucy
Reverse k-Ranks	Linear Model	Jim, Joe	Jim, Joe, Lily, Mary
Reverse Top-k-Ranks	Extended Linear Model	Joe, Lily	Joe, Lily, Mary, Kate, Lucy

Fig. 1. Exmaple of Reverse Top-k-Ranks Query

query [7], needs to maintain two R-trees to index product datasets and user preference datasets to find the users who regard the query product as their top-k list. This algorithm cannot store the k users who have the minimal values. MPA, proposed in [6] to solve the problem of reverse k-ranks query efficiently, needs to construct a histogram for users previously. This method cannot be applied directly because our query employs an extended linear model. The primary contributions of our research are summarized as follows.

- We propose a novel rule to generate the user interest during the data pre-processing.
- We design a new query (Reverse Top-k-Ranks Query) to find the matching users for a given product, improving the coverage ratio of returned users and then propose the Batch-Based Pruning Approach (BBPA), which employs a pruning strategy from the perspective of both products and users, to handle Reverse Top-k-Ranks Query efficiently.
- We conduct extensive experiments to evaluate the performance of our query frame on real and synthetic data sets. The results show the superior of our proposal in finding customers for a given product.

The remainder of the paper is organized as follows. Section 2 details the problem statement. Section 3 and 4 introduce data preprocessing and query processing respectively. Section 6 reviews the related work. We report experimental results in Section 5 and conclude the paper in Section 7.

2   Problem Statement

2.1   Preliminary

Consider a Web 2.0 Application about online consumption platform including many user reviews (e.g. contents and ratings) and products, such as Dianping [1] and Yelp [2], the users like to rate each feature of the product. In the application that we target, we mainly employ the rating information and then define the following objects to illustrate the problem.

Definition 1 (Product). *A product p can be represented as the tuple $(\vec{p}, B_p)$. $\forall i \in [1, ..., d]$, $p^{(i)}$ denotes the value of the i-th numeric attribute of $\vec{p}$, and $p^{(i)} \geqslant 0$. B_p denotes the boolean attributes of the product p.*

Definition 2 (User Rating). *A user rating record u_r is a triple (uid, pid, $\vec{s}$). $\forall i \in [1, ..., d]$, $s^{(i)}$ denotes the user's rating score of the i-th numeric attribute of product $\vec{p}$, and $s^{(i)} \geqslant 0$. The pid and uid are the identity number of the product p and the user u respectively.*

Definition 3 (User). *A user u is a triple (uid, $\vec{w}$, V_u). It means a user u with uid has the weight preference vector $\vec{w}$ and V_u represents the vote for the boolean attributes. $\forall i \in [1, ..., d]$, w^i denotes the user's preference weight of the i-th numerical attribute of product $\vec{p}$ and $\sum_{i=1}^{d} w^i = 1$.*

For any user u and product p, the utility function $f(u, p)$ computes the degree how the user likes the product p. It is defined as the score function.

Definition 4 (Score Function, $f(u, p)$). *The score function $f(u, p)$ of product p for a user u is defined as the combination of the inner product of $u.\vec{w}$ and $p.\vec{p}$ and $\frac{1}{AND(u.V_u, \ p.B_p)}$ between p and u: $f(u, p) = \frac{1}{AND(u.V_u, \ p.B_p)} \sum_{i=1}^{d} u.w^i \cdot p.p^i$, where $u.w^i \geqslant 0$ $(1 \leqslant i \leqslant d)$, $\sum_{i=1}^{n} u.w^i = 1$ and AND represents the logical AND operator, whose result is 0 or 1. When $AND(u.V_u, \ p.B_p)$ is 0, it is an exception and we set the score value as infinity.*

We use rank value to measure relative preference of a user for product p compared based on the score value in this study.

Definition 5 (rank value, rank(u, p)). *Given a product set D, a user u, and a query point q, the rank of q for u is rank$(u, q) = |R|$, where $|R|$ is the size of R, a subset of D. For each $p \in R$, we have $f(u, p) < f(u, q)$; and for each $s \in D - S$, we have $f(u, s) \geqslant f(u, q)$.*

2.2   Problem Formulation

Based on the definition 5, Reverse Top-k-Ranks Query is defined. By comparison, we also give the definition of Reverse Top-k-Scores Query.

---

[1] www.dianping.com
[2] www.yelp.com

Definition 6 (Reverse Top-k-Scores Query). *Given a product set D, a user set S, Reverse Top-k-Scores Query returns the user set R, $R \subseteq S$, such that $\forall u \in R$, u is in the top-k list with $min(score\ value, f(u, p))$.*

Definition 7 (Reverse Top-k-Ranks Query). *Given a product set D, a user set S, Reverse Top-k-Ranks Query returns a set R, $R \subseteq S$, such that $\forall u \in S$, $\mathcal{P}(rank(u, p))$, where $\mathcal{P}$ is a conditional predicate. The conditional predicate $\mathcal{P}$ constructs a condition to filter the user based on $rank(u, p)$, the user result set that this query returns is the union of the users who have the rank value $rank(u, p) < k$ and are in the top-k list with $min(rank(u, p))$.*

In the example 1, the reverse top-2-ranks query will return Joe and Lily and the reverse top-2-scores query will return Joe and Mary. Actually, Reverse Top-k-Scores Query can be solved directly in a way similar to traditional top-k query and it is less reasonable for the biased product set [6]. Hence, our focus is on Reverse Top-k-Ranks Query. In this study, we proposed a framework, including data preprocessing and query processing, to find the matching users for a given product. As shown in Figure 2, During the data preprocessing phase, we devise a rule to process the data from the online consumption platform, including the user review dataset, user dataset and product dataset, etc. By executing the rule, we can generate the user interest for numeric attributes and the users' votes for the boolean attributes of products. The detailed procedure will be introduced in section 3. During the query processing phase, we execute the Reverse Top-k-Ranks Query to find the matching users.

3   Data Preprocessing

In this part, the generation of user interest for numeric attributes and the users' votes for products' boolean attributes are completed.

The Generation of User Interest. For the products' numeric properties, the user interest can be represented by the vector $\vec{w}$ in a quadruple u (see definition 3). In the online consuming platform, the user's interest can be learned from his (her) rating scores. Let's take the data from Dianping as the example.

Example 2. *Assume a restaurant has four numeric attributes:* average spend (as), taste (ta), environment (en) *and* service (se). *In general, a user prefers a restaurant with lower as, but higher ta, en and se. The weight of a dimension should be set low if a user does not care about that dimension. For example, it is reasonable to set a low weight on average spend for a user who often choose expensive restaurants. Let S_{as}, S_{ta}, S_{en}, and S_{se} denote the average scores of a user in four attributes, and ϕ_{as}, ϕ_{ta}, ϕ_{en} and ϕ_{se} denote the proportions of restaurants lower than the corresponding score. The weight vector of the user is represented as $(\frac{1-\phi_{as}}{sum}, \frac{\phi_{ta}}{sum}, \frac{\phi_{en}}{sum}, \frac{\phi_{se}}{sum})$, where $sum = 1 - \phi_{as} + \phi_{ta} + \phi_{en} + \phi_{se}$.*

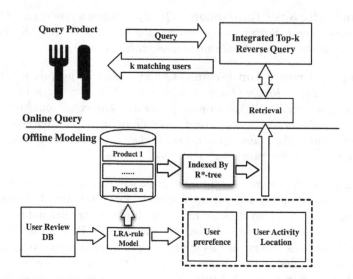

Fig. 2. The specification of Product-Customer Matching Framework

Algorithm 1 illustrates the main steps of the rule to learn the user's preference. The set D and S include all the products and users respectively. The set U_r represents all the users' rating records, in which S_r includes the rating records of a user. Assume $\vec{as}$ represents the vector of users's average scores in each attribute of the products. In the first phase, it checks each S_r in the dataset U_r to compute the vector $\vec{as}$ of a user. In the second phase, it will compute the rank value for each entry of $\vec{as}$ in the product dataset D (at lines 7-9). At last, the algorithm normalizes the rank values (at lines 10 and 11).

As for the collection of boolean properties, we can extract the features from the product profile. The users' votes can be processed by employing the technology of Nature Language Processing.

4   Query Processing

In the query part, we first devised two simple methods to find the matching users: the naive approach (NA) and the extension of RTA (ERTA). RTA is first employed in the literature [5]. NA needs to evaluate every product-user pair, which will consume huge amounts of computation resources in the datasets with a large number of user and products. ERTA first employs the RTA algorithm to get the result set S_1 of reverse top-k query. Assuming the size of S_1 is m. If m is smaller than the parameter k, then we continue to compute the remained product-user pair to find the $k - m$ users. This algorithm only uses the threshold method in the phase of RTA and has not any pruning strategies for products to reduce the computation. Due to the limitation of the above methods, we devise a

---

Algorithm 1. User Interest Generation (D, U_r)

---

1 Let L denote a set to store the users with average rating score for products;

2 **foreach** S_r in U_r **do**

3   compute $\vec{as}$;

4   Let u_s denote $(uid, \vec{as})$; L.add(u_s);

5 $\vec{rv} \leftarrow$ a new vector to record the rank value for each feature of the products;;

6 **foreach** l in L **do**

7   **foreach** as^i in $l.\vec{as}$ **do**

8 $rv^i \leftarrow$ the rank of of as^i in D;

9   Let u'_r is $(uid, \vec{rv})$;

10   **foreach** rv^i in $u_r'.\vec{rv}$ **do**

11 $u.w^{(i)} = \frac{rv^{(i)}}{\sum_{j=1}^{d} rv^j}$; $/ * Normalization * /$:

12   add u to S;

13 **return** *the user interest set S;*

---

novel algorithm, named Batch-Based Pruning Approach (BBPA), to reduce the computation of product-user pair from the perspective of both user and product.

4.1 Reverse Top-k-Ranks Query Processing Technology

In this section, we first analyze the query technology and then describe the algorithm in detail.

4.2 Analysis

In fact, based on the boolean attributes we can execute a pruning computation for the users. There exists the following fact between a user and a product.

Fact 1. *Given a user u and a query product q, if the user vote reject for the product in terms of the boolean attributes, then the value of $\frac{1}{AND(u.V_u,\ p.B_p)}$ tend to infinity and the score value $f(u,p)$ also is infinity (see definition 4).*

According to the Fact 1, the users who do not vote accept for the query product can be pruned safely since their score value are infinity and their rank value are also the largest. For the users that cannot be pruned, BBPA employs a R-tree for the product datasets to reduce the computation of product-user pair. At first, we review the dominance relationship between two points, which has been adopted in many literatures, including reverse top-k query processing [5].

Definition 8 (Dominance). *Given two points p_1 and p_2, we say p_1 dominates p_2, denoted as $p_1 \prec p_2$, if and only if the following two conditions hold: (i) $\forall l$, $p_1^{(l)} \leq p_2^{(l)}$; (ii) $\exists j$, $p_1^{(j)} < p_2^{(j)}$.*

We build an R-tree to index all products previously. Let r denote a minimal bounding rectangle (MBR) in R-tree. Let $r.L$ and $r.U$ denote the bottom left and top right points of r respectively. We have two facts and one theorem below.

Fact 2. *Given a query point q, a user u, and an MBR r in R-tree, if $f(u, q) < f(u, r.L)$, $\forall p \in r$, we have $f(u, q) < f(u, p)$.*

Fact 3. *Given a query point q, a weight vector w_i, and an MBR r in R-tree, if $f(u, r.U) < f(u, q)$, $\forall p \in r$, we have $f(u, p) < f(u, q)$.*

Based on the definition 7, we can maintain a buffer S_{result} to store the users who have the k minimal rank currently during the processing. This buffer strategy can reduce the space consumption and also can help prune the user by cooperating with the R-tree. Here is the pruning property.

Theorem 1 (Pruning Strategy). *Given a user u, a existing set of users $S_{result} \subseteq S$, and a query point q, if k data items (MBRs or data points) dominate p based on u and the current rank value of u is larger than the kth minimal rank value of a buffer set S_{result}, then u can be safely discarded.*

Proof 1. *By contradiction. Let us assume that the user u belongs to the reverse ranking result of q. Since k data items dominate q based on u (a MBR r encloses k data points) and the current rank value of u is larger than the kth minimal rank value of S_{result}, it holds that $\exists p_i \subseteq r$, $1 \leqslant i \leqslant k$ such that $f(u, r.U) < f(u, q)$. This means that there exist at least k data points with better scores and ranks than q for u. This leads to a contradiction, since by definition if u belongs to the reverse ranking result of q, then at most k-1 data points p_i have better scores and ranks than q based on u.*

4.3   Algorithm Description

BBPA employs two pruning strategies based on the users and a pruning strategy based on the products. For a given product, BBPA first checks whether the current user vote accept for the boolean properties. If not, this user can be pruned safely. Then, BBPA employs a result reusing strategy to prune the current user by maintaing a Top_k list of the previous user (details will be introduced in the following). If the above strategies cannot prune the current user, then the algorithm *compRank* will be invoked to compute the user's rank value. In this algorithm, we can prune some products based on the R-tree. Now, we first describe the *compRank* algorithm, which is the main component of BBPA and is also the most time-consuming during the data query.

The compRank Algorithm. In this algorithm, we use a buffer set S_{result} to store the users whose rank is less than k and k users who have the minimal rank currently. The value of *minRank* records the kth minimal value in the set S_{result}. Algorithm 2 llustrates the main steps of the *compRANK* algorithm about how to compute the rank value of a user. Assume all products have been indexed by an R-tree R during the data preprocessing phase. Initially, a queue Q is created. The set S_{result} and *minRank* are two parameters. At first, we add the root of R-tree to the queue Q (at line 4). Then for each node r in the tree, the algorithm computes the ranking scores of $r.L$ and $r.U$. According to Fact 2 and 3, the nodes satisfying $f(u, q) < f(u, r.L)$ or $f(u, r.U) < f(u, q)$ are in the candidate list of pruned users. Thereafter, we need to check If the rank of a user is larger than k and $minRank$ and the size of S_{result} is larger than k. If the above two conditions are made, based on the Theorem 1 we can exclude the user u from the result set (at lines 8 and 9). Otherwise, we need to traverse the R-tree and the subroutine enqueue appends an entry to the tail of a queue. If the tree reaches the leaf node, we need to compute the score value of each point. Here, we also use the Theorem 1 to exclude the user u from the result set (at lines 17 and 18). Finally, the algorithm returns -1 if the user can be discarded safely or the rank value if the user may belong to the result set.

Algorithm 2. compRANK $(u, q, k, minRank, S_{result})$

---

1 **Input:** user u, query product q, parameter k
2 **Output:** -1: discard u, *other*: the rank value of u
3 Let Q denote a queue;
4 Clear Q;   enqueue(Q, R.root);
5 **while** $((r = dequeue(Q)) \neq \emptyset)$ **do**
6 | **if** $(f(u, r.U) < f(u, q))$ **then**
7 | | $rank \leftarrow rank + |r|$;
8 | | **if** $(|r| > k \text{ and } rank > minRank \text{ and } S_{result}.size > k))$ **then**
9 | | | return -1;
10 | **else if** $(f(u, q) < f(u, r.L))$ **then**
11 | | continue ;
12 | **else**
13 | | **if** (r *is a leaf node*) **then**
14 | | **foreach** *point* r *in* r **do**
15 | | | **if** $f(u, r) < f(u, q)$ **then**
16 | | | | $rank \leftarrow rank + 1$;
17 | | | | **if** $(rank > k \text{ and } rank > minRank \text{ and } S_{result}.size > k)$ **then**
18 | | | | | return -1;
19 | | **else**
20 | | | $\forall r'$ (r' is r's child), enqueue(r', Q) ;

21 return $rank$;

Batch-Based Pruning Approach (BBPA). Clearly, the performance of query processing depends on the number of invocations of the *compRANK* algorithm, which is the cause of I/Os on the index of data set D. To improve the performance of query processing, it is beneficial to avoid *compRANK* invocations (and the respective I/Os) when possible. Therefore, we employ two methods to reduce the accesses on the R-tree index. First, we employ the boolean properties to prune a user according to the fact 1. Then, we employ a result reusing method that reduces the accesses on the R-tree index. The details are depicted in Algorithm 3.

Algorithm 3. Batch-Based Pruning Approach (BBPA) (D, S, q, k)

---

1 Let S_{result} denote an set to store the result set;
2 $minRank \leftarrow$ the kth minimal value in S_{result};
3 **foreach** u *in* S **do**
4 $\quad$ Compute $AND(u.V_u, q.B_p)$;
5 $\quad$ **if** $f(u, p) \rightarrow \infty$ **then**
6 $\quad\quad$ continue;
7 $\quad$ **foreach** r *in* TOP_k **do**
8 $\quad\quad$ **if** $(f(u, r.U) < f(u, q))$ **then**
9 $\quad\quad\quad$ $count + +$;
10 $\quad$ **if** $(count > k$ *and* $S_{result}.size > k$ *and* $count > minRank)$ **then**
11 $\quad\quad$ continue;
12 $\quad$ $rank \leftarrow$ compRANK$(u, q, k, minRank, S_{result})$;
13 $\quad$ **if** $(rank <> -1)$ **then**
14 $\quad\quad$ $S_{result}.add(u, rank)$;
15 $\quad\quad$ update S_{result};
16 $\quad\quad$ $minRank \leftarrow$ the kth minimal value in S_{result};
17 $\quad$ **else**
18 $\quad\quad$ update TOP_k;
19 **return** *the result set* S_{result};

For each user in the dataset S, BBPA first computes the logical value of $AND(u.V_u, q.B_p)$ to check if the score value of u reaches the infinity (at lines 4 and 5). For the users who cannot be pruned, BBPA exploits previously computed results to avoid redundant processing. As explained in the above section, Algorithm 3 achieves to discard a user due to the retrieval of k data items (data points or MBRs) of the index on D that dominate q on u. These data items can potentially lead to discard subsequent users, i.e., if a MBR dominate q based on the next entry u . Therefore, BBPA maintains this set of MBRs in a list of bounded size k, denoted as Top_k, and uses this list for avoiding invocations of the $compRANK$ algorithm. In the pseudocode of Algorithm 3, each time a user accessed (at line 3), a test is conducted between q and MBRs maintained in the list Top_k (at lines 7-9). If k points dominate q and the rank value of u is larger than the k-th minimal rank value in the buffer S_{result} in the TOP_k list (at line 10), then the $compRANK$ invocation based on u is avoided (at line 11). The list Top_k is updated each time the $compRANK$ algorithm results in discarding an user u (at line 18).

5   Experimental Results

In this section, we conduct extensive experimental evaluations on the matching frame. All codes are written in JAVA, and run on a stand-alone computer with an Intel CPU/2GHz and 8GB memory.

5.1   Data Description

In the experiment, we use two kinds of data sets in this study, including product set D and user data set S. We regard the location as the boolean attribute of products.

Product Data Set D. We use a real dataset (from Dianping). In order to test the scalability of Reverse Top-k-Ranks Query, we also use synthetic datasets.

- *Real product dataset.* We use a real dataset (from Dianping), which contains millions of ranking records about tens of thousands of restaurants in Shanghai from Jan. 2009 to Dec. 2013. Each record has four numeric scores in four dimensions: average cost per person (ac), taste (ta), environment (en) and service (se). The product dataset consists of 9,800 restaurants ranked at least ten times. The value in each dimension is computed as the average of all corresponding records.
- *Synthetic product dataset.* We generate three synthetic datasets, which follow uniform (UN), correlated (CO) and anti-correlated (AC) distributions respectively. See [8] for detailed generation methods. The value of the location is also shanghai.

User Data Set S. We employ the rule of the data preprocessing to generate users' interest and then extract users' votes from Dianping website. Except that, we also generate two synthetic data sets to test the efficiency of Reverse Top-k-Ranks Query, which are also used in [5, 9]. In the experiment, assume that the users who are in the same location as the product will vote accept. Otherwise, the users will vote reject. In the synthetic user preference dataset, we employ a random strategy to populate users' votes.

- *Real Preference Dataset:* it refers to 3,096, users who have rated at least ten times in the Dianping website.
- *Synthetic Preference Dataset:* There are two synthetic datasets: Uniform distribution (UNI) and Clustered distribution (CLU). For the UNI set, we repeatedly select one vector from d-dimensional space, and then normalize it to a standard form. The CLU set is created based on two parameters: g and σ^2. We first randomly select g cluster centroids in d-dimensional space. Then we generate some weights around the centroids with a variance of σ^2 in each dimension.

5.2   Effectiveness Study

Precision Test. We evaluate the precision of Reverse Top-k-Ranks Query and Reverse Top-k-Scores Query upon all hot restaurants with at least 100 ranking records, where precision is defined as the proportion of users who have visited the query restaurant. We use the data from Jan. 2009 to Dec. 2012 as the training data and the data from Jan. 2013 to Dec. 2013 as the testing data. Figure 3(a) shows Reverse Top-k-Ranks Query outperforms Reverse Top-k-Scores Query significantly in all situations: the precision of Reverse Top-k-Ranks Query is at least

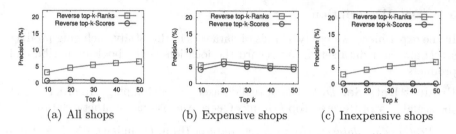

(a) All shops   (b) Expensive shops   (c) Inexpensive shops

Fig. 3. Precision Evaluation

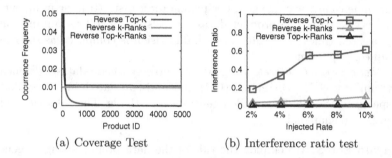

(a) Coverage Test   (b) Interference ratio test

Fig. 4. Coverage and Interference Evaluations

3.8 times greater than that of Reverse Top-k-Scores Query. Figure 3(b) and 3(c) report the effect after dividing all restaurants into two groups: expensive (with average cost per person $\geqslant$ 100 RMB) and inexpensive (with average cost per person < 100 RMB). For the expensive restaurants, Reverse Top-k-Ranks Query and Reverse Top-k-Scores Query behave similarly, but for the other group, Reverse Top-k-Ranks Query is significantly better than Reverse Top-k-Scores Query. The reason is that Reverse Top-k-Scores Query only works well for a small part of restaurants (especially expensive ones), while Reverse Top-k-Ranks Query works well for all restaurants. In fact, Reverse Top-k-Ranks Query focus more on the users' preference by employing the rank value as the metric.

Coverage Test. Coverage indicates the proportion of products that are recommended to at least one customer in multiple queries. In other words, this metric reflects the ability of recommendation for long-tail products [10]. We employ reverse k-ranks query and reverse top-k query as the baseline to test coverage upon real dataset *Dianping*. Here, we set $k = 50$ and choose 5000 query points randomly. Figure 4(a) shows the coverage testing results of three queries. The x-axis represents product ID and the y-axis represents the coverage ratio of each product. We observe that the recommendation chance of each product in reverse top-k query varies differently. Only a small number of products have a high probability to be recommended, while a large proportion of products can't be recommended. As for reverse k-ranks query, each product can get the same recommendation ratio, but the ratio is low. For our Reverse Top-k-Ranks Query, it has the better coverage.

Interference Ratio Test. As we all know, a good recommendation system should keep relatively good stability when new products are added into the system [11], which has important implications on users' trust and acceptance of recommendations. In other words, for a given query type, if we add some other products randomly to the original product set, the query result should not be affected greatly. Based on this standard, we continue our experimental research in a given product set by adding new products which can dominate the most of original product sets. Here, we define this "add" action as injection, and the proportion of added products as injected rate. Then, we use *interference ratio* to measure the stability of a query type, defined by $\frac{|R_{before} - R_{after}|}{|R_{before} \cup R_{after}|}$, where R_{before} is the query result on D before injection, and R_{after} is the query result on D after injection. Large interference ratio is not preferable. Figure 4(b) illustrates the interference ratio's variation with the increment of injection percentage upon real Dianping dataset. The x-axis represents the the proportion of the original products which new products accounted for. The y-axis represents the *interference ratio* under the certain percentage. For a given data set D, we can observe that our query has better stability for the lower interference ratio.

5.3   Efficiency Study

NA, ERTA vs. BBPA for Varying d. Figure 5 illustrates the performance of three methods (NA, ERTA and BBPA) when varying the number of dimensions. We use UNI weight set and three product sets, including UN, CO and AC. We set $|D| = 20K$, $|S| = 400K$ and $k = 10$. Each query product is randomly selected in the domain. The y-axes of Figure 5(a) and (b) represent the executing time and the total number of pairwise computations (refers to computing the ranking score for a product-customer pair) respectively. The longest, the second longest and the shortest bars represent the performance of NA, ERTA and BBPA respectively.

ERTA and BBPA outperform NA for at least one order of magnitude, since NA has the time complexity of $O(|D| \cdot |S|)$, while the other two algorithms can reduce the number of comparison operations efficiently by employing different pruning rules. ERTA prunes some unnecessary pairwise computations from the perspective of users based on the threshold, while BBPA simultaneously prune unnecessary pairwise computations from the perspectives of both products and users, avoiding scanning all $u \in S$ and $p \in D$. Figure 5(b) shows the pairwise computations are insensitive to the number of dimensions because the number of pairwise (u, p) is almost unchanged when d increases.

Scalability of NA, ERTA and BBPA. Figure 6 reports the scalability of NA, ERTA and BBPA under different $|D|$, $|S|$ and k. The metrics are time cost. By default, $|D| = 20K$, $k = 10$ and $|S| = 400K$. Figures 6(a) shows the performance change when $|S|$ increases from 100K to 500K. For all methods, the processing

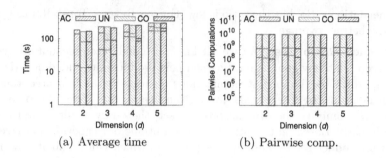

(a) Average time   (b) Pairwise comp.

Fig. 5. Performance by varying d [NA (the longest bar) vs. TPA (the shortest bar)]

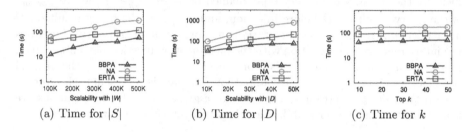

(a) Time for $|S|$ (b) Time for $|D|$ (c) Time for k

Fig. 6. Scalability of four approaches [NA vs. ITRA]

time will rise when $|S|$ increases. We also observe that NA is more sensitive to the value of $|S|$, because ERTA and BBPA have pruning rules on S. Figures 6(b) shows the performance change when $|D|$ increases from 10K to 50K. For all methods, the processing time will rise when $|D|$ increases. Meanwhile, BBPA outperforms NA and ERTA since it have the pruning rules for D. Figures 6(c) shows the time consumption of NA remains almost unchanged when k increases, because the time complexity of NA is $O(|D| \cdot |S|)$ for any k. ERTA and BBPA are also insensitive to k, since in general $k \ll |S|$.

6   Related Work

This section shows the work related to ours in two parts, including the generation of user interest and Reverse Top-k-Ranks Query. [12] employs a LDA method to generate user interest based on the content in the location-based social network. In [13], it makes full use of folkowees' tweets to generate user preference in the microblog. However, we try to find user preference by the users' rating information in the online consuming platform (e.g. Dianping and Yelp). [14] employs a personalized, content-based recommendation approach for high-involvement furniture products which adopts linguistic information to express customer preferences and product evaluations.

Our work in the query processing phase is considered as one of the reverse ranking query. As a principle query type in the database management field, it has been widely adopted in many applications, and it has multiple variants nowadays. It is first proposed by Li et.al [15]. And then Xiang Lian et.al. [16] studied reverse ranking queries in uncertain database, Ken C.K.Lee et.al. [17] discussed reverse ranking query over imprecise spatial data.

Reverse top-k query is recently proposed by Akrivi Vlachou et.al. [5] [9] [7] to find customers where a given data point belongs to its top-k query result set. Shen Ge et.al [4] proposed methods for batch evaluation of all top-k queries by the block indexed nested loops and view-based algorithm. It is another approach to answer reverse top-k query by efficiently computing multiple top-k queries. Thereafter, reverse k-ranks is recently by [6], which can return k customers who like the given product mostly, no matter it is hot or niche. However, this query cannot cover the customers who regard the query product as his or her top-k list. The above literatures compute the rank of a given point based on the original linear model. We try to find the rank of a given point by the extended linear model after considering the boolean properties of products.

7 Conclusion

In this paper, we propose a novel product-customer matching framework to find matching users for a product. During the data preprocessing phase, a weight generation rule is proposed to learn the users preference. In the query processing phase, we first propose a new query named Reverse Top-k-Ranks Query to find some users to match the query product, and then devise an efficient method (BBPA) to handle this new query. The experimental results show the efficiency and effectiveness of our framework.

Acknowledgement. Our research is supported by the 973 program of China (No. 2012CB316203), NSFC (No. 61370101), Shanghai Knowledge Service Platform Project (No. ZF1213), Innovation Program of Shanghai Municipal Education Commission(14ZZ045) and the Natural Science Foundation of Shanghai (14ZR1412600) and a fund of ECNU for oversea scholars, international conference and domestic scholarly visits.

References

1. Chang, Y.-C., Bergman, L.D., Castelli, V., Li, C.-S., Lo, M.-L., Smith, J.R.: The onion technique: Indexing for linear optimization queries. In: Proc. of ACM SIGMOD, pp. 391–402 (2000)
2. Chester, S., Thomo, A., Venkatesh, S., Whitesides, S.: Indexing reverse top-k queries. CoRR abs/1205.0837 (2012)
3. Hristidis, V., Koudas, N., Papakonstantinou, Y.: Prefer: A system for the efficient execution of multi-parametric ranked queries. In: Proc. of ACM SIGMOD, pp. 259–270 (2001)
4. Ge, S., Leong Hou, U., Mamoulis, N., Cheung, D.: Efficient all top-k computation - a unified solution for all top-k, reverse top-k and top-m influential queries. IEEE Trans. Knowl. Data Eng. 25(5), 1015–1027 (2012)

5. Vlachou, A., Doulkeridis, C., Kotidis, Y., Nørvåg, K.: Reverse top-k queries. In: Proc. of ICDE, pp. 365–376 (2010)
6. Zhang, Z., Jin, C., Kang, Q.: Reverse k-ranks query. PVLDB 7(10), 785–796 (2014)
7. Vlachou, A., Doulkeridis, C., Nørvåg, K., Kotidis, Y.: Branch-and-bound algorithm for reverse top-k queries. In: SIGMOD Conference, pp. 481–492 (2013)
8. Börzsönyi, S., Kossmann, D., Stocker, K.: The skyline operator. In: Proc. of ICDE, pp. 421–430 (2001)
9. Vlachou, A., Doulkeridis, C., Kotidis, Y., Nørvåg, K.: Monochromatic and bichromatic reverse top-k queries. IEEE Trans. Knowl. Data Eng. 23(8), 1215–1229 (2011)
10. Shani, G., Gunawardana, A.: Evaluating recommendation systems. In: Recommender Systems Handbook, pp. 257–297. Springer (2011)
11. Adomavicius, G., Zhang, J.: On the stability of recommendation algorithms. In: Proceedings of the Fourth ACM Conference on Recommender Systems, pp. 47–54. ACM, New York (2010)
12. Yin, H., Sun, Y., Cui, B., Hu, Z., Chen, L.: Lcars: a location-content-aware recommender system. In: KDD, pp. 221–229 (2013)
13. Xu, C., Zhou, M., Chen, F., Zhou, A.: Detecting user preference on microblog. In: Meng, W., Feng, L., Bressan, S., Winiwarter, W., Song, W. (eds.) DASFAA 2013, Part II. LNCS, vol. 7826, pp. 219–227. Springer, Heidelberg (2013)
14. Gerogiannis, V.C., Karageorgos, A., Liu, L., Tjortjis, C.: Personalised fuzzy recommendation for high involvement products. In: SMC, pp. 4884–4890 (2013)
15. Li, C.: Enabling data retrieval: by ranking and beyond. PhD thesis, University of Illinois at Urbana-Champaign (2007)
16. Lian, X., Chen, L.: Probabilistic inverse ranking queries in uncertain databases. VLDB J. 20(1), 107–127 (2011)
17. Lee, K.C.K., Ye, M., Lee, W.C.: Reverse ranking query over imprecise spatial data. In: COM. Geo. 17:1–17:8 (2010)

Rapid Development of Interactive Applications Based on Online Social Networks

Ángel Mora Segura, Juan de Lara, and Jesús Sánchez Cuadrado

Modelling and Software Engineering Group
Department of Computer Science
Universidad Autónoma de Madrid, Spain
http://www.miso.es
{Angel.MoraS,Juan.deLara,Jesus.Sanchez.Cuadrado}@uam.es

Abstract. Online social networks, like Twitter or Google+, are widely used for all kind of purposes, and the proliferation of smartphones enables their use anywhere, anytime. The instant messaging capabilities of these services are used in an ad-hoc way for social activities, like organizing meetings or gathering preferences among a group of friends, or as a means to contact community managers of companies or services.

Provided with automation mechanisms, posts (messages in social networks) can be used as a dialogue mechanism between users and computer applications. In this paper we propose the concept of post-based application, an application that uses short messages as a medium to obtain input commands from users and produce outputs, describing several scenarios where these applications are of interest. In addition, we provide an automated, *Model-Driven Engineering* approach (currently targeting Twitter) for their rapid construction, including dedicated Domain-Specific Languages to express the interesting parts to be detected in posts; and query matched posts, aggregate information or synthesize posts.

Keywords: Social Networks, Post-based Application, Model-Driven Engineering, Domain-Specific Languages, Social Applications.

1 Introduction

Online social networks (OSNs) based on microblogging are booming nowadays, thanks in part to the proliferation of smartphones and mobile devices. Hence, services like Twitter or Google+[1] are extremely used nowadays to connect with friends, or to organize social activities. These services are not only used for leisure, but most companies and brands recognise the reach and importance of OSNs today, and use these services to keep in contact with clients [18].

In this setting, we observe a growing need to automate social activities, leveraging on popular OSN platforms, like Twitter. On the one hand, users of OSNs – possibly lacking any programming skills – may wish to define simple applications involving the participation of a community of users. On the other hand,

---

[1] http://www.twitter.com, https://plus.google.com

B. Benatallah et al. (Eds.): WISE 2014, Part II, LNCS 8787, pp. 505–520, 2014.

companies may like to open their information systems to OSN platforms, but this integration effort needs to be done by hand. Related to this last issue, companies frequently interact with potential customers via the *online community manager*, in charge of managing and moderating the relationships with their clients through posts (messages in OSNs). Even though this role is of crucial importance nowadays, many of his tasks are performed manually, in ad-hoc ways.

Our thesis is that OSNs based on short, instant messages, are suitable as front-ends for computer applications. We call them *post-based* applications, and present many advantages in some scenarios. First, OSNs are designed to support a high load of users and posts, serving as a robust front-end, which could be difficult to achieve for companies or end users. Second, many people are familiar with specific OSNs and have it already installed in their devices. Hence, they do not need to learn a new application, or install a new one. Finally, applications can leverage from the social network structure provided by the OSN platform.

Several scenarios benefit from post-based applications. In the first one, small, simple, self-contained applications can be designed by unexperienced end users. For example, for mobile learning games, or to organize votings. The second one involves the rapid construction of applications to coordinate a large amount of people upon unexpected events, like natural disasters or strikes in airports. Finally, specific OSNs can be used as a front-end for existing information systems. For example, an airport may send notifications with flight information, or with status updates via *tweets* (posts in the Twitter platform) to interested users.

These scenarios present several challenges. First, relevant information needs to be extracted from posts. Posts are unstructured, and users cannot be expect to follow a tight syntax. Hence, we need a simple way to express and detect the interesting information. Second, a mechanism is needed to specify simple actions, like querying the extracted information, or synthesizing posts with collected information. Finally, a quick, easy way for constructing this kind of applications is needed, enabling their use by non-experts, but supporting also their deployment into servers, and their integration with existing information systems.

Contributions. We introduce the concept of *post-based user interface* and *post-based application*, proposing a Model-Driven Engineering (MDE) approach for their automated rapid construction. The solution includes: (a) A Domain Specific Language (DSL) for expressing patterns that is connected with WordNet [16], a lexical database for the English language; (b) a DSL for describing actions; and (c) an Eclipse-based prototype environment to model and deploy post-based applications, which currently targets Twitter. This paper extends [21], a short tool-demo paper which focussed on the tooling aspect of the solution.

Paper Organization. Section 2 describes several motivating scenarios for *Post-based applications*, showing their benefits. Section 3 proposes our architecture. Section 4 describes a DSL for expressing patterns on posts. Section 5 presents a DSL for describing simple actions. Section 6 describes tool support. Section 7 analyses related research, and Section 8 ends with conclusions and future work.

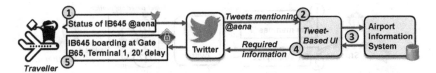

Fig. 1. A tweet-based user interface for an airport information system

2 Motivating Examples

In this section, we provide some motivating scenarios for post-based applications. The underlying concepts are applicable to different OSNs, but for several reasons (popularity, availability of an API, the availability of a social network structure, possibility of both public and private messages), we currently consider Twitter.

Twitter is an OSN based on microblogging, which permits a direct communication among users via posts, short messages limited to 140 characters called *tweets*. Tweets are public by default, and searchable by any person not necessarily a Twitter user. There are also private messages, directed to a specific user. Twitter users may follow a number of other users. A follower receives any tweet of the users they follow. User names start by '@'. Tweets may contain user names, and the mentioned users get notified whenever this happens. Tweets may also contain hashtags (a word preceded by '#'), names agreed by community convention, which facilitate the searching of tweets. Finally, tweets may contain links (in particular to pictures), and be geolocalized (i.e., contain information on the position on earth where the tweet is being sent).

Next, we describe some scenarios where post-based applications are of interest.

1. **Tweet-based user interfaces for Enterprise Systems.** In this scenario a company decides to use Twitter as a means to provide access to its information system. The advantage is that users do not need to install a new software for interacting with the information system. They use their Twitter accounts for the interaction, so that they can access the information systems either from smartphones, laptops or desktop computers in a unified way. Moreover, additional information can be obtained from users if tweets are geolocalized. The social network structure of Twitter can also be exploited, e.g., to broadcast messages, received by all followers.

   As an example, Figure 1 shows the integration of a tweet-based user interface with an airport information system. The purpose of such system is to inform travellers of the status of their flights, among other services. In a first step (label 1), a person about to travel sends a tweet requesting information about a certain flight. Such tweet mentions the account of the airport or the management entity, like *@aena* for the case of Spanish airports. In step 2, the tweet-based interface receives a mentioning tweet, and matches such tweet accross a collection of relevant requests. In this case, the tweet request information for a flight, which is directed to the information system (step 3). The flight number is extracted and sought in the airport's database. Then, a

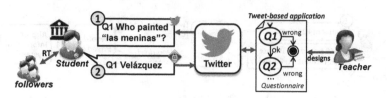

Fig. 2. Tweet-based outdoor quizz

private message for the particular user is automatically synthesized (label 4) by the interface, which is received by the user via Twitter (label 5). In this case, the initial message is public, but for privacy issues, a private message could have been used instead.

2. ***Ad-hoc* Tweet-Based Collaborative Applications.** In some cases, groups of end-users are interested in using Twitter as a means to collect structured information that could be later automatically processed. However, they may lack the technical abilities to develop such an application from scratch. These situations include all sorts of votings, organization of sport events, educational outdoor games, and many others.

   As an example, Figure 2 shows the working scheme of an outdoor educational game, where the teacher designs a questionnaire, to be answered by groups of students during a visit to a museum. Hence, questions are answered by exploring the museum. Students organize in groups and can use the social network structure of Twitter to collaborate among them (e.g., via retweets and tweets) so that they can search for the answer in parallel. The system provides new questions when a question is answered, and can finally provide some statistics. In this scenario, the teacher may design the questionnaire, not needing to be proficient in programming languages or the Twitter API. Along the paper, we will use a simpler example in this scenario, consisting in a simple yes/no voting system among a community of users.

3. **Massively Collaborative Application for Unexpected Situations.** Information technologies play a crucial role for emergency response (so called peace technologies) [9]. In particular in [15] it is reported that Twitter was an important medium for the Japanese government to distribute information to millions of people during the Fukushima nuclear radiation disaster. In this type of applications, the geoposition information offered by geolocalized tweets can be valuable. As an additional example, we may consider a Twitter front-end to help in reassigning flights in case of a strike. This is useful, as the airplane company web site may not support the high peak load if thousands of users try to access the system at a time. Hence, each passenger may provide its flight, which is saved into a database. Once the flight is reassigned (manually by some operator), the user may receive a notification, or he may be asked to contact a service desk. Hence, in this scenario, we profit from the high load supported by Twitter, and its popularity, so that users do not need to install a new application.

3   Architecture

The working scheme of our solution for tweet-based applications is shown in Figure 3, where the numbers illustrate a typical interaction.

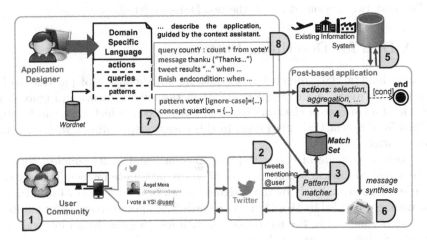

Fig. 3. Architecture provided by our solution

In the first place (label 1), users send tweets or private messages via Twitter. Then, the relevant information in tweets needs to be extracted from the application. Our solution relies on the definition of patterns, expected to be found in tweets. Not every tweet is sought, but only those mentioning the user associated to the application, or private messages directed to it (label 2). The patterns (label 3) are defined by the application designer using a dedicated DSL. A typical application may include different queries, selecting the relevant concepts in matching tweets, or calculating different aggregation values from them (label 4). In addition, data can be obtained or sent to existing information systems (label 5). The data extracted from queries, or provided by the information system can be used to synthesize tweets or private messages, directed to the users (label 6). Finally, conditions can be defined to signal the end of the execution.

In order to facilitate the construction of such system, we provide an MDE solution, based on two domain-specific languages (DSLs). The first DSL helps in the definition of relevant patterns, and concepts to be found in them (label 7) in the figure. The latter are sets of relevant words, or fragments, and sets of synonyms can be automatically extracted from WordNet [16].

The second DSL (label 8), is targeted to the description of the processing logic of post-based applications. Both languages are defined in a modular way, permitting extensions for the peculiarities of different OSNs, but we currently target Twitter. The DSL allows defining queries on posts matching some pattern, using an SQL-like syntax. Queries can be used to select relevant information from posts, or to calculate aggregated information from a set of posts. The DSL also provides commands to synthesize private messages and tweets, and to signal the end of the application execution. Finally, it is also possible to define what we call *data hooks*, as a way to push extracted data into an existing information system, or to gather data from it. Being an MDE framework, we can profit from code generation for different technologies, like REST or SOAP web services.

The next two sections describe the two DSLs in detail.

4   A DSL for Describing Patterns

We have built a DSL, directed to facilitating the definition of patterns to be identified in tweets. A small excerpt of its meta-model is shown in Figure 4. It is made of a core set of elements (package PatternCore), which then can be specialized for different OSN platforms. In this case, we specialize it with concepts from Twitter.

A pattern is made of concepts (class Concept), and in its simplest form, a concept (called ExplicitConcept) is a set of words (attribute tokens). This set can be either defined explicitly by the designer, or can be automatically taken from a synonym set provided by WordNet. We have defined special concepts for numbers and letters (omitted in the meta-model), with length given by an interval, and permit defining sequences of these.

We have also included specific Twitter concepts (TwitterPatterns package), like patterns to detect user names, URLs (specially pictures), and to define collections of interesting hashtags. As we will discuss in the next section, the meta-data information present in tweets, like the originator, date or geoposition can be retrieved and does not need to be explicitly declared in patterns.

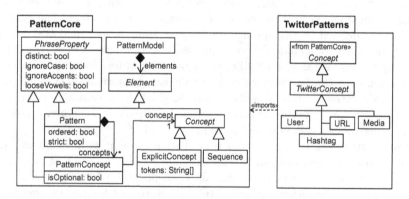

Fig. 4. Simplifed excerpt of the meta-model to describe patterns

The class Concept actually inherits from ResultSet (not shown in the figure), which allows matched concepts to be sent as external data (see next section), and referenced in queries. As can be observed, we have separated the Concept definition from their usage within Patterns, which permits their reuse in different contexts. An intermediate class PatternConcept enables the configuration of how the concept is to be used, specifying for example whether it is optional in the pattern. Patterns also indicate if concepts have to appear in some specific order, or allow the interleaving of concepts with other words. It is also possible to specify that some concept cannot occur in a pattern, and whether concepts are to be sought ignoring upper/lower case, ignoring accents, and permitting missing vowels, as this is a usual idiom in posts, due to their restrictive length in some cases (especially in tweets).

The DSL has been provided with a simple textual syntax. Figure 5 shows some example patterns and concepts for the voting application mentioned in Section 2. The goal of the application is to detect positive or negative votings to simple questions, and the patterns detecting it are defined in lines 1–2. They are tagged as ordered meaning that concepts should appear in order, but are not strict, so that other elements may occur in the tweet, possibly interleaved with the concepts of the pattern. Only the yes and the no concepts are mandatory, while the question concept is optative as indicated by the ? symbol. Moreover, for all concepts in the pattern we ignore whether they are in upper/lower case (flag ignore-case), and admit variations with missing vowels (flag loose-vowels). The patterns make use of the concepts of lines 4–6. The words making concept *question* were actually taken from WordNet synonyms, as we will see in Section 6.

```
1   pattern voteY [ordered loose−vowels ignore−case] = {question?, yes}
2   pattern voteN [ordered loose−vowels ignore−case] = {question?, no}
3
4   concept question = {question, inquiry, enquiry, query, interrogation}
5   concept yes = {yes, affirmative, 1, true, y}
6   concept no = {no, negative, 0, false, n}
```

Fig. 5. Patterns for the voting example

Altogether, *voteY* matches tweets like *"My vote 4 the qstn is YES!"*. In this case, the actual value of concept question is qstn while the value of yes is YES.

5   A DSL for Queries and Actions

Our approach considers the description of actions by means of another DSL. Its meta-model is split in a core package which includes those elements typically applicable to OSNs, and a package specific to Twitter. An excerpt of both packages is shown in Figure 6 (many concrete classes have been omitted, leaving only the base abstract classes of the main hierarchies). The action DSL uses the patterns DSL to perform queries on the concepts found in posts matching a certain pattern, and in addition supports other actions, namely:

– **Querying.** Queries can be issued using an SQL-like syntax. They may refer to a set of matches of a pattern, as if they formed an SQL table, and

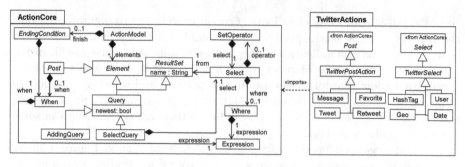

Fig. 6. An excerpt of the actions meta-model with the Twitter extension

```
1    query countY : count yes from voteY;
2    query countN : count no from voteN;
3    @newest query users_voting : users from voteY union users from voteN;
4    message thanku ("Thanks for your vote.") to users_voting;
5
6    tweet presults ("Partial results: (%s) yes, (%s) no", countY, countN) when (countY + countN) = 15;
7    tweet results ("Results: (%s) yes, (%s) no", countY, countN) when (countY + countN) = 30;
8    finish end: when (countY + countN) = 30;
```

Fig. 7. Actions for the voting example

the concepts in the pattern, as if they were SQL columns (as explained before, Concept inherits from ResultSet to this end). Three kinds of specialized queries can be issued. *Pattern queries* to select some concepts from a set of posts matching a pattern (represented as SelectQuery, which takes data from a result set obtained in the pattern matching phase), *aggregation queries* (class AddingQuery, to perform some arithmetical operation on result sets, like counting, and *metadata queries* (class TwitterSelect in the case of the Twitter package), to obtain a result set made of some tweet metadata, like its users, geopositions, images, or hashtags. Aggregation queries can calculate the maximum, minimum, average, count or add elements in a result set. Every query inherits from ResultSet and has a name, so that its results can be obtained from other actions. We currently support three kinds of operations (subclasses of SetOperator) with result sets from queries: union, intersection and substraction.

For example, Figure 7 shows two *adding* queries (lines 1–2) and one metadata query (line 3). The former counts the tweets matched by the voteY or voteN patterns. The later gathers the users that issued a tweet matching pattern, performed by a union operator.

While queries are similar to SQL queries, the data gathered from Twitter is dynamic. Hence, similar to data stream management systems [3] we may query using temporal windows [11]. Currently, we support two kinds of temporal windows, one considering all data, and another one with the last tweet. The two queries in lines 1 and 2 simply consider all data from the beginning of the execution. The users_voting, being labelled with @newest, discards an incoming tweet as soon as considers it has been processed by a dependent action. Other temporal windows could be interesting as well, and we will consider them in future work.

- **Composing and Sending Messages**. Once data becomes available from queries, messages can be composed and sent to a collection of users. This is reflected by class Post and its specialization TwitterPostAction. In the case of Twitter, messages can be public (class Tweet), or private, directed to a certain user (class Message). In addition, received tweets can also be retweeted, and be categorized as favorite. Public messages can be sent to a number of users (obtained through a query), in which case the messages contain a mention to those users. Messages may have a trigger (class When), so that the message is sent when the trigger becomes true.

  As an example, in Figure 7, line 4 sends a private message to the user that has voted. The user is obtained from query users_voting, which obtains the user of the last positive or negative vote. Lines 6 and 7 show the construction of two tweets. Similar to C's printf function, the tweet is composed by inserting data (countY, countN) into a string, in the places indicated using '%s' (independently of the variable type). Both tweets have trigger conditions, so that they are sent when the number of votes reach 15 and 30 respectively. In this case, they will not mention any user, but we could synthesize tweets mentioning users, in a similar way to private messages.

- **Execution End**. We also provide means to signal the end of the application execution (class EndingCondition and omitted subclasses). The condition may depend on several factors, like the number of tweets received that match a certain pattern, or time conditions. In Figure 7, line 8 shows a finish condition, when the number of votes reaches 30.

In this DSL, each action has a name, so that actions can refer to the data they produce simply by that name. The type of data does not need to be declared, but it is inferred by simple rules: data produced in a tweet match is considered String, and adding queries produce Integer data.

The execution model of the DSL is based on data flow, relying on data dependencies. This is the recommended execution model for reactive, event-driven, scalable applications [19]. In this way, an action is performed as soon as its data becomes available, unless it contains an explicit trigger, in which case it is executed when the data is available *and* the trigger becomes true. Figure 8 shows the data dependencies obtained from the example of Figure 7. The squares refer to the different actions (patterns, queries, messages, tweets and finish). Every action may produce some data, as indicated by solid arrows, and may depend on some data for their execution (shown by dashed arrows). The data produced by queries is assigned the same name as the query. Actions may have triggers, depicted as black circles (for simplicity we omit the actual trigger condition). For example, the thanku message can be sent when the users_voting data becomes available, which is produced by the users_voting query. Such query depends on the user metadata produced by tweets matching patterns voteY or voteN. Hence,

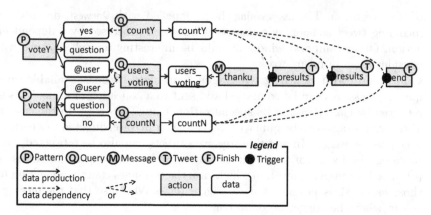

Fig. 8. Data flow execution model for the example of Figure 7

users_voting needs to be reexecuted whenever a tweet matching voteY or voteN arrives. We forbid cycles of dependencies to avoid deadlock situations.

Our prototype does not make use of any data stream management system or stream processing library, but as future work we plan to improve its efficiency and scalability by compiling our DSL programs to a selected system.

Finally, in order to facilitate the connection with external information systems, the DSL permits declaring external data dependencies (expected data from an existing information system), and also data that is pushed into the information system. The former are asynchronous events, triggered by the external source, which provide some data to the model. Currently, we generate service interfaces from those two descriptions, one for the existing information system (to obtain the data from), and another for the tweet processing system (so that the information system can notify our system). The former service interface needs to be manually integrated with the existing information system.

6 Tool Support

In this section, we present a prototype modelling environment, built as an Eclipse plugin. The environment permits describing patterns (Section 4) and actions (Section 5) using both DSLs, test the application in the environment, and then deploy it on a server. More information about the tool, including a video of the tool in action can be found at: http://www.miso.es/tools/twiagle.html.

The tool supports an agile development method, as during development, the user may test the patterns against tweets received in real-time, and send testing tweets using the tool itself. Figure 9 shows the tool being used to define some patterns corresponding to the running example. The question concept is being defined through the use of WordNet, and a contextual window offers the different synonym sets (i.e., the different *senses* for a given concept) for the given concept. Selecting one synonym set makes the tool copy all words into the concept.

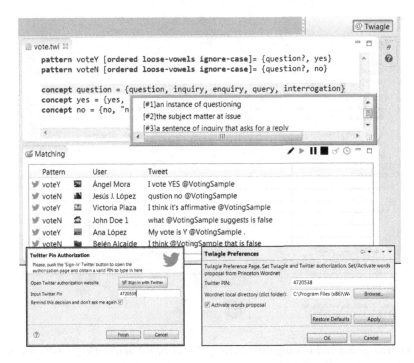

Fig. 9. Defining and testing patterns against live tweets

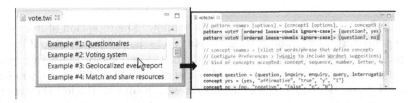

Fig. 10. Twiagle assistant suggesting different application kinds

The tool is designed for its use by both software engineers and non-experts. For this purpose, it incorporates an assistant that suggests typical examples for a range of applications, as Figure 10 shows. Upon selection of one application kind, a skeleton example is presented, with explanations in the form of comments. The range of examples have been extracted from an analysis of current web applications, details of which are available from the tool web site.

The tool can be used for testing purposes, and the left-bottom of Figure 9 shows a connection window, where the user needs to authorize the tool to access the Twitter account associated to the application (@VotingSample in the figure). The tool enables storing such authorization pin in the tool preferences registry, so that it does not need to be entered for every execution. Once Twiagle is authorized, tweets can be obtained and matched against the defined patterns, as

Fig. 11. Executing and tracing actions (up). Tweet-based application results shown in the Twitter console (bottom).

shown in the *"Matching"* view of the main window. The reception of live tweets can be paused, resumed, and testing tweets can be issued from the tool itself.

Figure 11 shows the tool being used to define the actions for the voting example, and test them over live tweets, so that the results of the different queries are visualized. In particular, the tool offers two trees, the one on the left showing the results of the different queries, while the one to the right indicates the actions being executed. The latter capability is very useful for debugging purposes, and as an execution log for analysis. The bottom of the figure shows the Twitter web console, showing the emitted tweet with the partial results (left), and a private message confirming the vote for a user (right).

While applications can be executed in the development environment, for integration purposes, they can also be deployed on a server. For this purpose, the tool generates skeleton service interfaces from the external data dependencies, which need to be completed manually. We consider both REST and SOAP web services, but this feature is currently under development.

7   Related Work

This section compares related work along several axes: (i) approaches to build social applications over OSNs, (ii) analysis of social interactions in OSNs, (iii) applications for OSNs, built manually, (iv) processing posts, and (v) approaches to create specific kinds of social applications, e.g. for crowsourcing.

(i) Tools for Creating Social Applications. As we have seen, a social application uses the infrastructure of OSNs to enhance their reach and dissemination [13]. We find several approaches to build such kind of applications: OSN-specific, multi-channel, or based on web-engineering systems.

In the first kind, in [13] the authors propose a flow metaphor to concatenate actions to be performed on resources of an OSN, like posting a comment, or uploading a photo. Similar to ours, that framework is amenable for non-experts, as it abstracts from low level programming tasks based on APIs. However, we are directed to tweet-based applications, having a dedicated DSL to detect patterns, and we allow the connection with external sources.

Concerning multi-channel approaches, IFTTT[2] is a mashup approach to construct simple "recipes" to automate tasks from different *channels* (like Facebook or Twitter), like "tweet my facebook status updates". In a similar vein, in Zapier[3] users write graphical transformation rules between different web services, in which the input data may come from a general-purpose web service or OSN, and the output data can be another web service. Compared to our work, we enable the definition of more complex tasks, involving e.g., queries, and we make available a dedicated DSL for pattern-match and consider mechanisms for connection with other sources, hence enabling the integration of existing information systems with social applications. A more advanced way to integrate web services is proposed by the MuleSoft commercial tool[4], however it is a heavyweight solution, requiring from expert knowledge in software architecture and design.

Finally, BPM4People [5] extends the BPM language for enabling to model complex data flow, coming from a social domain. The approach is integrated with WebRatio[5]. This is useful because, ever more frequently, web applications need to be extended with features enabling their interaction with OSNs. For this purpose, in [6] an extension of WebML is proposed, to incorporate social primitives, permitting either cross-platform operations (login and search) applicable to several social networks, and specific actions, e.g., to send a tweet.

All these tools differ from our approach in the level of expertise required from the user, since being generic web-engineering tools, they are generally expert-user, or IT staff, oriented. Our proposal enables non-expert - or with a really short training - users to collect data from OSNs immediately for using in it in a domain-specific field that they can define by themselves. This is facilitated by our use of DSLs specific for OSN applications (for pattern definition, actions).

(ii) Analysis of Social Interactions in OSNs. Twitter has been used in numerous recent academic works, studying its network structure [12,24], or its use as social medium for communication. As an example of the latter, in [23] a classifier was developed to detect messages contributing to situational awareness,

---

[2] https://ifttt.com/
[3] http://zapier.com/
[4] http://mulesoft.com/
[5] http://www.webratio.com/

using a combination of manually annotated and automatically-extracted linguistic features. In [15] an analysis of tweet content and retweet behaviour during the Fukushima nuclear disaster is performed. In our case, our goal is to detect simple patterns in text, so for now we did not use sophisticated natural language processing (NLP) techniques. We believe frameworks similar to ours could be very valuable as a basis to build this kind of applications.

(iii) Applications for OSNs. Some works aim at using Twitter to create new applications, but they are normally built manually with no automated support from frameworks like ours. This has the drawback that developers need to deal with complicated programming concepts, directly manipulating the OSN API. For example, in [20], tweets are used to detect earthquakes in Japan, by classifying tweets according to whether they convey the occurrence or not of an earthquake. While this application was built ad-hoc, it could have benefited from an automated framework, like ours, for their construction.

(iv) Processing Posts. Tweets contain unstructured information, and in order to make them computer-processable, some works [22,8] have proposed the incorporation of structured information in tweets, while others use NLP techniques. In the first group of works, in [22], a simple workflow language is proposed, which combines SOA principles within Twitter. Therefore, SOA primitives, e.g., for service discovery or service binding are embedded in tweets, and Twitter is used as a means to reuse existing infrastructure. In this way, tweets may be used to invoke services and coordinate crowd-sourced activities that are required to fulfill a certain task. However, tweets need to follow a strict syntax, and are not suitable for a natural human-machine or machine-human communication. HyperTwitter [8] proposes the use of Twitter for collaborative knowledge engineering. For this purpose, it defines RDF-like syntaxes to be detected on tweets, which are converted into RDF statements. As our goal is not knowledge engineering, but building tweet-based applications, we support the definition of a richer syntax for tweets, and also provide the machinery for the rapid creation of tweet-based applications, as well as an advanced MDE-based developing environment.

Regarding the use of NLP, EquatorNLP [7] uses deep NLP and machine learning to extract relevant facts of posts in social networks during an emergency situation; and [14] proposes an hybrid technique to recognise named entities. Other works, like [2] are concerned with efficient processing of streams of tweets. We will consider in future work the possibility to integrate some of these techniques in our framework. Tweets are limited to 140 characters, and Twitter users tend to use slang, or abbreviations. Hence, in order to use NLP techniques in unrestricted domains, normalization techniques, like those in [17], are targeted to ammend words with missing vowels. In our case, the recognision task is much simpler, because we define patterns a priori, and enable variations of their concepts (e.g., missing vowels, or distinct letter case).

(v) Crowdsourcing Applications. There are approaches to facilitate the construction of crowdsourcing (or human-computation) applications [10,4,1] a kind of social application, based on the distribution of tasks among a high number of human participants. For example, *we flow* [10] is an approach to facilitate the creation of human computation applications, based on the availability of a coordination language, and a generator that synthesizes a collaborative web application from such specification. Similar to our work, *We Flow* is directed to empower users with the ability to create their own applications. However, we use the OSN capabilities of Twitter, make available DSLs, and enable the connection with existing information systems. In [4] a model-based approach to systematize the definition of crowdsourcing applications is proposed. It is based on modelling task types and the interaction with performers, so that the system guarantees certain properties. For some applications, their models could be compiled to our DSLs to obtain a Twitter-based crowdsourcing applications.

Altogether, we are witnessing an increasing interest in both, the construction of all sorts of social applications, and in the analysis of the interactions produced in ODNs. We believe that an MDE framework, like ours would greatly help in these two aspects, as otherwise expert programmers are needed to deal with the intricacies of the OSN API and pattern detection.

8   Conclusions and Future Work

In this paper, we have introduced the concept of post-based applications: applications whose inputs and outputs are extracted and produced from OSNs, like Twitter. We have shown some scenarios where those applications are useful, and demonstrated the feasibility of their construction through an MDE approach. We have presented a prototype realization, targeting Twitter.

Even though we target short messages, we would like to increase the expressiveness of our pattern DSL, considering more advanced NLP techniques for pattern match. We are currently improving our action language with new primitives (e.g., for presenting outputs in maps or in charts) and taking inspiration from data-stream systems for tweet querying. We are currenlty working on improving our tool, in particular the deployment mode, and considering a web-based version of the tool. We are integrating other OSNs, in addition to Twitter, which would allow from inter-platform applications. Finally, we are designing more specific languages for certain applications on top of our DSLs, like for mobile learning.

Acknowledgements. Work supported by the Spanish Ministry of Economy and Competitivity with project Go-Lite (TIN2011-24139).

References

1. Ahmad, S., Battle, A., Malkani, Z., Kamvar, S.: The jabberwocky programming environment for structured social computing. In: UIST, pp. 53–64. ACM (2011)
2. Asadi, N., Lin, J.: Fast candidate generation for real-time tweet search with bloom filter chains. ACM Trans. Inf. Syst. 31(3), 13 (2013)
3. Babcock, B., Babu, S., Datar, M., Motwani, R., Widom, J.: Models and issues in data stream systems. In: PODS, pp. 1–16. ACM (2002)

4. Bozzon, A., Brambilla, M., Ceri, S., Mauri, A.: Reactive crowdsourcing. In: Proc. 22nd Int. Conf. on World Wide Web, pp. 153–164 (2013)
5. BPMN4People., http://www.bpm4people.org
6. Brambilla, M., Mauri, A.: Model-driven development of social network enabled applications with webml and social primitives. In: Grossniklaus, M., Wimmer, M. (eds.) ICWE Workshops 2012. LNCS, vol. 7703, pp. 41–55. Springer, Heidelberg (2012)
7. Döhling, L., Leser, U.: EquatorNLP: Pattern-based information extraction for disaster response. In: Foundations, Technologies and Applications of the Geospatial Web, pp. 127–138 (2011)
8. Hepp, M.: Hypertwitter: Collaborative knowledge engineering via twitter messages. In: Cimiano, P., Pinto, H.S. (eds.) EKAW 2010. LNCS, vol. 6317, pp. 451–461. Springer, Heidelberg (2010)
9. Hyman, P.: 'Peace technologies' enable eyewitness reporting when disasters strike. Commun. ACM 57(1), 27–29 (2014)
10. Kokciyan, N., Üsküdarli, S.M., Dinesh, T.B.: User generated human computation applications. In: SocialCom/PASSAT, pp. 593–598. IEEE (2012)
11. Krämer, J., Seeger, B.: Semantics and implementation of continuous sliding window queries over data streams. ACM Trans. Database Syst. 34(1) (2009)
12. Kumar, S., Morstatter, F., Liu, H.: Twitter Data Analytics. Springer Briefs in Computer Science. Springer (2014)
13. Jara, J., Daniel, F., Casati, F., Marchese, M.: From a simple flow to social applications. In: Sheng, Q.Z., Kjeldskov, J. (eds.) ICWE Workshops 2013. LNCS, vol. 8295, pp. 39–50. Springer, Heidelberg (2013)
14. Li, C., Sun, A., Weng, J., He, Q.: Exploiting hybrid contexts for tweet segmentation. In: SIGIR, pp. 523–532. ACM (2013)
15. Li, J., Vishwanath, A., Rao, H.R.: Retweeting the Fukushima nuclear radiation disaster. Commun. ACM 57(1), 78–85 (2014)
16. Miller, G.A.: Wordnet: A lexical database for english. CACM 38(11), 39–41 (1995)
17. Porta, J., Sancho, J.-L.: Word normalization in twitter using finite-state transducers. In: Tweet-Norm@SEPLN, CEUR, vol. 1086, pp. 49–53 (2013)
18. Qualman, E.: Socialnomics: How Social Media Transforms the Way We Live and Do Business, 2nd edn. Wiley, Chichester (2012)
19. Reactive manifesto, http://www.reactivemanifesto.org
20. Sakaki, T., Okazaki, M., Matsuo, Y.: Tweet analysis for real-time event detection and earthquake reporting system development. IEEE Trans. Knowl. Data Eng. 25(4), 919–931 (2013)
21. Segura, Á.M., de Lara, J., Cuadrado, J.S.: twiagle: A tool for engineering applications based on instant messaging over twitter. In: Casteleyn, S., Rossi, G., Winckler, M. (eds.) ICWE 2014. LNCS, vol. 8541, pp. 536–539. Springer, Heidelberg (2014)
22. Treiber, M., Schall, D., Dustdar, S., Scherling, C.: Tweetflows: Flexible workflows with twitter. In: PESOS 2011, pp. 1–7. ACM (2011)
23. Verma, S., Vieweg, S., Corvey, W., Palen, L., Martin, J., Palmer, M., Schram, A., Anderson, K.: Natural language processing to the rescue? extracting "situational awareness" tweets during mass emergency. In: ICWSM. AAAI Press (2011)
24. Zhao, F., Tung, A.K.H.: Large scale cohesive subgraphs discovery for social network visual analysis. PVLDB 6(2), 85–96 (2012)

Introducing the Public Transport Domain to the Web of Data

Christine Keller, Sören Brunk, and Thomas Schlegel

TU Dresden - Junior Professorship in Software Engineering
of Ubiquitous Systems, Germany
{christine.keller,soeren.brunk,thomas.schlegel}@tu-dresden.de

Abstract. The public transport domain generates large amounts of structured data. Making that information available on the Web of Data and linking it to other data sources can enable new services and applications for the benefit of passengers as well as public transport providers. Most standard data models in the public transport domain lack explicit semantics and interoperability because they are modeled in an informal or implementation-centric way. In this paper, we describe the development process of an OWL-ontology based on existing data models and standards in the domain. We show that our ontology enables the development of advanced passenger information systems and we briefly illustrate the application in a tourism-themed prototype.

Keywords: ontology, public transport, linked data.

1 Introduction

The public transport domain is a field that generates and uses large amounts of structured data. This includes timetables, infrastructure data and real time data concerning vehicles and events, such as disruptions or delays. However, we are convinced that the full potential of the data has not been utilized yet. Our goal is to unlock that potential by modeling the data in an ontology and linking it to the Web of Data. On top of this data, we are developing advanced passenger information systems using Semantic Web technologies. One of the big challenges faced when designing innovative passenger information systems is the heterogeneity of the data. There have been several efforts to standardize data models for the public transport domain to support interoperability and facilitate passenger information. In the European Union there are standards like the Reference Data Model for Public Transport, **Transmodel** [2] or the Service Interface for Real Time Information **SIRI** [4]. The General Transit Feed Specification Reference (GTFS) is a data model developed by Google to utilize public transport data [10]. These approaches use very different modeling techniques ranging from graphical notations to XML Schema models. However, due to the different level of abstraction, varying, partly competing standards and different, often proprietary implementations, the integrated usage of public transport data still proves difficult. Combining it with other data sources, as found in the Web of Data, is even more complicated. In order to provide an integrated view on the public

B. Benatallah et al. (Eds.): WISE 2014, Part II, LNCS 8787, pp. 521–530, 2014.

transport domain, we have created an OWL-ontology that builds upon the standards in the field and leverages the semantic potential of public transport data. We have developed a prototype that integrates public transport data and other linked data sources using our ontology. It provides a smart mobile application that computes interesting routes for tourists, relying on public transport [12].

This paper is structured as follows: In Section 2, we describe the data models in the field that form the basis of our ontology. We also look at existing ontologies for the public transport domain as well as linked data sources that contain public transport information. In Section 3, we describe our development process for the ontology in several steps and present some core concepts of our ontology. We also discuss the application of the ontology in our prototype. We conclude the paper with a discussion and an outlook in Section 4.

2   Related Work

In this section, we introduce existing standards and data models for the public transport domain. We also describe existing ontologies that model public transport entities and present linked and open data approaches in the domain.

2.1   Data Models in Public Transport

Numerous standards for various areas of the public transport domain exist mainly to improve interoperability of public transport systems. In the following, we focus on Europe for its dense and highly connected network of public transport services. The Transport Protocol Expert Group (**TPEG**) as a European institution, defined a standard for traffic and travel information in broadcasting that includes classifications of modes of transport, event reasons and other entities of public transport [19]. It consists of a terminology in textual form and tables, including translations into different languages, and was applied in other data models for public transport. **Transmodel** is one of the earliest standard data models for public transport, resulting from a joint European effort in the 1990s and standardized by the European Committee for Standardization (CEN) [2]. It is the foundation of many other standardization efforts in Europe and was revised and complemented in 2001 [3]. It describes a conceptual model of the entities and their relations in public transport and covers, among others, tactical planning, personnel disposition, passenger information and fare collection. Transmodel mainly consists of a textual or tabular description of the defined entities, supplemented by entity-relationship diagrams and later complemented by UML diagrams. The "National Public Transport Access Node" (**NaPTAN**) is a UK based standard, which identifies and models access points to public transport [6]. Its terminology refers to Transmodel. Using NaPTAN, it is possible to describe the type of a stop point (e.g., bus stop) and its attributes (e.g. location, entrances). However, it does not cover other aspects of the public transport domain, such as realtime information. The schema is based on UML, also providing an XML Schema Definition for data exchange as well as a CSV schema for tabular representation. Similar to NaPTAN, there are other national standards, partly based on Transmodel, for example the french standard "Norme d'Echange Profil Transport collectif utilisant la Normalisation Européenne" (**NEPTUNE**) [5] or German standards such as the "ÖPNV Datenmodell" (**VDV 450**) [1]. On European level, there have been several efforts to unify and combine these standards as described in the following.

The "Identification of Fixed Objects in Public Transport" (**IFOPT**) is a CEN EN Technical Standard that extends Transmodel and is based on NaP-TAN, modeling access points to public transport in greater detail [7]. IFOPT describes fixed objects such as stop places, station buildings, points of interest and their physical properties and parts, such as entrances, quays or stairs. It is mainly focused on providing data for interchanges on trips. The IFOPT specification includes UML diagrams as well as an XML Schema Definition. The "Service Interface for Real Time Information" (**SIRI**) standard is also standardized by CEN and describes an interface and protocol for real time information in public transport [4]. It is based on Transmodel as well and adds means to describe real-time information about timetables, delays, situations and vehicle locations. SIRI focuses on providing a protocol that allows the exchange of realtime information as update to a planned timetable that is used as reference. For this purpose, it defines Web Services in WSDL and a data model as XML Schema Definition. One of the newest efforts and currently in development as CEN standard is the "Network and Timetable Exchange" schema (**NeTEx**) [8]. The main goals of NeTEx are to facilitate the exchange of the network topology and timetable data of public transport networks. NeTEx incorporates Transmodel and IFOPT data structures and provides mappings to other standards, including SIRI and several national standards. It is also modeled as XML Schema Definition.

Google launched its own effort to create a unified data model for public transport data, in particular for public transport schedules and realtime information. The "General Transit Feed Specification Reference" (**GTFS**) models static data such as routes and schedules, whereas the General Transit Feed Specification-realtime Reference (GTFS-realtime) covers the realtime components [10,9]. The GTFS standard defines a data format based on text files with comma separated values. While GTFS is simple and has been adopted by many public transport providers to publish their data, it only describes a small subset of the public transport domain. Furthermore, the CSV based data model makes extension to other parts of the public transport domain or integration with other schemas and datasets difficult.

Although standardization efforts have brought many advantages for the field, (e.g. a unified terminology and exchange protocols), the existing standards are not well suited for the integration of the public transport domain with the Web of Data, mainly because none of the existing models build upon Semantic Web standards, limiting their interoperability with linked data severely. Some of the standards are implementation templates but cannot be used as a formal data model, others describe more formal models, but are intended for a very specific purpose. They are also usually developed by and for public transport providers, thus exposing a lot of the complexity of the domain.

2.2 Ontologies for the Public Transport Domain

We found very few ontologies covering the public transport domain. Houda et al. have developed a public transport ontology to support multi-modal transportation planning [11]. While they describe some of the main concepts of public transport, they focus on planning aspects and public transport infrastructure concepts for their use case. Similar to Houda et al., Mykowiecka et al. cover only small aspects of the public transport domain, in their case to build a dialog system for public transport [13]. Both ontologies do not refer to existing public transport standards. Plu and Scharffe address a workflow and framework for publishing transport data as linked data and also describe two ontologies they developed for that purpose [16]. The first is mainly concerned with road traffic and the second is based upon the french NEPTUNE standard mentioned above. It covers a basic network description of public transport systems, but passenger information is only included to a very limited extent.

2.3 Public Transport as Linked Open Data

There are some datasets available as linked open data, that contain public transport data. NaPTAN data, for example, has been converted to and made available as linked data[1], based on an RDFS vocabulary[2]. The instance data is also available online as open data and can be queried via SPARQL. The vocabulary covers only public transport stops and is specific to the UK transportation system. For GTFS, there exists an RDFS vocabulary called TRANSIT, which models the GTFS terms as classes and their connecting properties[3]. Some vocabularies covering broader domains include a number of public transport concepts. The schema.org[4] vocabulary as an example, contains several public transport infrastructure terms such as BusStop or TrainStation. Likewise, the LinkedGeoData ontology contains public transport infrastructure concepts [18]. It has a focus on geospatial information and comes with a huge amount of community created instance data based on data of the OpenStreetMap[5] project.

3   An Ontology for Public Transport Services and Data

In the following sections, we describe the development process of our ontology, based on the analysis of existing data models. We then present some core concepts of our ontology and discuss our prototype.

---

[1] http://transport.data.gov.uk/

[2] Available at http://transport.data.gov.uk/def/naptan

[3] Available at http://vocab.org/transit/terms/

[4] http://www.schema.org/

[5] Available at http://www.openstreetmap.org/

3.1 Ontology Development

We roughly followed the methodology outlined by Noy and McGuiness in their "Ontology Development 101" Guide [14]. They propose an iterative process in seven steps. We adapted each step to benefit from domain knowledge in existing standards and iterated several steps, particularly regarding reviews and input from domain experts, provided by project partners.

Step 1: Scope of the Ontology and Competency Questions

First, we conducted a general requirement analysis for passenger information systems in public transport jointly with our project partners. Based on interviews with domain experts, reviews of statistical information and focus groups, a catalog of requirements was established. In order to outline the scope of our ontology, we phrased competency questions, for example *"How should a user travel from A to B so she may get her shopping done on the trip?"* or *"Where are restaurants for a romantic dinner in medium price range, open tonight, that are reachable using the bus lines 7 or 10?"*. We determined the scope of our ontology as passenger information in public transport.

Step 2: Consider Existing Ontologies

In our survey of existing transport ontologies described in section 2.2, we did not find ontologies that covered our requirements. We decided to rely upon existing ontologies for some concepts not specific to public transport. We integrated the geographical schema of GeoSPARQL [15] and used the vCard Ontology[17] to describe postal addresses.

Step 3: Terminology and Conceptualization

In order to model the central concepts of the ontology in a class hierarchy, they first have to be selected and named. In this step, we extracted common concepts and terminology from Transmodel, TPEG, NaPTAN, VDV 452, IFOPT, SIRI as well as NeTEx. We then identified the concepts and data structures relevant for advanced passenger information systems.

Step 4: Classes and Class Hierarchies

Based on the identified concepts, we modeled the central taxonomy of our ontology by creating the first OWL classes. We then followed a top down approach to define their subclasses. For each class, we examined corresponding concepts in the different standards. In those data models, concept hierarchies can be found as flat hierarchies in tables, expressed in UML or as entity-relationship diagrams and in XML schemas. We selected the subclasses that are relevant for the scope of our ontology and added them to our class hierarchy.

Step 5 and 6: Properties and Constraints

In the next two modeling steps, properties and constraints are to be defined. For datatype properties, we were able to rely on the XML schemas that contained datatype properties for core elements. We also extracted and applied some properties from UML and entity-relationship diagrams of existing data models. Based on an additional analysis of real data from public transport systems, we complemented our ontology with relevant object properties. For further refinement, we added domain and range to the properties as well as cardinalities. In case of datatype properties, those constraints are often included in the XML schemas or in the UML and entity-relationship diagrams of the standard data models. However, most of the constraints are only modeled implicitly in the domain standards and we relied on domain knowledge of experts in reviews and the examination of data from public transport systems to extract them.

Step 7: Instances

The last step in the development process is the integration of instances into the ontology. First, there are some instances that are fixed for the domain, for example, reasons for situations. We took these from existing enumerations, in this case from TPEG tables. The second type of instances is application specific, for example, trips or lines. Depending on the application that uses the data, scope and detail level of the required instances may vary.

3.2   Core Concepts

The basic structure in public transport is a **trip**. Our ontology provides a class Trip that is shown in Figure 1. The properties hasOrigin and hasDestination define origin and destination of a trip as a Site. Usually, a trip takes a person

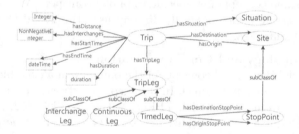

Fig. 1. Trips in public transport

from one point to another, but origin and destination are rarely public transport stops. In most cases, people will have to walk some distance or even drive by car before and after the journey in public transport vehicles. A trip therefore not only consists of possibly several passages in public transport vehicles, but also of routes to, from or between stop places. These are called legs and a trip can have several trip legs. The class TripLeg has the subclasses TimedLeg, ContinuousLeg and InterchangeLeg, whereas the last represents the path of an interchange. A timed leg is covered by a public transport vehicle and a continuous leg is covered on foot or by other vehicles (e.g. driving by car or bicycle). A timed leg therefore always has a stop point as origin and destination and the subproperties hasOriginStopPoint and hasDestinationStopPoint of hasOrigin and hasDestination represent this relation. Several properties of a trip, such as the covered distance or the overall duration, are modeled as DatatypeProperty.

Changing one's **location** is the purpose of transportation and therefore location is a key concept. To integrate GeoSPARQL into our ontology, we created a subclass of GeoSPARQLs geo:Feature class, Location, in order to describe locations in public transport, as shown in Figure 2 on the left. Using geo:Geometry, it is possible to describe the location as, for example, a GPS coordinate or polygon.

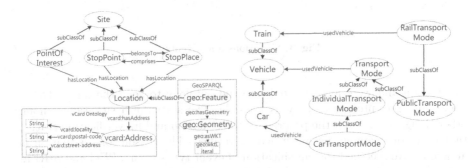

Fig. 2. Geographical Entities and Means of Transportation

Additionally, we use the class vcard: Address and properties such as vcard:-locality and vcard:street-address from the vCard Ontology[17] to represent postal addresses. As geographical object, the ontology defines a Site to which a location can be assigned by hasLocation. The class StopPlace, as subclass of Site, represents all kinds of stops for public transport services. A stop place in public transport can consist of several stop points. Those are the points at which vehicles stop and passengers board or alight from vehicles (cf. definition of "stop point" in Transmodel, [3]). In the ontology, the class StopPoint is also subclass of Site and the properties comprises and belongsTo are used to assign stop points to a stop place (see also Figure 2 on the left). Another specialization of Site is a PointOfInterest. The class TransportMode describes **means of transportation**. Figure 2 on the right shows that transport modes are distinguished in IndividualTransportMode and PublicTransportMode. These can be further specified as, for example, CarTransportMode for individual transportation, or RailTransportMode for rides by train. The property usedVehicle denotes the type of Vehicle for a transport mode. The specified transport modes are then defined using owl:Restriction to determine that, in the given example, for RailTransportMode, only Trains are used as vehicles.

In case of **delays or disruptions**, passenger information is crucial in public transport. Corresponding concepts are modeled using the Situation class, shown in Figure 3. A situation in public transport is classified by a Situation-Classifier, which includes an assessment of the severity of the situation (SituationSeverity), as well as the information whether it was planned, which can be true for drills, for example. The encapsulation of some situation properties in a Classifier and Status class stem from the data models that exist in

Fig. 3. Situations and Events

the domain and were incorporated into the ontology to keep the link between standards and ontology as easy as possible. The status of a situation, among other things, consists of the progress of the situation and its verification status, that may, for example, be unverified or verified. The reason for a situation is an Event. We included the classification of event types from TPEG and SIRI in our event model. Such events may be, for example, derailment or fuel shortage, both subclasses of EquipmentEvent. The property affects is used to identify the entities that are affected by the situation. Subproperties are specify affected elements, for example affectsStopPlace. They can be used to determine which particular trips have to be rerouted.

3.3 Prototype

In order to test the applicability of our ontology, we developed a prototypical system that implements an adaptive application for tourism. Our prototype consists of a server component and a mobile application, which is shown in Figure 4. The main features of the mobile application are to suggest points of interest in several categories, that are reachable in a given time using public transport. The determined points of interest are rated and filtered based on the user's preferences and context. Furthermore, the application allows the user to compute different themed tours for a given time frame, using public transport. The tours are also adapted to user preferences and context, such as weather conditions and daytime. Our mobile application is presented in greater detail in [12]. We request the Web service of a public transport provider and write its XML responses as RDF instances in our triplestore using an ontology-based conversion component. For data on points of interest, we extended our ontology with a tourism-themed ontology based on the schemas of LinkedGeoData and schema.org and combined data from the LinkedGeoData project in our triplestore with public transport instance data.

Our concept for context-adaptation and filtering of data relies on the exploitation of semantic data, describing different facets of the user's situation and information needs: the public transport situation, weather conditions and points of interests but also user preferences. Its ontological description also enables flexible and adpative presentations of relevant information. Additional features of the prototype are still work in progress, but it already shows that our ontology can be applied for enhanced passenger information systems in public transport and is suitable for integration of additional data sources with public transport data on an ontological level.

Fig. 4. Screenshots of the mobile application of our prototype

4 Conclusion

In this paper, we have presented an ontology for public transport built to integrate the public transport domain with the Web of Data and to make use of existing datasets for advanced passenger information systems. We described our approach to benefit from the domain knowledge contained in existing standards and data models. We have implemented a prototype that shows the application of our ontology for advanced passenger information systems. We have successfully integrated existing passenger information systems and have added data sources from the Web of Data to exploit common links and structures. We will further refine and improve the ontology while developing additional features of our prototype. Taking advanced passenger information systems into everyday life reveals new opportunities for context-adaptive and intelligent support of passengers in public transport.

Acknowledgments. Part of this work has been executed under the project IP-KOM-ÖV funded by the German Federal Ministry for Economic Affairs and Energy (BMWi) under the grant number 19P10003O as well as within the SESAM project under the grant number 100098186 by the European Social Fund (ESF) and the German Federal State of Saxony.

References

1. ÖPNV-Datenmodell Version 4.1. VDV Schrift 450, Verband Deutscher Verkehrsunternehmen (VDV), Version 4.1 (1996)
2. Road Transport and Traffic Telematics - Public Transport Reference Data Model (Transmodel). Technical Report prENV 12896, Technical Committee CEN/TC 278 (May 1997)
3. Reference Data Model For Public Transport (Transmodel). Technical Report ENV12896, Technical Committee CEN/TC 278 (June 2001)
4. Service Interface for Real Time Information (SIRI). Technical Report prCEN/TS 15531, Technical Committee CEN/TC 278 (2006)
5. Télématique de la circulation et du transport routier - profil d'échange neptune. Technical Report P99-506PR, AFNOR Group (December 2009)
6. National Public Transport Access Node (NaPTAN). Technical report, Technical Committee CEN/TC 278 (December 2010)
7. Technical Committee CEN/TC 278. Identification of Fixed Objects in Public Transport (IFOPT). prCEN/ Technical Standard prCEN/TS 28701, European Committee for Standardization, CEN (2008)
8. Technical Committee CEN/TC 278. Network Exchange - NeTEx. prCEN/ Technical Standard pre/CEN/TS-xxx, European Committee for Standardization (CEN) (July 2012)
9. Google. General transit feed specification-realtime reference. Creative Commons Attribution 3.0 (August 2011),
   https://developers.google.com/transit/gtfs-realtime/reference
10. Google. General transit feed specification reference (June 2012),
   https://developers.google.com/transit/gtfs/reference
11. Houda, M., Khemaja, M., Oliveira, K., Abed, M.: A public transportation ontology to support user travel planning. In: Loucopoulos, P., Cavarero, J.-L. (eds.) RCIS, pp. 127–136. IEEE (2010)
12. Keller, C., Pöhland, R., Brunk, S., Schlegel, T.: An adaptive semantic mobile application for individual touristic exploration. In: Kurosu, M. (ed.) HCI 2014, Part III. LNCS, vol. 8512, pp. 434–443. Springer, Heidelberg (2014)
13. Mykowiecka, A.: Time expressions ontology for information seeking dialogues in the public transport domain. In: Loftsson, H., Rögnvaldsson, E., Helgadóttir, S. (eds.) IceTAL 2010. LNCS, vol. 6233, pp. 257–262. Springer, Heidelberg (2010)
14. Noy, N.F., McGuinness, D.L.: Ontology development 101: A guide to creating your first ontology. Technical report, Stanford University (March 2001)
15. Open Geospatial Consortium. OGC GeoSPARQL - a geographic query language for RDF data. Technical report (2012)
16. Plu, J., Scharffe, F.: Publishing and linking transport data on the web. CoRR, abs/1205.1645 (2012)
17. R. Iannella, J. McKinney. vCard Ontology - for describing People and Organizations. Technical report, W3C (2014)
18. Stadler, C., Lehmann, J., Höffner, K., Auer, S.: Linkedgeodata: A core for a web of spatial open data. Semantic Web Journal 3(4), 333–354 (2012)
19. Transport Protocol Expert Group (TPEG). Traffic and travel information (tti), public transport information (pti) application. Technical Report CEN ISO/TS 18234-5:2006, European Committee for Standardization (CEN) - International Organization for Standardization, ISO (2006)

Measuring and Mitigating Product Data Inaccuracy in Online Retailing

Runhua Xu[1] and Alexander Ilic[2]

[1]ETH Zurich, Zurich, Switzerland
[2]University of St.Gallen, St.Gallen, Switzerland
rxu@ethz.ch,alexander.ilic@unisg.ch

Abstract. Driven by the proliferation of Smartphones and e-Commerce, consumers rely more on online product information to make purchasing decisions. Beyond price comparisons, consumers want to know more about feature differences of similar products. However, these comparisons require rich and accurate product data. As one of the first studies, we quantify how accurate online product data is today and evaluate existing approaches of mitigating inaccuracy. The result shows that the accuracy varies a lot across different Web sites and can be as low as 20%. However, when aggregating product information across different Web pages, the accuracy can be improved on average by 11.3%. Based on the analysis, we propose an attribute-based authentication approach based on Semantic Web to further mitigate online data inaccuracy.

Keywords: Data Quality, Data Accuracy, Product Data, Linked Data.

1 Introduction

Product information on the Web is gaining more importance to help consumers choose the right products from an overwhelming number of offerings. Thanks to the widespread availability of Smartphones, this is not only a phenomenon for online retailing but also in the rapidly growing category of online-influenced offline sales [7]. Consumers quickly want to understand the price and feature differences of similar products. In the European Union, there is also a regulatory component that forces manufacturers and retailers to publish rich information for all food products sold online from 2014 onwards [6].

However, there is currently no standard for product identification on the Web and also the standards for describing product attributes in a machine-readable form such as schema.org are still in their infancy. Product data is mainly published by merchants in a human readable form and is often inaccurate [4], which leads to consumer confusion and lose of sales [13]. While data quality has been intensively studied within organizations and also some studies have provided approaches for analyzing and consolidating data online, there is little insight into the quantitative dimension of the online product data quality. With fast development of Semantic Web and linked data technologies, we see the potential to leverage such technologies to mitigate online data inaccuracy.

B. Benatallah et al. (Eds.): WISE 2014, Part II, LNCS 8787, pp. 531–540, 2014.

Our work comprises three main contributions. First, we quantify product data inaccuracy on the Web and describe the extent of the problem by analyzing data of e-Commerce Web sites. In addition, we review and compare different mitigation approaches to improve the data accuarcy by aggregating product information across the Web. Finally, we propose a new approach based on Semantic Web, linked data and digital signature to mitigate online product data inaccuracy.

2   Related Work

2.1   Data Quality Dimensions

Data quality (DQ) is commonly regarded as a multi-dimensional concept and its definition differs depends on the context. Wang and Strong [15] categorized DQ into 16 dimensions like accuracy, completeness, and security. However in a quality model developed by Bovee et al. [3], different essential attributes, namely accessibility, interpretability, relevance and integrity, were used to evaluate DQ. Although researchers evaluated DQ from different aspects, they used some common dimensions. Knight and Burn [11] and Barnes and Vidgen [2] compared different DQ dimensions and found that data accuracy was the most important and frequently used dimension. Thus, data accuracy was used to evaluate data quality in the study.

2.2   Data Quality on the Web

Previous research on online data quality assessment mainly focuses on evaluating the quality of Web sites. Eppler and Muenenmayer [5] presented a methodology to measure DQ on the Web by using Web mining tools and site analyser. Such tools helped to quantitatively evaluate quality criteria like accessibility, consistency and speed in a Web site. But data accuracy can only be measured qualitatively based on user surveys. Barnes and Vidgen [2] developed a framework to assess quality of Web sites. Based on 376 online questionnaires, they evaluated and compared the quality of three largest online bookshops from aspects like usability, information and trust. Different from survey-based DQ assessment, Frber and Hepp [8] developed a framework to evaluate DQ quantitatively in the Semantic Web context. They calculated data quality scores according to the four chosen dimensions: accuracy, completeness, timeliness and uniqueness. The framework was applied on 1.3 million semantic triples of BestBuy. However, more than 90% of the triples are related to geographic data. The remaining triples are email addresses, phone numbers and opening hours instead of product data.

2.3   Mitigation Strategies to Improve Data Quality

Consumers get confused when they find multiple values on the Web for one product attribute. In previous research, mitigation strategies have been developed by aggregating data from multiple sources. In a framework developed by Mendes et al. [12], different approaches were used to pick up the correct value for each attribute. These approaches were 1) taking average, maximum or minimum values, 2) taking first, last or random values, 3) taking most frequently used values, and 4) taking values from specific URLs. Based on these approaches, the framework decided which values to keep, discard or transform when multiple values exist for a single attribute.

As people know the DQ problem on the Web is common and Internet data is never perfect [4], to the best of our knowledge, no research has quantitatively measured online product data accuracy and has shown how big the problem is. Furthermore, current mitigation approaches can help to choose the most possible correct value when data conflict happens. But no research has evaluated and compared their effectiveness on improving data accuracy. Consequently, our research tries to address these two research gaps.

3 Research Method

To analyze online product data accuracy, 15 products on major e-Commerce Web sites in Switzerland were selected. Information extraction technologies are available to retrieve product information automatically on the Web [1], however, in addition to extraction accuracy, it is still difficult to use them to link and compare data from different sources. One reason is that some attributes have different names or they are even presented in different languages on different data sources, and there is no simple way to semantically link them. Another reason is that product data is presented in pictures on some Web sites, which makes it difficult for current information extraction algorithms to work. Therefore, product data used in this study were extracted manually. To reduce the possible manual mistakes, all the data was cross-verified by two people.

Based on discussions with industry experts, electronics and high value goods from category Smartphone, laptop, digital camera, coffee machine and printer were selected as target products because 1) they are of high interest to consumers, 2) they have high relevancy for rich product data, and 3) most of their official specs are available online to serve as ground truth values. An official spec is defined as a product's spec sheet that is found on the product's official Web page. In each category, the top three most popular products on a countrywide well-known product comparison Web site were chosen. Product data that came from popular online retailers who sold the product was gathered. In the study, data was defined as accurate when an attribute value is exactly the same as the one in the official spec. It was defined as similar if an attribute value rounds the official one. Otherwise, it was regarded as wrong. For instance, if the screen resolution of a digital camera was 1.2 MP on its official spec, then 1.0 MP would be regarded as similar while 2.0 MP would be considered as wrong. Based on the above definitions, data accuracy was measured and the results were compared in Section 4.1.

Furthermore, mitigation approaches introduced in Section 2.3 were applied on the dataset. However, not all of them are suitable in online retailing environment.

First, the timeliness of product data on the Web is difficult to be identified. Second, based on our definition, taking average value actually moves data from accurate to similar. For non-numerical attributes, it is impossible to average out data with different values. Accordingly, four mitigation approaches were selected: taking maximum values (MAX), taking minimum values (MIN), taking most frequently used values (FREQ) and taking values from a specific URL (SURL). For non-numerical attribute values, MAX and MIN algorithms would take values according to their string lengths or alphabetical orders depends on the context. One of the most well-known e-Commerce shops, www.digitec.ch, was selected as the specific Web site in the SURL approach. Based on the result, we analyzed possible causes for online product data inaccuarcy and proposed an attribute-based authentication approach to mitigate the inaccuracy.

4   Result Analysis

4.1   Data Accuracy

We collected 3000 attribute values on 173 product Web pages from 66 distinct online retailers in November 2013. Attribute values on each Web page were compared with the ground truth values and then labelled as accurate or inaccurate. In this stage, both similar and wrong data were regarded as inaccurate. The accuracy of each product Web page was calculated by dividing the number of accurate values by the number of available attributes on that Web page.

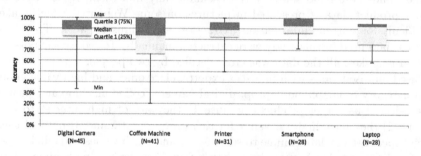

Fig. 1. Data accuracy of seleted products grouped by categories

Fig. 1 shows the overall data accuracy for each category. Smartphone has the highest data accuracy with a median at 93%. Laptop has a similar median accuracy (92%) but its first quartile accuracy is only 75%. Digital camera and printer have almost the same accuracy except that the minimum accuracy of digital camera is only 33%. Coffee machine has the worst accuracy: It has the lowest minimum accuracy (20%), first quartile accuracy (67%) and median accuracy (83%). Overall, 88.9% of all the attribute values were accurate.

Data inaccuracy existing in different attributes has various levels of impact on manufacturers, retailers and consumers. Therefore, attributes were further

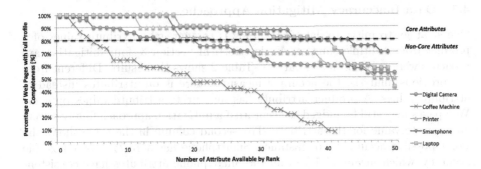

Fig. 2. Ranking attributes based on the percentage of Web pages that publish them

divided into core and non-core attributes based on popularity. A more sophisticated approach of categorization is required in further research.

As shown in Fig. 2, attributes of all products in each category were ranked according to the percentage of Web pages that publish them. Taking coffee machines for example, the first attribute has 100% on the y-axis, which means all the Web pages involved in the study that sell the product publish values for this attribute. As most Web pages publish values for the attribute, we assume it to be a core attribute. For the last attribute, conversely, only 7% of all the Web sites that sell the product publish values for it, which makes it to be a non-core attribute. According to Pareto principle, the boundary between core and non-core attributes is set to 80% in the study.

Table 1. Data accuracy of core attribute values gathered from 173 Web pages

Product Category	Core Values M(SD)	Accurate Values M(SD)	Pct.	Similar Values M(SD)	Pct.	Wrong Values M(SD)	Pct.
Digital Camera	102.3 (93.2)	85.7 (74.0)	83.7%	11.7(11.5)	11.4%	5.0 (8.7)	4.9%
Coffee Machine	21.0 (6.6)	18.7 (5.8)	88.9%	0.7 (0.6)	3.2%	1.7 (2.1)	7.9%
Printer	91.0 (20.0)	75.3 (8.3)	82.7%	9.4 (6.8)	10.3%	6.3 (5.5)	7.0%
Smartphone	125.3 (43.0)	116.0 (43.3)	92.6%	8.7 (3.1)	6.9%	0.6 (0.6)	0.5%
Laptop	119.3 (40.0)	100.3 (24.9)	84.1%	7.3 (3.2)	6.1%	11.7 (15.0)	9.8%
All Products	91.8 (57.4)	79.2 (44.4)	86.3%	7.5 (6.6)	8.2%	5.1 (8.0)	5.5%

After identifying core attributes, data accuracy was measured on core attributes as shown in Table 1. Taking the first row for example, each digital camera selected in the study has on average 102.3 attribute values gathered from different Web pages, and 85.7 or 83.7% of these values are accurate, 11.4% are similar and 4.9% are wrong. Smartphone has the highest accuracy with 92.6% while printer has the lowest accuracy with 82.7%. Laptop has the highest percentage of wrong values (9.8%). Overall, 86.3%, 8.2% and 5.5% of all the core attribute values are accurate, similar and wrong, respectively.

4.2   Data Inaccuracy Mitigation Approaches

One strategy to mitigate data inaccuracy is to aggregate attribute values of a product from multiple sources. As described in Section 3, four mitigation approaches were applied on the dataset. Table 2 shows the results. Different from the previous analysis, the accuracy measured here is on attributes instead of attribute values. A product attribute has inaccurate attribute values on some Web pages may still be judged as accurate if a mitigation approach finally selects an accurate value for the attribute. The second column in the table shows the worst possible accuracy. For instance, Smartphone has a 75.7% worst possible accuracy, which means 75.7% of all the Smartphone attributes have consistent and accurate values; only the remaining 24.3% attributes have multiple values on different Web sites. In the worst case, a mitigation approach chooses inaccurate values for all the attributes with multiple values, which makes the data accuracy to be 75.7%. However, if MAX is used, the accuracy will be improved to 82.4%. Compared to the worst case, the four mitigation approaches can on average improve the accuracy by 11.3%. FREQ performs best but it still leaves around 9% inaccurate attributes.

Table 2. Improvement of algorithms on mitigating data inaccuracy

Product Category	Worst Possible Data Accuracy	Improved Data Accuracy with Mitigation Approaches			
		MAX	MIN	FREQ	SURL
Digital Camera	88.2%	93.3%	88.9%	97.0%	89.1%
Coffee Machine	61.9%	73.8%	66.7%	80.9%	77.8%
Printer	80.5%	92.2%	84.4%	87.0%	91.4%
Smartphone	75.7%	82.4%	86.5%	93.2%	91.3%
Laptop	71.9%	82.5%	87.7%	85.9%	89.1%
Total Products	79.0%	87.3%	84.9%	90.9%	88.6%

4.3   Potential Causes for Data Inaccuracy

The results of the preliminary study provide insights into the potential causes for data inaccuracy. Thus, we generated an explanatory model to describe the information flow and impact of errors, as shown in Fig. 3. Some retailers get product data from manufacturers through distributors and publish it online, while others just copy and publish the existing online data. In the end, consumers face data from different retailers. We suspect four main types of errors in the information flow:

1) Data Input Errors: Retailers publish product data imprecisely on the Web mainly due to manual mistakes. Although such input errors were mostly found on retailers' Web site in the study, it is possible that they also occur within the manufacturers when data is being typed into databases.

2) Data Lack Errors: When official information is not available, some retailers use wrong numbers for all the products in the same category. For instance,

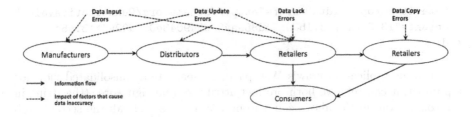

Fig. 3. Information flow and impact of factors that cause online product data inaccuracy

one retailer in the study set 10kg as the weight for several coffee machines from different brands, ignoring the fact that their actual dimensions and weights differ a lot.

3) Data Update Errors: In some cases, manufacturers update product information to correct previous errors. But such information is not always conveyed to consumers. It is possible that manufacturers, distributors and retailers do not replace the wrong attribute values with the updated ones, or they fail to notify their partners in the information chain of the updates.

4) Data Copy Errors: Instead of publishing product data, some retailers just copy the HTML source file from other retailers that sell the same product. Consequently, if an error exists on a Web page, it will be amplified by other retailers who copy the page. This explains why FREQ works best among all the approaches but still leaves around 9% inaccurate attribute values in the result.

In the next step, 20 expert interviews will be conducted on manufacturers, distributors and retailers to validate and further develop the proposed model. Afterwards, simulation modelling methods will be applied to measure the over product data inaccuracy in online retailing. The default values in the model come from both literatures like corporate data quality research and the data collected in our expert interviews.

5   New Approach to Mitigate Data Inaccuracy

We developed an attribute-based authentication approach to mitigate data inaccuracy through automatically differentiating manufacturer data and open data on the Web. Based on the Public-Key Cryptography [14], two Web tools were developed: The first one was used by manufacturers to generate a digital signature for each authentic attribute value. The second one was used by anyone on the Web to verify these signatures. Together with attribute values, online retailers can publish the corresponding signatures they get from manufacturers as meta-data in products' HTML files, as shown in the example code below.

```
<div>Width:
 <meta itemprop="width" href="http://schema.org/QuantitativeValue"
 content="13.55cm | 311b09f8ad88aaa83d549043e037a02a">13.55cm
</div>
```

When an application crawls Web pages to generate a consolidated view for a product, it can retrieve both product attributes and signatures (the string in the content tag in the above code) from a Web page and call the second Web tool API to verify whether the attribute values are authentic or not. Authentic values will always overwrite inauthentic ones when multiple values exist for a single attribute.

When manufacturers update product information, the first Web tool will make sure that old signatures become invalid and new ones come into force. Thus, only the latest attribute values will be verified as authentic by the second tool. Furthermore, retailers who copy online product data will also copy the signatures since the signatures are integrated in HTML files. Therefore, the new approach can contribute to reduce the data update and copy errors. Regarding data input error, if a retailer gets a correct signature but a wrong value for an attribute from a manufacturer, the approach will still solve the problem by verifying this attribute as inauthentic. However, if data input error occurs in the process of generating a signature, the proposed approach cannot contribute to solve the problem.

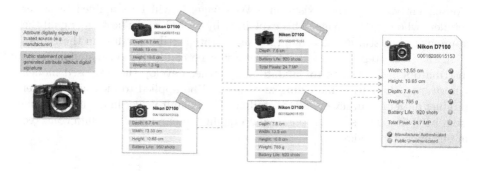

Fig. 4. An attribute-based authentication approach to mitigate data inaccuracy

Fig. 4 demonstrates how the approach works. In the demo, Retailer3 and Retailer4 publish accurate product data with digital signatures on their product Web pages, whereas Retailer1 and Retailer2 publish data without signatures or with invalid signatures. When the demo application crawls the four Web pages, it verifies each attribute signature and decides which value to trust when conflict occurs. The consolidated result is shown in the rightmost picture, where manufacturer authentic data with a digital signature is always selected when data conflict happens, thereby improving data accuracy. For attribute values without signatures, FREQ is used to select a proper value.

6 Conclusion, Limitation and Further Work

The study revealed a significant problem of product data inaccuracy on the Web. On average over 11% all attribute values and 13% core attribute values were inaccurate. We showed that the data accuracy can be increased by crawling the Web and consolidating data with several approaches. We investigated four popular approaches and showed that they were able to increase the accuracy on average by 11.3%. However, the most frequently appeared attribute value is not always accurate. Thus, we suspect that the dynamics of copying information from one page to another amplifies in several cases the problems. Based on the findings, we proposed an attribute-based authentication approach to mitigate online product data inaccuracy.

Our study is not without limitation and provides several opportunities for further research. First, product data was extracted manually and the number of samples used in the study is relatively small. This prevents us from drawing scientific and robust conclusions although it provides with an intuitive understanding about current data quality problem on the Web. It would be interesting to leverage Semantic Web and linked data technologies to automatically extract and compare online data of a larger number of products from different categories. Second, we categorized core and non-core attributes only based on their popularity. In future study, we plan to conduct a survey study to better understand what attributes are more important to consumers. Third, we proposed an explanatory model and suspected four potential causes for online data inaccuracy. We acknowledge that small sample size might lead to biased understanding, therefore, 20 expert interviews will be conducted on manufacturers, distributors and retailers to validate the suspected causes. Furthermore, we developed a prototype to demonstrate our attribute-based authentication approach. In the next step, we will collaborate with online retailers and manufacturers to implement the approach in a field trial and then evaluate to what extent it can improve data accuracy in online retailing.

References

1. Bădică, C., Bădică, A.: Rule learning for feature values extraction from HTML product information sheets. In: Antoniou, G., Boley, H. (eds.) RuleML 2004. LNCS, vol. 3323, pp. 37–48. Springer, Heidelberg (2004)
2. Barnes, S., Vidgen, R.: An Integrative Approach to the Assessment of E-commerce Quality. Journal of Electronic Commerce Research 3(3) (2002)
3. Bovee, M., Srivastava, R.P., Mak, B.: A Conceptual Framework and Belief-function Approach to Assessing Overall Information Quality. Int. J. Intell. Syst 18, 51–74 (2003)
4. Cho, V.: Data Quality on the Internet. Services and Business Computing Solutions with XML: Application for Quality Management and Best Processes, pp. 171–176. Information Science Reference, New York (2009)
5. Eppler, M., Muenzenmayer, P.: Measuring Information Quality in the Web Context: A Survey of State-of-the-Art Instruments and an Application Methodology. In: 7th International Conference on Information Quality, Cambridge, pp. 187–196 (2002)

6. European Union.: Regulation (EU) No 1169/2011 of the European parliament and of the council of 25 October 2011. Official Journal of the European Union (2011)
7. Evans, P.F., Sehgal, V., Bugnaru, C., McGowan, B.: US Online Retail Forecast, 2008 to 2013. Forrester Inc. Report (2009)
8. Fürber, C., Hepp, M.: A Semantic Web Information Quality Assessment Framework. In: 25th European Conference on Information System, Helsinki (2011)
9. Klein, B.D.: User Perceptions of Data Quality: Internet and Traditional Text Sources. The Journal of Computer Information System 41(2001), 9–18 (2001)
10. Knap, T., Michelfeit, J., Necasky, M.: Linked Open Data Aggregation: Conflict Resolution and Aggregate Quality. In: Computer Software and Applications Conference Workshops (2012)
11. Knight, S., Burn, J.: Developing a Framework for Assessing Information Quality on the World Wide Web. Informing Science Journal 8, 159–172 (2005)
12. Mendes, P.N., Mühleisen, H., Bizer, C.: Sieve: Linked Data Quality Assessment and Fusion. In: Joint EDBT/ICDT Workshops, Berlin, pp. 116-123 (2012)
13. Redman, T.C.: The Impact of Poor Data Quality on the Typical Enterprise. Magazine Communications of the ACM 41, 79–82 (1998)
14. Rivest, R.L., Shamir, A., Adleman, L.: A Method for Obtaining Digital Signatures and Public-key Cryptosystems. Magazine Communications of the ACM 21, 120–126 (1978)
15. Wang, R., Strong, D.: Beyond Accuracy: What Data Quality Means to Data Consumer. Journal Management Information System 12, 5–33 (1996)

WISE 2014 Challenge: Multi-label Classification of Print Media Articles to Topics

Grigorios Tsoumakas[1], Apostolos Papadopoulos[1], Weining Qian[2],
Stavros Vologiannidis[3], Alexander D'yakonov[4], Antti Puurula[5], Jesse Read[6],
Jan Švec[7], and Stanislav Semenov[8]

[1] Aristotle University of Thessaloniki, Thessaloniki 54124, Greece
{greg,papadopo}@csd.auth.gr
[2] East China Normal University, China
wnqian@sei.ecnu.edu.cn
[3] DataScouting, Greece
svol@datascouting.com
[4] Lomonosov Moscow State University, Russia
djakonov@mail.ru
[5] The University of Waikato, New Zealand
asp12@students.waikato.ac.nz
[6] Aalto University, Finland
jesse.read@aalto.fi
[7] University of West Bohemia, Czech Republic
honzas@kky.zcu.cz
[8] Higher School of Economics and the Yandex School of Data Analysis, Russia
stasg7@gmail.com

Abstract. The WISE 2014 challenge was concerned with the task of multi-label classification of articles coming from Greek print media. Raw data comes from the scanning of print media, article segmentation, and optical character segmentation, and therefore is quite noisy. Each article is examined by a human annotator and categorized to one or more of the topics being monitored. Topics range from specific persons, products, and companies that can be easily categorized based on keywords, to more general semantic concepts, such as environment or economy. Building multi-label classifiers for the automated annotation of articles into topics can support the work of human annotators by suggesting a list of all topics by order of relevance, or even automate the annotation process for media and/or categories that are easier to predict. This saves valuable time and allows a media monitoring company to expand the portfolio of media being monitored. This paper summarizes the approaches of the top 4 among the 121 teams that participated in the competition.

1 Introduction

In the past, gathering information was paramount only for top-tier companies. In the information age, mining and categorization of relevant information is necessary for all companies. Media monitoring - the activity of monitoring the

B. Benatallah et al. (Eds.): WISE 2014, Part II, LNCS 8787, pp. 541–548, 2014.

output of the print, online and broadcast media - allows every company to search a wide range of media, from print media to internet publications, and be informed on their area of expertise and remain competitive.

Media monitoring companies rely on human experts that watch/read/listen to media clips and manually index them using concepts from a predefined ontology. This task requires significant amount of time and money to be accomplished. Machine learning can be employed to construct systems for assisting the human annotators by suggesting a list of all concepts by order of relevance, or even for automating the annotation process for media and/or concepts that are easier to predict. This would allow a media monitoring company to invest the saved resources towards expanding the portfolio of media being monitored.

The WISE 2014 challenge was concerned with the task of multi-label classification of articles coming from Greek print media. Data was collected by scanning a number of Greek print media from May 2013 to September 2013. Articles were manually segmented and their text extracted through OCR (optical character recognition) software. The text of the articles is represented using the bag-of-words model and for each token encountered inside the text of all articles, the tf-idf statistic is computed and unit normalization is applied to the tf-idf values of each article. There are therefore 301561 numerical attributes corresponding to the tokens encountered inside the text of the collected articles. Articles were manually annotated by a human expert with one or more out of 203 labels ranging from specific persons, products, and companies that can be easily categorized based on keywords, to more general semantic concepts, such as environment or economy. 99780 articles were collected. The chronologically first 64857 form the training set, and the following 34923 form the test set. The goal is to predict the relevant labels in the test set, where the labels of the articles are withheld. The evaluation metric was the mean F_1 score, also known as example-based F_1 score [1].

This paper discusses the approaches of the top four teams, which were also the ones that provided a summary of the solution. These teams, in order of their final ranking are:

1. Alexander D'yakonov, from the Lomonosov Moscow State University, Russia.
2. Antti Puurula and Jesse Read from University of Waikato, New Zealand and Aalto University, Finland respectively.
3. Jan Švec, from the University of West Bohemia, Czech Republic.
4. Stanislav Semenov, from the Higher School of Economics and the Yandex School of Data Analysis, Russia.

The rest of this paper is organized as follows. Section 2 discusses the learning approaches employed by the teams to obtain a vector of numerical scores for the labels and Section 3 discusses the thresholding approaches that were used to obtain bipartitions of the set of labels into relevant and irrelevant ones from the numerical scores. Finally, Section 4 presents the conclusions of the WISE 2014 challenge.

2   Learning Methods

The 1st team employed binary relevance instantiated with an ensemble of 10 classifiers:

- Four kNN classifiers, with $k \in \{1, 2, 3, 50\}$ respectively. For $k > 1$ the neighbors are averaged according to their cosine distance. For $k = 50$ initial experiments measured an F_1 score of around 0.68. This was the least accurate family of models.
- Three ridge regression classifiers, with ridge parameter taking values from $\{0.4, 0.8, 1.2\}$. For ridge parameter equal to 0.8, initial experiments measured an F_1 score of around 0.76.
- Three logistic regression classifiers, with L1 regularization and regularization parameter taking values from $\{2, 6, 10\}$. For regularization parameter value equal to 6, initial experiments measured an F_1 score of around 0.78. This was the most accurate family of models.

The ensemble was combined via Stacking [2] using ridge regression. Training data for the meta-model were produced by training the ensemble on the first 50,000 examples of the training set and obtaining its predictions on the rest 14,857 examples. The same last set of examples was used for initial parameter exploration of the base classifiers.

Feature engineering investigations (singular value decomposition, transforming the features to polynomial, adding the features a 2nd time sorted by value) did not lead to significant improvements in the best case.

A distinctive aspect of the approach of the 2nd team was that it reverse-engineered the word counts in the documents from the provided tf-idf vectors. This clever hack allowed the construction of two additional feature vectors: word pairs [3] and 50-300 topics extracted via LDA (Latent Dirichlet Allocation).

The 2nd team created a much larger ensemble, of over 200 classifiers, by employing a variety of multi-label classification algorithms (binary relevance, classifier chains, (pruned) label powerset, random k-labelsets and others) instantiated with a subset of a variety of base classifiers (centroid classifier, multinomial naive bayes, random forest, c4.5, support vector machines) using combinations of the 3 different feature sets. Depending on the size of the feature set and the complexity of the base classifiers different multi-label classifiers could be afforded. In other words not all the above combinations were realized. The final ensemble of the 2nd team consisted of about 50 models selected by a hill-climbing search attempting additions, removals and replacements of the classifiers in the ensemble.

The ensemble of the 2nd team was combined by a variant of Feature-Weighted Linear Stacking [4,5], using the first 59857 documents for developing base classifiers and the following 5000 documents for optimizing the ensemble. This improves simple majority voting by predicting for each instance optimal vote weights based on meta-features computed from all available information. Their variant approximates optimal weights for each training instance, and uses the approximated weights as targets for regression models. Linear regression was used to develop the ensemble, but for the best submission a random forest with 40 trees was used to predict the vote weights. The set of meta-feature included document features, training set frequencies of the predicted labels and labelsets, as well as correlations between the classifier outputs. One new type of meta-feature that proved useful was the labelset predictions for neighboring documents: since the data was organized in time order, labels occurred often in sequences, and predictions for neighboring documents could be used to improve the classifier vote weight prediction. Two windows of neighboring documents were used: one with 650 documents and one with 6. The score for each label was the sum of the weighted classifier outputs.

The 3rd team employed binary relevance instantiated with a simple linear model trained using stochastic gradient descent (SGD) [6] with modified Huber loss and elastic net regularization. For predicting the posterior probability the method described in [7] was used. The regularization parameters of the binary models were tuned iteratively based on the mean F_1 score: models were tuned one after the other in descending order of frequency. Multiple iterations over all labels were conducted.

In addition, a distinctive aspect of the approach of the 3rd team was that it employed semi-supervised learning. In specific, the aforementioned supervised model tuned using three iterations over all labels was used to give predictions in the test set and then these predictions were taken as ground truth. The same SGD models with another three tuning iterations were applied to this expanded training set. After each tuning iteration, the test data were re-labeled.

The final model arbitrated among: 1) the supervised models, 2) the semi-supervised models, and 3) two additional tuning iterations of the semi-supervised models. This can be seen as a classifier selection approach per label, choosing among 1 supervised model and 3 semi-supervised models, with 3, 4 and 5 tuning iterations respectively.

The 4th team used a linear SVM with L1 regularization for each label. This shows how far one can get by focusing on the design of a powerful thresholding method.

3   Thresholding Methods

Let $\mathcal{Y} = \{\lambda_1, \ldots, \lambda_q\}$ be a set of q labels. Let g_j, $j = 1 \ldots q$ be the predicted score of label λ_j for a given test instance. Let p be a threshold.

The 1st team explored 4 different thresholding rules, according to which a label λ_j was included in the final output when the following corresponding expressions were true:

$$g_j \geq \min(p, \max(g_1, \ldots, g_q)), \tag{1}$$

$$\frac{g_j}{\max(g_1, \ldots, g_q)} \geq p, \tag{2}$$

$$\frac{g_j}{g_1 + \cdots + g_q} \geq p, \tag{3}$$

$$\frac{g_j - (g_1 + \cdots + g_q)/q}{\max_i (g_i - (g_1 + \cdots + g_q)/q)} \geq p. \tag{4}$$

Figure 1 shows the F_1 score of these rules when logistic regression is used as a binary classifier for each label for different values of the threshold p ranging from 0 to 1. The second and forth decision rules appear to be more effective. In its final solution the 1st team used the second rule with $p = 0.55$.

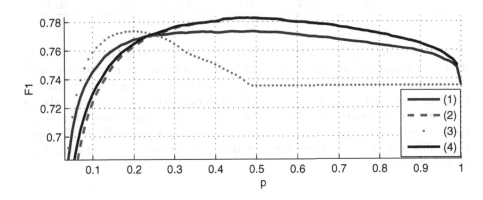

Fig. 1. Performance of decision rules

The 2nd team followed the same simple, but obviously effective, rule, but with $p = 0.5$. This team had successfully used this rule again in the past as part of its winning solution for the LSHTC4 competition [8].

The approach of the 3rd team outputs all labels with probability higher than 0.5. The label with the highest probability is given in the output even if its probability, p^*, is lower than 0.5. In addition, the approach outputs the 4, 3 or 2 labels with the highest probability when the probability of the 4th, 3rd and 2nd label correspondingly are larger than $t_C \cdot p^*$, $t_B \cdot p^*$ and $t_A \cdot p^*$ respectively, where the values of t_A, t_B and t_C are tuned using grid search to optimize the mean F_1 measure on the training data. In cases where multiple conditions are satisfied (e.g. it is possible to assign both two and four labels) the set with the larger cardinality is used.

The approach of the 4th team follows the paradigm of [9,10], which train a regression model to predict a separate threshold for each test instance. This is based on obtaining the predicted probabilities for (a subset of) the training data. Consider for example the predicted probabilities (sorted in descending order) and the corresponding ground truth for an instance of a multi-label learning task with 5 labels that are given in the first two rows of Table 1.

In [9] the regression target was the threshold that minimized the number of labels with wrong prediction, i.e. a threshold in (0.7,0.9) or in (0.3,0.6) in the above example. In [10] the regression target was the threshold that maximized the F_1 score, i.e a threshold in (0.3,0.6) in the above example. Usually, the mean

Table 1. Number of wrong labels (3rd row) and F_1 score (4th row) for 6 different threshold ranges for an instance of a multi-label learning task with 5 labels. The first two rows show the predicted probabilities sorted in descending order and the corresponding ground truth for the 5 labels.

predictions	0.9		0.7		0.6		0.3		0.1	
ground truth	1		0		1		0		0	
wrong labels	2	1		2		1		2		3
F_1	0	0.67		0.5		0.8		0.67		0.57

of the lower and upper boundaries is taken, e.g. for a chosen range of (0.3,0.6), the target would be 0.45. In the approach of the 4th team, two regression models are built, one for predicting the upper and one for the lower boundary of the range that optimizes the F_1 score. A linear combination of the two predicted boundaries can then be taken in the form of $\alpha l + (1 - \alpha)u$, where l and u are the predicted lower and upper boundaries respectively. Figure 2 shows the Mean F_1 for different values of α as investigated by the 4th team. A value around 0.5 leads to best results.

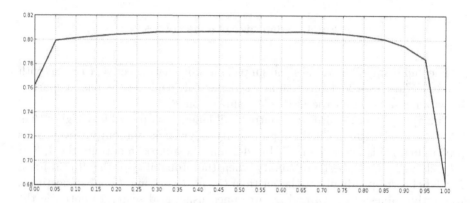

Fig. 2. Mean F_1 for different values of combining the predicted upper and lower boundary

The predicted probabilities were used as input features for the regression in [9], while the features of the instance itself were used in [10]. The 4th team used a versatile set of input features for the two regression tasks: the predicted probabilities, the sorted predicted probabilities, the differences among two consecutive values of the sorted predicted probabilities, the differences among all pairs of the top 10 probabilities and the 20 first principal components extracted by applying PCA to the input space of the main learning task.

4   Conclusions

Examining the top 4 solutions in the WISE 2014 challenge, a number of conclusions can be drawn. A first well-known and expected conclusion is that ensemble methods do well. Indeed, the top 3 solutions employed more than one models (10, 50, 5 respectively). On the other hand, the single model solution of the 4th team shows the importance of a strong thresholding technique. The top 3 solutions used simple, yet effective thresholding techniques, based on dividing each score with the maximum score for a given instance. The 4th team employed a more elaborate thresholding technique that involved learning and appears to be more successful. Another conclusion, more or less also known, is that linear models do well in text classification. Linear models were important components in the solutions of all the top 4 teams. Furthermore, from the approach of the 3rd team, it seems that semi-supervised approaches can go that extra mile compared to supervised approaches, especially for multi-label data where scarcity of labeled examples often arises for some labels.

Another issue worth mentioning is that of overfitting, as it most probably played a decisive role in the final standing. Both the first two teams were well behind in the leaderboard until the private results came out. Both of these teams submitted less than half than the 3rd and 4th teams: (26, 21) vs (52, 72). The first two teams also divided the dataset according to the time-order, instead of doing cross-validation, so that the models were fitted to data closer in time to the test set, and not across the whole dataset. This is always a *wise* choice for data streams.

Comparing the top 2 solutions, we can see evidence in favor of Occam's razor [11]. The solution of the 2nd team involved three different feature representations and over 200 classifiers that resulted from the combination of many different multi-label and single-label classification algorithms, combined by a very powerful stacking variant [4]. The solution of the 1st team involved the original feature vectors and 10 different classifiers from 3 standard families (kNN, logistic regression, ridge regression), combined using standard stacking with ridge regression. Given that they both employed almost the same thresholding strategy, we can say that perhaps simple solutions are still worth being considered first.

Finally, the clever hack of the 2nd team, teaches us that if privacy of the sources has to be protected due to copyright or other issues, then more careful pre-processing has to be applied to the data, such as adding noise, adding bigrams, removing frequent/rare words and disclosing as few details as possible for the actual pre-processing steps.

References

1. Tsoumakas, G., Katakis, I., Vlahavas, I.: Mining multi-label data. In: Maimon, O., Rokach, L. (eds.) Data Mining and Knowledge Discovery Handbook, 2nd edn., pp. 667–685. Springer, Heidelberg (2010)
2. Wolpert, D.H.: Stacked generalization. Neural Networks 5, 241–259 (1992)

3. Lesk, M.E.: Word-word associations in document retrieval systems. American Documentation 20(1), 27–38 (1969)
4. Sill, J., Takács, G., Mackey, L., Lin, D.: Feature-weighted linear stacking. CoRR abs/0911.0460 (2009)
5. Puurula, A., Bifet, A.: Ensembles of sparse multinomial classifiers for scalable text classification. In: ECML/PKDD - PASCAL Workshop on Large-Scale Hierarchical Classification (2012)
6. Zhang, T.: Solving large scale linear prediction problems using stochastic gradient descent algorithms. In: Proceedings of the Twenty-First International Conference on Machine Learning, ICML 2004, p. 116. ACM, New York (2004)
7. Zadrozny, B., Elkan, C.: Transforming classifier scores into accurate multiclass probability estimates. In: Proceedings of the eighth ACM SIGKDD International Conference on Knowledge Discovery and Data Mining, KDD 2002, pp. 694–699 (2002)
8. Puurula, A., Read, J., Bifet, A.: Kaggle LSHTC4 winning solution. CoRR abs/1405.0546 (2014)
9. Elisseeff, A., Weston, J.: A kernel method for multi-labelled classification. In: Advances in Neural Information Processing Systems 14 (2002)
10. Nam, J., Kim, J., Gurevych, I., Fürnkranz, J.: Large-scale multi-label text classification - revisiting neural networks. CoRR abs/1312.5419 (2013)
11. Domingos, P.: The role of occam's razor in knowledge discovery. Data Min. Knowl. Discov. 3(4), 409–425 (1999)

Author Index

Aberer, Karl   I-276
Achilleos, Achilleas P.   II-304
Agathangelou, Pantelis   I-47
Agrawal, Shradha   I-135
Aksoy, Cem   I-448
Alabau, Vicent   II-460
Alba, Alfredo   II-17
Albitar, Shereen   I-105
Aldalur, Iñigo   I-293
Almulla, Mohammed   II-32
Altingovde, Ismail Sengor   II-78
Andreae, Peter   I-512, I-523
Arellano, Cristóbal   I-293

Bai, Changqing   II-381
Bass, Len   II-425
Bell, David A.   I-115
Berberich, Klaus   I-156
Bernardino B. de Campos, Gilda Helena
    II-351
Bianchini, Devis   I-218
Bieliková, Mária   I-372
Blazquez, Desamparados   II-435
Bochmann, Gregor V.   II-365
Boon, Ferry   I-433
Bornea, Mihaela   I-480
Bouzidi, Sabri   I-433
Brunk, Sören   II-521
Burger, Roman   I-372

Cao, Longbing   I-1
Cao, Yu   II-246
Casanova, Marco Antonio   I-324, II-351
Ceroni, Andrea   II-90
Chaudry, Rabia   II-425
Chawda, Bhupesh   II-278
Chen, Fengjiao   I-95
Chen, Wei   I-170
Chen, Xi   II-336
Chowdhury, Israt J.   I-146
Chung, Tonglee   I-308

Dao, Bo   I-398
Dass, Ananya   I-448
De Antonellis, Valeria   I-218

de Boer, Nienke   I-357
de Lara, Juan   II-505
Dey, Akon   II-262
Díaz, Oscar   I-293
Dietze, Stefan   I-324
Dimitriou, Aggeliki   I-448
Ding, ZhiJun   I-79
Di Pietro, Roberto   I-15
Djafari Naini, Kaweh   II-90
Dolby, Julian   I-480
Domenech, Josep   II-435
Drews, Clemens   II-17
Duong, Thi   II-474
Dustdar, Schahram   II-415
D'yakonov, Alexander   II-541

Enoki, Miki   II-395
Erdmann, Maike   II-109
Espinasse, Bernard   I-105

Fei, Jinlong   II-336
Feigenbutz, Florian   II-294
Fekete, Alan   II-319, II-425
Fetahu, Besnik   II-351
Firmenich, Sergio   I-293
Fisichella, Marco   II-90
Fokoue, Achille   I-480
Fournier, Sébastien   I-105
Frasincar, Flavius   I-357, I-418, I-433,
    I-534

Gan, Zaobin   I-63
Gao, Xiaoying   I-512, I-523
Gao, Yang   I-186
Georgescu, Mihai   II-90
Geva, Shlomo   I-125
Giannopoulos, Giorgos   II-189
Gil, Jose A.   II-435
Gong, Zhiguo   II-62
Gruhl, Daniel   II-17
Gu, Jun   II-246
Guabtni, Adnene   II-425
Gunopulos, Dimitrios   II-178
Guo, Xiaohui   I-464
Gupta, Himanshu   II-278

Hattori, Gen   II-109
Hirzel, Martin   II-395
Hogenboom, Frederik   I-418
Holzmann, Helge   II-47
Hong, Jun   I-115
Horii, Hiroshi   II-395
Hrgovcic, Vedran   I-496
Hu, Weishu   II-62
Husmann, Maria   II-199

IJntema, Wouter   I-418
Ikeda, Kazushi   II-109
Ilic, Alexander   II-531
Ishizaki, Hiromi   II-109

Jabeen, Shahida   I-512, I-523
Jiajie, Xu   I-170
Jin, Cheqing   II-489
Jourdan, Guy-Vincent   II-365
Joy, Mike   II-158

Kang, Qiangqiang   II-489
Kapitsaki, Georgia M.   II-304
Katakis, Ioannis   I-47
Kau, Chris   II-17
Kawase, Ricardo   II-351
Keller, Christine   II-521
Kementsietsidis, Anastasios   I-480
Kermarrec, Anne-Marie   I-276
Kliem, Andreas   II-294
Kokkoras, Fotios   I-47
Kompatsiaris, Ioannis   II-168
Kondylakis, Haridimos   I-496
Koniaris, Marios   II-189

Lee, Kevin   II-319
Leiva, Luis A.   II-460
Lewis, Neal   II-17
Li, Fangfang   I-1
Li, Guohui   I-244
Li, Jianjun   I-244
Li, Kan   I-31, I-95
Li, Victor O.K.   II-125
Li, Xue   II-1
Li, Xueming   II-1
Li, Yuefeng   I-186, I-408
Liu, An   II-141
Liu, Anna   II-319, II-425
Liu, Guanfeng   II-141
Liu, Shengli   II-336
Liu, Xudong   I-464

Liu, Yongbin   I-308
Lofi, Christoph   I-340
Long, Yi   II-125
Lu, Hongwei   I-63
Luo, Changyin   I-244
Luo, Wei   I-266

Ma, Jiangang   I-256
Ma, Yuanchao   I-388
Maamar, Zakaria   II-32
Macha, Meghanath   I-135
Medina, Haritz   I-293
Melchiori, Michele   I-218
Mendes, Pablo N.   II-17
Miller, James   II-445
Mirtaheri, Seyed M.   II-365
Mora Segura, Ángel   II-505
Móro, Róbert   I-372

Nagarajan, Meena   II-17
Nayak, Richi   I-125, I-146
Nebeling, Michael   II-199
Nguyen, Thin   I-266, I-398, II-474
Nieke, Christian   I-340
Niu, Guolin   II-125
Norrie, Moira C.   II-199
Ntonas, Konstantinos   I-47

Olteanu, Alexandra   I-276
Onal, Kezban Dilek   II-78
Onut, Iosif-Viorel   II-365, II-445
Ozsoy, Makbule Gulcin   II-78

P. Paes Leme, Luiz André   I-324
Pai, Deepak   I-135
Panev, Kiril   I-156
Panwar, Abhimanyu   II-445
Papadopoulos, Apostolos   II-541
Papadopoulos, Symeon   II-168
Peng, Shu   II-246
Pereira Nunes, Bernardo   I-324, II-351
Petrocchi, Marinella   I-15
Phung, Dinh   I-266, I-398, II-474
Plexousakis, Dimitris   I-496
Pongelli, Stefano   II-199
Pont, Ana   II-435
Premm, Marc   I-496
Puurula, Antti   II-541

Qadan Al Fayez, Reem   II-158
Qian, Weining   I-234, II-541

Rabello Lopes, Giseli   I-324
Rajani, Meena   II-262
Rao, Weixiong   II-246
Read, Jesse   II-541
Risse, Thomas   II-47
Roels, Reinout   II-215
Röhm, Uwe   II-262

Sánchez Cuadrado, Jesús   II-505
Saravanou, Antonia   II-178
Scekic, Ognjen   II-415
Schlegel, Thomas   II-521
Schouten, Kim   I-357
Schuele, Michael   I-496
Sellis, Timos   II-189
Semenov, Stanislav   II-541
Seyfi, Majid   I-125
Sharaf, Mohamed   II-1
Sharang, Abhijit   I-135
Shemshadi, Ali   I-202, I-256
Sheng, Quan Z.   I-202, I-256, II-405
Shuai, Kaiyan   I-308
Signer, Beat   II-215, II-231
Siméon, Jérôme   II-395
Sladescu, Matthew   II-319
Spognardi, Angelo   I-15
Srinivas, Kavitha   I-480
Stanik, Alexander   II-294
Sun, HaiChun   I-79
Sun, Hailong   I-464
Švec, Jan   II-541
Szabo, Claudia   II-405

Takishima, Yasuhiro   II-109
Tao, Han   II-336
Tayeh, Ahmed A.O.   II-231
Theodoratos, Dimitri   I-448
Tian, Nan   I-408
Tran, Truyen   I-266
Tran, Tuan   II-90
Truong, Hong-Linh   II-415
Tsakalidis, Adam   II-168
Tsoumakas, Grigorios   II-541

Unankard, Sayan   II-1

Valkanas, George   II-178
Vandic, Damir   I-357, I-418, I-433, I-534
van Leeuwen, Marijtje   I-357
van Luijk, Ruud   I-357
Vassiliou, Yiannis   II-189
Venkatesh, Svetha   I-266, I-398, II-474
Vermaas, Raymond   I-433, I-534
Vologiannidis, Stavros   II-541

Wang, PengWei   I-79
Wang, Ting   II-381
Wang, X. Sean   II-246
Wang, Yanghao   I-464
Welch, Steve   II-17
Weng, Daiyue   I-115
Woitsch, Robert   I-496
Wu, Chao   I-388

Xu, Bin   I-308, I-388
Xu, Chen   I-234
Xu, Guandong   I-1
Xu, Runhua   II-531
Xu, Yong   I-256
Xu, Yue   I-186, I-408

Yahyaoui, Hamdi   II-32
Yang, Min   II-246
Yao, Lina   I-256
Yuan, Bozhi   I-308, I-388
Yuan, Ji   I-464
Yuefei, Zhu   II-336

Zhang, Liang   I-234
Zhang, Richong   I-464
Zhang, Wei Emma   I-202
Zhang, Yihong   II-405
Zhang, Zhao   II-489
Zhao, Lei   I-170, II-141
Zhao, Qian   I-63
Zheng, Kai   I-170
Zhong, Jiang   II-1
Zhong, MingJie   I-79
Zhou, Aoying   I-234, II-489
Zhou, Xiaofang   I-170, II-141
Zhu, Feng   II-141
Zhu, Guangyao   I-31